International Conference on Environment, Renewable Energy and Green Engineering (EREGCE 2022)

E3S Web of Conferences Volume 350 (2022)

Online
22-24 April 2022

Editors:

Chunming Yang
Haoming Chen
Peigao Duan
Feipeng Jiao
Chang Wen

ISBN: 978-1-7138-5245-2

Printed from e-media with permission by:

Curran Associates, Inc.
57 Morehouse Lane
Red Hook, NY 12571

proceedings.com

Some format issues inherent in the e-media version may also appear in this print version.

For additional information, please contact EDP Sciences – Web of Conferences
at the address below.

EDP Sciences – Web of Conferences
17, Avenue du Hoggar
Parc d'Activité de Courtabœuf
BP 112
F-91944 Les Ulis Cedex A
France

Phone: +33 (0) 1 69 18 75 75
Fax: +33 (0) 1 69 28 84 91

contact-edps@webofconferences.org

Additional copies of this publication are available from:

Curran Associates, Inc.
57 Morehouse Lane
Red Hook, NY 12571 USA
Phone: 845-758-0400
Fax: 845-758-2633
Email: curran@proceedings.com
Web: www.proceedings.com

TABLE OF CONTENTS

GREEN CHEMICAL ENGINEERING

RENEWABLE ENERGY

EREGCE

International Conference on Environment, Renewable Energy and Green Chemical Engineering

Editorial Board:

April 22-24, 2022

Organized by:
Hunan Normal University

Keynote Speakers

Zhanying Zhang
(Queensland University of Technology)

Shahid Hussain
(Jiangsu University)

Peigao Duan
(Xi'an Jiaotong University)

Sponsors

Conference Program
Sponsored by

Committee

General Chairs

Prof. Chunming Yang
Hunan Normal University, China
Prof. Peigao Duan
Xi'an Jiaotong University, China
Prof. Feipeng Jiao
Central South University, China
Prof. Juai Wu
Nanjing University of Posts and Telecommunications, China

Program Chairs

Prof. Zonglin Chu
Hunan University, China
Prof. Zhiguo He
Central South University, China
Prof. Weizhi Zhou
Shandong University, China
Assoc. Prof. Wenjihao Hu
Central South University, China
Prof. Mohd Hasmadi Ismail
Universiti Putra Malaysia, Malaysia
Prof. Muhammed Bashir Mu'azu
Ahmadu Bello University, Nigeria

Technical Chairs

Prof. Rajendra P. SHRESTHA
Asian Institute of Technology, Thailand
Prof. Chang Xu
Hohai University, China
Assoc. Prof. Haoming Chen
Nanjing University of Science and Technology, China
Assoc. Prof. Neslihan Kulözü Uzunboy
Atatürk University, Turkey
Assoc. Prof. Chang Wen
Huazhong University of Science and Technology, China

Technical Program Committees

Prof. Hanfeng Liang
Xiamen University, China
Prof. Lehua Zhang
East China University of Science and Technology, China

Preface

It was our great honor and pleasure to organize the International Conference on Environment, Renewable Energy and Green Chemical Engineering (EREGCE 2022) online on April 22nd to 24th, 2022.

As Chemical industry plays a vital role in the development process of national economy, a large number of poisonous and harmful substances pose a great threat to the ecological environment and human. Therefore, it is a general trend to promote the development of renewable energy and green chemical engineering towards a greener environment. The conference focused on the latest research fields of environment, renewable energy and green chemical engineering, presenting scientific research results and its practical solutions to these matter. It has taken into account the principles of sustainable development, which provides an international platform for experts, professors, scholars and engineers from universities, scientific research institutes, enterprises and institutions at home and abroad to share their professional experiences, expand their professional networks and display their research results.

The comprehensive content of the conference has attracted great attention due to its wealthy information that may be useful to professionals working in the related fields. Many researchers in the related field had participated in the conference and made oral presentations. The conference proceedings contain 50 papers divided into 3 sections: Environment, Renewable Energy and Green Chemical Engineering.

The keynote speeches on April 23, 2022, were given by three scholars, Zhanying Yang, Shahid Hussain and Peigao Duan, and an invited lecture by Chunming Yang. In addition to the presentations brought by the authors, there were three parallel sessions. Session I: Ecosystem sustainability was hosted Haoming Chen (Nanjing University of Science and Technology, China) and Yuanming Guo (Nanjing University of Science and Technology, China), Session Ⅱ: Remediation of water and soil pollution was hosted by Da Tian (Anhui Agricultural University, China) and Gongwen Luo (Key Laboratory of Soil Environment and Pollution Remediation, Institute of Soil Science, Chinese Academy of Sciences,China), and Session Ⅲ: Landscape ecological planning and development was hosted by Haoming Chen (Nanjing University of Science and Technology, China) and Yang Feng (Xijing University, China). Whether in the main session or in the parallel sessions, all scholars expressed their views to showcase their research, contributing to drive innovation and breakthroughs in the whole field.

Many thanks to all the authors who had submitted papers for their participation. It is they who have put a lot of effort and creativity to make this work complete. And we feel honored that they are willing to present their excellent work on this conference. Many thanks also to all the editors and reviewers who had contributed a lot of time in their busy schedules to carefully read and evaluate all the manuscripts.

Finally, due to the epidemic prevention policy, it's a pity that we could not have offline communication and discussion, but still we would like to thank all the supporting organizations for their hardwork and the organizing committee for the extensive preparation that made this online conference a great success. We look forward to having a live conference next year to share the results.

Guest Editors

Chunming Yang (Hunan Normal University)

Haoming Chen (Nanjing University of Science and Technology)

Peigao Duan (Xi'an Jiaotong University)

Feipeng Jiao (Central South University)

Chang Wen (Huazhong University of Science and Technology)

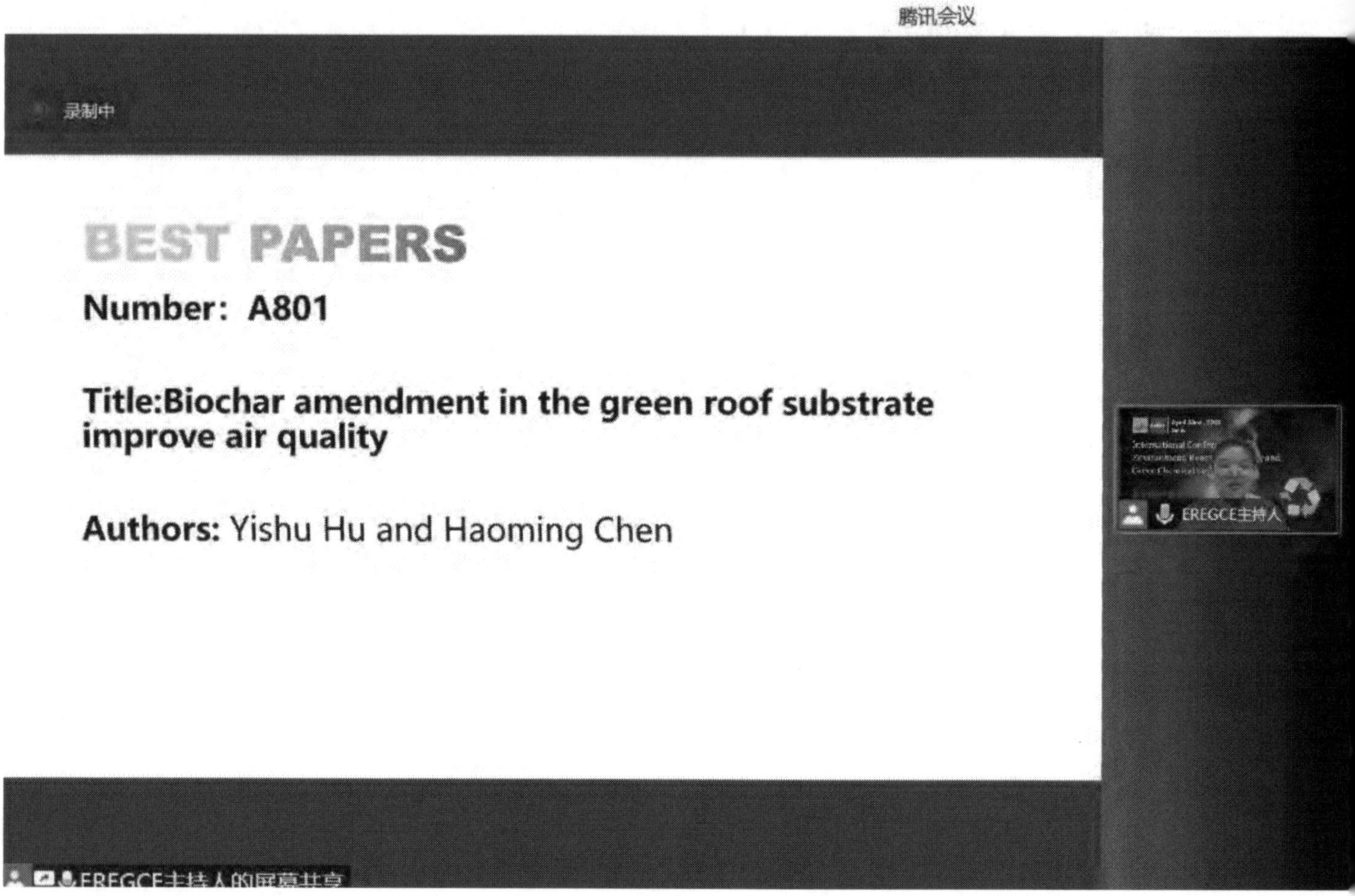
腾讯会议
录制中
BEST PAPERS
Number: A801
Title:Biochar amendment in the green roof substrate improve air quality
Authors: Yishu Hu and Haoming Chen
EREGCE主持人

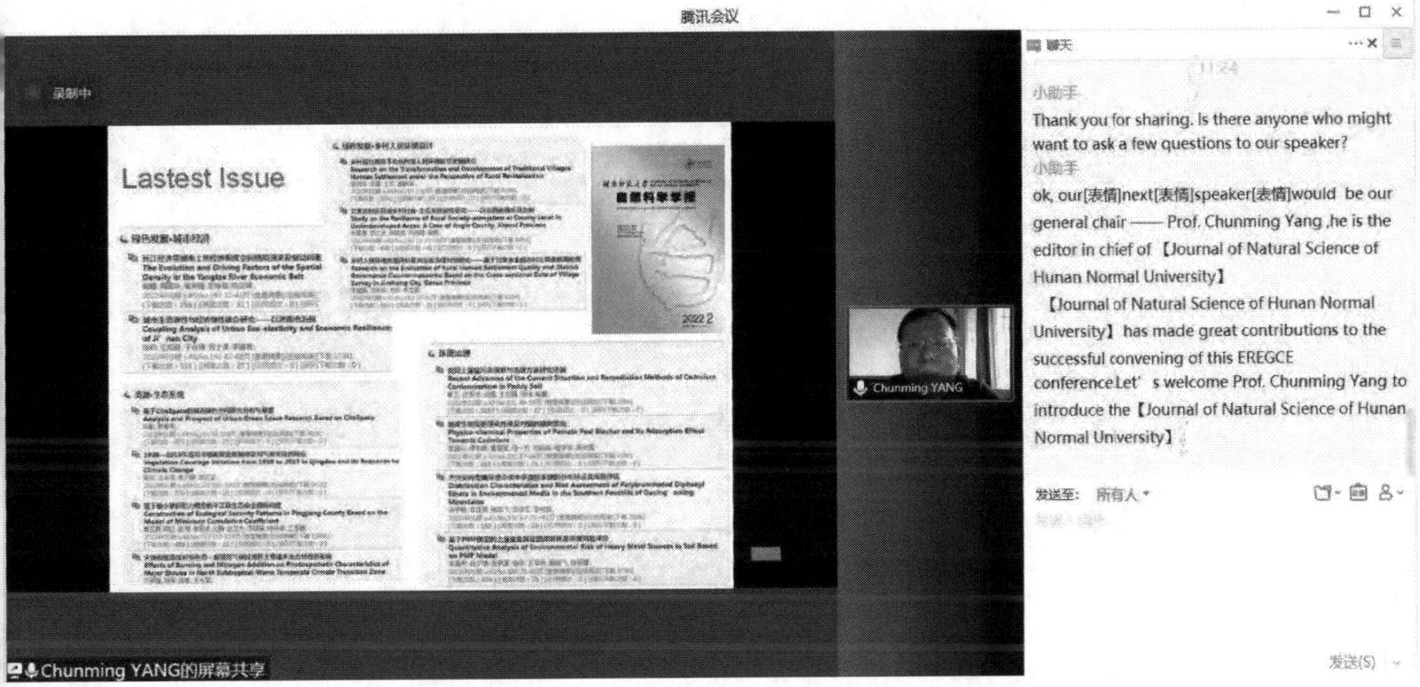
腾讯会议
录制中
Lastest Issue
2022 2
Chunming YANG
聊天
11:24
小助手
Thank you for sharing. Is there anyone who might want to ask a few questions to our speaker?
小助手
ok, our[表情]next[表情]speaker[表情]would be our general chair —— Prof. Chunming Yang ,he is the editor in chief of 【Journal of Natural Science of Hunan Normal University】
【Journal of Natural Science of Hunan Normal University】 has made great contributions to the successful convening of this EREGCE conference.Let' s welcome Prof. Chunming Yang to introduce the【Journal of Natural Science of Hunan Normal University】
发送至: 所有人
发送(S)
Chunming YANG的屏幕共享

腾讯会议
录制中
EREGCE主持人
Zhanying Zhang
陈颢明njust
A203+RUIMAO
Shuaijuan Bian
张清凌
A210-Sichun Ding
Junqi Wang
A808JinxingCai
Chunming YANG
A179 Qian Deng
Shahid Hussain
A214-Lipeng WANG
A209 Hao Chen
A189-Kai-Chen
Baishan Li
蒋好
HaoyuTang
安农田达
A187-Yangwei Wu
A221-Yang Feng
A196/A202-张德龙
A804Fangfang Min
A188+Yuhao Liu
合肥学院-陈俊

腾讯会议
06:10:36
宫格视图
录制中
EREGCE主持人
A111+Jian Yang
A113 - Hao Wu
A112-XIAOKUN
JunqiWang
A187+Yangwei Wu
A203+RUIMAO
A192+Leiqiong CAI
A218 Zijian Zhou
JinshuoYan
A209 Hao Chen
陈锴
A179 Qian Deng
Haoyu Tang
Shuaijuan Bian
张清凌
张德龙
A226YueZhou
小助手
ESBK国际学术交流中心
说点什么...

腾讯会议
录制中
Acknowledgements
• EREGCE 2022 - conference organisers, committee and staff
• Funding
Australian Government
Department of Agriculture and Water Resources
Australian Government
Australian Research Council
Queensland Government
syngenta
UPALA
• Colleagues in China
• QUT colleagues
Zhang's Group (Dec 2021)
QUT a university for the real world
Zhanying Zhang
Zhanying Zhang的屏幕共享

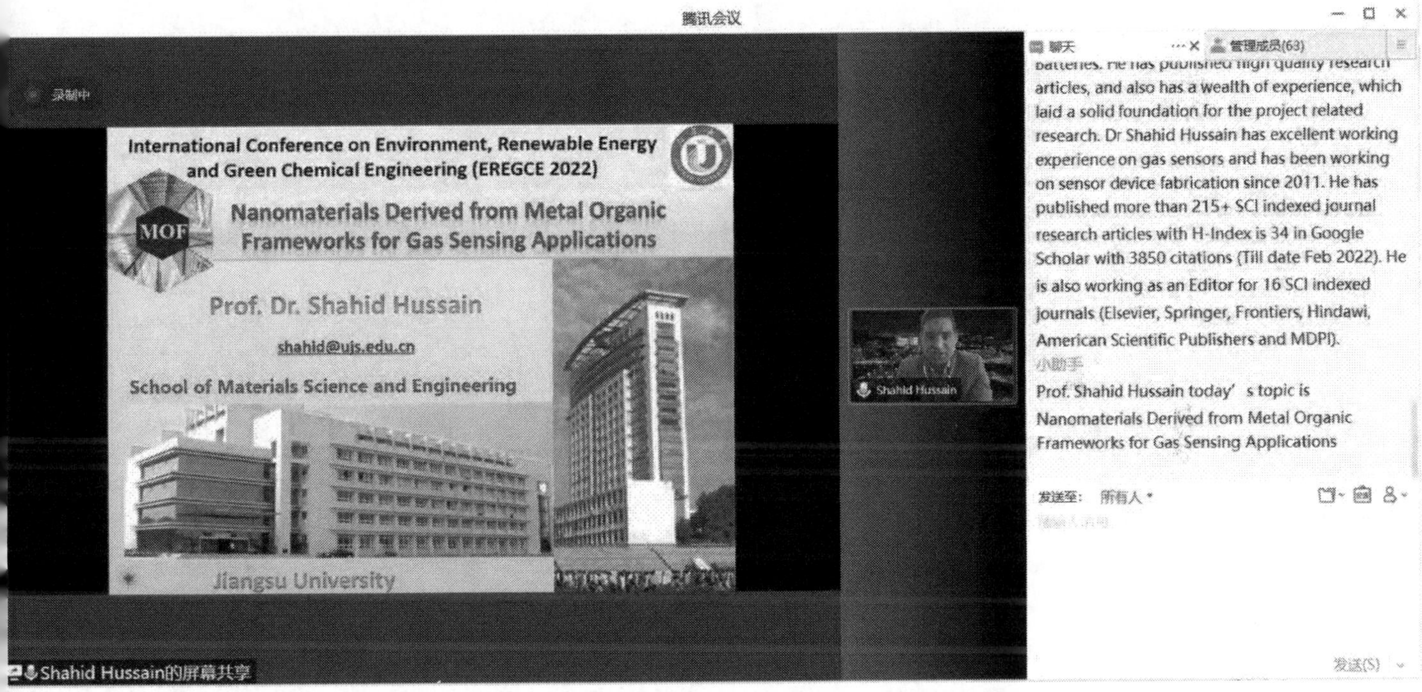
腾讯会议
录制中
International Conference on Environment, Renewable Energy and Green Chemical Engineering (EREGCE 2022)
MOF
Nanomaterials Derived from Metal Organic Frameworks for Gas Sensing Applications
Prof. Dr. Shahid Hussain
shahid@ujs.edu.cn
School of Materials Science and Engineering
Jiangsu University
Shahid Hussain
Shahid Hussain的屏幕共享
聊天
管理成员(63)
articles, and also has a wealth of experience, which laid a solid foundation for the project related research. Dr Shahid Hussain has excellent working experience on gas sensors and has been working on sensor device fabrication since 2011. He has published more than 215+ SCI indexed journal research articles with H-Index is 34 in Google Scholar with 3850 citations (Till date Feb 2022). He is also working as an Editor for 16 SCI indexed journals (Elsevier, Springer, Frontiers, Hindawi, American Scientific Publishers and MDPI).
小助手
Prof. Shahid Hussain today' s topic is Nanomaterials Derived from Metal Organic Frameworks for Gas Sensing Applications
发送至: 所有人
发送(S)

您正在观看EREGCE主持人的屏幕
01:05:09
演讲者视图
录制中
锁定画面
Keynote Speaker
Assoc. Prof Zhanying Zhang
Assoc. Prof Zhanying Zhang is the leader of Bioprocessing and Biorefining Group at the Centre for Agriculture and the Bioeconomy (CAB), Queensland University of Technology (Brisbane, Australia). He is also an Advance Queensland Industry Research Fellow. Assoc. Prof. Zhanying Zhang graduated from the University of Adelaide in 2008. He joined the CAB, QUT after he completed his PhD study. He has been working on lignocellulose pretreatment and biorefining, cellulosic biofuels and biochemicals, biomass-based materials since 2008. He developed several solvent-based innovative processes for pretreatment and fractionation of lignocellulosic biomass, which led to a number of patent applications and journal publications. His research is important to solve the real world challenges in energy, waste management and climate change. Assoc. Prof. Zhang has generated 102 publications, including 87 journal articles, 10 conferences papers and 5 book chapters. In addition, Assoc. Prof. Zhang has been granted 11 patents worldwide in the area of biomass pretreatment and biorefineries.
EREGCE主持人
合肥学院-纵诚
王KM
Zhanying Zhang
小助手
解除静音
开启视频
共享屏幕
安全
邀请
管理成员(68)
聊天
结束录制
分组讨论
应用
设置
结束会议

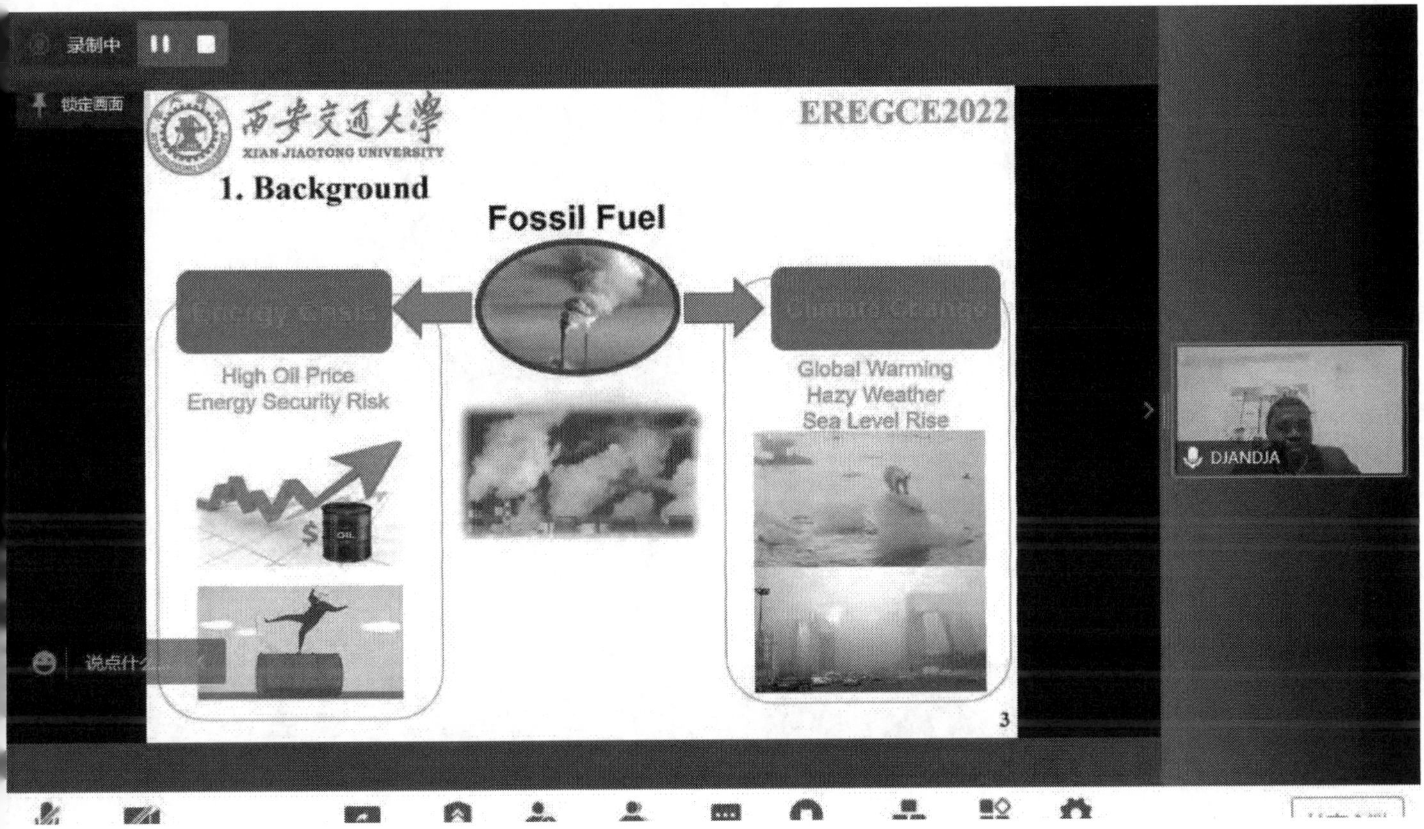
录制中
锁定画面
西安交通大學
XIAN JIAOTONG UNIVERSITY
EREGCE2022
1. Background
Fossil Fuel
Energy Crisis
Climate Change
High Oil Price
Energy Security Risk
Global Warming
Hazy Weather
Sea Level Rise
OIL
3
DJANDJA
说点什么...

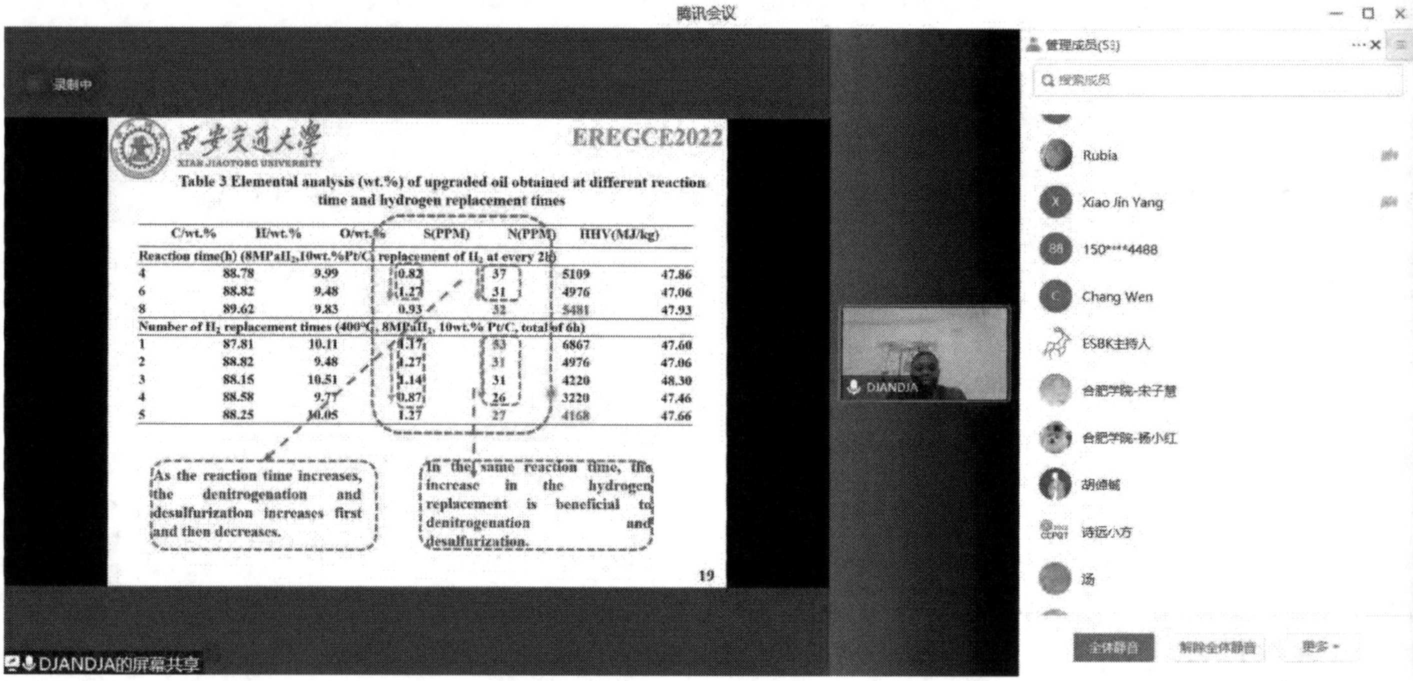
腾讯会议
录制中
西安交通大學
XIAN JIAOTONG UNIVERSITY
EREGCE2022
Table 3 Elemental analysis (wt.%) of upgraded oil obtained at different reaction time and hydrogen replacement times
C/wt.% H/wt.% O/wt.% S(PPM) N(PPM) HHV(MJ/kg)
Reaction time(h) (8MPaH2,10wt.%Pt/C, replacement of H2 at every 2h)
4 88.78 9.99 0.82 37 5109 47.86
6 88.82 9.48 1.27 31 4976 47.06
8 89.62 9.83 0.93 32 5481 47.93
Number of H2 replacement times (400°C, 8MPaH2, 10wt.% Pt/C, total of 6h)
1 87.81 10.11 1.17 53 6867 47.60
2 88.82 9.48 1.27 31 4976 47.06
3 88.15 10.51 1.14 31 4220 48.30
4 88.58 9.77 0.87 26 3220 47.46
5 88.25 10.05 1.27 27 4168 47.66
As the reaction time increases, the denitrogenation and desulfurization increases first and then decreases.
In the same reaction time, the increase in the hydrogen replacement is beneficial to denitrogenation and desulfurization.
19
DJANDJA
DJANDJA的屏幕共享
管理成员(53)
搜索成员
Rubia
Xiao Jin Yang
150****4488
Chang Wen
ESBK主持人
合肥学院-宋子慧
合肥学院-杨小红
胡德斌
诗远小方
汤
全体静音
解除全体静音
更多

腾讯会议

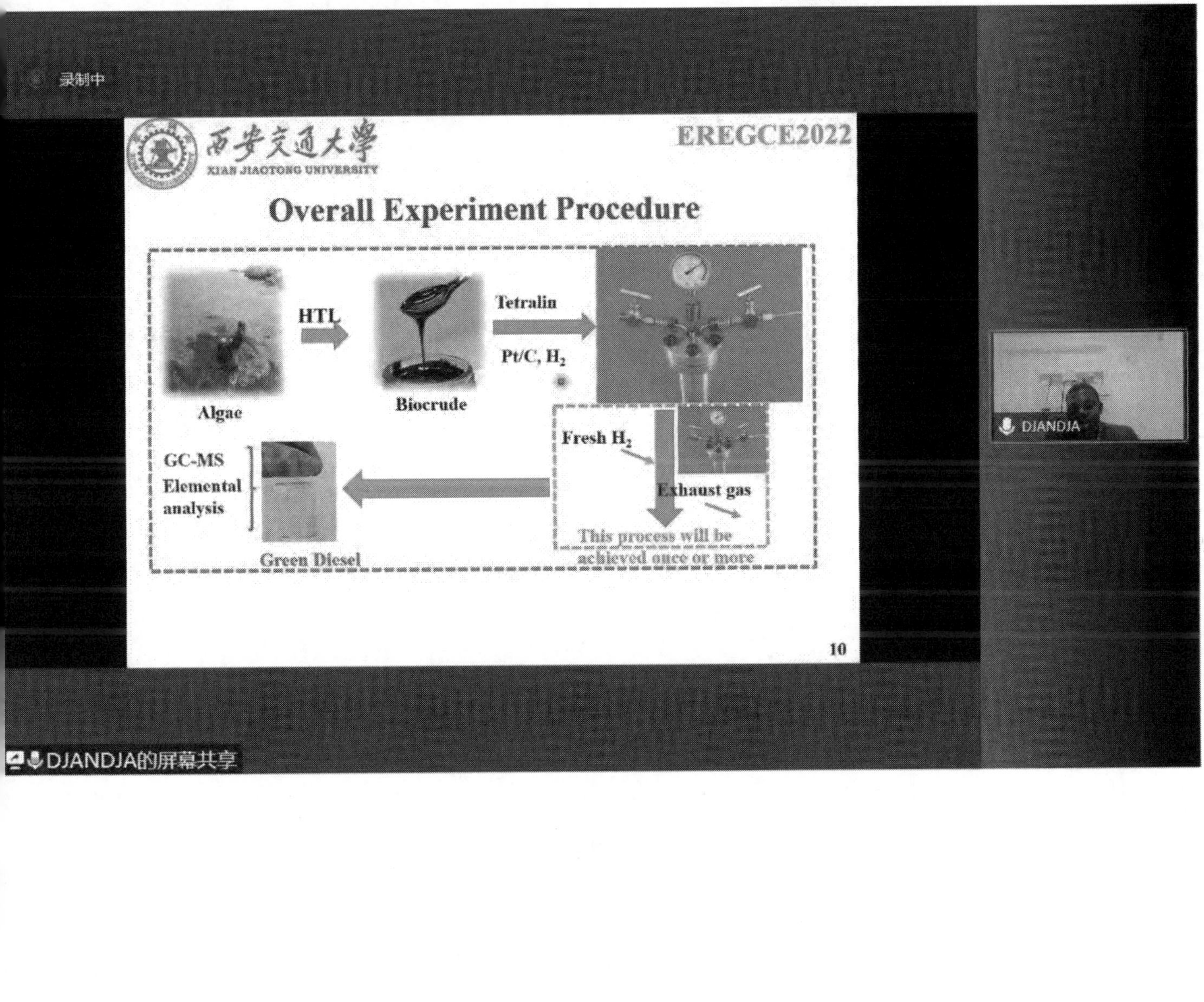
录制中
西安交通大學
XIAN JIAOTONG UNIVERSITY
EREGCE2022
Overall Experiment Procedure
HTL
Tetralin
Pt/C, H2
Algae
Biocrude
Fresh H2
Exhaust gas
GC-MS
Elemental analysis
This process will be achieved once or more
Green Diesel
10
DJANDJA
DJANDJA的屏幕共享

Statement of Peer Review

In submitting conference proceedings to Web of Conferences, the editors of the proceedings certify to the Publisher that

1. They adhere to its Policy on Publishing Integrity in order to safeguard good scientific practice in publishing.
2. All articles have been subjected to peer review administered by the proceedings editors.
3. Reviews have been conducted by expert referees, who have been requested to provide unbiased and constructive comments aimed, whenever possible, at improving the work.
4. Proceedings editors have taken all reasonable steps to ensure the quality of the materials they publish and their decision to accept or reject a paper for publication has been based only on the merits of the work and the relevance to the journal.

Title, date and place of the conference

International Conference on Environment, Renewable Energy and Green Chemical Engineering
Online, China | April 22-24, 2022

Proceedings editor(s):

Chunming Yang (Hunan Normal University)
Haoming Chen (Nanjing University of Science and Technology)
Peigao Duan (Xi'an Jiaotong University)
Feipeng Jiao (Central South University)
Chang Wen (Huazhong University of Science and Technology)

Date and editor's signature

Chunming Yang
Haoming Chen
Peigao Duan
Feipeng Jiao
Chang Wen

Extraction of Refined Crop Type Over Agriculture Region of Heilongjiang

Yanmin Shuai[1,2,3,4], *Hao* Wu[1,*], *Congying* Shao[1], and *Kexin* Ning[1]

[1] Liaoning Technical University, College of Surveying and Mapping and Geographic Science, 123000 Fuxin, China
[2] Xinjiang Institute of Ecology and Geography, CAS, 830011 Urumqi, China
[3] Research Center for Ecology and Environment of Central Asia, CAS, 830011 Urumqi, China
[4] University of Chinese Academy of Sciences, 100049 Beijing, China

Abstract. The spatial distribution of fine crop types at regional scale is required by numerous research communities. The traditional approach with limited time-phases is hard to capture the signatures presented within different growth period of various crop types. With the improvement of understanding on the phenology feature of major crops and the accumulation of satellite-based observations, there is a chance to distinguish detailed crop clusters with elevated accuracy. In this work, we investigated the phenological feature of four dominant crops (soybean, wheat, maize, and paddy) in multi-spectrum space through ~800 representative crop samples within typical agriculture regions of Heilongjiang using MODIS daily surface reflectance product covering related growth period of 2005-2018. Features with the high degree of separation among land cover clusters are screened out to construct the model in identifying typical crop types in terms of weighted temporal features and classification scheme, which is applied to extract the crop map of Heilongjiang province in 2019. The results show higher accuracy achieved over main agriculture region of soybean, wheat, maize, and paddy, and reduced accuracy over field of wheat or other mixed crops at MODIS pixel scale. Our validation shows the overall accuracy of 0.9816 and kappa coefficient of 0.9702 through the comparison with ~3000 random selected ground sites. The preliminary application of the presented approach performs well via the capture on valid phenology features of major crops within dominant agriculture region of Heilongjiang, with the potential to serve the extraction of fine crop types over wide agriculture regions.

1 Introduction

Accurate distribution of crop types is one of important datasets for food security, agricultural management and policy development, environmental sustainability and land surface ecosystem monitoring [1-2]. With the accumulation of satellite-based observation in medium or high resolution, there is a chance to develop effective methods in retrieving crop information and distinguishing various crop types, thus provide the potential refined crop map for user communities [3]. Numerous studies demonstrated the importance of multi-temporal information in crop identification. Supervised methods have been used most widely in multi-temporal classification, like decision tree [3], random forest [4], support vector machine (SVM) [5] and other nonparametric methods [6]. These methods can process multiple time-phase images to identify various land cover clusters. However, there is still a lack of the investigation on phenology features of different crops, especially in the satellite-based thematic retrieval.

Time series classification method has great potential in crop identification, due to the ability to capture unique growth rhythm and intrinsic phenological signatures [7]. In this study, we used MODIS data to extract a variety of time series features, and established a time series feature classification model to improve the accuracy of crop identification.

2 Study Area and Data

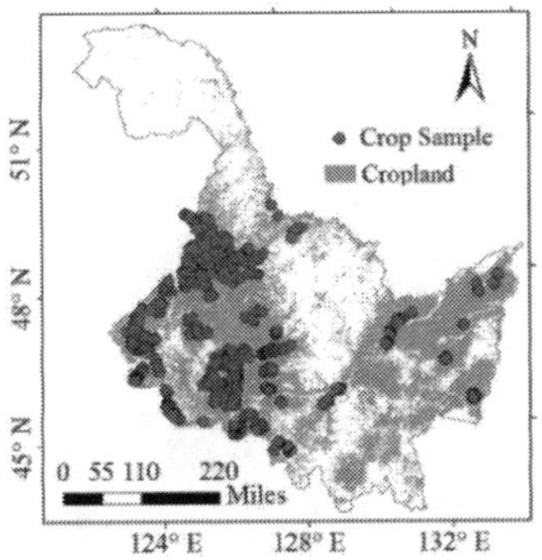

Fig. 1. The study area and crop samples in HLJ from 2005 to 2018 (the cropland is from GlobeLand30 in 2020 [8]).

* Corresponding author: 471920558@stu.lntu.edu.cn

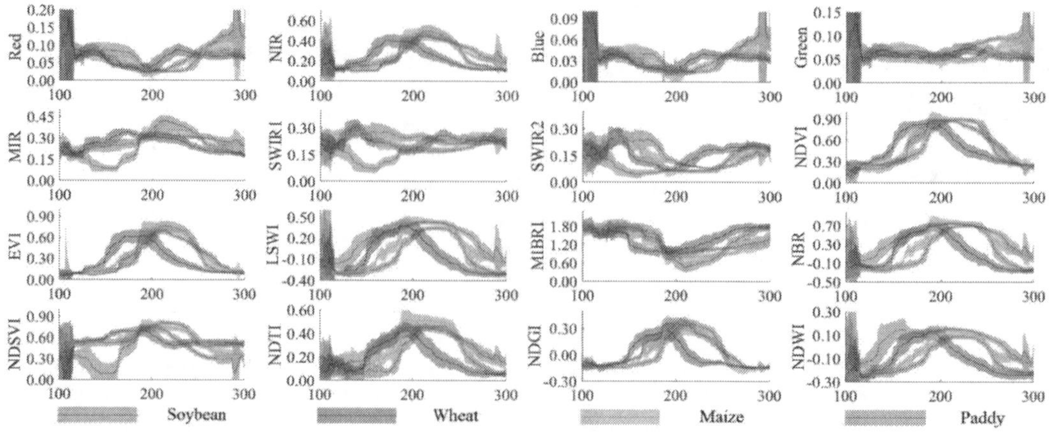

Fig. 2. The spectral reflectance and vegetation index (VI) of crops during the growth period of DOY 100-300 (x-axis), which the solid lines represent the mean value of features, and the translucent zones represent the standard deviation of features.

Heilongjiang (HLJ) is the northernmost province of China, plays the significant role in the supply of grains and agriculture management. The total sown area is 147,700 km^2, which is mainly distributed in Songnen Plain (SNP) and Sanjiang Plain (SJP). Now, percentages of plant area for soybean, maize and paddy respectively have occupied up to 40% ,13% and 12% of the total grains' amount within China. Wheat was widely planted in northern HLJ before 2010, but the current planting area has been reduced to less than 1% of the total planting area of the province [9-10]. HLJ has a continental monsoon climate. The accumulated temperature (>10℃) is 1800-2800 ℃, and the frost-free period is between 100-150 days. The annual rainfall averages 300-700mm, concentrated mainly from June to August [10].

MODIS provides operational standard products at 500m in daily, 8-day, and monthly data suits. We adopted the daily MODIS surface reflectance dataset MCD43A4 (collection 6) Nadir Bidirectional Reflectance Distribution Function (BRDF)-Adjusted Reflectance (NBAR) dataset during the growth period of 2005-2019, and as well as the related quality assessment layer to screen clear observations with less aerosol contamination.

We collected 3880 representative crop samples over the main agriculture region of HLJ from 2005 to 2018, according to the ground true data in 2018 and high-resolution remote sensing images (Fig. 1). In this study, we randomly selected 20% crop (~900) samples to construct the fuzzy membership model. The remaining 80% (~3000) samples were used to verify the classification accuracy. Due to the total sown area of HLJ almost unchanged in 2019-2020 [10], the GlobalLand30 of 2020 was used to screen out the non-agricultural region to tighten this work on the identification of fine crop types. In addition, the 10-m crop maps [11] in Northeast China in 2019 was used for cross-comparison.

3 Method

3.1 Time series Vegetation Index

Vegetation Index is frequently considered as the enhanced signature of crop types than individual reflectance band, which combines multiple bands of reflectance to highlight vegetation information and suppress disturbance from non-vegetation surface clusters. We investigated phenology features of soybean, maize, paddy, and wheat through seven MODIS spectral bands and nine selected indices (shown in Table 1) that have been widely used in detecting different biophysical characteristics of vegetation, especially on the change of chlorophyll content, water content and canopy structure.

Table 1. The calculation formulas of the VI in this study (The B1-B7 respectively represent the reference bands 1-7 of MCD43A4)

VI	Calculated Formula of MCD43A4
Normalized Differenced Vegetation Index, NDVI [10]	$\frac{B2 - B1}{B2 + B1}$
The Enhanced Vegetation Index, EVI [5]	$2.5 \times \frac{B2 - B1}{B2 + 6 \times B1 - 7.5 \times B3 + 1}$
Land Surface Water Index, LSWI [11]	$\frac{B2 - B6}{B2 + B6}$
Mid-IR Bi-spectral Index, MIBRI [12]	$10 \times B7 - 9.8 \times B6 + 2$
Normalized Burn Ratio, NBR [12]	$\frac{B2 - B7}{B2 + B7}$
Normalized Difference Senescent Vegetation Index, NDSVI [12]	$\frac{B6 - B1}{B6 + B1}$
Normalized Difference Tillage Index, NDTI [10]	$\frac{B6 - B7}{B6 + B7}$
Normalized Difference Green Index, NDGI [10]	$\frac{B4 - B1}{B4 + B1}$
Normalized Difference Water Index, NDWI [13]	$\frac{B2 - B5}{B2 + B5}$

3.2 Separability Index (SI)

We adopted SI to examine the separability of above selected factors in different growth period among crops. Somers defined SI as the ratio of inter-class spectral heterogeneity to intra-class spectral heterogeneity in 2010 [14]. For the spectrum of the two object types, SI increases when the intra-class spectral variance ($\Delta intra$) decreases and the inter-class distance ($\Delta inter$) increases, whereas decreases [10]. We calculated SI for each pair of crop types during DOY 100-300, and a global SI using the average method to describe the separability of time series features among all types. The formulas are as followed:

$$SI_{ij}(v,d) = \frac{\Delta inter}{\Delta intra} = \frac{|\mu_i - \mu_j|}{1.96 \times (\sigma_i + \sigma_j)} \tag{1}$$

$$SI_{gloval} = 1/M \times \sum_{i=1}^{n-1} \sum_{j=i+1}^{n} SI_{ij} \tag{2}$$

where SI_{ij} is the SI of type i and type j ($i, j = 1, 2, \ldots,$ n); n is the number of classes; v refers to a band or a VI in this study; d refers DOY; μ is the mean value of features; σ is the standard deviation of features; n is the number of all types; M is the number of class pairs.

3.3 Fuzzy Membership Model (FMM)

The FMM was used to evaluate the similarity of features between unclassified sample to target type. Fuzzy membership degree, from 0 to 1, is used to describe the degree of similarity between a sample to target class [15]. The formula is as followed:

$$s_{ik}(v,d) = e^{-(\frac{x_k - \mu}{2\sigma})^2} \tag{3}$$

where x_k is the feature value of sample k; s_{ik} means the fuzzy membership of sample k to class i; v, d, μ and σ are similar as the above formulas.

3.4 Multi-stage Weighting Model (MWM)

The MWM was used to screen out the features with the high degree of separation among land cover clusters, and identify the crop types according to the multi-stage results. In each stage, a normalized SI matrix was used to weight the FMM, and returned membership degrees of the unclassified sample belonging to target types. The weighting formulas are as followed:

$$p_{ik}^{(m)} = \sum_v \sum_d s_{ik}(v,d) \times SI_N^{(m)}(v,d) \tag{4}$$

$$SI_N(v,d) = SI(v,d) / \sum\sum SI(v,d) \tag{5}$$

where p_{ik} means the probability of sample k belonging to class i; SI_N is the normalized matrix of all SI (removing the values when the feature of sample i miss observation or the value is below the threshold of t_1); m means the m-th (the m times) judgment currently in progress; s, v and d are similar as the above formulas.

Firstly, we used the global SI to weight fuzzy membership matrix and summed up the weighted matrix for identifying the sample type as the first classification stage (m=1). The first stage returns the weighted membership degree which means the similarity degree of sample k belongs to each type. When the membership degree of class i is higher than t_2 (t_2 is the value of s when x_k-μ=2σ) and higher t_1 than the membership degree of other types, it is considered that sample k belongs to the type i. When the membership degree of type i is greater than t_3, the possibility of sample k belonging to type i the is marked +1, and the possibility belonging other type is marked 0. Otherwise, demanding the first stage results invalid.

Secondly, we evaluated the possibility of belonging to all paired crop types (m>1). When the membership degree of class i is higher t_3 than the membership degree of another, the possibility with the higher membership type is marked as +1, and the possibility with the lower membership type is marked as -1 Otherwise, the possibility of the judgment in m-th stage is marked as 0.

Finally, the possibilities of sample k belonging to each type are summed together. When the weighted membership degree returned in any stage is less than t_4 (t_4 is the value of s when x_k-μ=3σ), the possibility of sample k belonging to this type will become 0. When there is a possibility greater than 0, take the category with the maximum possibility as sample k. If the minimal probability is less than or equal to 0, marked sample k as non-crop type. In this study, the parameters t_1, t_2, t_3 and t_4 in classification were set as 0.7788, 0.3000, 0.1000 and 0.3679.

4 Results & Discussion

4.1 Crop separability

Fig.3. shows a global SI chart (a) for multiple crops and six SI charts (b-g) for each pair of crops (soybean, wheat, maize, and paddy) during the growth period of DOY 100-300 (x-axis) at seven spectral bands and nine indices (y-axis). The higher SI value means the higher separability for target crop types over 16 indices and 201 time points, and the corresponding grid becomes increasingly red. In this study, most of features with high SI values exhibit as the NIR- or SWIR- related indices, which are respectively linked with crop biomasses at the green-up and maturing stages, or the water content.

Fig. 3. (a) shows the global SI of four crop types with highest SI value of 1.4. The global SI higher than 1.0 is mainly during DOY 210-280, especially in NIR, SWIR2 and nine VI. However, the highest SI value during 140-160 (0.8-1.0) is much lower than the value of some pairwise SI, especially to Fig.3. d and g (>1.5).

The pairwise SI between wheat and other crop types (Fig.3. b, e, f) perform highest value about 1.8-2.4, especially to soybean and maize (b, e), and almost all of

VI can effectively distinguish wheat from other crops, during DOY160-210. To reflectance features, the highest SI of reflectance features is only concentrated in low vegetation cover periods (DOY140-160), and the highest value is about 1.5, which is still much higher than the maximum SI between soybean and maize.

The SI between soybean and maize (Fig.3. c) are relatively lower than other crop pairs with maximum 1.3. NDVI is the most widely used VI in crop classification, but the highest SI value is only 0.8. The higher SI between soybean and maize (1.0-1.3) is mainly showed in SWIR- related bands and VI (SWIR1, SWIR2, LSWI,

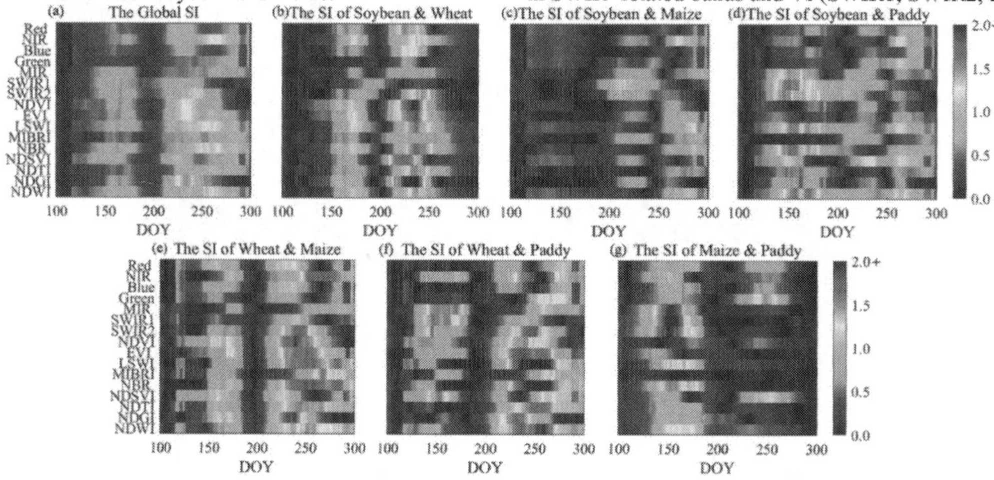

Fig. 3. Estimated SI values of each pair of crops during the growth period of DOY 100-300 (x-axis) at seven spectral bands and nine indices (y- axis).

NBR, MIBRI and NDWI) during DOY200-240.

The paddy presents the remarkable separability (SI>1.5) to maize and soybean (Fig.3. d, g) during transplanting period (DOY130-160). For paddy and soybean (Fig.3. d), the SI of three spectral bands (MIR, SWIR1 and SWIR2) and three SWIR- related indices (LSWI, NBR and NDMI) is similar (1.5-1.7). For paddy and maize (Fig.3. g), the SI of three spectral bands (1.7-2.1) is even higher than the SI of the three SWIR- related indices (1.5-1.7).

In this part, the four crop types have showed a great possibility to distinguish each other by the time series features. There is the highest separability between wheat and other types, due to the growth and the dry leaf period of wheat are earlier 30 days approximately than other crops. Similarly, paddy is extractable, due to the spectral features during irrigation and transplanting period (DOY140-160). The SI between soybean and maize is slightly lower, but the time series features with relatively high SI last for more than 40 days, which may give helps to effectively distinguish them.

4.2 Classification results

We applied the above presented method to classified 3103 random selected ground sites during 2005-2018 and the overall agriculture region of HLJ in 2019 using MODIS daily data of DOY 100-300. Table 2. shows the classification results of 3103 random selected ground samples. Each column of the matrix represents the indicates in actual crop types, and the row of the matrix represents the indicates in predicated crop types. In case of the classification results during 2005-2018, the overall accuracy is 0.9816, and the kappa coefficient is 0.9702. Almost all samples of wheat and paddy were correctly classified. Soybean and maize samples were slightly misclassified. The user accuracy of soybean (0.9330) is lower than the producer accuracy (0.9953), means the soybean area extraction by the model may be overestimated. Relatively, the maize area extraction by the model may be underestimated. In this study, about 40% agricultural samples were correctly classified by the first classification stage in MWM, and 60% agricultural samples were correctly classified by the 2-7 classification stage in MWM.

Table 2. The confusion matrix of classification results for samples in HLJ during2005-2018 (the Soy- means Soybean).

		Actual				
		Soy-	**Wheat**	**Maize**	**Paddy**	**ALL**
Predicted	**Soy-**	640	0	46	0	686
	Wheat	0	50	0	0	50
	Maize	3	0	1608	8	1619
	Paddy	0	0	0	748	748
	ALL	643	50	1654	756	3103

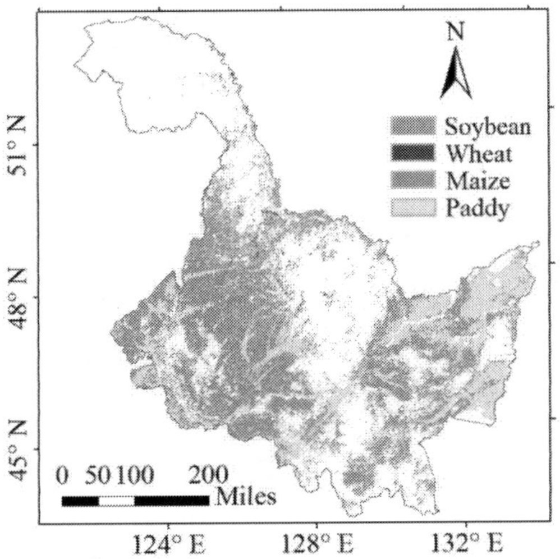

Fig. 4. The extracted crop map (500m) of HLJ in 2019

Fig.4. shows the extraction results for four crop types over the agriculture region of HLJ in 2019. Maize is the most sown crop in HLJ, followed with paddy, and they are mainly distributed in SJP and the south of SNP. Soybean is mainly distributed in the north of SNP. Wheat is the least sown crop and sporadically distributed in the north of HLJ. The 10-m crop maps in Northeast China in 2019 was used for cross-comparison, as showed in Fig.5. In the soybean dominant planting region (Fig.5. a1, a2), the extracted soybean area of MODIS-based 500-m crop map was underestimated compared to the soybean area of 10-m crop map. It has the same performance in maize dominant planting areas. For most of paddy dominant planting regions, the extracted paddy distribution in this study is highly consistent with the paddy distribution of 10m reference map. In some boundary of maize field and paddy field, numbers of MODIS pixels with mixed crop types were divided into soybean, although soybean is rarely planted in this area (Fig. 5. b1, b2). Besides, the classification results performed weak over the field of wheat or other mixed crops at MODIS pixel scale, because of the increase of fragmentation crop fields which are far smaller than 500m MODIS pixel.

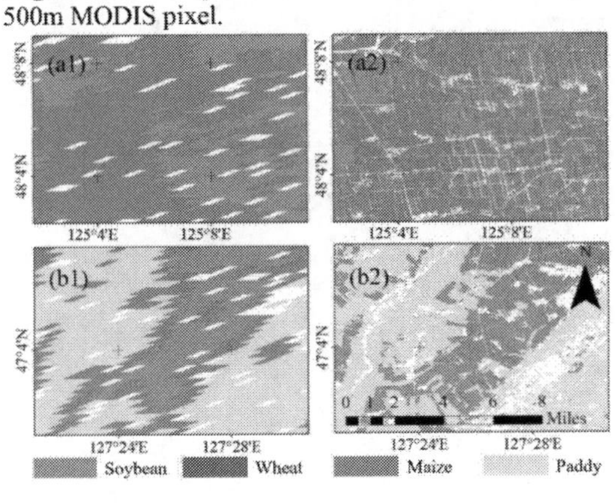

Fig. 5. Comparison between MODIS-based 500-m crop map (a1, b1) and the sentinel-based 10-m crop map (a2, b2) in case of two typical crop regions of HLJ in 2019.

5 Conclusions

We developed a physiological feature determined classification model at daily intervals during the growing season, and we classified four major crops (soybean, wheat, maize, and paddy) in case study of agriculture region of HLJ. The results show higher accuracy achieved over main agriculture region of soybean, wheat, maize, and paddy, and reduced accuracy over field of wheat or other mixed crops at MODIS pixel scale. Compared with the traditional multi temporal classification method, our model potentially improves the degree of automatic extraction of fine crop categories in the large region scale, due to the classification mechanism is interpretable and flexible. In the following study, we are going to continuously contribute efforts to improve the model and the identification of the misclassified targets.

This work was supported by the National Natural Science Foundation of China (No. 42071351), the National Key Research and Development Program of China (No. 2020YFA0608501, No. 2017YFB0504204), the Liaoning Revitalization Talents Program (No. XLYC1802027), talent recruitment program of CAS (No. Y938091, No. Y674141001), Project supported discipline innovation team of Liaoning Technical University (No. LNTU20TD-23).

References

1. B. Wu, J. Meng and Q. Li, China crop watch system with remote sensing. Journal of Remote Sensing, **8** 481-497. (2004)
2. V.S. Konduri, J. Kumar, J.W.W. Hargrove, et al. Mapping crops within the growing season across the United States. Remote Sensing of Environment, **251** 112048. (2020)
3. Z. Chen, J. Ren, H Tang, et al. Progress and perspectives on agricultural remote sensing research and applications in China. Journal of Remote Sensing, **20(5)**: 748-767 (2016)
4. L. Zhong, P. Gong and G. Biging, Efficient corn and soybean mapping with temporal extend ability: A multi-year experiment using Landsat imagery. Remote Sensing of Environment, **140** 1-13 (2014)
5. H. Thi, T Nguyen, L.V. Nguyen, et al. Mapping Maize Cropping Patterns in Dak Lak, Vietnam Through MODIS EVI Time Series. Agronomy, 10: 1-16. (2020)
6. Q. Hu, W. Wu, Y. Shao, et al. Advances in remote sensing extraction of crop planting structure. Scientia Agricultura Sinica., **48** 1900-1914. (2015)
7. Q. Xu, G. Yang, H. Long, et al. Crop information identification based on MODIS NDVI time-series data. Transactions of the Chinese Society of Agricultural Engineering, **30** 134-144. (2014)

8. C. Jun, Y. Ban and S. Li, China: Open access to Earth land-cover map. Nature, **514**: 434-434. (2014)
9. National Bureau of Statistics. Statistics of main crop sown areas in each province of China. (2021). http://www.stats.gov.cn/tjsj/
10. Q. Hu, Research on crop remote sensing recognition method based on time series MODIS images. Chinese Academy of Agricultural Sciences. (2018)
11. N. You, J. Dong, J. Huang, G. Du, et al. The 10-m crop type maps in Northeast China during 2017–2019. Scientific Data, **8** 1-11. (2021)
12. L. Gu, Y. Shuai, C. Shao, et al. Angle Effect on Typical Optical Remote Sensing Indices in Vegetation Monitoring. Remote. Sensing, **13**: 1699. (2021)
13. B. Gao, NDWI—A normalized difference water index for remote sensing of vegetation liquid water from space. Remote Sensing of Environment, **58**: 257-266. (1996)
14. B. Somers, S Delalieux, W.W. Verstraeten, et al. An automated wave band selection technique for optimized hyperspectral mixture analysis. International Journal of Remote Sensing. **31** 5549-5568. (2010)
15. Z. Peng and Y. Sun, Fuzzy Mathematics and its Applications. Wuhan University Press, Wuhan. (2007)

Analysis of Energy Flow in the Sea Area of Zhongjieshan Islands and Assessment of Ecological Capacity of Sepiella Maindroni Based on Ecopath Model

Wenjing Shi[1], *Feng* Gui, *Manyao* Zhang, *Xianmin* Zeng, *Kaida* Xu and *Sheng* Zhao[2,*]

Zhejiang Ocean University, School of Marine Science and Technology, 316000, Zhejiang Ocean University, Zhoushan, Zhejiang. Peoples R China.

[1]shiwenjing01@163.com.

[2]zhaosh@zjou.edu.cn,13454069742.

Abstract. Sepiella maindroni was once one of the four traditional marine products in our country. In order to evaluate the ecological capacity of Sepiella maindroni in the sea area of Zhongjieshan Islands, this study used Ecopath with Ecosim 6.5 (EwE) software to construct an Ecopath model composed of 17 functional groups in the sea area of Zhongjieshan Islands. The nutrient structure and energy flow characteristics of the sea area of Zhongjieshan Islands are analyzed from the ecosystem level, and the reasonable ecological capacity of the sea area of Zhongjieshan Islands is estimated by the model.The results show that the trophic level of each functional group in the marine ecosystem of the Zhongjieshan Islands is between 1-4.648, total primary production/total respiration is 4.539, the system Connectance Index is 0.478, the System Omnivory Index is 0.389, The trophic level of Sepiella maindroni is 3.562, and the Ecotrophic Efficiency is 0.306, indicating that the Ecotrophic Efficiency of each trophic level in the system is low and the energy is not fully utilized. There is still room for breeding and exile for the Sepiella maindroni. The model estimates that the ecological capacity of Sepiella maindroni is about $2t\cdot km^2$.

1 Introduction

Sepiella maindroni, commonly known as cuttlefish, belongs to the class Cephalopods. It is widely distributed along the coast of our country and is one of the four major fishery products in the East China Sea. But due to human large-scale fishing and unreasonable use, the resources of Sepiella maindroni decreased sharply in the early 1980s and almost disappeared in the 1990s. The resource situation is worrying. In recent years, in order to restore the resource amount, Zhejiang Fisheries has actively carried out resource enhancement measures combining artificial breeding of seedlings and release, and carried out large-scale reproduction and release work, but it still cannot meet the requirements of the recovery of the Sepiella maindroni resources. The current research on the resource recovery of Sepiella maindroni mainly focuses on the biological characteristics and seed breeding [1-6]. There are relatively few studies on the evaluation of its ecological capacity. In previous studies on the Sepiella maindroni, many experts and scholars have carried out some studies on its biology, reproductive habits, migration distribution, resource status, etc. Tang Yimin and others investigated the species of phytoplankton in the spawning ground area of Zhongjieshan. Zooplankton species, benthic species composition and main species distribution and changes.

At present, most of the assessment for ecological capacity is done by models. Some models require too many parameters and the operation of the model is more complicated. Therefore, relatively speaking, the Ecopath model constructed by EwE software is an ideal tool for studying biological ecological capacity because of its convenient operation and its ability to cover all limiting factors of ecological capacity. Ecological carrying capacity assessment is the basis and prerequisite for carrying out the conservation of biological resources and maintaining the health of the ecosystem.

The sea area of Zhongjieshan Islands is located in the east of Zhoushan Islands New Area. The superior geographical location and marine environmental conditions make it an important habitat for the reproduction, feeding and growth of many marine organisms. The biodiversity of this sea area is relatively high, which has high research value, economic value and protection value. The Sepiella maindroni is also one of the most representative biological species in the sea area, and it is also an important resource protection and restoration species. In this study, by constructing the Ecopath model in the sea area of Zhongjieshan Islands, the nutrient structure and energy flow characteristics in the sea area of Zhongjieshan Islands were analyzed from the ecosystem level. The model is used to estimate the release flow, determine the reasonable flow and scale of the biological resources, and evaluate the ecological capacity of the growth of Sepiella maindroni, in order to provide help for the subsequent construction of the Sepiella maindroni Habitat. In this study, when using the

* Corresponding author: zhaosh@zjou.edu.cn

Ecopath model to estimate the ecological capacity of the proliferation species, the Ecotrophic Efficiency can be maximized from the perspective of the ecosystem food web and food chain, and the number of released species can be maximized without destroying the own ecosystem. , To utilize the energy of the ecosystem as much as possible, to ensure the survival rate of proliferation species, and to promote the development of "productive" fisheries.

2 Materials and Methods

2.1 Model principle

The Ecopath model is an important module in the Ecopath with Ecosim (EwE) software. It is an important method for studying the energy flow and material circulation of the ecosystem. Its main content is to quantitatively describe the flow of energy between the biological components of the ecosystem by determining the important ecological parameters of the ecosystem such as biomass, production, consumption, and ecological nutrient conversion efficiency according to the law of conservation of energy.

The definition of the Ecopath model requires that the ecosystem is composed of a series of interrelated functional groups, and that all functional groups must cover the entire process of the energy flow of the ecosystem. The system function group can include organic debris, plankton, and a function group consisting of fish species of the same size, age, or ecological characteristics. According to the principle of thermodynamics, the Ecopath model defines the energy output and input of each functional group in the system to maintain a balance, that is, production-predator death-other natural death-output = 0. The model defines the ecosystem with a set of simultaneous linear equations, where each linear equation represents a functional group in the system, expressed by the following mathematical formula:

$$P_i = Y_i + B_i + B_i + M_{2i} + E_i + B_{Ai} + M_{0i} + B_i$$

$$Q_i = P_i + R_i + U_i$$

Among them: $M_{0i} = \frac{P_i \times (1 - EE_i)}{B_i}$,

$$M_{2i} = \sum_{j=1}^{n} \frac{Q_j \times DC_{ji}}{B_i}$$

P_i is the production of organisms in functional group$i$, Y_i is the total catch, B_i is the biomass, E_i is the net migration, B_{Ai} is the biomass accumulation, R_i is the respiration, U_i is the amount of undigested food, and Q_i is Consumption, M_{0i} is other mortality, M_{2i} is predation mortality, EE_i is the Ecotrophic Efficiency of functional group$i$, which refers to the proportion of production used in the system,. See Table 1 for specific definitions. In the Ecopath model, at least 3 of the 4 parameters of B_i, $(P/B)_i$, $(Q/B)_j$ and EE_i must be known, and one unknown parameter can be calculated by the model inferred.

Table 1. The interpretation of each parameter in the Ecopath model

parameter	definitions
P_i	production of organisms in functional group$i$
Y_i	Catch of organisms in functional group$i$
B_i	Biomass of organisms of functional group$i$
E_i	Net migration of organisms of functional group$i$
B_{Ai}	Biomass accumulation
M_{0i}	Other biological mortality of functional group$i$
M_{2i}	Mortality of predator in functional group$i$
EE_i	Ecotrophic Efficiency of functional group$i$
DC_{ji}	The biological composition ratio of predator functional group$i$ to predator functional group$j$

2.2 Functional group division

Based on the existing biological community structure and model calculation rules in the sea area of Zhongjieshan Islands, this study will analyze the species with a high degree of overlap in the niche (food composition, age composition, feeding method, individual size, and statistical classification method of catches). Merged to simplify the food web and divide organisms with similar functional status into a functional group. According to the existing survey data, in this study, the marine ecosystem in the sea area of Zhongjieshan Islands is divided into 17 functional groups.

The functional groups basically cover the entire process of energy flow of the marine ecosystem in the sea area of Zhongjieshan Islands, including Detritus, phytoplankton, zooplankton, meiobenthos, Benthic crustaceans, shrimps, crabs, cephalopods, Sepiella maindroni, Larimichthys polyactis, Anchovy, Small benthic fishes, other pelagic fishes, Large yellow croaker, Large demersal fishs, Trichiurus lepturus, Other piscivorous fishes.

Since this article mainly focuses on the breeding and release of Sepiella maindroni, we set it as a functional group separately. See Table 2 for the division of specific ecosystem model functional groups.

Table 2. The functional groups and main biological species of the Ecopath model in the sea area of Zhongjieshan Islands

Serial number	Functional groups	Main species
1	Other piscivorous fishes	Lophius litulon Champsodon snyderi
2	Trichiurus lepturus	
3	Large demersal fishes	Muraenesox cinereus Paralichthysolivaceus
4	Large yellow croaker	
5	other pelagic fishes	Pampus argenteus Coilia mystus
6	Small benthic fishes	Priacanthus tayenus Acropoma japonicum
7	Anchovy	Setipinna taty anchovy
8	Larimichthys polyactis	
9	Sepiella maindroni	
10	cephalopods	Uroteuthischinensis Uroteuthisduvauceli Sepia aculeata
11	crabs	Charybdis japonica Portunus trituberculatus Charybdis bimaculata
12	shrimps	Palaemon gravieri Oratosquilla oratoria Alpheus distinguendus
13	Benthic crustaceans	shellfish snails
14	meiobenthos	Granulifusus kiranus Polychaeta
15	zooplankton	siphonohora Chinese stinging algae Sagitta
16	phytoplankton	Noctiluca scintillans coscinodiscus jonesianus prorocentrum donghaiense
17	Detritus	

2.3 Model data and parameter debugging

2.3.1 Model data

In the Ecopath model, the movement of energy in the system can be expressed in the form of energy (for example, biological wet weight t/km^2), and the length of time is generally one year. The biomass (B) data in the model is generally estimated from survey data, and the phytoplankton biomass can be calculated by converting the ratio of the chlorophyll a concentration to the carbon content of phytoplankton. The biomass of organic detritus can be estimated according to the empirical formula of organic detritus and primary production carbon [7]; Production/biomass (P/B) data is also called biomass turnover rate, which refers to the ratio of production to biomass per unit time. The unit is 1/y. Generally speaking, in a balanced and stable ecosystem, the P/B value of fish is equal to the total instantaneous death Z [8], Calculate according to the formula Z=M+F. M is the natural mortality rate, which can be calculated using the empirical formula of the growth function [9], F is the fishing mortality rate. Consumption/Biomass (Q/B) refers to the ratio of a certain animal's food intake to its biomass per unit time (usually one year). The unit is 1/y. The P/Q coefficient refers to the ratio of a certain animal's production to food intake, that is, the reciprocal of the food coefficient. When Q/B is difficult to calculate, if you know P/B and P/Q, you can also calculate Q/B. The Q/B value of fish can be calculated according to the multiple regression model of the fin profile ratio [10].

The fishery resource data of the ecosystem in the sea area of Zhongjieshan Islands are mainly derived from the collection and analysis of environmental field data and historical data, and are obtained through the analysis and collation of historical literature data on fishery resources and ecological environment. As part of the data of P/B and Q/B is difficult to obtain, this model refers to the model data with similar ecosystem characteristics in the sea area of Zhongjieshan Islands, such as the relevant system parameters of the Ecopath model in the East China Sea.

2.3.2 Parameter debugging

The debugging process of the Ecopath model is the process of keeping the ecosystem and the input and output of each functional group in balance [11]. The basic condition of ecosystem balance is that the EE of each functional group is between 0 and 1. When EE reaches 1, it means that the functional group is under predation pressure from other functional groups or fishing pressure is overly heavy [12]. In this study, when inputting the raw data in the sea area of Zhongjieshan

 https://doi.org/10.1051/e3sconf/202235001002

Islands, the EE value (Ecotrophic Efficiency) estimated by the input parameters is greater than 1, indicating that the model is unbalanced, and it is necessary to repeatedly adjust the unbalanced functional groups and the related food habits. P/B, Q/B values and food composition of functional groups. The debugging results make the EE of all functional groups in the model ≤ 1, and the input and output are kept in balance, so as to obtain reasonable values of other ecological parameters of the ecosystem [13].

3 Results and Discussion

3.1 Analysis of functional group trophic level and Ecotrophic Efficiency

In the Ecopath model of the sea area of Zhongjieshan Islands, the input and output parameters after debugging are shown in Table 3. The EE value of each functional group of the marine ecosystem is less than 1 (0.066-0.729). Among them, the EE value of the Sepiella maindroni is 0.306, which is at a medium level in the ecosystem. According to the diet analysis matrix, the trophic level of the functional group of the marine ecosystem is 1-4.648, and the entire ecosystem can be divided into 5 trophic levels. The P/Q values of all functional groups are 0.268-0.476, and the results are reasonable. In addition, some parameters missing from the model can be estimated by the model.

Table 3. Basic input parameters and estimated parameters of Ecopath model in the sea area of Zhongjieshan Islands

Serial number	group	trophic level	biomass ($t{\cdot}km^2$)	Production /Biomass (P/B)	Consumption /Biomass Q/B	Ecotrophic Efficiency (EE)	Production /consumption (P/Q)
1	Other piscivorous fishes	4.648	0.19	1.1	4.1	0.382	0.268
2	Trichiurus lepturus	4.407	0.104	2.2	5.9	0.268	0.373
3	Large demersal fishs	4.317	0.524	1.525	4.7	0.154	0.324
4	Large yellow croaker	4.080	0.21	2.42	6.46	0.557	0.375
5	other pelagic fishes	3.843	0.25	2.61	8.7	0.590	0.300
6	Small benthic fishes	3.734	0.56	6.6	15.4	0.729	0.429
7	Anchovy	3.453	0.39	4.9	10.8	0.488	0.454
8	Larimichthys polyactis	3.284	0.45	7.05	14.8	0.452	0.476
9	Sepiella maindroni	3.562	1.3	4.59	15.55	0.306	0.295
10	cephalopods	3.085	2.8	6.9	17	0.335	0.406
11	crabs	3.025	3.21	9.1	21	0.396	0.433
12	shrimps	2.833	4.68	9.9	23	0.532	0.430
13	Benthic crustaceans	2.762	6.89	10.65	26.9	0.582	0.396
14	meiobenthos	2.442	12.96	12	35	0.651	0.343
15	zooplankton	2	25.85	45	150	0.263	0.3
16	phytoplankto n	1	150.8	70		0.351	
17	Detritus	1	500			0.066	

3.2 Trophic level and food web structure

3.2.1 trophic level analysis of functional groups

The trophic level distribution of the ecological function group in the sea area of Zhongjieshan Islands shows in Figure 1. The lowest trophic level is detritus and phytoplankton, the trophic level is 1, and the highest trophic level is other piscivorous fishes. The trophic level of the Sepiella maindroni is 3.562, which belongs to the middle trophic grade in the surveyed sea area. It can be seen from Figure 1 that there are 5 trophic levels in the ecosystems in the sea area of Zhongjieshan Islands. The first trophic level is organic detritus and phytoplankton; the second trophic level includes zooplankton, meiobenthos, etc.; the third nutrition level Levels include Larimichthys polyactis, Small benthic fishes, etc. the fourth trophic level is mostly omnivorous fish; the fifth trophic level is mostly Other piscivorous fishes.

 https://doi.org/10.1051/e3sconf/202235001002

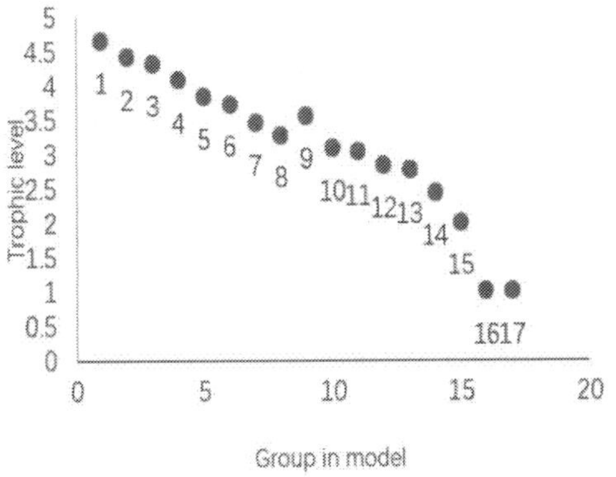

Fig.1 The trophic level of the ecological function group in the sea area of Zhongjieshan Islands,
Refer to Table 2 for the number of function groups

3.2.2 Food web structure analysis

Figure 2 is the food web structure of the ecosystem in the sea area of Zhongjieshan Islands. This figure shows the nutrient flow relationship of all functional groups in the entire study area and the influence of food competition and mutual predation between biological functional groups in the ecosystem. The circles and sizes in the figure represent the size of each functional group and biomass in the system. The lines represent the transfer of energy between the functional groups. The thickness and color of the lines represent the amount of the energy flow. It can be seen from Figure 2 that the Zhongjieshan Islands of the energy flow in the seas is mainly concentrated between trophic levels I to IV, along two paths, one is a detrital food chain starting from organic detritus, and the other is a pastoral food chain starting from phytoplankton. The energy of the system is provided by detritus and phytoplankton, and the energy gradually decreases from low trophic level to high trophic level.

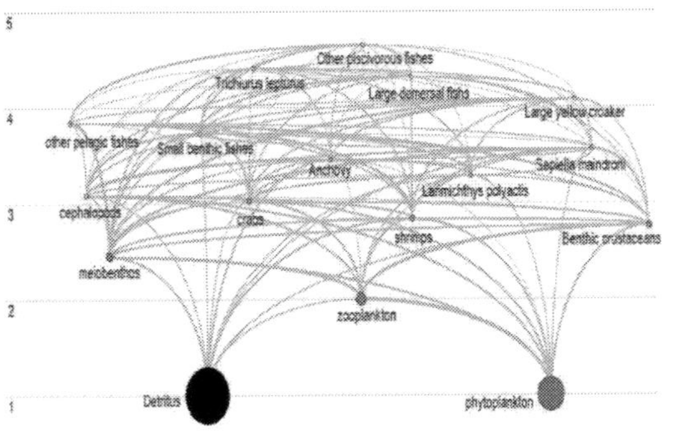

Fig.2 Schematic diagram of the energy channel of the ecosystem in the sea area of Zhongjieshan Islands

3.3 Nutrient structure and energy flow

In the Ecopath model, each functional group species is allocated to different integrated trophic levels according to the proportion of foods of different trophic levels in their food sources. The distribution of the energy flow of each functional group in the integrated trophic level is expressed in terms of relative flow and absolute flow. The integrated trophic level of the marine ecosystem of the Zhongjieshan Islands obtained by the model operation is composed of I-XII. Since the flow of the trophic level VII-XII is relatively small, it can be ignored in this study. Therefore, it can be considered that the marine ecosystem of the Zhongjieshan Islands consists of VI trophic level. Table 4 shows the energy flow of each integrated trophic level in the ecosystems to the sea area of Zhongjieshan Islands. The sea ecosystems of the Zhongjieshan Islands include VI trophic levels. We can see from Table 4 to Table 6 that the distribution of energy flow is mainly concentrated in the first IV trophic levels. The total energy flow of the system is 24154t/km^2/y. The total flow of the first trophic level is 19362t/km^2/y, accounting for 80.16% of the total energy flow of the ecosystem; the predation is 4286t/km^2/y, accounting for 89.55% of the total food intake. The total energy flow of the second trophic level is 4286 t/km^2/y, accounting for 17.74% of the total energy flow of the system, of which the energy flows for predation, inflowing debris and breathing are 395.1t/km^2/y,1776t/km^2/y,2115t/km^2/y, respectively, accounted for the total prey, total inflow debris are 8.26%, 20.16%, and 90.97% of the total respiration. The to34tal energy flow of trophic level III is 395.1t/km^2/y, which accounts for 1.6% of the total energy flow of the system. The energy flow of trophic levels IV to VI is relatively small, accounting for 0.35%, 0.07%, and 0.01% of the total energy flow of the system, respectively.

The flow from each integrated trophic level to the debris in the system is 8809t/km^2/y, accounting for 36.47% of the total energy flow of the system. In the total energy flow of the ecosystem, the flows from primary producers and detritus are 14666 t/km^2/y and 9481 t/km^2/y, respectively, accounting for 60.73% and 39.27%, indicating that the energy flow of the grazing food chain is slightly higher than that of the detritus food chain.

Tab.4 The total energy flow of the integrated trophic levels of the ecosystems in the sea area of Zhongjieshan Islands

unit：t/km²/y

Trophic level	Consumption by predators	Export	Flow to detritus	Respiration	Throughput
VI	0.395	0.000	1.156	1.166	2.717
V	2.714	0.000	7.218	6.752	16.68
IV	16.68	0.000	34.65	33.54	84.87
III	84.85	0.000	141.8	168.4	395.1
II	395.1	0.000	1776	2115	4286
I	4286	8221	6848	0.000	19362
sum	4786	8221	8809	2325	24148
Extracted to break cycles					2.713
Input					0.000
Total throughput					24154

Table 5. Energy flow from primary producers to different trophic levels

unit：t/km²/y

Trophic level	Consumption by predators	Export	Flow to detritus	Respiration	Throughput
VI	0.333	0.000	0.970	0.973	2.276
V	2.274	0.000	6.005	5.590	13.87
IV	13.87	0.000	28.22	27.44	69.52
III	69.51	0.000	114.1	138.3	321.9
II	321.9	0.000	1547	1837	3705
I	3705	0.000	6848	0.000	10553
sum	4113	0.000	8544	2009	14666

Table 6. Energy flow from debris into different trophic levels

unit：t/km²/y

Trophic level	Consumption by predators	Export	Flow to detritus	Respiration	Throughput
VI	0.0623	0.000	0.186	0.193	0.441
V	0.441	0.000	1.213	1.162	2.816
IV	2.815	0.000	6.431	6.100	15.35
III	15.34	0.000	27.75	30.11	73.21
II	73.20	0.000	229.5	277.8	580.6
I	580.5	8228	0.000	0.000	8809
sum	672.4	8228	265.1	315.4	9481

3.4 Analysis of mixed nutrition effect

The Ecopath model describes the nutritional relationship between the functional groups of the ecosystem through the analysis of mixed nutrition effects, and judges the impact on other functional groups through the change of the biomass of the functional groups. The positive effect means that the increase of biomass in one functional group promotes the increase in biomass of another functional group. The negative effect is the opposite, which means that the increase in biomass of one functional group has an inhibitory effect on the biomass from another functional group. The white circle reflects that the increase in biomass of the functional group has a positive or promoting effect on the biomass of another functional group, while the black circle has a negative or inhibitory effect. The larger the circle, the more significant the effect.

The mixed nutrient effect of the ecosystem in the sea area of Zhongjieshan Islands is shown in Figure 3. The mixed nutrient effect of each functional group on itself is generally negative. Detritus and phytoplankton, as primary producers, which have positive effect on almost all functional groups. Among them, phytoplankton has a more obvious positive effect on zooplankton and benthic species; small benthic animals have a significant negative effect on zooplankton, while have positive effect on the shrimps and small benthic fishes. Sepiella maindroni has obvious negative and inhibitory effects on cephalopods and small benthic fishes, mainly due to the competition relationship between Sepiella maindroni and these functional groups. It has a positive effect on Anchovy and small yellow croaker. Other piscivorous fishes have a negative effect on Large demersal fishs and other pelagic fishes

https://doi.org/10.1051/e3sconf/202235001002

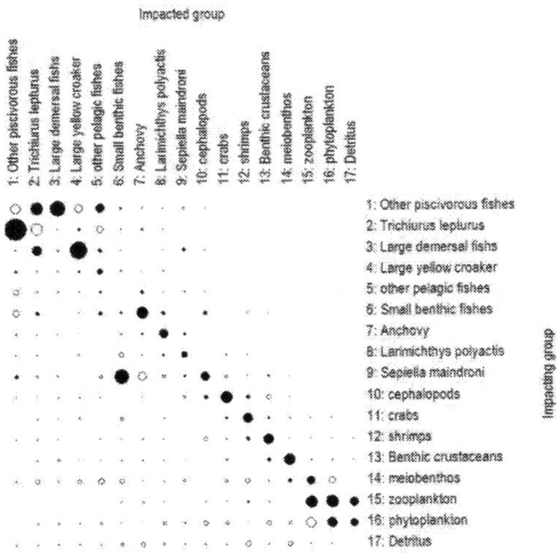

Fig.3 Mixed nutrition effects among functional groups of ecosystems in the sea area of Zhongjieshan Islands

3.5 Niche overlap analysis

The niche overlap analysis of functional groups means that in the ecosystem involved in the study, certain biological species play similar roles in the ecosystem where they are located. so the niches are relatively similar, the competition relationship is more obvious, such as similar bait species, living environment, as well as the predation habits. In the ecosystem we study, the more similar conditions the two functional groups have, the more obvious the niche overlap and the higher the value of ecological overlap. In natural environmental conditions, environmental resources are limited, so it is impossible for organisms to obtain all the food and energy they need from the environment blindly. Therefore, organisms in the ecosystem must have a competitive relationship through intra-species and inter-species competition, to obtain the required food and habitat for growth and development and reproduction of offspring. After competition, the species will inevitably lose to make concessions. By migrating to other environments and choosing to replace other organisms as food sources or choosing other time to prey, the species cross the ecological niche with the competitors to stop competition. [14-15].

The Ecopath model can calculate the food overlap index and predator overlap index between functional groups by analyzing the food web of the ecosystem in the sea area of Zhongjieshan Islands, reflecting the niche overlap between functional groups. The food overlap index indicates whether the two functional groups have similar food sources. When the value is 0, it means that the two functional groups do not have the same food source. When the value is 1, it means that the two functional groups compete fiercely for the same prey object. The predator overlap index indicates the similarity of pressure from the predator and prey between the two functional groups. A value of 0 indicates that there is no same predator, and a value of 1 indicates that the two functional groups have the same predator. The analysis result is shown in Figure 4. It can be seen from Figure 4 that cephalopods and crabs have the highest niche overlap index, and the overlap is above 0.8. The niche overlap index of Anchovy and small yellow croaker is also above 0.8. Other piscivorous fishes and Anchovy have the lowest overlap index, around 0.1.

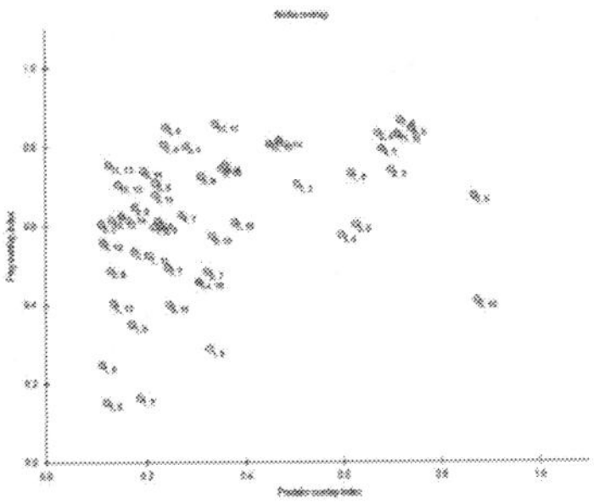

Fig.4 Analysis of Niche Overlapping of Ecosystems in the sea area of Zhongjieshan Islands

4 Conclusions

4.1 Overall characteristics of the ecosystem

Table 7 shows the total parameters of the ecosystem characteristics in the sea area of Zhongjieshan Islands after the establishment of the Ecopath model. It can be seen from Table 7 that in the current system state, the total consumption is 4786.189t/km^2·year, the total output is 8228.404 t/km^2·year, the total breathing is 2234.796t/km^2·year, and the total amount of debris flowing is 8808.951 t/km^2·year, total biomass (excluding debris) is 211.128t/km^2·year. From the model data, the sea area of Zhongjieshan Islands is rich in nutrition and bait resources, and the productivity level is at a relatively high level. When the Sepiella maindroni reaches the ecological capacity, the variation of the characteristic parameters of the ecosystem is subtle, so the ecosystem is basically stable, and the organisms of other functional groups in the sea area have no significant impact on.

Table 7. Ecosystem characteristic parameters in the sea area of Zhongjieshan Islands

parameters	Value	Units
Sum of all consumption	4786.189	t/km^2/y
Sum of all exports	8228.404	t/km^2/y
Sum of all respiratory flows	2324.796	t/km^2/y
Sum of all flows into detritus	8808.951	t/km^2/y
Total system throughput	24148.34	t/km^2/y

Sum of all production	12057.35	t/km²/y
Calculated total net primary production	10553.2	t/km²/y
Total primary production/total	4.539	
Net system production	8228.403	t/km²/y
Total primary production/total biomass	49.985	
Total biomass/total throughput	0.009	
Total biomass (excluding detritus)	211.128	t/km²
Connectance Index	0.478	
System Omnivory Index	0.389	

4.2 Estimation of Growth Capacity of Sepiella maindroni

To estimate the growth capacity of Sepiella maindroni in the sea area of Zhongjieshan Islands based on the Ecopath model, it is necessary to maximize the biomass of Sepiella maindroni in the Ecopath model. In order to maintain the stability of the ecosystem, the model is continuously adjusted to other functional groups. When the ecological efficiency (EE) ≥ 1, the biomass of Sepiella maindroni is the maximum ecological capacity. The maximum growth capacity is the value of the maximum ecological capacity minus the original biomass. The highest annual production of Sepiella maindroni in Zhejiang Province once reached 60,000 tons. Taking into account that the sea area of the Zhongjieshan Islands is a suitable habitat for the Sepiella maindroni, after using the sea area to estimate,

We assume that in this model, the raw biomass of Sepiella maindroni in the sea area of the Zhongjieshan Islands is 1.3 t·km². In the model, the biomass of the Sepiella maindroni is compensated by 0.2 t·km² based on the original data. When the biomass of the Sepiella maindroni is increased to 2.1 t·km², the EE value of the small demersal fish in the model is 1.035, the EE values of other functional groups are all less than 1. Therefore, it can be concluded from the model that the ecological capacity of Sepiella maindroni at this time is about 2 t·km². The specific debugging process is shown in Table8.The Zhongjieshan Islands Marine Special Protection Area covers an area of 202.9km². Considering that the sea area of the Zhongjieshan Islands is a suitable habitat for the Sepiella maindroni, it is estimated that the ecological capacity of Sepiella maindroni in the Zhongjieshan Islands Marine Special Reserve is 405.8t.

Table 8. The variation of the releasing capacity of Sepiella maindroni in the model

Add value /t·km²	biomass /t·km²	Change in model
0(current)	1.3	balance
0.2	1.5	balance
0.2	1.7	balance
0.2	1.9	balance
0.1	2	balance
0.1	2.1	Small demersal fish EE=1.035

5 Summary and Prospect

5.1 The main conclusions and results of this research

In this study, the sea area of the Zhongjieshan Islands was used as the research object, and the EwE6.5 version model software was used to construct the Ecopath model of the ecosystem in the sea area of Zhongjieshan Islands. According to the running results of the model, the structure and function changes of the ecosystem in the sea area of Zhongjieshan Islands are discussed, and the energy flow of the ecosystem is evaluated. Which provides guidance for the estimation of the growth capacity of Sepiella maindroni in the sea area of Zhongjieshan Islands.

The main conclusions of this study are as follows:

(1) Constructed an Ecopath model of the ecosystem in the sea area of Zhongjieshan Islands including 17 functional groups. The effective trophic level of each functional group ranges from 1-4.648. Functional groups such as shrimps, small demersal fish, pelagic fish and other functional groups have higher Ecotrophic Efficiency, while detritus, phytoplankton, zooplankton and other functional groups have lower Ecotrophic Efficiency.

(2) The total energy flow of the ecosystem in the sea area of Zhongjieshan Islands is 24154 t/km²·year. The total energy flow of the ecosystem from primary producers and debris accounted for 60.73% and 39.27% of the total energy flow of the system, respectively, which indicates that the energy flow of the grazing food chain is slightly higher than that of the detrital food chain.

(3) The TPP/TR value of the ecosystem in the sea area of Zhongjieshan Islands is 4.539; the SOI value is 0.389; the CI value is 0.478; and the TB/TT value is 0.009. In the current system status, the Sum of all consumption is 4786.189t/km²·year; the Sum of all exports is 8228.404 t/km²·year; the Sum of all respiratory flows is 2324.796t/km²·year; and the total flow of debris is 8808.951 t/km²·year. Year, the total biomass (excluding debris) is 211.128t/km²·year. From the model data, the sea area of Zhongjieshan Islands is rich in nutrition and bait resources, and the productivity level is at a relatively high level.

(4) Based on the Ecopath model analysis of the sea area of the Zhongjieshan Islands, it is concluded that the trophic level of the Sepiella maindroni in this sea area is 3.562, which belongs to the middle trophic level. The Ecotrophic Efficiency is 0.306, and the ecological capacity is 2 t·km^2.

5.2 The shortcomings and prospects of this research

(1) Ecopath is a static model. The model presents an idealized state. It only inputs data from the perspective of material energy balance to estimate the proliferation capacity, but cannot evaluate the process of biological growth. Subsequent studies can add Ecosim dynamic simulation of fishery policy and environmental changes and other parameters to evaluate the impact of observations on biomass and catches within a certain period of time, and conduct in-depth spatial analysis in conjunction with Ecospace[16].

(2) In addition, in the process of constructing the Ecopath model, some data such as gastric content analysis data, biomass data, P/B, Q/B, etc. are difficult to obtain. Therefore, in this study, we often refer to the literature of neighboring similar sea area models for data integration and reference. However, various factors such as insufficient data accuracy will lead to deviations in the model results. In subsequent studies, the data of the study area needs to be further obtained and studied, and more accurate collection of survey data of various types of operations will be considered, and various influencing factors such as climate and environment will be considered comprehensively, so as to achieve the ecological assessment of the study area. The accuracy of the analysis of the energy flow characteristics of the system further improves the accuracy and reliability of the Ecopath model.

(3) In addition, this study can also consider the habitat environment, food habits and other factors of the Sepiella maindroni to be more scientifically analyzed to improve the accuracy of the evaluation of the breeding and release effects of Sepiella maindroni. In this study, due to the incompleteness of some important basic data, such as the food matrix of some functional groups, these unknown data can only be estimated, so it will also have a certain impact on the research results. Which will also cause some errors with the actual local conditions.

This work was funded by National Key R&D Program of China (2019YFD0901204), Fundamental Research Funds for the Provincial Universities(2021JD006), The Key R&D Program of Zhejiang Province(2019C02056)

References

1. Z.Y. Ni, H.X. Xu. A preliminary study on the age and growth of Sepiella maindroni in coastal waters of Zhejiang.J. Ma-rine Fisheries, **7(3)**: 102-105 (1985).
2. J.X. Huang, Y.S. Yu. Preliminary investigation on biologicalbasis of Sepiella maindroni offshore Zhejiang in fishing grounds.J. Zhejiang Agricultural Science, **13(4)**: 186-190 (1962)
3. C.W. Wu, C. Zhou, B.Y. Guo, et al. Study on changes in re-productive biology characteristics of Sepiella maindroni (Rochebrune) offshore Zhejiang.J. Oceanologia et Limnolo-gia Sinica, **43(4)**: 689-694 (2012),
4. F. Fan, F. Yin and J. Wang, Status and Prospects on biological research of Sepiella maindroni. J. Modern Fisheries Infor-mation, 26(6): 6-9 (2011)
5. J.S. Zhang, Culture biological characteristics and haemocytes immune function of Sepiella maindroni. D. Xiamen: Xiamen University, 21-28 (2007)
6. F. Fan, F. Yin, S.M. Peng, et al. Impacts of starvation on bio-chemical indices and behaviors of Sepiella maindroni juve-niles.J. Chinese Journal of Ecology, **30(10)**: 2262-2268 (2011).
7. D. Pauly, M.L. Soriano-Bartz, M.L.D. Palomares. Improved construction, parametrization and interpretation of steady-state ecosystem models. In: Proceedings of the ICLARM Conference on Trophic Models of Aquatic Ecosystems, Metro Manila, Philippines. **26**: 1-13 (1993)
8. J.A. Gulland, The Fish Resources of the Ocean. Fishing News (Books) Ltd, Surrey (1971)
9. D. Pauly, On the interrelationships between natural mortality, growth parameters, and mean environmental temperature in 175 fish stocks.J. ICES Journal of Marine Science, **39(2)** 175-192 (1980)
10. M.L.D. Palomares and D. Pauly, Predicting food consumption of fish populations as functions of mortality, food type, morphometrics, temperature and salinity.J. Marine and Fresh-water Research, **49(5)** 447-453 (1998)
11. Z.Z. Chen and Y.S. Qiu, Assessment of the food-web structure, energy flows, and system attribute of northern South China Sea ecosystem.J. Acta Ecological Sinica, **30(18)** 4855-4865 (2010)
12. V. Christensen and D. Pauly, A guide to the ECOPATH II Soft-ware System (version 2.1). International Center for Living Aquatic Resources Management,.Manila (1992)
13. S.N. Xu, Z.Z. Chen, X. W. Zheng, et al. Assessment of eco-logical carrying capacity of intertidal mangrove planting-aquaculture ecological coupling system.J.Journal of Fishery Sciences of China, **17(3)** 393-403 (2010)
14. M. Liang, Q. Jiang, L.Y. Sun, et al. Study on the spatial niche of dominant species of large and medium-sized zooplankton in coastal waters.J. Ecology and Environment, **27(07)** 1241-1250 (2018)
15. X.B. He, J. Li, Z. Sheng, et al. Niche breadth and overlap of the main fish caught in the Minjiang Estuary. J. Chinese Journal of Applied Ecology, **29(09)**:3085-3092 (2018),

16. B. Song, L.Q. Chen, et al. Ecopath with Ecosim in the application of aquatic ecosystem research. J. Marine Sciences, **2007(01)** 83-86 (2007)

New insight into the Bibliometric Analysis on the Topic of Environmental Kuznets Curve

Libo Xia[1*], *Zhiliang* Wang[2], *Shuang* Du[1], *Decun* Tian[2] and *Sujun* Zhao[3]

[1]North China University of Water Resources and Electric Power, School of Environmental and Municipal Engineering,450045, Zhengzhou, China
[2]North China University of Water Resources and Electric Power, College of Water Resources,450045, Zhengzhou, China
[3]The People's Hospital of Jianshui,654300, Honghe, China

Abstract. The topic of the environmental Kuznets curve (EKC) was searched on Web of Science and 2,747 documents were obtained from 2010 to 2020. Among them, 2616 documents were selected and analysed using the bibliometrix package in the R language. Results showed that Environmental Science and Pollution Research were the most literature sources, and the country with the most researches was China. Keywords analysis revealed that most researches were focused on the EKC, economic development, and carbon dioxide (CO2) emission, and the last item was the main environmental pollution problem. Most researches were focused on income distribution, benefits, ecological footprint, globalization, and degradation evidence. Dividing these documents into three stages with 2014 and 2018 as the nodes, economic growth was focused from 2010 to 2014, following was focused on emissions and then the international tourism from 2019 to 2020. EKC was easily visualized among the co-citation network of authors, the co-occurrence network of keywords and the cooperation network of countries. EKC historical graph showed that the global environment should be considered as a whole to research environmental pollution and economic development, and more research should be focused on the EKC related to clean energy and sustainable development. Therefore, by analysing a large number of documents and using scientific measurement methods, it is convenient for future research on the EKC hypothesis.

1 Introduction

The problem between environmental pollution and economic development has always existed. It was proposed that as the labour force shifted from low-income agriculture to high-income industries, the process of industrialization led to increased income inequality, and at a more advanced level of development, inequality began to decrease [1]. The data from multiple countries were used to find that the per capita emissions of pollutants have an inverted-U relationship with per capita GDP [2]. According to the relationship between various environmental indicators and the level of per capita income, the environmental conditions brought an initial stage of deterioration with economic growth and then began to improve. Later, this situation was called environmental Kuznets curve [3]. The environmental Kuznets curve (EKC) related literature was summarized, including theoretical development and empirical research on the EKC phenomenon, and found that EKC had problems [4]. The new model helps to understand the true relationship between economic development and the environment and may lead to the demise of the classic EKC [5]. The results showed that energy consumption and control of carbon dioxide emissions may not have an impact on Turkey's economic growth, and used a linear logarithmic model to point out that the EKC hypothesis is invalid [6]. Carbon dioxide emissions have a strong relationship with energy consumption and foreign investment which, also supports the EKC hypothesis [7]. The panel data such as carbon dioxide emissions, per capita GDP, population density, land, and industry's share of GDP were studied the EKC hypothesis of 14 Asian countries, which was an EKC hypothesis that provided empirical support [8]. The causal relationship among energy consumption, carbon dioxide emissions, economic growth, trade opening, and urbanization in EU states were investigated and provided evidence to support the EKC hypothesis [9]. The relationship between environmental pollution and economic development is still under study.

Bibliometric analysis is to scientifically determine the relationship between related fields of thesis research by analysing a large number of related documents. According to the defined coupling standards, the automatic processing results of a large number of scientific papers were described, showing the logic between the papers [10]. The network of co-cited papers and co-occurrence data can become a specific scientific discipline and provide a new method for article indexing. [11]. A method that analyses the evolution of topics in a specific research field can allow us to quantify and visualize the evolution of topics in a specific research field [12]. There are already some bibliometric analyses on environmental pollution issues. The relationship between energy growth and discuss future research directions and emerging trends was analysed by the

* Corresponding author: 201514321@stu.ncwu.edu.cn

CiteSpace (https://citespace.podia.com/) software [13]. And the VOS viewer software (https://www.vosviewer.com/) and meta-analysis methods were used to track the historical trends of the topic studied the author's contribution, and transformed the network analysis into a visual form [14]. There were 4,225 articles on haze and economic growth as the research object, and the CiteSpace software was used to explore the spatial and temporal distribution of literature. [15]. By using the bibliographic analysis of the Web of Science (https://www.webofscience.com/) to give a detailed overview of the EKC hypothesis. [16]. The bibliometrix package in the R language [17] for the research of comprehensive bibliometric analysis [18]. This article will use the method of bibliometric analysis to analyse 2,616 articles from the Web of Science, and conduct co-citation, co-occurrence and co-word analysis on the articles. Through the use of scientific methods to analyse the relationship between the documents, the change trend of keywords, and cluster analysis of these documents, and finally study the historical citation of the EKC. This article can help researchers who study the curve to provide a direction.

2 Materials and Methods

By searching the topic "Environmental Kuznets Curve" in the Web of Science, selecting publication years from 2010 to 2020, and the database as the Web of Science core collection, 2747 articles were obtained. Export documents, selecting all records in the output section, including Author(s)/Editor(s), Title, Source, Conference Information, Abstract, Cited References, Document Type, Conference Sponsors, Addresses, Times Cited, Keywords, Publisher Information, ISSN / ISBN, Cited Reference Count, Source Abbrev, Page Count, IDS Number, Language, Web of Science Categories, Research Areas, Funding Information, Accession Number, Author Identifiers, Usage Count, PubMed ID, Open Access, Hot Paper, Highly Cited. After pre-processing the literature, the actual publication year of the other 131 articles is 2021, it is found that the actual number of documents published from 2010 to 2020 is 2,616, which are the main parts of the bibliometric analysis of this article.

The 2616 documents were analysed by using the open-source software R language, which is a language and environment for statistical computing. The method is comprehensive scientific measurement analysis, using the bibliometrix package in R. The main analysis is divided into three parts, main information about the collection, analytics and plots for three different level metrics, and analysis of three structures of knowledge and data visualization. There are three analyses, co-citation analysis B_{cocit}, collaboration analysis B_{coll}, co-word analysis B_{coc}.

$$\mathrm{B}_{cocit} = A \times A^{'} \quad (1)$$

A is a document cited reference matrix, $A^{'}$ is the transposed matrix of A, element a_{ij} indicates how many co-citations exist between documents i and j.

$$B_{coll} = B \times B^{'} \quad (2)$$

B is a document author matrix, $B^{'}$ is the transposed matrix of B, element b_{ij} indicates how much collaboration exist between authors i and j.

$$B_{coc} = C \times C^{'} \quad (3)$$

C is a document word matrix, C_{ij} is the transposed matrix of C, element c_{ij} indicates how many co-occurrences exist between words i and j. Figure 1 shows the workflow of this research.

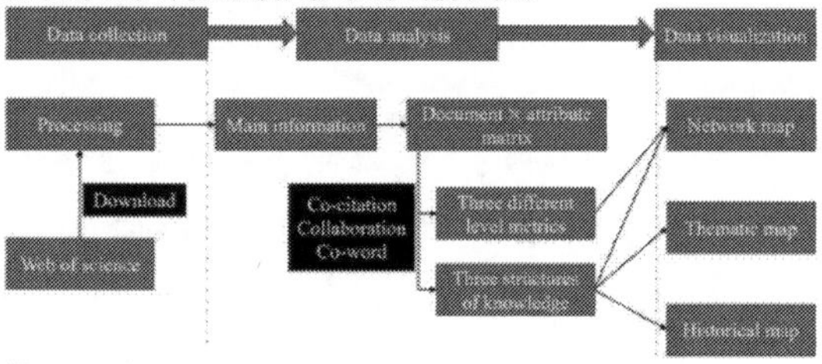

Fig. 1. The bibliometric analysis workflow, which contains data collection, date analysis and data visualization.

3 Results & Discussion

3.1 Main information about the collection

Table 1 shows, the main information of the published papers obtained through statistics of the literature data of the EKC. There are 2,616 articles and 64,946 references for these texts. The data sets of these documents are from 2010 to 2020, there are 389 literature sources and 2,616 articles. And the keywords plus is the calculating keyword by using the author's keywords [19], the author's keywords become the keywords plus from 4887 to 2593. Few articles were completed by a single author, only 312, and most of the articles were completed by multiple people. And co-authors per documents is 3.06.

Table 1. Main information about data

Main information		Document and authors	
Timespan	2010:2020	Keywords plus	2593
Sources (Journals, Books)	389	Author's keywords	4887
Documents	2616	Authors	4334
Average years from publication	3.69	Author appearances	7994
Average citations per document	34.39	Authors of single-authored documents	247

Average citations per year	7.013	Authors of multi-authored documents	4087
References	64946	Single-authored documents	312
Article	2451	Documents per author	0.604
Review	134	Authors per document	1.66
Collaboration index	1.77	Co-Authors per documents	3.06

As shown in Figure 2, according to the graph of article changes over time, it can be seen that the number of articles is increasing every year, with a growth rate of 25.28%, of which there is not much growth from 2010 to 2014, and more growth from 2018 to 2020. From the number of articles changes, it can be seen that the research on the EKC is attracting people's attention greatly in recent years.

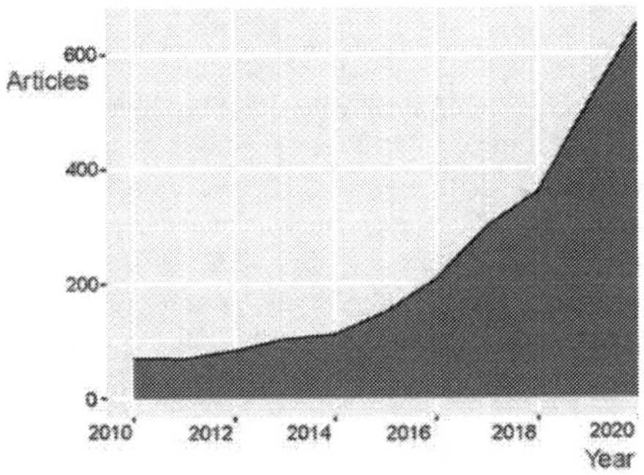

Fig. 2. Annual production of EKC.

3.2 Analytics and plots for three different levels of metrics

The main researches in this part are the three indicators of different levels: the source of the documents, the author, and the documents. According to the sources of these documents and references, we know that Environmental Science and Pollution Research (ESPR) has the most articles, with 375 articles. The second is Journal of Cleaner Production (JCP), with 198 articles. The third is Sustainability, with 150 articles. According to Figure 3, there are five core journals related to EKC research: ESPR, JCP, Sustainability, Renewable and Sustainable Energy Reviews (RSER), and Energy Policy. According to the Brasdord's Law, sources of information on specific subjects [20], these 5 journals provide the articles that accounted for 35.47% of the total number of 2616 articles. There are 64,946 references citing the studied 2616 articles. The most cited journal is Energy Policy. The journals in it have been cited 12,329 times during the studied timespan. Ecological Economics is the second journal, which has been cited 8713 times. The third one is RSER. In the next few years, these documents will become indispensable reference tools and have been cited 7,450 times in the field.

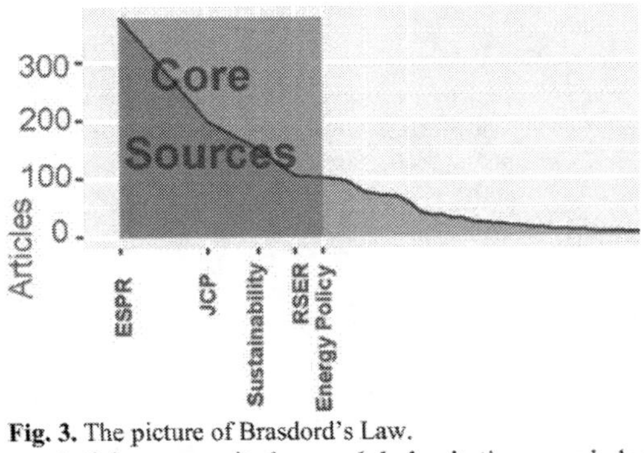

Fig. 3. The picture of Brasdord's Law.

Articles are cited as global citations and local citations. Global citations refer to these references being cited by articles in all fields, and local citations refer to those cited by people in the research field of the EKC, so it is more reasonable and accurate to analyse local citations. Regarding the four indexes cited H, G, total cited (TC), and the number of publications (NP). The index of H is an index to quantify an individual's scientific research output [21]. The G-index is introduced as an improvement of the H-index of Hirsch to measure the global citation performance of a set of articles [22]. Through comparing these source journals, it will be found that the number of publications of ESPR is the most. The G' index of RSER and Energy Policy is 100, but their NP is only 104 and 101. Combining these indexes, it can be concluded that JCP is the most suitable for the EKC.

This part mainly analyses the number of articles by authors, and we can see which authors have studied more on the EKC. It can be concluded that the author with the most articles is SHAHBAZ M, with 66 articles, the second is OZTURK I with 48 articles, and the third is Wang Z with 40 articles. It can also be found that OZTURK I, the most cited research field of the EKC, has been cited 1821 times, followed by SHAHBAZ M with 1707 times, and AL-MULALI U with 792 times. It can be seen from Figure 4 that SHAHBAZ M has been studied the EKC from 2012 to the present, and since 2013, more articles have been published every year. OZTURK I has been researched this issue since 2010. It may be earlier because the record in the Web of Science started in 2010 and published the most articles in 2015. Wang Z also started in 2012. There were no articles from 2013 to 2015 and most articles were published in 2019.

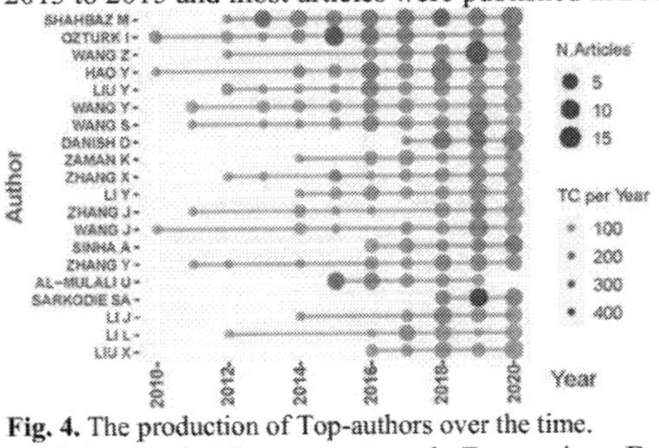

Fig. 4. The production of Top-authors over the time.

School of Management and Economics, Beijing Institute of Technology, and Eastern Mediterranean University, are top three affiliations for publishing which are 190, 100, and 88, respectively. Figure 5 shows the situation of the countries where the authors of these articles are located, and whether the authors are in a single country or multiple countries. China has the most articles, with 968 articles, and 293 articles in cooperation with other countries, of which the ratio of cooperation between multiple countries is 0.306.

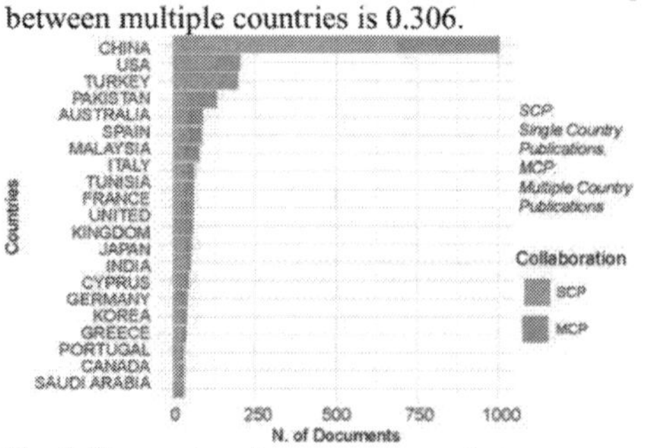

Fig. 5. The country of corresponding author.

After the analysis of the literature, it is mainly the citation of the literature and the change of the keywords. Among the 2616 documents, the most cited is an article named the lancet commission on pollution and health [23], which mainly talks about pollution and health issues, and has little to do with the EKC we are studying. Maybe the problems caused by environmental pollution are closely related to people's health. With the improvement of the economic level, health problems are also important, so it has been cited 1061 times. However, the most cited in this research field, the topic of the EKC, is named testing environmental Kuznets curve hypothesis in Asian countries [8], 230 times, mainly. Among all the references, the most cited is economic growth and the environment [3]. This document belongs to a source of the EKC, so this reference is cited 839 times in the field.

Adding algorithm-derived indexes based on the author's keywords, such keywords can better represent the status of the literature in the bibliometric analysis. After the keywords of these documents are converted by keywords, it can be found through analysis that the EKC appears the most times, with 1220 times, followed by economic growth, and the third is CO_2 emissions. Therefore, the most researched on the EKC is economic growth and carbon dioxide emissions. Through the changes in the subject headings of the literature, in Figure 6, it can be seen that the problem of income balance was studied from 2010 to 2016, 2013 was the focus of Kuznets curves, scenarios, benefits, and the focus of 2018 was economic growth, EKC and Kuznets curve. By 2020, the focus has become on ecological footprint, degradation evidence, and globalization.

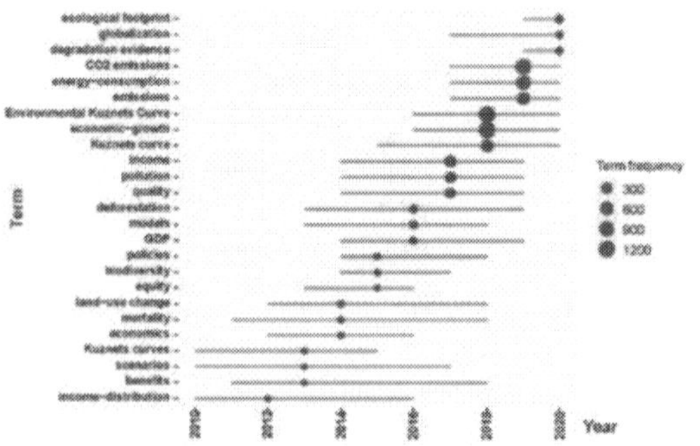

Fig. 6. The trend topic all over the word.

3.3 Analysis of two structures of knowledge

Structures of knowledge are also the science mapping. A general framework for creating a large-scale document space is proposed, as well as some methods of performing certain individual processing steps [24]. It is mainly used for citation data but can be used for other types of documents. The overall picture of scientific knowledge has always been desirable. Scientific mapping attempts to find intellectual connections in the dynamically changing scientific knowledge system and aims to show the structural and dynamic aspects of scientific research [25], and the evolution of theory and technology. After that, each researcher will give a comprehensive overview of the main findings related to his particular field, and scientific mapping allows the investigation of scientific knowledge from a statistical perspective.

3.3.1 Conceptual structure

This part will analyse what science talks about, the main themes, and trends. The conceptual structure represents the relationship between the concepts or words in a set of publications: the words appearing together in the document will be related in the network. It is also called the common word network. This structure is used to understand the topics covered by the research field and to define what is the most important and up-to-date issue also helps to study the evolution over time. Similar to network analysis, the use of clustering algorithms can also be applied to factor analysis. Clustering them can help reduce the dimensionality of elements. By analysing them, the topic network can be drawn, the axis is the centre and density function of the topic network, the time is divided into time slices, and the topic evolution of a specific research field can be represented by the alluvial diagram. When only a small number keywords together of articles were used, they are far away from each other, and larger number articles put keywords together, indicating that they are close to each other.

In factor analysis, the multidimensional scaling method [26] was used, as shown in the Figure 7, the keywords of the article can be seen divided into two categories. When using the clustering method, these keywords are also clustered according to the citations of the article and the keywords used in the article, and up to

9 levels are used. And the keywords of cointegration, tests, error-correction, time-series, and unit-root are the research methods of the EKC, so there are divided into one category. The other category specifically introduces the actual problems of this curve and related influencing factors. These two categories can be simply understood as methodology and practice. Through the clustering and visualization of keywords, the research of these documents can be clearly understood.

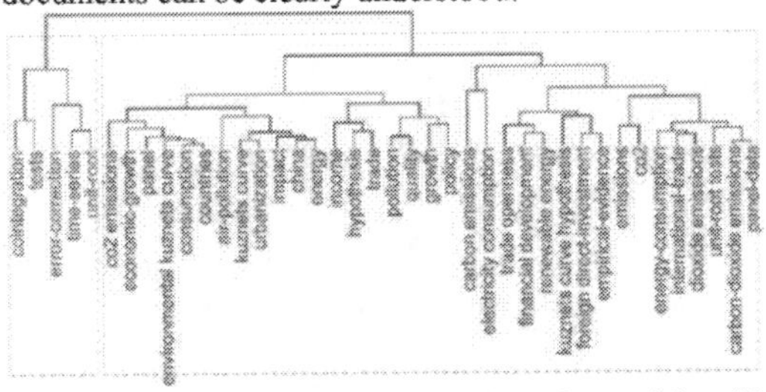

Fig. 7. The topic dendrogram of clustering articles. Blue indicates the method used, red indicates the research question.

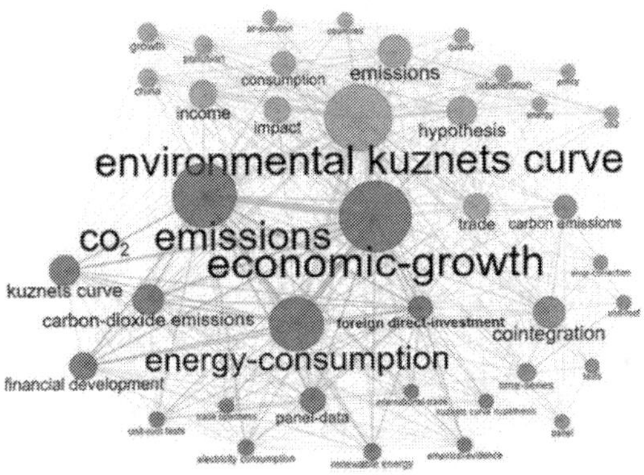

Fig. 8. The Co-occurrence networks of the keywords.

The co-occurrence network analysis of articles by using a clustering algorithm [27] and regularization [28] also is used. Through this co-occurrence network, the mathematical structure between articles can be simulated. The network is composed of lines and points. The vertices represent the keywords of the analysis. The size of the vertices is the number of occurrences. The colours of the vertices are of different types. The lines indicate the connection between them, and the thickness of the line indicates the strength of the connection. As shown in the Figure 8, economic growth and CO_2 emissions fall into one category, and their betweenness is 2.87 and 2.23, in Table 2, cointegration is the second category, and betweenness is 0.42; the EKC is the third category, and betweenness is 2.17. In this classification we can conclude that economic growth, cointegration, and EKC represent the main parts of these three categories. Betweenness for a given node is the number of shortest paths between two nodes. The value of closeness is the mean length of the shortest paths between the node and all other nodes in the network. PageRank for a given node is a metric used to estimate the importance of the node. If the nodes with a higher rank have links to the node, the node has a higher PageRank [29].

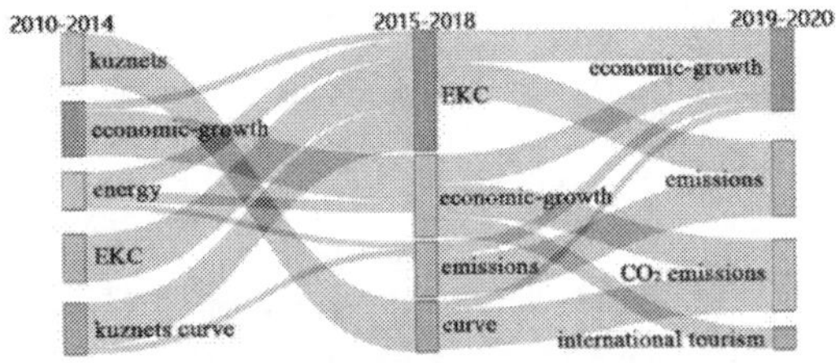

Fig. 9. Thematic evolution and (b) thematic maps.

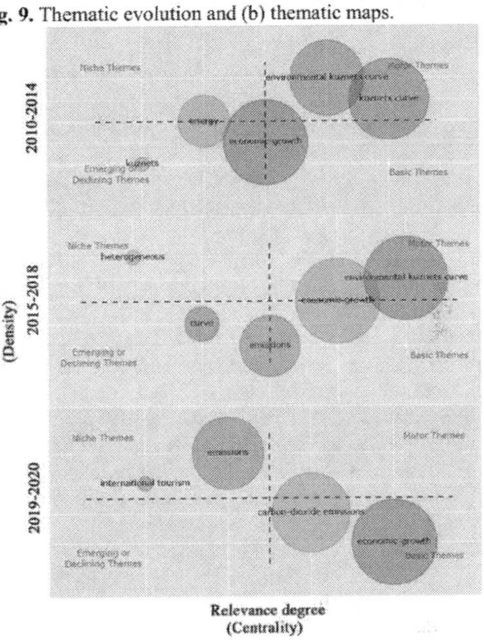

Fig. 10. Thematic maps.

Table 2. The Co-occurrence Network Analysis

Node	Cluster	Betweenness	Closeness	PageRank
Economic-growth	1	2.873	0.0256	0.0795
CO2 emissions	1	2.234	0.0256	0.0721
Energy-consumption	1	1.533	0.0263	0.0588
Cointegration	2	0.425	0.0263	0.0339
Carbon emissions	2	0.220	0.0256	0.0256
Time-series	2	0.162	0.0263	0.0203
Environmental Kuznets Curve	3	2.178	0.0250	0.0716
Emissions	3	0.499	0.0263	0.0333
Hypothesis	3	0.253	0.0263	0.0303

Thematic maps [12] are also used in the analysis. Each cluster can be represented on the map and different colours represent different topics. Callon centrality is represented as the importance of the research field, Callon density is a measure of theme development [30]. According to the number of articles per year, three-time periods from 2010 to 2014, 2015 to 2018, and 2019 to 2020 are selected, as shown in Figure 9. By analysing the changes of the theme maps in different years, we can further look at the theme evolution of the entire field. In the Figure 6 (b), the upper left part is a highly developed and isolated theme, the upper right part is motor themes, the lower left is an emerging or declining theme, and the lower right part is a basic theme. The study found that the theme of aging in the field of the EKC is cointegration, and the theme of the motor is economic growth. The topic maps of different years and their clusters are shown in the Figure 10.

3.3.2 Intellectual structure

The intellectual structure represents the relationship between the citations of the article. Citation analysis is the most common bibliographic analysis in common citations between authors or documents [11]. Based on the analysis of these citations, the historical trend of the literature was obtained, and a certain historical change map was formed, a new way for the future professional structure was provided [31]. As shown in Figure 11, from the historical direct citation network, the research content of EKC can be found from 2010 to 2020, and the citation rules in it. To study the problem of the EKC, the autoregressive distributed lag (ARDL) bounds testing method was used [32], and the long-term and causal relationship between economic growth, carbon emissions, energy consumption and employment was examined in Turkey [6]. Some studies have questioned the existence of EKC to determine the relationship between environmental quality and economic growth [33]. Because the causal relationship between income and environmental quality cannot be proved, it is necessary to use more representative data and more appropriate statistical techniques [34]. When attracting foreign direct investment, developing countries should strictly review foreign investment qualifications to avoid environmental damage and the management methods of foreign direct investment [35].

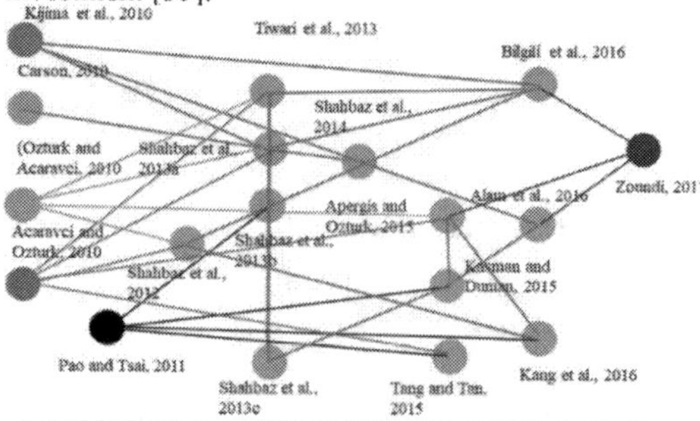

Fig. 11. Historical co-citation network about the EKC.

The research showed that there was a long-term relationship between variables, supporting EKC hypothesizes that energy consumption will increase carbon dioxide emissions [7]. Whether economic development reduces carbon dioxide emissions, the research results verify the link among carbon dioxide emissions, economic development, energy consumption, and economic growth [36]. Indonesia's economic growth, energy consumption, trade openness, and carbon dioxide emissions, and the economic growth and carbon dioxide emissions are also interrelated with a two-way causal relationship [37]. The results of the study confirmed that energy consumption was the main factor causing energy pollutants. Policy and financial development improved the environment and redirect resources to environmentally friendly projects, making a significant contribution to reducing carbon dioxide emissions [38]. The coal consumption and trade openness led to carbon dioxide emissions and trade openness leads to economic growth and coal consumption [39]. By using annual time series data from 1971 to 2010 to investigate the existence of the Tunisian EKC. The results shew that the existence of EKC [40]. The method using panel data was used in a multivariate framework to test the EKC hypothesis [8]. In terms of the inverted U-shaped correlation between emissions and per capita income, the estimated value has expected signs and is statistically significant to provide empirical support for the existence of the EKC hypothesis. At this stage, some technologies can improve energy efficiency, energy conservation, and renewable energy, but low-income countries cannot obtain these technologies due to high costs [41]. The causal relationship between energy consumption, carbon dioxide emissions, economic growth, trade opening, and urbanization was investigated in the new EU states. The estimated coefficients of the lag correction period in the equations of, trade opening is statistically significant [9]. The energy consumption, foreign direct investment, and income are the key determinants of Vietnam's carbon dioxide emissions. It also believes that foreign investment should adopt clean technologies [42]. The annual time series data of India, Indonesia, China, and Brazil were researched during the period 1970-2012 by using the ARDL boundary test method, considering the main linear and non-linear assumptions for short-term and long-term relevant time-series data to carbon dioxide [41]. By applying the spatial panel data model to study China's carbon dioxide EKC hypothesis, the results show that spatial effects affect the shape of the EKC, and the spatial structure of China's provincial carbon dioxide emissions has a clear distribution [43]. All countries should use renewable energy to obtain electricity fairly, and improve renewable energy technologies to increase the supply of renewable energy policies [44]. Research shows that renewable energy has a positive effect on reducing carbon dioxide emissions and is an effective alternative to traditional fossil energy [45].

By analysing the historical map of this curve, it can be found that the problems of environmental pollution and economic development are changing. From 2010 to 2020, the first is simple EKC, followed by carbon dioxide and energy consumption, and sustainable development. It also used different data, different research methods, and studied different pollution. With the development of time, the economy of different places, the pollution of different places, and the spatial impact are also taken into account, so in the end, the EKC problem will be analysed as a whole.

4 Conclusions

The study found that the key issue in EKC is carbon dioxide emissions in environmental pollution, which is resulted from the continuous use of energy. Different countries or cities makes their own local EKC present an inverted U-shape, but destroys the environment of another place. Therefore, when the economic development reaches a certain level and more investment technologies, more renewable energy and clean energy will be used to reduce the overall environmental pollution and reach the real inverted U-shaped EKC. Bibliometric analysis revealed that the NO.1 cited article in the EKC research field is testing EKC hypothesis in Asian countries published by Nicholas Apergisa and Ilhan Ozturkb in 2015, which was published in Ecological Indicators, 230 times. NO.1 cited reference is economic growth and the environment published by Grossman, Gene M and Krueger, Alan BG in NBER WORKING PAPER SERIES in 1995, 839 times. Through the clustering of keywords, the relevant factors of the EKC and the methods to research environmental pollution can be clearly understood. In the following research, more databases should be selected, and all relevant documents should be taken into consideration.

Acknowledgments

This paper thanks the data from the Web of Science and the bibliometrix package in R language.

References

1. S. Kuznets, Economic Growth and Income Inequality Simon. Econ. Growth Income Inequal. Simon **45** 1–28 (1955)
2. T.M. Selden and D. Song, Environmental quality and development: Is there a kuznets curve for air pollution emissions? J. Environ. Econ Manage. **27** 147–162 (1994)
3. G.M. Grossman and A.B. Krueger, Economic Growth and the Environment*. Q. J. Econ. **110** 353–377 (1995)
4. S. Dinda, Environmental Kuznets Curve hypothesis: A survey. Ecol. Econ. **49** 431–455 (2004)
5. D.I. Stern, The Rise and Fall of the Environmental Kuznets Curve. World Dev. **32** 1419–1439 (2004)
6. I. Ozturk and A. Acaravci, CO2 emissions, energy consumption and economic growth in Turkey. Renew. Sustain. Energy Rev. **14** 3220–3225 (2010)
7. M. Shahbaz, H.H. Lean and M.S. Shabbir, Environmental Kuznets Curve hypothesis in Pakistan: Cointegration and Granger causality. Renew. Sustain. Energy Rev. **16** 2947–2953 (2012)
8. N. Apergis and I. Ozturk, Testing environmental Kuznets curve hypothesis in Asian countries. Ecol. Indic. **52** 16–22 (2015)
9. A. Kasman and Y.S. Duman, CO2 emissions, economic growth, energy consumption, trade and urbanization in new EU member and candidate countries: A panel data analysis. Econ. Model. **44** 97–103 (2015)
10. M.M. Kessler, Bibliographic coupling between scientific papers. Am. Doc. **14** 10–25 (1963)
11. H. Small, Co-citation in the scientific literature: A new measure of the relationship between two documents. J. Am. Soc. Inf. Sci. **24** 265–269 (1973)
12. M.J. Cobo, A.G. López-Herrera, E. Herrera-Viedma and F. Herrera, An approach for detecting, quantifying, and visualizing the evolution of a research field: A practical application to the Fuzzy Sets Theory field. J. Informetr. **5** 146–166 (2011)
13. M.A. Anwar, R. Zhou, F. Asmi, D. Wang and A. Hammad, Mapping the Evolution of Energy-Growth Nexus: Synergies and Trade-Offs. J. Econ. Surv. **33** 968–998 (2019)
14. S.A. Sarkodie and V. Strezov, A review on Environmental Kuznets Curve hypothesis using bibliometric and meta-analysis. Sci. Total Environ. **649**, 128–145 (2019)
15. H. Xiong and Z. Zhao, The correlation between haze and economic growth: Bibliometric analysis based on wos database. Appl. Ecol. Environ. Res. 18, 59–75 (2020)
16. M.F. Bashir, B. Ma, M.A. Bashir, L.B. Shahzad, Scientific data-driven evaluation of academic publications on environmental Kuznets curve. Environ. Sci. Pollut. Res. **28** 16982–16999 (2021)
17. R Core Team R: A Language and Environment for Statistical Computing 2021.
18. M. Aria, Cuccurullo, C. bibliometrix: An R-tool for comprehensive science mapping analysis. J. Informetr. **11** 959–975 (2017)
19. J. Zhang, Q. Yu, F. Zheng, C. Long, Z. Lu and Z. Duan, Comparing keywords plus of WOS and author keywords: A case study of patient adherence research. J. Assoc. Inf. Sci. Technol. **67** 967–972 (2016)
20. Geophysics, A. Sources of information on specific subjects 1934. J. Inf. Sci. **10** 176–180 (1985)
21. J.E. Hirsch, An index to quantify an individual's scientific research output. Proc. Natl. Acad. Sci. U. S. A. **102** 16569–16572 (2005)
22. L. Egghe, Theory and practise of the g-index. Scientometrics **69** 131–152 (2006)
23. P.J. Landrigan, R. Fuller, N.J.R. Acosta, O. Adeyi, R. Arnold, N. Basu, A.B. Baldé, R. Bertollini, S. Bose-O'Reilly, J.I. Boufford, et al. The Lancet Commission on pollution and health. Lancet **391** 462–512 (2008)
24. H. Small, Update on science mapping: Creating large document spaces. Scientometrics **38** 275–293 (1997)
25. S.A Morris, B. Van Der Veer Martens, Mapping research specialties. Annu. Rev. Inf. Sci. Technol. **42** 213–295 (2008)

26. C. Cuccurullo, M. Aria and F. Sarto, Foundations and trends in performance management. A twenty-five years bibliometric analysis in business and public administration domains. Scientometrics **108** 595–611 (2016)
27. A. Lancichinetti and S. Fortunato, Community detection algorithms: A comparative analysis. Phys. Rev. E **80** (2009)
28. N.J. Eck and L. Waltman, How to normalize cooccurrence data? An analysis of some well-known similarity measures. J. Am. Soc. Inf. Sci. Technol. **60** 1635–1651 (2009)
29. A.K.W. Lau, Y. Kajikawa, N. Sharif, The roles of supply network centralities in firm performance and the moderating effects of reputation and export-orientation. Prod. Plan. Control **31** 1110–1127 (2020)
30. M. Callon, J.P. Courtial and F. Laville, Co-word analysis as a tool for describing the network of interactions between basic and technological research: The case of polymer chemistry. Scientometrics **22** 155–205 (1991)
31. E. Garfield, Historiographic mapping of knowledge domains literature. J. Inf. Sci. **30** 119–145 (2004)
32. A. Acaravci and I. Ozturk, On the relationship between energy consumption, CO2 emissions and economic growth in Europe. Energy **35** 5412–5420 (2010)
33. M. Kijima, K. Nishide and A. Ohyama, Economic models for the environmental Kuznets curve: A survey. J. Econ. Dyn. Control **34** 1187–1201 (2010)
34. R.T. Carson, The Environmental Kuznets Curve: Seeking Empirical Regularity and Theoretical Structure. Rev. Environ. Econ. Policy **4** 3–23 (2010)
35. H.T. Pao and C.M. Tsai, Multivariate Granger causality between CO2 emissions, energy consumption, FDI (foreign direct investment) and GDP (gross domestic product): Evidence from a panel of BRIC (Brazil, Russian Federation, India, and China) countries. Energy **36** 685–693 (2011)
36. M. Shahbaz, S.A. Solarin, H. Mahmood and M. Arouri, Does financial development reduce CO2 emissions in Malaysian economy? A time series analysis. Econ. Model. **35** 145–152 (2013)
37. M. Shahbaz, M. Mutascu, P. Azim, Environmental Kuznets curve in Romania and the role of energy consumption. Renew. Sustain. Energy Rev. **18** 165–173 (2013)
38. M. Shahbaz, Q.M.A. Hye, A.K. Tiwari and N.C. Leitão, Economic growth, energy consumption, financial development, international trade and CO2 emissions in Indonesia. Renew. Sustain. Energy Rev. **25** 109–121 (2013)
39. A.K. Tiwari, M. Shahbaz, Q.M. Adnan Hye, The environmental Kuznets curve and the role of coal consumption in India: Cointegration and causality analysis in an open economy. Renew. Sustain. Energy Rev. **18** 519–527 (2013)
40. M. Shahbaz, N. Khraief, G.S. Uddin and I. Ozturk, Environmental Kuznets curve in an open economy: A bounds testing and causality analysis for Tunisia. Renew. Sustain. Energy Rev. **34** 325–336 (2014)
41. M.M. Alam, M.W Murad, A.H.M. Noman and I. Ozturk, Relationships among carbon emissions, economic growth, energy consumption and population growth: Testing Environmental Kuznets Curve hypothesis for Brazil, China, India and Indonesia. Ecol. Indic. **70** 466–479 (2016)
42. C.F. Tang and B.W. Tan, The impact of energy consumption, income and foreign direct investment on carbon dioxide emissions in Vietnam. Energy **79** 447–454 (2015)
43. Y.Q. Kang, T. Zhao and Y.Y. Yang, Environmental Kuznets curve for CO2 emissions in China: A spatial panel data approach. Ecol. Indic. **63** 231–239 (2016)
44. F. Bilgili, E. Koçak and Ü. Bulut, The dynamic impact of renewable energy consumption on CO2 emissions: A revisited Environmental Kuznets Curve approach. Renew. Sustain. Energy Rev. 54, 838–845 (2016)
45. Z. Zoundi, CO2 emissions, renewable energy and the Environmental Kuznets Curve, a panel cointegration approach. Renew. Sustain. Energy Rev. **72** 1067–1075 (2017)

Study on Bibliometric Visualization of Sustainable City Based on VOSviewer (2008-2021)

Yang Feng[1,*], *Xinyuan* Gu[1], *Jianqing* Ye[1], *Xiaolin* Jia[1], *Hongchen* Zhang[1], *Sirong* Wang[1] and *Jianfeng* Yang[1]

[1]Xijing University, School of Art and Design, 710123, No.1 Xijing Road, Xi'an, China

Abstract. In order to clarify the co-research, research hotspots and knowledge evolution in the field of sustainable city, a visualization analysis of 1816 articles about sustainable city in the Web of Science was conducted by using bibliometrics method based on VOSviewer. The conclusions are listed as follows. (1) The citable documents of sustainable city have shown a rapid upward trend, with increasing attention and rising research hot spot, year by year, since 2015. (2) Scholars should strengthen the construction of global academic community, and China should be more extensive and in-depth cooperation with and scientific research institutions in European and American countries or regions. (3) The research dimension of sustainable city has been greatly expanding under the continuous development of complex algorithm, data science and artificial intelligence, Hence, the interdisciplinary characteristics of this research field, such as intersections, comprehensiveness and systematicness, are becoming increasingly prominent.

1 Introduction

As city is the crystallization of human civilization, people live in a city to enjoy its attribute of providing a better life. However, since the industrial revolution, rapid urbanization has brought a series of problems, such as environmental pollution, heat island effect, urban waterlogging, traffic congestion and social differentiation. In 1987, the report, *Our Common Future*, of the World Commission on Environment and Development (WCED) put forward the concept of Sustainable Development for the first time, emphasizing that development "should meet the demands of contemporary people without endangering the ability of future generations to meet their demands"[1]. Sustainable city is proposed based on sustainable development as a contemporary paradigm for building an ideal city in the future, that is, " a city sustainable in the three dimensions of environment, society and economy, with sustainable use of resources and have high risk resistance". The early definition of sustainable city pays more attention to environmental issues. Although the concept includes social dimension, it is still not prominent. With the promotion of urbanization and the enhancement of public participation awareness, the definition focuses on balancing the concepts of green and economy while achieving the balance among various development objectives. In recent years, information and communication technologies based on data science have become more and more mature, which can provide new ideas and technical support for solving complex and diversified problems in urban development[2].

Sustainable city research mainly focuses on quantitative model, evaluation index, spatial planning and design. In terms of quantitative models, scholars have developed such models as extended metabolism[3], comprehensive conceptual framework[4], future urban development[5], micro-urban ecosystem sustainability index and the comprehensive sustainable assessment of land[6]. Sustainable city assessment indicators include city scale and community scale. In terms of city scale, Ecological Footprint (EF), City Development Index (CDI), City Sustainability Index (CSI) are single assessment indicators of sustainable city, but the index system, which is constructed through environmental, economic, social and ecological aspects, is a composite assessment indicator of sustainable city[7-9]; In terms of community scale, LEED-ND, BREEAM and CASBEE-UD are the community-scale assessment indicator tools for sustainable city with a far-reaching impact[10], with some other scholars who have used eco-label to measure the environmental performance of buildings[11]. In terms of the spatial planning and design of sustainable city, Pincetl have tried to connect the three research paths of urban ecosystem services, urban metabolism and urban political ecology to form a new idea for sustainable urban spatial planning[12]. Some scholars have discussed the impact of urban intensive and compact development on urban green space and its planning, and promoted sustainable urban construction through network optimization layout strategy[13]. Pearson et al. believe that urban agriculture can promote the benign development of sustainable city, and have reviewed the studies on urban agriculture from three dimensions of society, economy and environment[14]. Other scholars consider that the spatial planning and design of sustainable city should be a complex and interdisciplinary comprehensive design, and the future

* Corresponding author: 20200118@xijing.edu.cn

sustainable city should fully consider the integration of historical heritage, land mix, ecological architecture, public space at different spatial scales, so as to make the city more resilient, attractive and livable [15-16].

After decades of development, sustainable city has been promoted in theory and tried in practice in many countries around the world. With the deepening of cross and interdisciplinary research, the scientific research achievements in the field of sustainable city have seen explosive growth. Traditionally, for accurately grasping the knowledge evolution of a certain research field, scholars usually need to read a large number of literature materials and make professional judgments, but it is inevitable to be disturbed by subjective factors, such as scholars' research level, depth and preference. In order to further objectively, clearly and comprehensively present the development context, research hotspots and evolution characteristics of sustainable city, and further discuss the difficulties and opportunities encountered in the process of sustainable city development, this paper makes an in-depth study on the field of ecological agriculture by using bibliometrics and visualization tools, In order to provide meaningful reference for the theoretical development and practical ideas of sustainable cities this paper uses bibliometrics method and visualization tools to conduct in-depth research in the field of ecological agriculture, looking forward to providing meaningful reference for the theoretical development and practical thinking of sustainable city.

2 Materials and Methods

2.1 Data Sources

Bibliometric analysis should be based on high-quality sci-tech text data, while the content of Web of Science (WoS) database can cover the fields natural science, social science, engineering technology, art and humanities, with its full coverage, comprehensiveness and authority[17]. Therefore, this study uses the Web of Science (WoS) database as the retrieval source. On November 4, 2021, electronic resources provided by the Library of Xi'an University of Technology were retrieved into the Web of Science™ Core Collection, with All Years as the time span of the retrieved data and "TS: (Sustainable City) OR TS: (Sustainable Cities)" as the retrieval fields. A total of 1816 literatures were retrieved, with the time spanning from 2008 to 2021 in literature distribution.

2.2 Analysis Tools and Methods

It is necessary to rely on the assistance of computers for mining valuable information from massive Sci-Tech literature. Since the introduction of auxiliary visualization in scientometrics analysis, a number of valuable scientific knowledge atlas tools, such as VOSviewer, CiteSpace and CitNetExplore, have appeared in the field of scientific knowledge atlas[18]. In view of the characteristics of Web of Science (WoS) database and the powerful scientific metrology and visual analysis function of VOSviewer, the software is selected in this study as a tool for bibliometric analysis to convert a large number of literature data into a visual atlas to discover the laws and imperceptible things hidden in a large number of historical literature data, so that researchers and practitioners can understand knowledge in a more direct way. In addition, the calculation and ordering of measurement indicators are completed in Excel 2013, and the drawing of data presentation in Origin Pro 9.2.

3 Results and Discussion

3.1 Analysis of Annual Literature Volume and Annual Cumulative Citation Frequency

As can be seen from Fig. 1, the annual literature volume of sustainable city shows an overall upward trend from 2008 to 2021. The volume of literature in 2008 reached the lowest in the past years, with only 15 articles. The volume of literature in 2021 was in the highest level during 14 years, with 319 articles, accounting for 17.6% of the total. The literature distribution of sustainable city research can be divided into three stages: low-speed growth period (2008-2010), fluctuation stagnation period (2011-2014), and high-speed growth period (2015-2021). From 2008 to 2010, the annual increase of literature was less than 20, with a slow growth. During the period of fluctuation stagnation (2011-2014), the volume of literature in 2012 decreased by 7 compared with 2011, and there was a slight increase in 2013 and 2014, but the increase was so slight and almost stagnated. After 2015, the annual literature volume showed a trend of rapid growth, and especially the literature volume in 2019 increased the most (60 more than that in 2018). In addition, according to the annual cumulative citation frequency, the literature was cited the most frequently in 2017 (2701 times), but the least frequently in 2008 (243 times). The citation frequency per document was the highest in 2011 (26.9 times), with 2014 (25.1 times) and 2015 (23.9 times) ranking the second and third, respectively. In short, the field of sustainable city still has a high degree of attention and research heat.

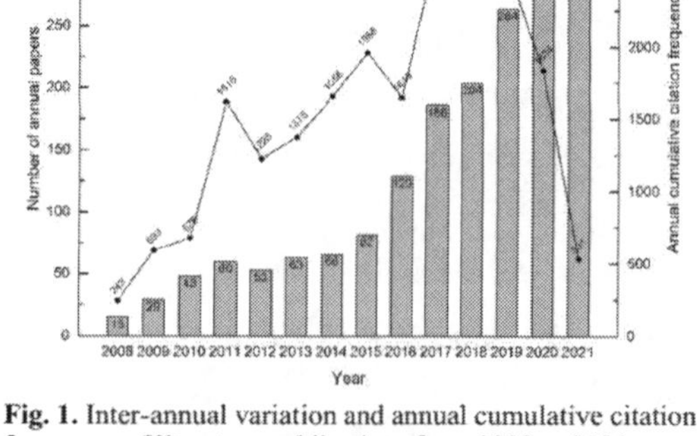

Fig. 1. Inter-annual variation and annual cumulative citation frequency of literature publications from 2008 to 2021.

3.2 Network Analysis of Co-research

Co-research refers to the behavior of scholars working together for the common purpose of producing new scientific knowledge. In practical work, co-research can also be presented as co-authorship, co-institution and co-countries (co-region). Price, the father of scientometrics, believes that with the in-depth development of interdisciplinary and marginal disciplines, the average number of collaborators on a collaborative paper will increase[19]. After analysis, the co-author network in the research field of sustainable city is relatively scattered, indicating that the co-researcher in practical work of the field is not very close with each other, but the co-institution and co-country (co-region) are relatively close.

Fig. 2 shows the largest sub-networks of co-institution in the research field of sustainable city, with the size of nodes as the volume of publications of institutions, the color of nodes as the class group of institutions obtained by default clustering method, and the connection between nodes as the relationship and intensity of co-institutions. The wider the connection is, the stronger the cooperation intensity will be.

Among the largest sub-networks of co-institution in the research field of sustainable city, research institutions with publications over 400 include Chinese Academy of Sciences (CAS, Citations=720), Arizona State University (ASU, Citations=705), University of British Columbia (UBC, Citations=631), Humboldt State University (HSU, Citations=546), Delft University of Technology (TU Delft, Citations=466), The University of Tokyo (UTokyo, Citations=448), University of Wageningen (Wageningen UR, Citations=429), Stockholm University (Wageningen UR, Citations=418) and Fudan University (FDU, Citations=418).

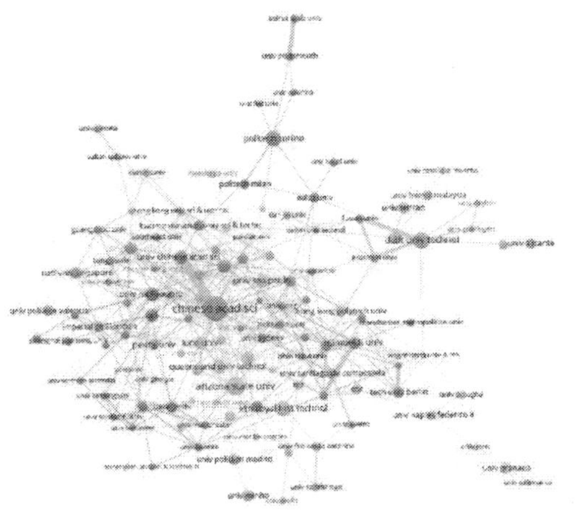

Fig. 2. The largest sub-networks of co-institution.

In terms of research categories, Chinese Academy of Sciences is similar to National University of Singapore, University of Chinese Academy of Sciences, City University of Hong Kong and Tsinghua University (Cluster=2). Arizona State University is similar to University of British Columbia, Beijing Normal University and University of Toronto (Cluster=4). Humboldt State University is similar to The University of Tokyo and Stockholm University (Cluster=3). Delft University of Technology is similar to Fudan University (Cluster=7). In terms of co-institution relation and strength, Chinese Academy of Sciences is closely cooperated with University of Chinese Academy of Sciences. There is also close cooperation among Delft University of Technology, Fudan University and Erasmus University Rotterdam.

It can be seen from Table 1 that, in terms of scientific co-country (or co-region) in the sustainable city field, China (251), America (205), Spain (165), England (146) and Italy (145) rank the top five co-countries in the number of published documents, and America (4028), China (3142), England (2923), Netherlands (2364) and Australia (2354) rank the top five in the cited documents, indicating that there is no lack of high-quality classic documents of the sustainable city field in America, compared with other countries (or regions).

Table 1. The top ranked items of co-country by documents.

Documents	Country	Citations	Cluster
251	China	3142	6
205	America	4028	5
165	Spain	1755	1
146	England	2923	6
145	Italy	1568	1
110	Australia	2354	7
101	Germany	1744	3
76	Netherlands	2364	5
74	Japan	1279	2
72	Sweden	1898	1

3.3 Analysis of Co-keywords

The word frequency analysis method in scientometrics research can be used to study the knowledge structure, development trend and research hotspot in a certain scientific field. Fig. 3 refers to the density diagram of co-keyword in the sustainable city field. The warmer the color is, the higher the frequency of the co-keywords will be. According to the 30 co-keywords with the strongest occurrences listed in Table 2 and their occurrence years, it can be seen that the four keywords, Sustainable Cities (236), Sustainability (172), Sustainable City (167) and Sustainable Development (110) in 2017, as well as the keywords, Smart City (103) and Smart Cities' (79) in 2019, have been the current research hotspots in the field of sustainable city.

From 2008 to 2010 (low growth period), there was no high-frequency co-keywords in the field of sustainable city, indicating that there was no research hotspot and no clear knowledge structure in this field. Research hotspots from 2011 to 2014 (fluctuation stagnation period) focused on sustainable aspects of urban design, urban architecture and eco-city, which can be seen from the four co-keywords, Eco-city (8) and Architecture (8) in 2013, Urban Design (9) and Sustainable (8) in 2014. From 2015 to 2021 (rapid growth period), in addition to the research hotspots on the high-frequency co-keywords such as Sustainable or Sustainability, there have also been changes in the knowledge structure of sustainable city, placing emphasis on Urban Environment (10) in 2015, focusing on Industrial Ecology (16) in 2016, paying attention to Urban Planning (58), Climate Change (45), Urbanization (38), Urban Agriculture (26) and Resilience (25) in 2017, probing into Sustainable Urban Development (27), Smart Sustainable Cities (24), and Green Infrastructure (22) in 2018, turning to Smart City (103)/Cities (79), Urban Heat Island (23), and Circular Economy (21) in 2019, and tending to be Sustainable Development Goals (36), Internet of Things (19) and Deep Learning (8) in 2020 and 2021 from the research perspective of sustainable city.

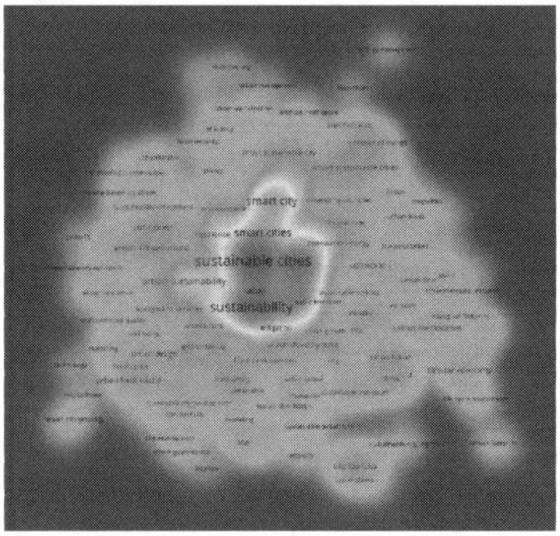

Fig. 3 Density graphic of co-keywords in the research field of sustainable city.

In summary, the evolution in the research field of sustainable city is characterized by those concepts related to ecology, environment and development are always running through the field. However, before 2017, more attention was paid to the planning and design of sustainable city in the context of GIS technology. From 2017 to 2019, emphasis was placed on how to maintain a city's own basic structure and function by preventing, absorbing and adapting to external interference[20]. The application of big data, Internet of Things and artificial intelligence has greatly expanded the research dimension of sustainable city for nearly two years[21].

Table 2. Top 30 keywords with the strongest influences.

Occurrences	Keywords	Year
236	sustainable cities	2017
172	sustainability	2017
167	sustainable city	2017
110	sustainable development	2017
103	smart city	2019
79	smart cities	2019
58	urban planning	2017
45	climate change	2017
41	urban sustainability	2018
38	urbanization	2017
36	sustainable development goals	2020
34	cities	2017
27	sustainable urban development	2018
26	urban agriculture	2017
25	resilience	2017
24	smart sustainable cities	2018
23	urban heat island	2019
22	green infrastructure	2018
21	circular economy	2019
19	internet of things	2020
19	environment	2016
16	industrial ecology	2016
15	gis	2016
14	urban design	2014
11	sustainable	2014
10	urban environment	2015
9	eco-city	2013
8	deep learning	2021
8	mitigation	2015
8	architecture	2013

3.4 Document Citation Network Analysis

In the VOSviewer, the correlation among documents is determined by the number of citations among them[22]. There are 21 documents with citations greater than 100 times in the research field of sustainable city (Fig. 4).

The most frequently cited document was written by Haase et al. in 2014 (Citations=429, Cluster=11), aiming to make people better understand and govern the complex urban system dominated by human beings by understanding the operation, service functions and internal changes of urban ecosystem[23]. In addition, Dobbs et al., under the same cluster (Cluster=11), published a document (Citations=429) discussing how to better plan sustainable city by optimizing landscape structure from the aspects of various ecosystem services and hazards of urban forests[24].

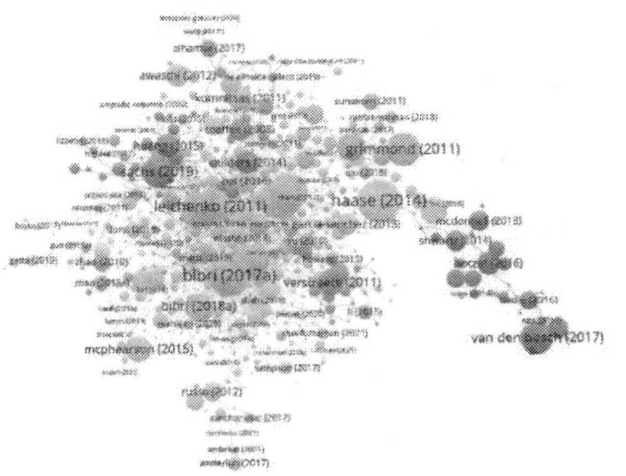

Fig. 4 Document citation analysis network of sustainable city.

The third most cited document (Citations=354, Cluster=28) counted the emerging concepts related to city in global policies and practices over the past few decades through comprehensive bibliometrics analysis. It has been found that Sustainable City is the largest node in the co-keyword network, and closely related to the concepts of Ecological City and Green City, but Smart City has gradually occupied a dominant position in urban planning policies in recent years[25]. Bibri and Krogstie wrote a document (Citations=378, Cluster=23) in 2017, emphasizing that the current difficulties faced by sustainable city construction and development can be solved by establishing a smart sustainable cities model or development framework with persuasive theory and practice, which can provide researchers and practitioners with valuable and pioneering references different from the past[26].

The realization of sustainable city requires not only the understanding of ecosystem, but also the systematic understanding of the relationship among urban ecosystem, urban residents and biodiversity (Cluster=1). Nature-based Solutions (NBS), as nature-inspired, nature-supported and nature-based dynamic solutions that address a variety of social challenges in a resource-efficient and adaptive manner while providing economic, social and environmental benefits, is playing an increasingly important role in urban and human health sustainability[27].

In short, the document citation network analysis in the research field of sustainable city can reflect the correlation among documents from the clustering characteristics and strength relationship. On the whole, there are many research dimensions of sustainable city. Under the background of globalization, this field has gradually shifted from the interpretation and analysis of concepts to the discussion of urban environment, ecology, livability and equity issues. Spatial planning tends to adopt a comprehensive method from a systematic perspective, and landscape design emphasizes the organic integration of geographical, ecological, cultural, aesthetic, social and economic dimensions in a reasonable way. Urban management supports public participation and actively explores the collaborative mechanism of different interest groups.

4 Conclusions

Based on the bibliometrics theory, 1816 documents in the field of sustainable city on the Web of Science database were visually analysed. From 2015 to now, the citable documents of sustainable city have shown a rapid upward trend, with increasing attention year by year and increasing research heat. In terms of building academic community of sustainable city, the scientific co-research of scholars needs to be strengthened. Although there has been a certain scale of institutional and national (or regional) academic communities, it is necessary to further enhance the in-depth cooperation among Chinese and European and American scientific research institutions (or countries/regions), and share the latest scientific research achievements and successful practical experience through international academic exchanges. The knowledge evolution in the research field of sustainable city always runs through such concepts as ecology, environment and development. However, in recent years, based on the development of complex algorithms, data science and artificial intelligence, the research dimension of sustainable city has been greatly expanded. In addition, according to the document citation network, the research in this field can increasingly reflect the interdisciplinary characteristics, intersection, comprehensiveness and systematicness. Sustainable city will be the focus of urban planning, landscape architecture and landscape ecology in a quite long time at present and even in the future, so that it can be conducive to providing reference for researchers, managers and practitioners with the use of visualization method.

Acknowledgments

This paper is one of the phased achievements of the National Social Science Fund Art Project, *Study on the Development of Rural Characteristic Culture Industry Enabled by Handicraft Design* (Grant No. 20BH158) and the Scientific Research Project of Education Department of Shaanxi Provincial Government, *Study on the Value Evaluation and Optimization Strategy of Cultural Space of Heritage Park under the Concept of Park City* (Grant No. 21JK0416).

References

1. WCED, World Commission on Environment and Development. Our Common Future. Oxford University Press, New York (1987)
2. L. Anthopoulos, Smart utopia VS smart reality: Learning by experience from 10 smart city cases. Cities, **63**, 128-148 (2017)
3. Egger, S. (2006) Determining a sustainable city model. Environ. Modell. Softw., 21(9): 1235-1246.
4. J.R. Kenworthy, The eco-city: Ten key transport and planning dimensions for sustainable city development. Environ. Urban, **18(1)**, 67-85 (2006)
5. S. Niza, L. Rosado and P. Ferrão, Urban metabolism: Methodological advances in urban material flow

accounting based on the Lisbon case study. J. Ind. Ecol., **13(3)**, 384-405 (2009)

6. J. Rajaonson and G.A. Tanguay, A sensitivity analysis to methodological variation in indicator-based urban sustainability assessment: A Quebec case study. Ecol. Indic., **83**, 122-131 (2017)
7. S.L. Huang, J.H. Wong and T.C. Chen, A framework of indicator system for measuring Taipei's urban sustainability. Landscape Urban Plan., **42(1)**, 15-27 (1998)
8. C. Turcu, Re-thinking sustainability indicators: Local perspectives of urban sustainability. J. Environ. Plann. Man., **56(5)**, 695-719 (2013)
9. T. Yigitcanlar and S. Teriman, Rethinking sustainable urban development: Towards an integrated planning and development process. Int. J. Environ. Sci. Te., **12**, 341-352 (2015)
10. A. Sharifi and A. Murayama, A critical review of seven selected neighborhood sustainability assessment tools. Environ. Impact Asses., **38**, 73-87 (2013)
11. J. Burnett, City buildings – Eco-labels and shades of green!. Landscape Urban Plan., **83(1)**, 29-38 (2007)
12. S. Pincetl, Nature, urban development and sustainability – What new elements are needed for a more comprehensive understanding?. Cities, **29**, S32-S37 (2012)
13. C. Haaland and C.K. van den Bosch, Challenges and strategies for urban green-space planning in cities undergoing densification: A review. Urban For. Urban Gree., **14(4)**, 760-771 (2015)
14. L.J. Pearson, L. Pearson and C.J. Pearson, Sustainable urban agriculture: Stocktake and opportunities. Int. J. Agr. Sustain., **8(1)**, 7-19 (2011)
15. C. Kennedy, S. Pincetl and P. Bunje, The study of urban metabolism and its applications to urban planning and design. Environ. Pollut., **159(8-9)**, 1965-1973 (2011)
16. K. Alawadi, Rethinking Dubai's urbanism: Generating sustainable form-based urban design strategies for an integrated neighborhood. Cities, **60**, 353-366 (2017)
17. Y. Feng, Y.W. Han, Q. Fan, X.P. Zhang and J.J. Liu, A visualized review of ecological planning and design based on bibliometrics from 1992 to 2017. Int. Rev. Spat. Plan. Su., **7(1)**, 101-116 (2019)
18. N.J. van Eck and L. Waltman, Software survey: VOSviewer, a computer program for bibliometric mapping. Scientometrics, **84(2)**, 523-538 (2010)
19. D.J. Price and D.D. Beaver, Collaboration in an invisible college. Am. Psychol., 21(11): 1011-1018.
20. Leichenko, R. (2011) Climate change and urban resilience. Curr. Opin. Env. Sust., **3(3)**, 164-168 (1966)
21. S.E. Bibri, The IoT for smart sustainable cities of the future: An analytical framework for sensor-based big data applications for environmental sustainability. Sustain. Cities Soc., **38**, 230-253 (2018)
22. L. Šubelj, N.J. van Eck and L. Waltman, Clustering scientific publications based on citation relations: A systematic comparison of different methods. PLOS ONE, **11(4)**, e0154404 (2016)
23. D. Haase, N. Frantzeskaki and T. Elmqvist, Ecosystem services in urban landscapes: Practical applications and governance implications. AMBIO, **43**, 407-412 (2014)
24. C. Dobbs, D. Kendal and C.R. Nitschke, Multiple ecosystem services and disservices of the urban forest establishing their connections with landscape structure and sociodemographics. Ecol. Indic., **43**, 44-55 (2014)
25. M. de Jong, S. Joss, D. Schraven, C.J., Zhan and M. Weijnen, Sustainable–smart–resilient–low carbon–eco–knowledge cities; making sense of a multitude of concepts promoting sustainable urbanization. J. Clean. Prod., **109**, 25-38 (2015)
26. S.E. Bibri and J. Krogstie, Smart sustainable cities of the future: An extensive interdisciplinary literature review. Sustain. Cities Soc., **31**, 183-212 (2017)
27. M. van den Bosch, A.O. Sang, Urban natural environments as nature based solutions for improved public health – A systematic review of reviews. Environ. Res., **158**, 373-384 (2017)

Biochar Amendment in the Green Roof Substrate Improve Air Quality

Yishu Hu[2] and *Haoming* Chen[1,2,*]

[1]School of Environmental and Biological Engineering, Nanjing University of Science and Technology, Nanjing 210094, China.
[2]College of Resources and Environmental Sciences, Nanjing Agricultural University, Nanjing, Jiangsu 210095, China.

Abstract. In this paper, biochar was added to the green roof in order to study the effect of biochar amendment on improving the air quality. Simulation experiments showed that biochar treatments can reduce the concentration of CO_2 by 7.3-28.8%, CO by 10.6-20.7%, SO_2 by 9.8-21.6%, and NO_2 by 2.1-11.5% in the micro-environment. Outdoor monitoring experiments showed that biochar can reduce 24.0-46.6% of PM2.5, 3.7-23.3% of PM1.0, and 17.3-38.4% of PM10. In addition, the high temperature is the main factor affecting the adsorption of small particles. Structural equation analysis showed that biochar mainly absorbed the particulate matters and harmful gases by reducing the surface temperature, increasing the humidity and promoting the plants growth.

1 Introduction

Seven million people die every year due to air pollution [1]. Urban air pollutants are mainly dust come from solid waste combustion, automobile exhaust emissions and tire friction. PM2.5 and PM10 have become the hot issues of the pollutants. Green roofs can absorb and filter particles and dust in the air. At the same time, they can absorb various harmful gases such as CO, NO and SO_2 in the air. However, because of the harsh roof environment, the green roof combined with soil and plants is often difficult to achieve the expected ecological benefits. In order to further improve air quality and repair air pollution by green roofs, biochar was used as a soil amendment in this study. Biochar has many merits such as rich pore structure, so biochar has the function of adsorbing air pollution. The raw materials of biochar come from solid organic waste. Therefore, the production and use of biochar can also contribute to solid waste resource utilization. Adding biochar into the green roof substrate is a potential method to purify the air pollution.

2 Materials and Methods

2.1 Materials

Sludge biochar (SB) and urban natural soil (Pukou District, Nanjing, 5-10cm above the ground) were mixed for roof greening research. SB with the treatment of pyrolysis and dehydration at 600°C is provided by Mississippi International Water (China) Co., Ltd. (Hangzhou, Zhejiang Province). The heavy metal indicators of biochar meet the national standards (AQSIQ and SAC, 2009) and can be used for roof greening. *Sedum lineare Thunb* was planted on the green roof, which seeds were purchased from the Jiangsu Academy of Agricultural Sciences.

2.2 Methods

2.2.1 Experimental Method

There are five treatments in the experiment, with different ratio of biochar and soil (CK, 5%SB, 10%SB, 15%SB, 20%SB). Each treatment has 3 repetitions. The test plants were traditional roof plants: *Sedum lineare Thunb*. The average biomass of plants during the monitoring period was CK (9.15±1.2 g/plant), 5%SB (11.93±0.9 g/plant), 10%SB (13.21±1.1 g/plant), 15% SB (15.65±0.8 g/plant) and 20% SB (15.37±1.2 g/plant). Air purification experiment (simulation): uses the simulation box shown in Figure 1. The hole above the analog box is sealed with tape, and to prevent gas loss the gas measuring port is plugged with a cotton plug when not detecting. The polluted air (the flue gas produced by the burning of 1g mixture of fallen leaves, dried branches and straw) is filled into the simulation box (Fig.1). The pollutant concentration of simulated gas is shown in Table 1. The simulation time of the experiment is from 10 am to 1 pm in the unobstructed outdoor space, so that the test can ensure that the plants can be illuminated fully by the sun and can carry out effective photosynthesis.

Environmental monitoring experiment (outdoor, Fig.2): A portable air quality detector (LB-S06, Lianyungang Lanbao Electronic Technology Co., Ltd., Fig.3A) is used for monitoring. The monitoring indicators are PM2.5, PM1.0, PM10, temperature, humidity and wind speed. The instrument is placed in the center of each treatment sample plot, 5cm high from the

* Corresponding author: chenhaoming89@hotmail.com

ground. The environmental climate during monitoring is shown in Table 2.

Table 1. Pollutant concentration of simulated gas

	PM2.5 μg/m³	PM1.0 μg/m³	PM10 μg/m³	CO_2 mg/m³	CO mg/m³	SO_2 mg/m³	NO_2 mg/m³
Pollutant concentration	356±87	178±85	381±68	21051±566	1277±37	15±7	92±21

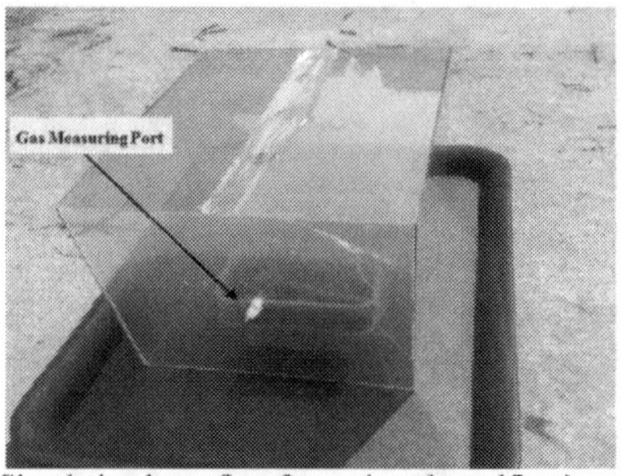

Fig. 1. Simulation box of roof greening air purification test

Fig. 2. Roof greening air purification test

Table 2. Environmental climate table

Date	Maximum temperature	Minimum temperature	Weather	Wind direction	Wind force
2016/12/8	16 ℃	5 ℃	Partly Sunny	Southwest	3-4
2016/12/19	16 ℃	9 ℃	Partly Rainy	Northeast	1-2
2017/1/2	14 ℃	5 ℃	Cloudy	North	3-4

2.2.2 Sample collection and analysis method

The air condition was detected by a portable gas detector (CD4(B) multi-parameter gas detector, Beijing Zhuoan Hengrui Technology Co., Ltd., Fig 3B). To prevent the loss of polluted gas, the air should be detected at 1 h, 2 h, 3 h after the injection of the polluted gas and be measured in the simulation box with an air intake of 500 ml every 30 minutes. The conversion is carried out according to the following formula: Mass concentration (mg/m³) = Substance molecular weight (M) ÷ 22.4× Volume concentration (ppm).

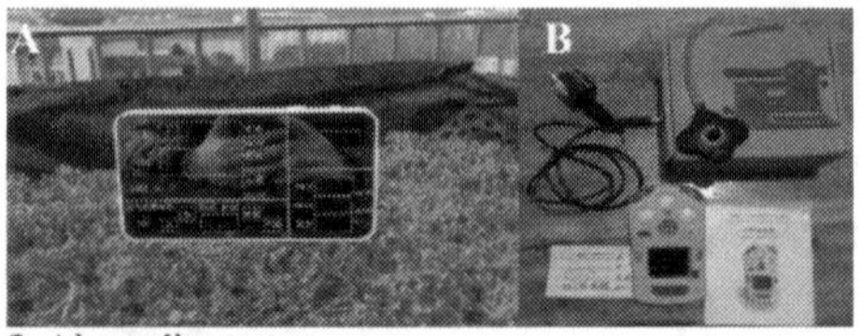

Fig. 3. Air quality tester

3 Results

3.1 The effect of air purification on green roof by biochar amendment

The simulation test results show that the green roof with biochar treatment has a stronger effect on air pollution elimination, which can reduce CO_2 by 7.3-28.8%, CO by 10.6-20.7%, SO_2 by 9.8-21.6% and 2.1 -11.5% of NO_2 (Fig 4). With the increase of time, the air purification capacity of the green roof with biochar treatments is always better than those without biochar. The decrease rate of CO concentration slows down with time, but the decrease rate of NO_2 concentration increases with time. The correlation shows that the concentration of the four gases is negatively correlated with water, but positively correlated with temperature (Table 3).

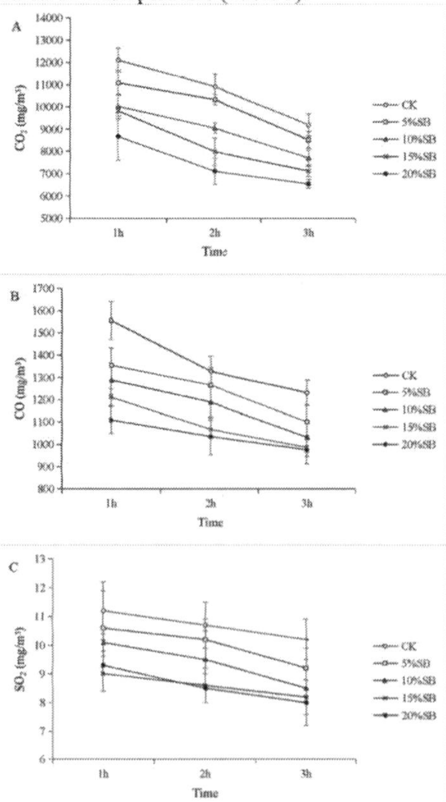

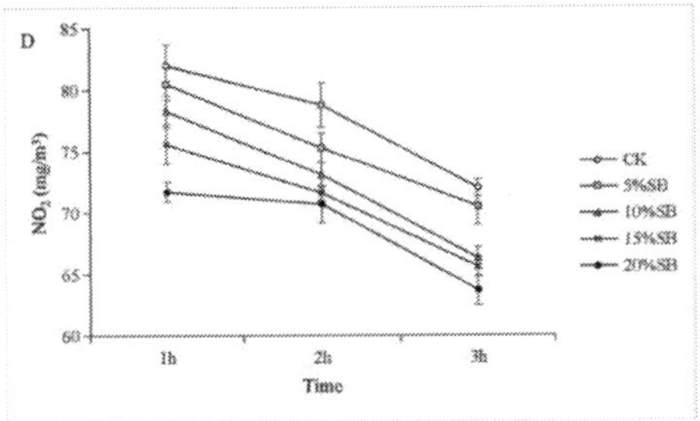

Fig. 4. Changes in the influence of roof greening on air pollution

(A) CO_2, (B) CO, (C) SO_2, (D) NO_2.

Table 3. Correlation analysis of substrate temperature and moisture and pollution gas in roof greening

		CO_2	CO	SO_2	NO_2	Temperature	Humidity
CO_2	Pearson's r	1.00	0.961**	0.873**	0.50	0.894**	-0.930**
	P Value		0.00	0.00	0.06	0.00	0.00
CO	Pearson's r	0.961**	1.00	0.921**	0.47	0.858**	-0.929**
	P Value	0.00		0.00	0.08	0.00	0.00
SO_2	Pearson's r	0.873**	0.921**	1.00	0.32	0.743**	-0.882**
	P Value	0.00	0.00		0.25	0.00	0.00
NO_2	Pearson's r	0.50	0.47	0.32	1.00	0.606*	-0.566
	P Value	0.06	0.08	0.25		0.02	0.04
Temperature	Pearson's r	0.894**	0.858**	0.743**	0.606*	1.00	-0.805**
	P Value	0.00	0.00	0.00	0.02		0.00
Humidity	Pearson's r	-0.930**	-0.929**	-0.882**	-0.566	-0.805**	1.00
	P Value	0.00	0.00	0.00	0.04	0.00	

* The correlation is significant at 0.05
** The correlation is significant at 0.01, n = 15

3.2 The effect of particle adsorption on green roof by biochar amendment

The test results show that the particle adsorption capacity of the green roof with biochar treatments is stronger than those without biochar, which can reduce PM2.5 by 24.0-46.6%, PM1.0 by 3.7-23.3%, PM10 by 17.3-38.4% and 13.6-50.9% AQI, respectively. (Fig 5). The analysis of variance showed that the purification efficiency of PM2.5 and PM10 treated by the green roof with 10-20% biochar has not much difference ($P > 0.05$).

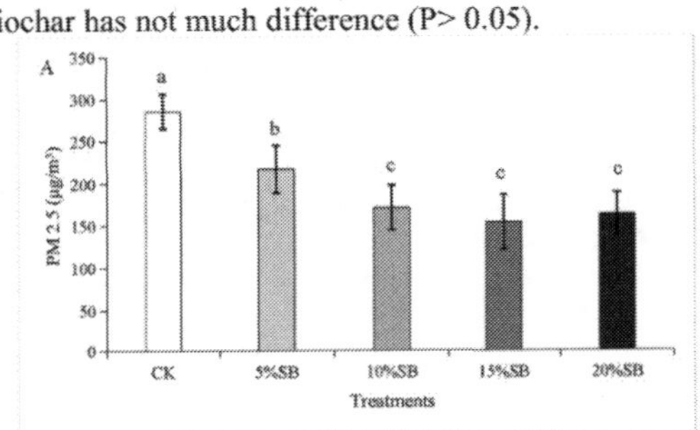

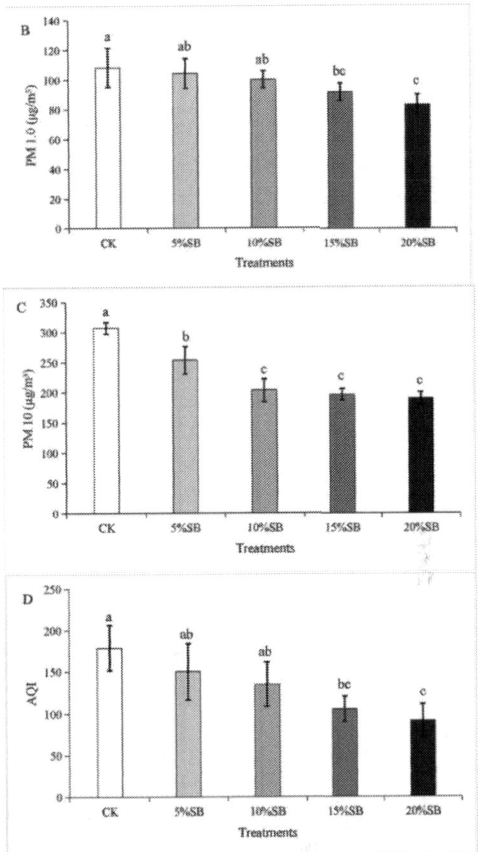

Fig. 5. Effect of biochar on purification of particulate pollutants by roof greening

(A) PM2.5, (B) PM1.0, (C) PM10, (D) AQI.

3.3 The effect of air quality on green roof by environmental factors

Outdoor test results show that the air quality level of roof greening has very strong correlation with temperature (R^2 = 0.637-0.757), humidity (R^2 = 0.516-0.684), wind speed (R^2 = 0.579-0.684) and botany (R^2 = 0.534-0.709) (Table 4). The higher the particle pollutant index is, the more serious the pollution will be. Therefore, the test results show that the purification effect of roof greening on particle is negatively correlated with temperature, and positively correlated with humidity, wind speed and botany (Table 4). Through structural equation analysis, it is found that biochar has no significant effect on PM2.5 and other particles (path coefficient = 0.11). It mainly enhances the adsorption of particles by influencing the near-ground temperature, humidity and vegetation, indirectly (Fig 6).

Table 4. Correlation analysis of particulate matter near the roof and roof environment in roof greening

		PM2.5	PM1.0	PM10	AQI	Temperature	Humidity	Wind Speed	Botany
PM2.5	Pearson's r	1	0.635*	0.860**	0.714**	0.801**	-0.827**	-0.812**	-0.761**

	P Value		0.011	0	0.003	0	0	0	0.001
PM1.0	Pearson's r	0.635*	1	0.639*	0.870**	0.809**	-0.809**	-0.809**	-0.731**
	P Value	0.011		0.01	0	0	0	0	0.002
PM10	Pearson's r	0.860**	0.639*	1	0.721**	0.870**	-0.807**	-0.830**	-0.842**
	P Value	0	0.01		0.002	0	0	0	0
AQI	Pearson's r	0.714**	0.870**	0.721**	1	0.798**	-0.718**	-0.761**	-0.804**
	P Value	0.003	0	0.002		0	0	0	0
Temperature	Pearson's r	0.801**	0.809**	0.870**	0.798**	1	-0.988**	-0.901**	-0.937**
	P Value	0	0	0	0		0	0	0
Humidity	Pearson's r	-0.827**	-0.809**	-0.807**	-0.718**	-0.988**	1	0.769**	0.931**
	P Value	0	0	0	0	0		0	0
Wind Speed	Pearson's r	-0.812**	-0.809**	-0.830**	-0.761**	-0.901**	0.769**	1	0.837**
	P Value	0	0	0	0	0	0		0
Botany	Pearson's r	-0.761**	-0.731**	-0.842**	-0.804**	-0.937**	0.931**	0.837**	1
	P Value	0.001	0.002	0	0	0	0	0	

* The correlation is significant at 0.05,

** The correlation is significant at 0.01, n = 15

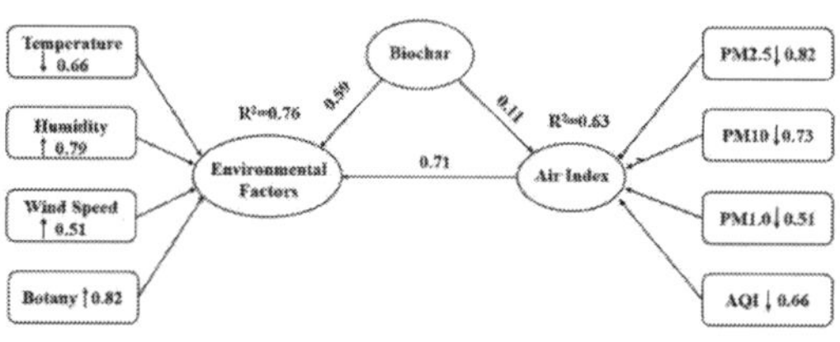

Fig. 6. Structure equation diagram of the effect of biochar on air particulate matter

4 Discussion

There are two main ways for biochar amendment to improve air quality. Firstly, it can help to reduce the emission of polluting gases in the roof matrix. Secondly, it can promote the growth of plants and strengthen the absorption of polluted gases by plants.

The results of this experiment show that the air pollution index of roof greening treated with biochar is lower than those without biochar. And the purification effect of CO_2 and SO_2 is the best (Fig 3). The reason why biochar can effectively reduce the emission of CO_2 and CO is that it can strengthen the photosynthesis of plants, which enhance plants to absorb a large amount of CO_2 and CO. In our previous study, it has been proved that biochar can improve plant growth [2]. And it can also regulate the soil temperature and moisture [3] so that the carbon sequestration capacity of the soil will be improved. Studies have shown that soil temperature is one of the factors that affect the improvement of CO_2 by green roofs [4]. Changes in soil temperature may also change the growth and community structure of microorganisms, thereby affecting soil CO_2 emissions. The water content of the substrate in this experiment is negatively correlated with the concentration of CO_2. This may be because substrate water can lead to the formation of an anaerobic environment, thus inhibiting the oxidation activity of microorganisms and reducing CO_2 emissions [5]. The results also show that temperature has a significant correlation with humidity (R^2 = 0.648). Because of the complexity of the roof environment, temperature changes are often affected by substrate humidity. Temperature and humidity will have a synergistic effect on CO_2 emissions in soil and the air [6].

The reason why the addition of biochar can reduce the concentration of NO_2 is that biochar increases soil porosity. Especially in the low moisture condition, it can increase the oxygen content of the soil and inhibit the anaerobic denitrification process. Biochar can also improve soil pH, thus reducing NO_2 emissions from soil [7]. In addition, biochar provides protection and nutrients for beneficial microorganisms related to the nitrogen cycle, and promotes their reproduction [8].

The addition of biochar significantly reduces the concentration of SO_2 in the air, since biochar increases the plants and leaf area. In addition, because biochar increases the water content of the substrate, the evaporation effect of the substrate is enhanced. So that SO_2 can be dissolved in the water vapor of the air, and then absorbed by the plant leaves and the soil.

Comparing the purification of three types of particles, it is found that the adsorption of green roof with biochar on PM2.5 and PM10 is stronger than that on PM1.0 (Fig 4). This phenomenon is due to the strong sunlight radiation on the roof, which rises the surface temperature of the substrate and causes dust pollution. In addition, the rise of hot air brings small particles higher, which intensifies the pollution of small particles in the roof environment. Besides, the results show that wind speed has a significant impact on PM2.5 in the roof air. The wind speed increases the settling rate, thereby improving the adsorption effect of plant leaves [9].

Plants have always been the main way for green roofs to purify the air due to the adsorption and transformation capacity. The results show that plants have a significant correlation with air quality (Table 4). Plant leaf morphology, leaf area, and leaf coverage can all affect the adsorption efficiency of particles. In addition, plants can also change the spatial pattern of pollutant on the roof by affecting air flow, increasing or decreasing the local pollutant concentrations [10]. Higher pollution concentration in the simulation box than that in the outdoor test is because the circulation limitation and the difficulty of pollutant diffusion [11].

In conclusion, biochar added in green roofs can not only improve the air quality, but also further promote the ecological benefits, such as purifying air. Green roof with biochar treatment can reduce the concentration of CO_2, CO, SO_2 and NO_2 in micro-environment. In addition, the results show that biochar can improve the adsorption of particles in the air by green roofs. Biochar can also improve the roof micro-environment by adjusting the soil environment and promoting plant growth.

References

1. B. Kuehn, WHO: More Than 7 Million Air Pollution Deaths Each Year. JAMA: the journal of the American Medical Association, **311**, 1486 (2014)
2. H.M. Chen, J.Y. Ma, X.J. Wang, P. Xu, S. Zheng and Y. Zhao, Effects of biochar and sludge on carbon storage of urban green roofs. Forests **9(7)**, 413 (2018)
3. J. Lehmann, Bio-energy in the black. Frontiers in Ecology and the Environment **5(7)**, 381-387 (2007)
4. X. Li, H. Fu, D. Guo, X. Li and C. Wan, Partitioning soil respiration and assessing the carbon balance in a Setaria italica (L.) Beauv. Cropland on the Loess Plateau, Northern China. Soil Biology and Biochemistry **42**, 337-346 (2010)
5. B. Liang, J. Lehmann, S. Sohi, J. Thies, B. O'Neill, L. Trujillo, J. Gaunt, D. Solomon, J. Grossman, E. Neves and F. Luizão, Black carbon affects the cycling of non-black carbon in soil. Organic Geochemistry. 206-213 (2010)
6. B. Gao, X. Ju, F. Su, Q. Meng, O. Oenema, P. Christie, X. Chen and F. Zhang Nitrous oxide and methane emissions from optimized and alternative cereal cropping systems on the North China Plain: A two-year field study. Science of The Total Environment **472**, 112–124 (2014).
7. Y. Wu, G. Xu, Y. Lu and H. Shao Effects of biochar amendment on soil physical and chemical properties. Adv. Earth Sci. **29**, 68-77 (2014).
8. S. Steinbeiss, G. Gleixner and M. Antonietti, Effect of biochar amendment on soil carbon balance and soil microbial activity. Soil Biology and Biochemistry **41**, 1301-1310 (2009).
9. P. Freer-Smith, A. El-Khatib and G. Taylor Capture of Particulate Pollution by Trees: A Comparison of Species Typical of Semi-Arid Areas (Ficus Nitida and Eucalyptus Globulus) with European and North American Species. Water, Air, and Soil Pollution **155**, 173-187 (2004)
10. Z. Tong, T. Whitlow, P. MacRae, A. Landers and Y. Harada, Quantifying the effect of vegetation on near-road air quality using brief campaigns. Environmental Pollution **201**, 141-149 (2015)
11. S.D. Sabatino, R. Buccolieri, G. Pappaccogli and L. Leo, The effects of trees on micrometeorology in a real street canyon: consequences for local air quality. International Journal of Environment and Pollution **58**, 100 (2015).

Overview of Phytoremediation Technology for Heavy Metal Contaminated Soil

Can Li[1], *Guomin* Yang[1], *Zhuojun* Liu[1] and *Jinxing* Cai[1,*]

[1]Nankai University, The College of Environmental Science and Engineering, 300350, Tongyan Road No.38, Jinnan District, Tianjin, P.R. China

Abstract. Phytoremediation of heavy metal polluted soil has become an efficient and environment-friendly method. The main mechanisms of phytoremediation consist of phytoextraction, phytovolatilization phytostabilization, and phytotransformation. But there are some limitations in the actual remediation effect of plants enriching heavy metals. At present, a wealth of joint remediation methods have been carried out. This paper reviewed the phytoremediation mechanisms and the research status of phytoremediation, such as Cd, As and Pb, and put forward the research direction of phytoremediation and the combination with other technologies.

1 Introduction

Soil heavy metal pollution is a global problem which poses acute threats to the environment and human beings. Their Their sources are divided into natural and anthropogenic sources. The former include volcanic eruptions and weathering of metal-bearing rocks, while the latter include mining and various agricultural or industrial activities. As non-degradable persistent pollutants, heavy metals in soil can reduce biological activity, soil fertility and endanger human health through accumulation in the food chain. Therefore, the remediation of heavy metal polluted soil is of great significance for reducing environmental risks and maintaining human health and ecological security. The phytotechnologies are environmentally safe and cost-effective, which can compensate for other soil remediation technologies. Thus the use of phytoremediation is gradually being accepted by the public.

Phytoremediation has some limitations, too. Phytoremediation relies on the biological cycle and requires multiple crop cycles to repair the soil; phytoremediation can be carried out on sites with little or moderate contamination to ensure normal plant growth; root depth, plant growth rates, low overground biomass and are also limiting factors for phytoremediation; heavy metals bound tightly to soil are difficult to be repaired by plants because of their low biological benefit and so on. Thus phytoremediation began to try to combine with other technologies to improve the phytoremediation effect of heavy metals. This paper summarized the mechanisms of phytoremediation as comprehensively as possible, lists several application examples of phytoremediation, and put forward the research direction of the combination of phytoremediation and other technologies.

2 Phytoremediation

Phytoremediation which uses plants to remove heavy metals from contaminated soil is considered to be a sustainable bioremediation process.

Phytoremediation of heavy metal contaminated soil has multiple mechanisms, which include phytoextraction (uptake, accumulation and translocation of heavy metals), phytostabilization (stabilization of heavy metals in the root zone), phytovolatilization (emission to atmosphere) and phytotransformation (plant degradation).[1]

In phytoextraction, heavy metals are absorbed by the plants together with nutrients and water. The heavy metals are then accumulated and translocated to the overground parts of the plant (leaves, bud, shoot, etc.). Phytoextraction is split into natural phytoextraction and chemical induced phytoextraction. Natural phytoextraction is referred to the extraction of heavy metals from soil by using the natural accumulation capacity of plants. Its main advantages are simple process and low cost, but low extraction efficiency of heavy metals. Compared with natural phytoextraction, chemically induced phytoextraction promotes the extraction of heavy metals by adding chemical modifiers in the rhizosphere of plants and forming chelates with heavy metals in the rhizosphere soil. Therefore, it has the advantages of high extraction efficiency and large removal of heavy metals.

Phytostabilization refers to in-situ deactivation or immobilization of heavy metals that decreases the bioavailability of heavy metals and prohibits their ectopic transportation. In this way, heavy metals are fixed at the roots of the plants by absorption. The benefit

* Corresponding Author: caijinxing666@163.com

of this technique is that it helps to protect the surface water and groundwater from heavy metal pollution. While, heavy metals are not really removed from the soil, which is the primary shortcoming of phytostabilization.

In phytovolatilization, heavy metals are absorbed by plants and pushed up through the xylem stream. They are finally converted into volatile forms entering the atmosphere via stomata. Phytovolatilization remediation only transfers pollutants from the soil to the atmosphere. Volatile pollutants are constantly dispersed and diluted in the atmosphere, and will not harm the soil in a short time. However, the state of pollutants cannot be tracked and controlled. So it is very important to monitor the migration of pollutants in the atmosphere for the application of phytovolatilization.

In phytotransformation, plants absorb heavy metals in soil and changed them into nontoxic or less toxic forms. Subsequently, metal ions are decomposed and mineralized by metabolism in plant cells or degradation of enzymes. The degradation of heavy metal pollutants in some plants is related to their own metabolic activities, which can degrade or mineralize heavy metals.The above processes are shown in Fig. 1.

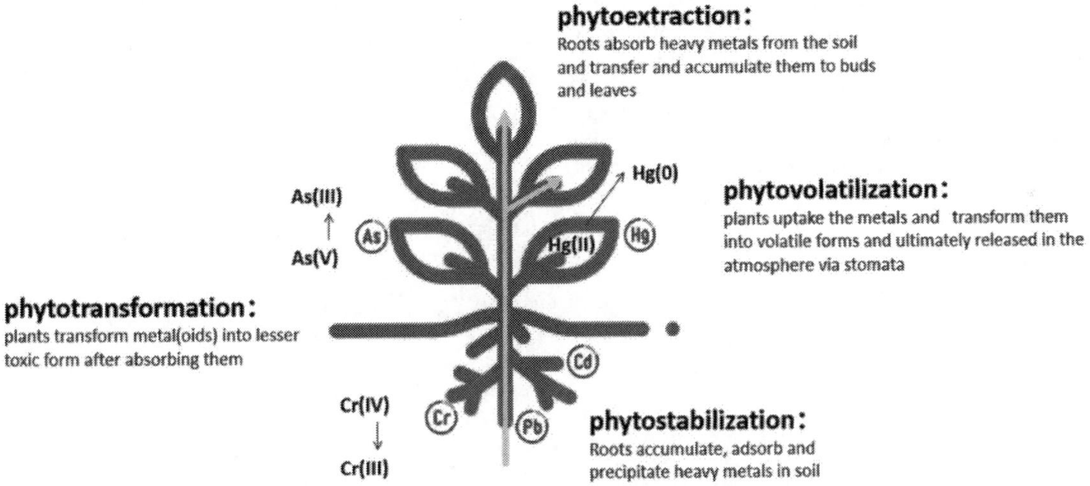

Fig. 1. Multiple mechanisms of phytoremediation.

3 The efficiency of phytoremediation of soils contaminated with different heavy metals

According to the National Soil Pollution Survey Bulletin, we selected a total of eight metals, including cadmium, chromium, zinc, arsenic, copper, mercury and lead, nickel to explore the restoration effect of which contaminated soil by different plants.

BCF (bioconcentration factor) and TF (translocation factor) are commonly applied to measure the efficiency of phytoremediation. TF is the ratio of the heavy metal element density in shoots to that in roots. And BCF is the ratio of heavy metal element density in roots to the that in soil[2]. Plants with both TFs and BCFs higher than 1 (TF > 1and BCF > 1) is promsing to be applied in phytoextraction.

In order to restore and reforest areas degraded by mining, Salas-Moreno, M.[3] planted *Paspalum fasciculatum Wild* in soils with three lead (Pb) concentrations (15, 30, 50mg/kg) and calculated the translocation factor (TF). TF value reached 1.68 after 30 days (Pb treatment concentration is 30mg/kg). They also found that the concentration of heavy metals in this plants declined as the extension of exposure duration, which demonstrates that the plant can be exposed to high concentration of Pb. Their tolerant mechanisms are accumulating the high concentration of Pb in the roots to avoid high concentration in the shoots and then minimize the lesion to the photosynthesis system. At the same time, this species can ameliorate soil conditions by reducing soil acidity and increasing the content of organics.

Azam, S.K. et al.[4] studied whether silicon affected the uptake and accumulation of As by *I. cappadocica*, an As hyperaccumulator, and how Si affected the physiological response under As stress. It indicated that 1 mM Si level enhanced the length of roots and shoots, photosynthetic pigment and biomass of *I. cappadocica*, and increased plant biomass contributed to improving the repair efficiency. Si reduces the concentration of bud as in *I. cappadocica* tissue, so Si addition can be used as a plant phytostabilizer in *I. cappadocica* as contaminated site. As the Si reduces the absorption and translocation of As, the use of Si cannot be used to promote the phytoextraction of *I. cappadocica*. Liu, X. et al.[5] found that P is an effective inhibitor of As uptake by plants. This is because As and P are chemical analogues, which are absorbed by P transporters in plants, and their affinity for P is higher than As (V). Inorganic phosphorus inhibits As uptake by *Pteris vittata*, which is not conducive to the phytoremediation of As contaminated soil, and phytate plays a beneficial role in enhancing As uptake.

Cu is a essential element needed in the growth of various organisms, but high concentration Cu has obvious toxicity to organisms. High concentration of Cu will inhibit plant photosynthesis, nutrient absorption and cell growth[6], and cause animal diseases, such as neurasthenia. Pan, X.C. et al.[7] proved that Cu accumulation in the shoot of *Elsholtzia haichowensis* was always higher than that in the root system due to the large shoot biomass accumulation. However, the concentration factors of roots was greater than that of shoot, and the tolerance of shoot was greater than that of root, indicating that *Elsholtzia haichowensis* mainly accumulated Cu into roots and precipitated to reduce the harm to the growth of shoot of *Elsholtzia haichowensis*.

Yu Xun et al. studied the mercury transport and bioaccumulation ability of *Cyrtomium macrophyllum* at different time (15d-60d) through pot experiments. A total of nine gradients of mercury contaminated soil samples were set.The experiments indicated that both TF and BCF decreased with the rise of the mercury concentration in soil, and BCF increased gradually with the extension of exposure duration. When it was exposed below 200 mg/kg, TF & BCF were greater than 1 in the whole duration of exposure, and when exposed above 500 mg/kg, TF & BCF were less than 1. *Cyrtomium macrophyllum* may be a potential plant for phytoremediation of mercury contaminated soil.

Medicago scutellata is a legume widely cultivated for which can produce high quality feed and has a wide range of uses. Shahrbanoo, P. et al. [8]found that *Medicago scutellata* can eliminate cadmium from polluted soil count on its strong absorption and accumulation ability of cadmium. In the research, the following two factors were mainly focused on, cadmium concentration and drought degree, respectively. TF was greater than 1 in all treatments, reflecting the high ability of *Medicago scutellata* to translocate cadmium from roots to shoots. The above results might be related to the differences in chemical functional groups of the roots and aerial parts of plants. It manifested that the above plant had an excellent ability of extracting cadmium from soil and could be regarded as a cadmium hyperaccumulator.

Vetiver, a perennial plant, which is known to prevent soil erosion can tolerate broad range of pH and withstand the increase of toxic metals. Jiang,Y.B. et al.[9] found that its processing effect is as follows: total absorption, rootsand shoots 248.3 (mg/kg dry soil),1.13 (BCF),0.796 (TF). In *Vetiver* plant, cadmium absorbed elements, remain attached to the cell wall or are stored in the roots vacuole. Another reason for the increase of cadmium in the roots of the *Vetiver* is its accumulation in the vacuoles; the accumulation of elements in cell vacuoles prevents their transmission to the aerial parts. Therefore, the value of these elements in the roots is far greater than the shoots. *Vetiver* showed the possibility for phytoremediation of heavy metal polluted lands.

Cassia tora, which possesses rapid growth, short lifetime and high biomass is greatly appropriate to grow and sustain in soil. Through research, Patra, D.K. et al.[10] found that the above plant in chrome polluted soil produced a wealth of biomass increasing with the absorption of heavy metals. In the study, the chrome stress improved the level of antioxidant enzymes and proline in plants. Besides, these plants protected themselves from oxidative stress by altering the sundry metabolic processes. Examples of the main phytoremediation of heavy metals are given in Table 1.

Table 1. TF or BCF of different plants in phytoremediation of heavy metal polluted soil.

Heavy metals	Plant(s)/Tree	TF, BCF
Cd	*M. Scutellata*[8]	1.76 (TF)
	Mulberry[11]	0.11 (TF),0.31 (BCF)
	Vetiver[9]	1.13 (BCF),0.796 (TF)
Cr	*Vetiver*[12]	0.53 (BCF),0.94 (TF)
	Cassia tora[10]	0.51 (BCF)
As	*Pteris vittata*[13]	7.99 (BCF)
	Pityrogramma calomelanos[14]	12.0 (TF)
Cu	*Ricinus communis*[15]	6.62 (TF)
	Commelina communis[16]	7.43 (TF)
Pb	*Paspalum fasciculatum Wild*[17]	1.68 (TF)
	Crambe abyssinica Hochst[18]	0.44 (BCF)
Zn	*Poinsettia*[19]	10.04 (BCF)
	Atriplex. hortensis var.purpurea[20]	2.35 (BCF)
Hg	*Cyrtomium macrophyllum*[21]	2.62 (TF)
	Jatropha curcas[22]	1.32 (BCF)
Ni	*Polypogon monspeliensis*[23]	1.07 (TF)
		1.85 (BCF)
	Rumex dentatus[23]	1.05 (TF)
		9.52 (BCF)

4 Conclusions and perspectives

Soil pollution caused by heavy metals is a global concern. Phytoremediation has been proved to be an important means to solve this problem. Its major strengths are feasible, cost-effective and eco-friendly. This paper reviewed the background and mechanisms of phytoremediation of heavy metal polluted soil, summarized the plant species and remediation effects of different heavy metal contaminated soil.

Combined with other remediation technologies, the phytoremediation effect of heavy metal pollution can be improved，such as plant microbial joint remediation, plant animal joint remediation, plant gene improvement remediation and chemical plant joint remediation. Plant microbial joint remediation provides favorable environment for plants to enrich heavy metals through bioaugmentation, and improves the enrichment effect of phytoremediation plants, For example, *Bacillus subtilis*-plant combined remediation of Cd contaminated soil can promote the accumulation of cadmium by plants and its transfer to the above ground. Plant animal combination technology refers to a remediation that uses the direct absorption and transformation of heavy metals by some lower animals (such as earthworms) in the soil, or improves soil physical and chemical properties to increase plant absorption of heavy metals. Plant gene improvement remediation is found on the over expression of specific genes participated in absorption, sequestration, translocation and plant tolerance of xenobiotic compounds in transgenic plants (Aken, B.V.)[24]. Chemical phytoremediation uses synthetic or organic chelating agents to assist plant extraction or induce plant extraction to improve their extraction ability. For example, some polymer materials can promote the absorption of cadmium by hyperaccumulators such as *Solanum nigrum* or *Amaranthus granatum*. The research on phytoremediation of heavy metal contaminated soil is still developing, whose existing limitations are shrinking with the research. Therefore, phytoremediation has broad development prospects.

This work was supported by Ningxia Hui Autonomous Region key research and development plan (2021BEG02002), Tianjin key research and development plan (21YFSNSN00030, 19YFZCSN00-450), Tianjin Agricultural Science and Technology Achievement Conversion and Extension Project (201901130).

References

1. V. Shah and A. Daverey, Phytoremediation: A multidisciplinary approach to clean up heavy metal contaminated soil. Environmental Technology & Innovation **18**, 100774 (2020)
2. S. Pasricha, et al. Molecular mechanisms underlying heavy metal uptake, translocation and tolerance in hyperaccumulators-an analysis. Environmental Challenges, **4**, 100197 (2021)
3. M. Salas-Moreno and J. Marrugo-Negrete. Phytoremediation potential of Cd and Pb contaminated soils by Paspalum fasciculatum Willd. ex Flüggé. International Journal of Phytoremediation, **22(1)**, 87-97 (2020)
4. S.K. Azam, N. Karimi, Z. Souri, et al. Multiple Effects of Silicon on Alleviation of Arsenic and Cadmium Toxicity in Hyperaccumulator Isatis Cappadocica Desv. J. Plant Physiology and Biochemistry, **168**, 177-187 (2021)
5. X. Liu, J.W. Fu, N., Tang, et al. Phytate Induced Arsenic Uptake and Plant Growth in Arsenic-Hyperaccumulator Pteris Vittata. J. Environmental Pollution, **226**, 212-218 (2017)
6. T. Ke, J. Zhang, Y. Tao, et al. Individual and combined application of Cu-tolerant Bacillus spp. enhance the Cu phytoextraction efficiency of perennial ryegrass. J. Chemosphere, 263 (2021)
7. X.C. Pan, J.Z. Xiao, Y.J. Bao, Effect of Cu on the Growth, Root Morphology and Copper Accumulation of Elsholtzia Haichowensis. J. Journal of Taizhou University, **39(06)**, 10-17 (2017)
8. P. Shahrbanoo and S. Hossein The phytoremediation effect of Medicago scutellata (L.) Mill. on soils under Cd-water stress: a good choice for contaminated dry lands.. Environmental science and pollution research international, **26**, 28 (2019)
9. G. Fourud, R. Aptin and P.G. Mohammad, Investigation of Vetiver Grass Capability in Phytoremediation of Contaminated Soils with Heavy Metals (Pb, Cd, Mn, and Ni) Soil and Sediment Contamination: An International Journal, **30**, 163-186 (2021)
10. D.K. Patra, A. Grahacharya, C. Pradhan, H.K. Patra, Phytoremediation potential of Coffe Pod (Cassia tora): An in situ approach for attenuation of chromium from over burden soil of Sukinda chromite mine, India, Environmental Progress & Sustainable Energy, 40 (2021)
11. Y.B. Jiang, S. Jiang, Z.B. Li, X.P. Yan, Z.X. Qin and R.Z. Huang,Field scale remediation of Cd and Pb contaminated paddy soil using three mulberry (Morus alba L.) cultivars Ecological Engineering, **129**, 38-44 (2019)
12. Y.S. Chintani, E.S. Butarbutar, A.P. Nugroho and T. Sembiring, Uptake and release of chromium and nickel by Vetiver grass (Chrysopogon zizanioides (L.) Roberty) SN Applied Sciences, **3**, 1-13 (2021)
13. Y.Q. Qiu, Y.H. Li, P.Q. Peng, et al. Effect of Chelating Agent on Pteris Vittata for Remediation of Arsenic-Contaminated Soil. J. Environmental Engineering, **39(03)**, 204-209, 119 (2021)
14. N.V. Campos, et al. Arsenic-induced responses in Pityrogramma calomelanos (L.) Link: Arsenic speciation, mineral nutrition and antioxidant defenses. Plant Physiology and Biochemistry, **97**: 28-35 (2015)
15. V. Kumar, L.F.R. Ferreira, M. Sonkar, and J. Singh, Phytoextraction of heavy metals and ultrastructural changes of Ricinus communis L. grown on complex organometallic sludge discharged from alcohol distillery. Environmental Technology & Innovation, 22 (2021)
16. J.Y. Ge, J.G. Zhang, Heavy Metal Contamination and Accumulation in Soil and Plant Species from the Xinqiao Copper Deposit, Anhui Province, China. J. Analytical Letters, **48(3)**, (2015)
17. M. Salas-Moreno, M.N. José, Phytoremediation potential of Cd and Pb-contaminated soils by Paspalum fasciculatum Willd. ex Flüggé. International Journal of Phytoremediation, **22(1)**, 87-97 (2020)
18. A.C. Gonalves, et al. Phytoremediation capacity, growth and physiological responses of Crambe abyssinica Hochst on soil contaminated with Cd and Pb.Journal of Environmental Management, **262**, 110342 (2020)
19. F. Xiao, et al. Phytoremediation of potentially toxic elements in a polluted industrial soil using Poinsettia.Physiology and Molecular Biology of Plants, **27(4)**, 675-686, (2021)

20. S.S. Kachout, A.B. Mansoura, R. Mechergui, et al. Accumulation of Cu, Pb, Ni and Zn in the halophyte plant Atriplex grown on polluted soil. Journal of the Science of Food and Agriculture, **92(2)**, 336-342 (2011)
21. X. Yu, F. Liu, Y.D. Li and H.C. Dong, Mercury accumulation plant Cyrtomium macrophyllum and its potential for phytoremediation of mercury polluted sites. J. Chemosphere, **189**, 161-170 (2017)
22. J. Marrugo-Negrete, J. Durango-Hernández, J. Pinedo-Hernández, J. Olivero-Verbel and S. Díez, Phytoremediation of mercury-contaminated soils by Jatropha curcas. J. Chemosphere, **127**, 58-63 (2015)
23. S. Samreen, A.A. Khan, M.R. Khan, S.A. Ansari and A. Khan, Assessment of Phytoremediation Potential of Seven Weed Plants Growing in Chromium- and Nickel-Contaminated Soil. J. Water, Air, & Soil Pollution, **232**, 209 (2021)
24. B.V. Aken Transgenic plants for phytoremediation: helping nature to cleanup environmental pollution. Trend Biotech, **26**, 225-237 (2008)

Effects of Smoothed BRDF on Multi-angle Albedo Retrievals

Yanmin Shuai[1,2,3,4], *Jian* Yang[1,*], *Junbo* Duan[1], *Latipa* Tuerhanjiang[2,3,5], *Chongyang* Wang[6] and *Yu* Ma[7]

[1] Liaoning Technical University, College of Surveying and Mapping and Geographic Science, 123000 Fuxin, China
[2] Xinjiang Institute of Ecology and Geography, CAS, 830011 Urumqi, China
[3] Research Center for Ecology and Environment of Central Asia, CAS, 830011 Urumqi, China
[4] University of Chinese Academy of Sciences, 100049 Beijing, China
[5] Xinjiang University, 830046 Urumqi, China
[6] Shenyang Gas Group Co. Ltd., 110000 Shenyang, China
[7] RIOH Automobile Testing & Certification Technology Co. Ltd, 101103 Beijing, China

Abstract. BRDF as the intrinsic feature of surface targets, is an important parameter required by albedo inversion from multi-angle observations, especially for satellite data suit with less directional measurements. Several studies have shown up to introduce BRDF priori knowledge into albedo retrievals at different scale by spatial or temporal smoothing. Thus, it is necessary to further understand what's the influence induced by BRDF smoothing on albedo retrieval. This work investigated effects of smoothed BRDF on albedo magnitude through case studies over North America region using operational MCD43A&C BRDF products respectively smoothed in spatial and temporal scales. Our results show that BRDF of seasonal DBF samples smoothed from daily to monthly can lead to apparent relative difference to smoothed values of 10.97%, 9.42%, 8.24% and detectable absolute differences of 0.0172, 0.0095 and 0.0035 on related albedo respectively at Near Infrared, Short Wave and Visible broadband. The spatial smoothing of BRDF from 500m to 5600m results in relative differences to smoothed values of 17.38%, 14.38%, 27.23% and absolute differences of 0.0250, 0.0139, 0.0052 for the inversed albedo at above three broadbands.

1 Introduction

Albedo is defined as the ratio of reflected radiation to received solar radiation over surface target, which is an important parameter affecting the energy budget, thus contributing to the change of climate, land surface model and earth-atmosphere system[1-2]. Satellite remote sensing provides an effective potential way to estimate continuous land surface albedo in global or regional scale. Compared with previous single-angle estimation models, modern albedo algorithm relies on multiple-angle observations to first build BRDF(Bi-directional Reflectance Distribution Function), then integrate over incident and view hemispheres to calculate albedo. Studies pointed out that the relative errors can reach up to 45% in the albedo estimation without the consideration of angle effects[3-4].

The requirement of albedo accuracy keeps increasing with the need from various user communities or with the increasing of our understanding on the process of Earth system. The albedo accuracy of 0.02-0.05 is required by the community of global climate change, but increased to 0.02 for regional investigations[5-6]. Studies point out the magnitude of albedo varies with the change of land surface, and as well as related surrounding parameters in fine resolution[7-8]. Existing satellite-based operational or sporadic albedo products span tens of meters to several kilometres in spatial scale and daily to annual phase in temporal scale, with different accuracy affected by inversion scheme, data quality, spatial or temporal resolution, etc[9-14]. Thus, it is important to further understanding the effect of factors induced the variation of albedo retrieval, for better service to user communities.

It is well known BRDF priori knowledge can improve the accuracy of albedo retrieval, especially for the estimation at high spatial resolution due to the lack of multi-angle observations. In recent years, BRDF smoothing process at spatial or temporal scale has been frequently used in satellite-based albedo retrieval from multi or single angle observations[15-16]. Therefore, it is an essential issue to understand what uncertainty induced by different BRDF smooth on albedo retrievals.

2 Data and Method

2.1 Data

Regions with rich land cover types and good public datasets in of North America was selected to serve the case study of this work. MCD12C1, MCD12Q1, MCD43A1, MCD43A3, MCD43C1, MCD43C3 and

* Corresponding author: 471910043@stu.lntu.edu.cn

other data sets of MODIS were used to provide classification, BRDF and albedo of land cover in 2019. These MODIS datasets were downloaded from https://ladsweb.modaps.eosdis.nasa.gov/search/.

This paper discussed the situation of land cover with no snow, as snow can not reflect the real situation of land cover and will cause great interference to spatiotemporal smoothing, so the DOYs(Day of year) with snow were removed from the data sets. In the operational products of MODIS albedo, there are two modes(full inversion and magnitude inversion) according to the observations. If it is magnitude inversion, the inversion quality of albedo is low, and there will be a certain difference from its true value, which will disturb the spatiotemporal smoothing. Therefore, the albedo with low quality should be removed through the QA data layer in the data sets.

Through IGBP data layer in MODIS-MCD12C1, we selected DBF sample(-94.88°W & 31.23°N) with high confidence(93%) as study site, whose average slope is 2.84° and altitude is 59.71m. Then, the DBF sample at 5600m was re-projected to IGBP data layer in MODIS-MCD12Q1(500m) to obtain more detailed classification result, and a representative sample at 500m spatial scale was selected from the homogeneous cushion. The MODIS pixel at 500m was used for temporal smoothing, while a spatial smoothing was carried out from 500m to 5600m. Fig. 1. shows the work follow of this study.

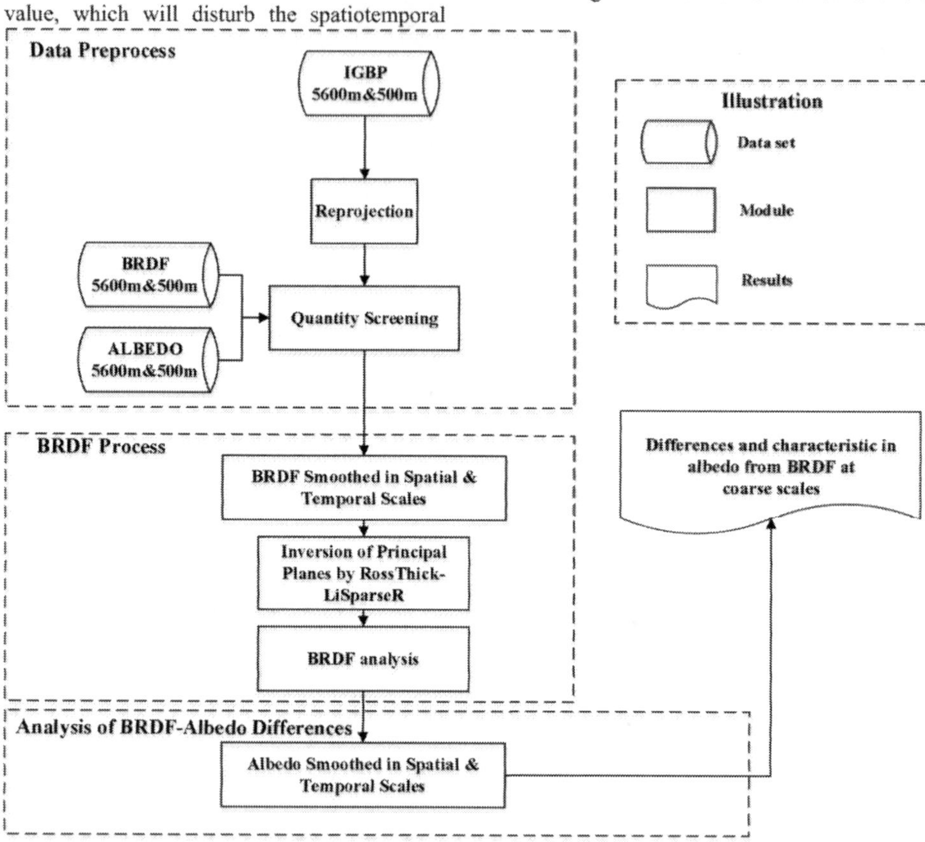

Fig. 1. Work flow chart

2.2 BRDF spatiotemporal smoothing

There are 3 work modules in Fig. 1: Data preprocess, BRDF process, Analysis of BRDF-Albedo differences. Data preprocess module has been discussed in section 2.1. In BRDF process module, as MODIS describes BRDF through a linear kernel driven model (RossThick-LiSparseR), a set of fixed kernel coefficients can determine a BRDF model and MODIS-BRDF products (MCD43C1 & MCD43A1) provide kernel coefficients, thus firstly BRDF parameters after spatiotemporal smoothing can represent the smoothed BRDF model; then, refletance with different incident-view geometry is retrieved by RossThick-LiSparseR model to obtain BRDF principle plane curve.

During the retrieval processing of albedo, albedo is a hemispherical integration of BRDF, so the magnitude and shape of BRDF principle plane curve have potential impact on albedo. In the reflecting hemisphere, the change of the principle plane (i.e. the plane determined by two lines from surface target respectively to observer and sun) is the most complex and representative. This principle plane can not only reflect the angle effect brought by the incident-view geometry, but also contains hot spot information, so it can best reflect the characteristics of BRDF. In this paper, negative values less than 0 degree represent backward and positive values greater than 0 degree represent forward; Then, based on the differences of the reflectivity in the curves at direction of 0°, ± 45°, ± 70°, the STD (Standard Deviation) of feature points is taken as the shape difference coefficient. If difference coefficient is small, the shape changes of principle plane curves are small; otherwise, the shape changes are large. The difference of BRDF magnitudes were used to investigate the dispersion of the all BRDF principle plane curves.

2.3 Absolute and relative differences of albedo

To explore the differences caused by BRDF temporal smoothing on the retrieval of albedo. Firstly, the smoothed monthly values of albedo in 12 months of the year 2019 were calculated, and then the smoothed WSA values at 16-days and 8-days scales within one month were calculated; Lastly, the daily albedo values were compared with the corresponding smoothed albedo values in three broad bands: Near-Infrared(NIR), Short Wave(SW), Visible(VIS) and the relative & absolute differences were calculated. When discussing the influence of BRDF spatial smoothing on the retrieval of albedo, we extracted albedo at 500m and 5600m from MCD43A3 and MCD43C3 products, then calculated the relative & absolute differences between them, lastly we counted the daily differences throughout the year 2019 to explore the magnitudes of the differences.

3 Results and Discussion

3.1 Temporal smoothing of BRDF and related albedo

In order to explore the difference of BRDF before and after multi-temporal smoothing, the BRDF parameters of DBF sample at 500m were extracted from MCD43A1 product, then BRDF was smoothed according to multi-temporal scales(8 days, 16 days, month). To obtain the smoothed principal plane curve and daily principal plane curves of BRDF, reflectance was calculated under different incident-view geometry based on RossThick-LiSparseR model and BRDF parameters before and after smoothing. Taking the growing season (March) and declining season (December) as an example (Fig. 2), this paper showed the differences of BRDF principle plane curves before and after temporal smoothing(day-month) in three broad bands of Near-Infrared(NIR), Short Wave(SW) and Visible(VIS). Meanwhile, the relative and absolute differences of albedo in March and December were counted (Table 1).

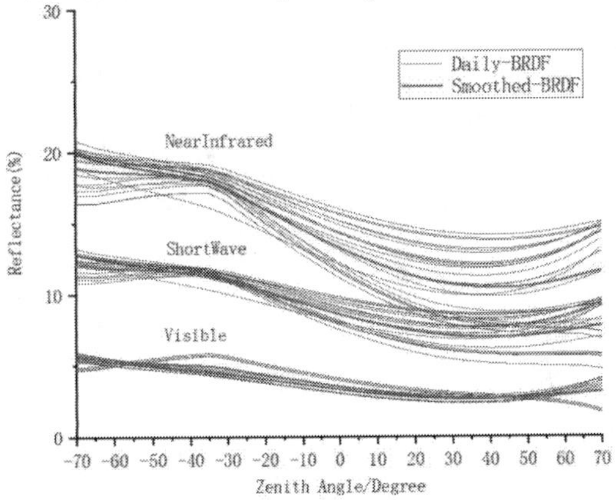

(a)March

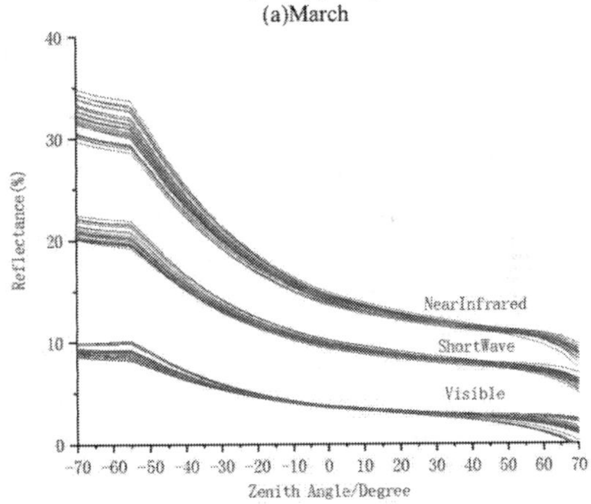

(b)December

Fig. 2. BRDF before and after temporal smoothing

Table 1. The differences of principle plane curves' magnitudes and feature points before and after BRDF temporal smoothing

Band	Month	Magnitude(%)	Hotspot(%)	-45°(%)	-70°(%)	45°(%)	70°(%)	0°(%)	STD
NIR	3	1.42	0.50	0.62	1.27	2.11	3.02	1.36	0.87
	12	0.38	1.19	0.89	1.20	0.10	0.31	0.24	0.45
SW	3	0.71	0.21	0.25	0.73	1.08	1.80	0.63	0.54
	12	0.22	0.69	0.51	0.66	0.12	0.31	0.15	0.23
VIS	3	0.17	0.62	0.35	0.39	0.10	0.96	0.35	0.27
	12	0.04	0.46	0.29	0.32	0.19	0.70	0.05	0.20

Table 2. Albedo differences caused by BRDF temporal smoothing

Band	Month	Relative difference (%)	Absolute difference	Month	Relative difference (%)	Absolute difference
NIR	3	10.97	0.0172	12	1.80	0.0029

SW	3	9.42	0.0095	12	1.51	0.0016
VIS	3	8.24	0.0035	12	1.67	0.0007

It can be seen from Fig. 2 and Table 1 that after temporal smoothing, the magnitudes and shapes of BRDF principle plane curves had changed to some extent. In three broad bands of NIR, SW and VIS, the overall magnitude differences of BRDF principle plane curves in March (1.42%, 0.71% and 0.17%) are significantly higher than those in December (0.38%, 0.22% and 0.04%), it means the dispersion of BRDF in March is higher than that in December; From the analysis of the differences dispersion of feature points(STD, Standard Deviation), the more discrete of feature points, the more intense changes of the shapes of BRDF, so the STD in March (0.87, 0.54, 0.27) are significantly higher than those in December (0.45, 0.23, 0.20).

Table 2 showed the magnitudes of albedo differences caused by BRDF temporal smoothing in March and December, and the relative and absolute differences of albedo in March are significantly higher than those in December. We selected the WSA(White Sky Albedo) from MCD43A3 data set as the processing object, as WSA is highly related to BSA(Black Sky Albedo) and has nothing to do with LSN(Local Solar Noon), so there is no need to consider the difference caused by LSN[17-18].

3.2 Spatial smoothing of BRDF and related albedo

To explore the differences of BRDF before and after spatial smoothing, the BRDF parameters of DBF sample at 5600m were extracted from MCD43C1 product, then BRDF parameters of land cover at 500m were extracted from MCD43A1 product. The BRDF principle plane curves and albedo values at two spatial scales were retrieved based on RossThick-LiSparseR model, then the daily differences of BRDF and albedo were counted. For the DOYs with extreme difference after albedo spatial smoothing (from 500m to 5600m) in 2019 (Doy1-365), statistics were made on the BRDF principle plane curves in three broad bands of NIR, SW and VIS(Fig. 3). We also made statistics of the albedo differences on the DOYs with extreme albedo difference(Table 3) and the changes of BRDF principle plane curves on the extreme DOYs(Table 4).

Table 3. Statistics of DOYs with extreme differences

Site	Band	Term	DOY	Relative Difference (%)	Absolute Difference
DBF	NIR	MAX	33	17.38	0.0250
		MIN	98	4.31	0.0104
	SW	MAX	32	14.38	0.0139
		MIN	99	4.22	0.0059
	VIS	MAX	158	27.23	0.0052
		MIN	93	8.31	0.0032

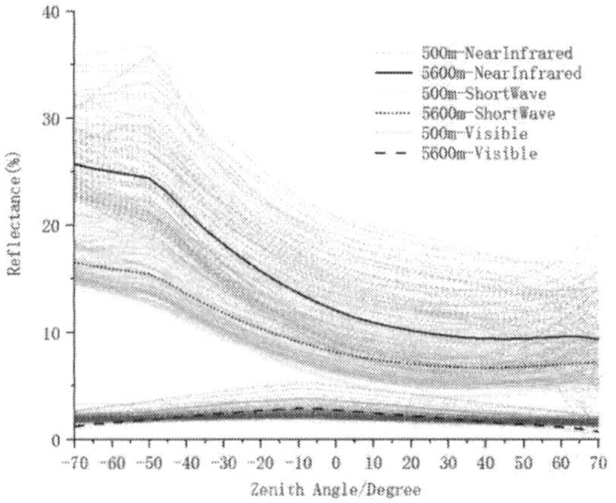

(a)Maximum DOYs

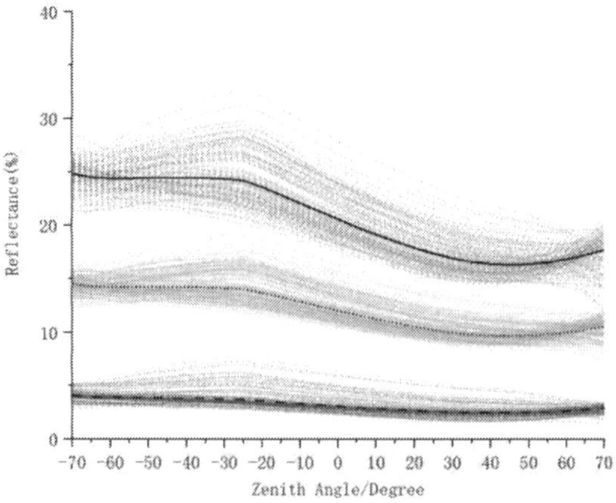

(b) Minimum DOYs

Fig. 3. Distribution of BRDF principle plane curves in extreme DOYs

Table 4. The differences of principle plane curves' magnitudes and feature points before and after BRDF smoothing

Band	Term	Magnitude(%)	Hotspot(%)	-45°(%)	-70°(%)	45°(%)	70°(%)	0°(%)	STD
NIR	MAX	2.90	3.87	3.72	2.71	2.83	3.23	3.22	0.42
	MIN	1.27	2.19	1.46	1.07	1.04	1.37	1.71	0.39
SW	MAX	1.63	2.21	2.13	1.59	1.54	1.86	1.81	0.25
	MIN	0.76	1.35	0.92	0.62	0.59	0.83	1.02	0.25
VIS	MAX	0.29	0.67	0.29	0.73	0.35	0.90	0.56	0.21
	MIN	0.39	0.60	0.47	0.36	0.35	0.28	0.47	0.11

It can be seen from Fig. 3 and Table 3&4 that after spatial smoothing, the magnitudes and shapes of BRDF had changed to some extent. The dispersion of feature points(STD) and overall magnitude differences are both impact factors to the differences. In the three broad bands of NIR, SW and VIS, the STD on maximum DOYs are almost larger than that on minimum DOYs; In the SW band where the maximum DOY and the minimum DOY are almost the same, but the magnitude of maximum DOY is significantly greater than that of minimum DOY.

4 Conclusion

The magnitudes and shapes of BRDF in principle planes are changed significantly after smoothing at multi-spatiotemporal scales, and it can induce detectable differences for the inversed albedo. As the case study of DBF sample, BRDF smoothed from day to month lead to relative differences to smoothed values of 10.97%, 9.42%, 8.24% and absolute differences of 0.0172, 0.0095 and 0.0035 on the estimated albedo respectively at Near-Infrared, Short Wave and Visible broad bands. The BRDF smoothed at spatial scale from 500m to 5600m results in relative differences to smoothed values of 17.38%, 14.38%, 27.23% and absolute differences of 0.0250, 0.0139, 0.0052 respectively on retrieved albedo at three above broad bands.

This work was supported by the National Natural Science Foundation of China (No. 42071351), the National Key Research and Development Program of China (No. 2020YFA0608501, No. 2017YFB0504204), the Liaoning Revitalization Talents Program (No. XLYC1802027), Talend recruited program of the Chinese Academy of Science (No. Y938091), Project supported discipline innovation team of Liaoning Technical University (No. LNTU20TD-23)

References

1. J.T. Kiehl, J.J. Hack, G.B. Bonan, et al. Description of the NCAR Community Climate Model (CCM3). National Center for Atmospheric Research Publishing, The Boulder (1996).
2. B. Pinty, T. Lavergne, T. Kaminski, et al. Partitioning the solar radiant fluxes in forest canopies in the presence of snow. Journal of Geophysical Research, **113** D04104 (2008)
3. D.S. Kimes and P.J. Sellers, Inferring Hemispherical Reflectance of the Earth's Surface for Global Energy Budgets from Remotely Sensed Nadir of Directional Radiance Values. Remote Sensing of Environment, **18**: 205–223 (1985)
4. D.S. Kimes, P.J. Sellers and W.W. Newcomb, Hemispherical Reflectance Variations of Vegetation Canopies and Implications for Global and Regional Energy Budget Studies. Journal of Climate and Applied Meteorology, **26**: 959–972 (1987)
5. A. Henderson and M. Wilson., Surface albedo data for climatic modeling. Reviews of Geophysics. **21**: 1743–1778 (1983)
6. P. Sellers, B. Meeson, F. Hall, et al. Remote sensing of the land surface for studies of global change: Models-Algorithms-Experiments. Remote Sensing of Environment, **51**: 3–26 (1995)
7. Z. Wang, C. Schaaf, Q. Sun, et al. Capturing rapid land surface dynamics with Collection V006 MODIS BRDF/NBAR/Albedo (MCD43) products. Remote Sensing of Environment, **207** 50-64 (2018)
8. Z. Wang, A. Erb, C. Schaaf, et al. Early spring post-fire snow albedo dynamics in high latitude boreal forests using Landsat-8 OLI data. Remote sensing of environment, **185**: 71-83 (2016)
9. P. Lewis, et al. The ESA globAlbedo project: Algorithm. In: 2012 IEEE International Geoscience and Remote Sensing Symposium. Munich. 5745-5748 (2012)
10. W. Lucht, C. Schaaf and A. Strahler, An algorithm for the retrieval of albedo from space using semiempirical BRDF models. IEEE TRANSACTIONS ON GEOSCIENCE AND REMOTE SENSING, **38** 977–998 (2000)
11. C. Schaaf, F. Gao, A. Strahler, et al. First operational BRDF, albedo nadir reflectance products from MODIS. Remote Sensing of Environment, **83**: 135–148 (2002)
12. C. Bacour and F. Bréon, Variability of biome reflectance directional signatures as seen by POLDER. Remote Sensing of Environment, **98** 80-95 (2005)
13. S. Liang, X. Zhao, S. Liu, et al. A long-term Global Land Surface Satellite (GLASS) data-set for environmental studies. International Journal of Digital Earth, **6** 5-33 (2013)
14. Q. Liu, L. Wang, Y. Qu, et al. Preliminary evaluation of the long-term GLASS albedo product. International Journal of Digital Earth, **6**: 69-95 (2013)
15. Y. Shuai, J. Masek, F. Gao, et al. An algorithm for the retrieval of 30-m snow-free albedo from Landsat

surface reflectance and MODIS BRDF. Remote Sensing of Environment, **115** 2204-2216 (2011)

16. Y. Shuai, J. Masek, F. Gao, et al. An Approach for the Long-Term 30-m Land Surface Snow-Free Albedo Retrieval from Historic Landsat Surface Reflectance and MODIS-based A Priori Anisotropy Knowledge. Remote Sensing of Environment, **152** 467-479 (2014)
17. Y.H. Zheng, L. Huang and J. Zhai, Impacts of land cover changes on surface albedo in China,the United States,India and Brazil. Journal of Remote Sensing(Chinese), **24**: 917-932 (2020)
18. H. Zhang, Z.T. Jiao, D.Y. Dong, et al. Albedo retrieved from BRDF archetype and surface directional reflectance. Journal of Remote Sensing(Chinese), **19** 355-367 (2015)

Near Real-time Fine-resolution Land Surface Phenological Prediction Using Convolutional Neural Network and Data Fusion

Kun Xiao [1,*], *Yidan* Wang[1], *Wei* Wu [1] and *Qinchuan* Xin [1]

[1] School of Geography and Planning, Sun Yat-sen University, Guangzhou 510275, China

Abstract. Near real-time fine-resolution land surface phenology (LSP) prediction is essential for understanding surface attributes and ecosystem functions, and solving important ecological processes related to phenology at the landscape scale. In this paper, we applied the Enhanced Spatial and Temporal Adaptive Reflectance Fusion Model (ESTARFM) to fuse image pairs of Landsat 8 and Moderate-resolution Imaging Spectroradiometer (MODIS) as train data, and then applied the first derivative method to retrieve phenophase transition dates from fused time series of satellite data as label data. The convolutional neural network (CNN) model was trained using fusion images as inputs and the label data as targets. The trained model was further used to predict LSP dates from individual Landsat images. As evaluated using the reference data, the predict land surface phenological dates and could match the reference well with the coefficient of determination of 0.77 and root mean squared errors of 3.535, and our study provides an alternative method to predict land surface phenological dates using individual Landsat images.

1 introduction

Land surface phenology (LSP) denotes the periodic patterns of variation and dynamics in vegetated land surface observed from remote sensing[1]. LSP derived from satellites reflects the responses of the land surface, mostly terrestrial ecosystems, to both climatic and anthropogenic forcings[2]. LSP information has also found helpful to a broad range of applications such as land use and land cover mapping, land surface change quantification, and land surface modeling[3].

As remote sensing could provide large-scale observations of the land surface at regular temporal steps, retrieving LSP information from satellite data such as MODIS data has received increasing interests in studies[4]. As this data missed spatial details on the landscapes, particularly in heterogeneous areas that include a mixture of multiple land cover types. As plant function types and vegetation communities have varied responses to climate variation and human disturbances there is a need to develop fine-spatial-resolution phenological products to meet both scientific and applicable demands. Fortunately, with the development of data fusion techniques, integrating data from different sensors has become feasible to obtain consistent data at fine spatial and temporal resolutions. Feng, Masek et al. (2006)[5] developed a spatial and temporal adaptive reflection fusion model (STARFM) to fuse both Landsat and MODIS data and obtained land surface reflectance data at the spatial resolution of Landsat and the temporal resolution of MODIS. Zhu, Chen et al. (2010)[6] developed an enhanced spatiotemporal adaptive reflection fusion model (ESTARFM) based on the STARFM algorithm with improved accuracy of data fusion. There are also a number of following studies that developed to fuse satellite data acquired from different sensors. These approaches provide key fused data for studying LSP at fine-spatial resolution.

A series of studies have shown that convolutional neural networks (CNN) have potentials and advances in remote sensing studies. CNN has been found effective in representing spatial patterns and extracting vegetation features from remote sensing images. Qi Yang et al. (2020)[7] extracted the near real-time phenology of rice from UAV images based on CNN. Cao et al. (2021)[8] retrieved phenological dates of deciduous broadleaf trees from near-surface camera images using CNN. It is worthy testing CNN on mapping LSP from remote sensing images.

The main purpose of this study are as follows: 1) proposes a model based on data fusion method and CNN to predict near real-time fine-spatial-resolution LSP; 2) Based on the specified individual Landsat 8 image, the deep learning model is used to predict the date far away from the phenophase transition date.

2 Study area and materials

2.1 Study area

We selected regions with a window size of 3000 × 3000 pixels (Fig.1) from Landsat images (Path 122 Row36,) as the study area. The latitude of the study area ranges from 34°N to 35°N. The study area mainly involves Shandong, Henan and Jiangsu provinces in the North

* Corresponding author: xiaok6@mail2.sysu.edu.cn

China Plain in China. It has typical monsoon climate and covers a variety of land cover types such as woodland, farmland, grassland, and wetland.

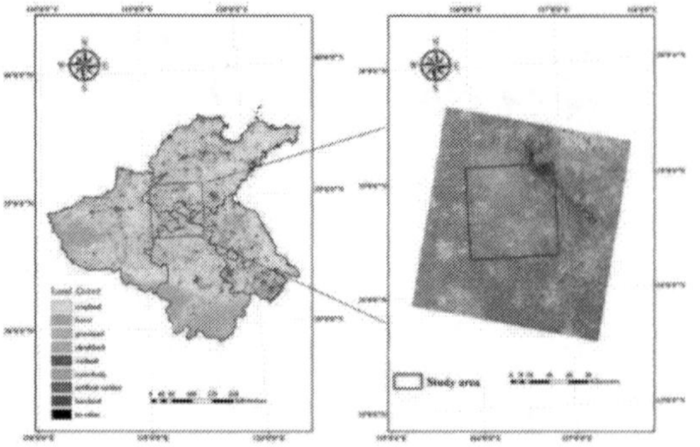

Fig.1. The location of the study area

2.2 Materials

We downloaded all available Landsat 8 images acquired in the year of 2018 for the studied four scenes from the website United States Geological Survey (https://earthexplorer.usgs.gov/). To ensure the accuracy and reliability of data fusion, we only used Landsat 8 OLI data with cloudiness less than 10% in each scene. We obtained MODIS Collection 6 land surface reflectance products (MOD09A1) at the spatial resolution of 500 m from https://ladsweb.nascom.nasa.gov/search, and the MODIS data that cover the studied Landsat scenes come from the tile of h27v05.

3 Methodology

3.1 Algorithm in prediction near real-time LSP

In the process of prediction near real-time LSP. Briefly, we pre-process the Landsat data and MOD09A1 data. since ESTARFM spatiotemporal fusion algorithm can effectively preserve the spatial details of non-heterogeneous landscapes so we applied the ESTARFM to fuse Landsat 8 and MOD09A1 data to obtain the training data. Then, calculate the EVI time series data through the fusion data, and use the EVI time series data to retrieving SOS and EOS, and calculate the days away from phenophase transition dates(eq.1) through SOS and EOS to obtain the label data corresponding to the training data. Finally, a convolutional neural network structure was designed to predict the LSP (Fig. 2), and the CNN model is trained through training data and label data, In training the model using fused data, the 23×3000×3000×6 images were cropped into 382,800 images with a size of 50×50×6, of which 2/3 were used as training and 1/3 were used for validation, and the ADAM optimization algorithm was used to set its batch size to 128, the learning rate to 0.01, and the loss function to mean square error (MSE), and the trained model is used to pediction near real-time LSP.

$$d = \begin{cases} 0, t < s \\ t - s, s < t < e \end{cases} \quad (1)$$

where d denotes the days away from phenophase transition dates, t denotes the fusion image time, s denotes the SOS, and e denotes is EOS.

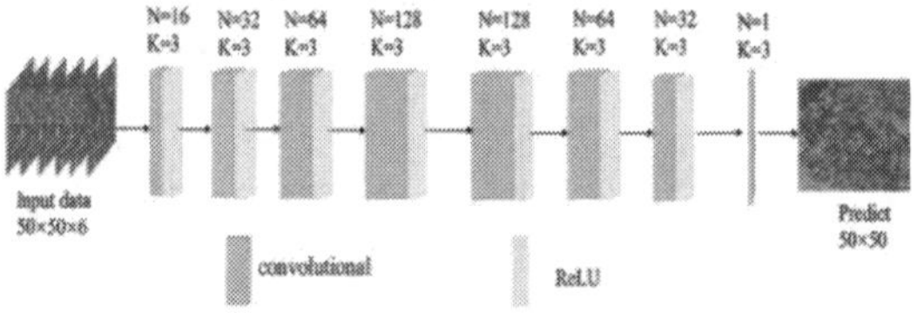

Fig. 2. The near real-time LSP prediction structure diagram, N and K denote the number of feature maps and the kernel size for the convolutional layer, respectively. Rectified and linear units (ReLU) denote the nonlinear activation layers.

4 Result

4.1 prediction near real-time LSP

Fig. 3 shows the spatial distribution of phenophase transition days on DOY (Day of year) 81， 113， 129， respectively. The phenophase transition days obtained using CNN and label have a similar spatial distribution across the study area, with approximately the same number of transition days for the same feature types and relatively low spatial heterogeneity. In different phenophase transition days on DOY, the phenophase transition days will also increase with the increase of DOY.

Fig. 4 shows the frequency statistics of the number of days difference between the CNN prediction results and the label, including only the image elements with valid retrievals in both results. The difference in phenophase transition days for DOY 81,113,129 are mainly concentrated within 50 days, and in fact, most of the differences are much smaller, with the number and frequency of pixels with differences of 20 days or less accounting for 80% of the valid retrieved pixels, which also indicates that most of the differences are mainly concentrated within 20 days.

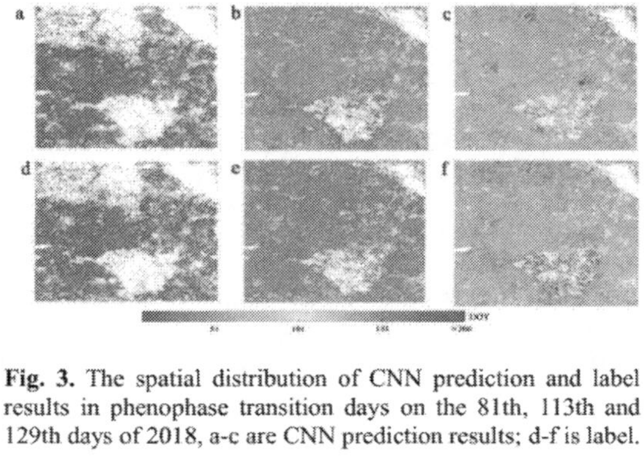

Fig. 3. The spatial distribution of CNN prediction and label results in phenophase transition days on the 81th, 113th and 129th days of 2018, a-c are CNN prediction results; d-f is label.

a and d) day 81, b and e) day 113 and c and f) day 129.

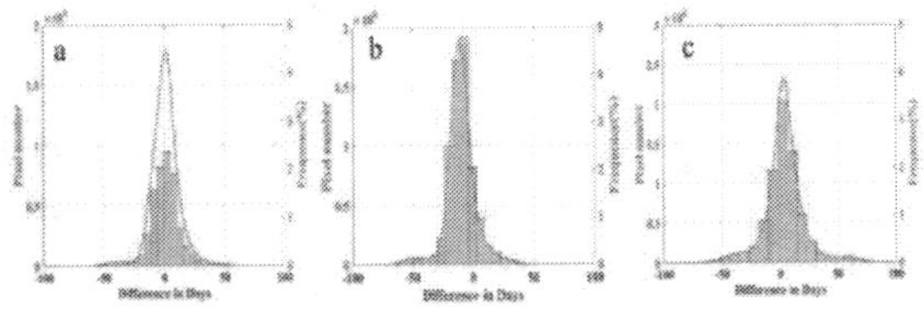

Fig. 4. The number and frequency of the days of difference between CNN prediction and label results in the transition days on the 81th, 113th and 129th days of 2018. a) day 81, b) day 113 and c) day 129.

4.2 Accuracy Evaluation of prediction near real-time LSP

The scatter plot compares the phenophase transition days obtained by CNN model prediction and label (Fig. 5). In each DOY we selected the effective retrieved pixels of a 30KM*30KM size arear to draw a scatter plot. For the phenophase transition days on DOY 81， 113,129, the prediction results and the label are generally correlated, and the scatter points are mainly distributed around the 1:1 line. The R^2 is 0.635, 0.7743, 0.72, with RMSE of 3.535, 3.58, 3.55, respectively

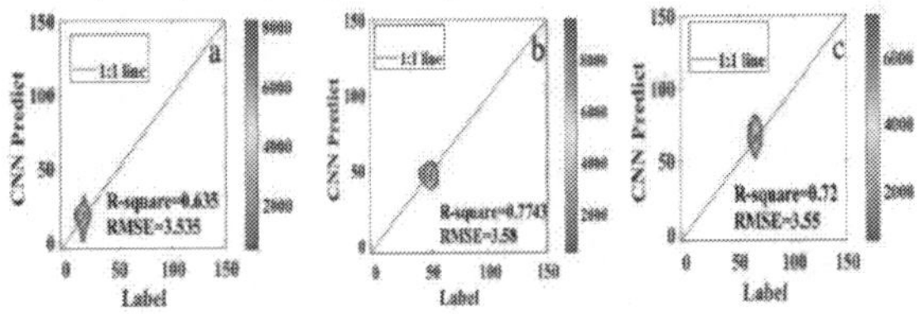

Fig. 5. Scatter plot of CNN prediction results and label results. a) day 81, b) day 113 and c) day 129.

5 Conclusions

This research proposes a method based on CNN to predict near real-time LSP using individual Landsat 8 images, which can directly extract advanced features from the reflectance data of Landsat 8 to predict the phenophase transition days of LSP, instead of the traditional SOS and EOS. However, it is worth noting that due to the lack of ground verification data, the method of this study is not compared with the ground measurement data.

This research is supported by National Key R&D Program of China (grant no. 2017YFA0604300), Natural Science Foundation of China (grant nos. U1811464 and 41875122), Natural Science Foundation of Guangdong Province (grant no. 2021A1515011429), Western Talents (grant no. 2018XBYJRC004).

Reference

1. A.D. Richardson, T.F. Keenan, M. Migliavacca, Y. Ryu, O. Sonnentag, M. Toomey, Climate change, phenology, and phenological control of vegetation feedbacks to the climate system. Agric. For Meteorol, **169** 156-173 (2013)
2. E.K. Melaas, D. Sulla-Menashe, J.M. Gray, T.A. Black, T.H. Morin, A.D. Richardson and M.A. Friedl, Multisite analysis of land surface phenology in North American temperate and boreal deciduous forests from Landsat. Remote Sens. Environ., **186**: 452-464. (2016)
3. Z. Zhu and C.E. Woodcock, Continuous change detection and classification of land cover using all available Landsat data. Remote Sens. Environ., **144**: 152-171. (2014)
4. X. Li, Y. Zhou, L. Meng, G.R Asrar, C. Lu and Q. Wu, A dataset of 30 m annual vegetation phenology indicators (1985–2015) in urban areas of the conterminous United States. Earth Syst. Sci. Data, **11**: 881-894. (2019)
5. G. Feng, J. Masek, M. Schwaller, F. Hall, On the blending of the Landsat and MODIS surface reflectance: predicting daily Landsat surface reflectance. IEEE Trans Geosci Remote Sens., **44**: 2207-2218. (2006)
6. X. Zhu, J. Chen, F. Gao, X. Chen and J.G Masek, An enhanced spatial and temporal adaptive reflectance fusion model for complex heterogeneous regions. Remote Sens. Environ., **114**: 2610-2623. (2010)
7. Q. Yang, L. Shi, J. Han, J. Yu and K. Huang, A near real-time deep learning approach for detecting rice phenology based on UAV images. Agric. For Meteorol., 287. (2020)
8. M. Cao, Y. Sun, X. Jiang, Z. Li and Q. Xin, Identifying Leaf Phenology of Deciduous Broadleaf Forests from PhenoCam Images Using a Convolutional Neural Network Regression Method. Remote Sens. 13. (2021)

Comprehensive Geochemical Evaluation of Land Quality in Strawberry Plant Area and Geoscience Suggestions in Chengde

Xia Li[1], *Xiaofeng* Wei[2], *Jin* Wu[3], *Liqin* Wan[1,*], *Yutong* Zhu[4], *Yuning* Ding[4]

[1] China Institute of Geo-Environment Monitoring, 100081 Beijing, China
[2] Beijing Institute of Geology for Mineral Resources, 100012 Beijing, China
[3] College of Architecture and Civil Engineering, Beijing University of Technology, 100124 Beijing, China
[4] China University of Geosciences, 100083 Beijing, China

Abstract. Strawberry Plant Area in Chengde is located in the Jingcheng Golden Tour Corridor. The main plantings in the area are strawberries, however in recent years, there have been problems such as insufficient supply of surrounding land and specialization of pesticide formulations. In this study, through testing and analysis of 30 nutrients and total metal elements, 7 efficient statuses of elements, pH and soil organic matter in surface soil samples, status of soil nutrients, soil environment and soil quality are comprehensive evaluated in the region. Based on the complete water quality analysis of irrigation water, the directional fertilization formula for the Quaternary accumulation area and the geological suggestion for the replacement of other high-quality soil for melioration are put forward to provide support for the sustainable development of Strawberry Plant Area.

1 Introduction

Strawberry has a very high nutritional value. Each 100g of fresh fruit pulp contains about 60-120 mg of vitamin C, which is more than 10 times higher than that of apples and grapes [1,2]. The pulp contains a lot of nutrients including sugar, protein, organic acid, pectin, and trace elements necessary for the human body such as calcium, phosphorus, ferrum, potassium, zinc, and chromium [3]. The Strawberry Plant Area in Qijia, Longhua County, Chengde City is located in the Jingcheng Golden Tour Corridor. The area spans 1 km from east to west and 2.8 km from north to south. The overall plan is 1.7 km2 (2560 Mu). The Wuliehe River is located in the west of the area and runs through. This survey involves the whole area of Strawberry Plant Area and the vast area in the south, with an area of 4.06 km2 (6084 Mu). The main plantings in the area are three types of characteristic agriculture including strawberry, economic forest begonia fruit, Chinese medicinal material yellow chrysanthemum.

A total of 45 surface soil samples were collected in this survey, with an average sampling density of 11 points per square kilometre. 32 types of elements are analysed including N, P, K, Mn, Mo, Zn, Cu, Se, Ge, B, soil organic matter (SOM), Cd, Cr, As, Hg, Ni, Pb, Sb, Sn, Co, V, Ti, Na, Al, TFe, F, Ca, Mg, S, Cl, Si, pH, and 7 types of element efficient statuses are also analysed including alkaline hydrolysis of nitrogen, available phosphorus, available potassium, available boron, available molybdenum, available ferrum, available manganese. At the same time, 2 samples of irrigation water were collected for complete water quality analysis. The soil quality of strawberry cultivation is mainly analysed, to put forward suggestions for comprehensive utilization of facility agriculture concentrated areas.

2 The status of land use

According to the latest third national land survey, there are 12 types of land use in the area (Figure 1). The area is 3050943.11 m2 of arable land, 33053.58 m2 of garden, 297832.09 m2 of woodland, 2388.32 m2 of grassland, 5673.69. m2 of Shangfu land, 168010.4 m2 of residential land, 18353.51 m2 of public land, 65298.15 m2 of other land, 280.39 m2 of special land, 162162.39 m2 of transportation land, 252927.48 m2 of water area, respectively. The area is mainly arable land, accounting for as high as 75.2%.

* Correspondence author: 313273589@qq.com

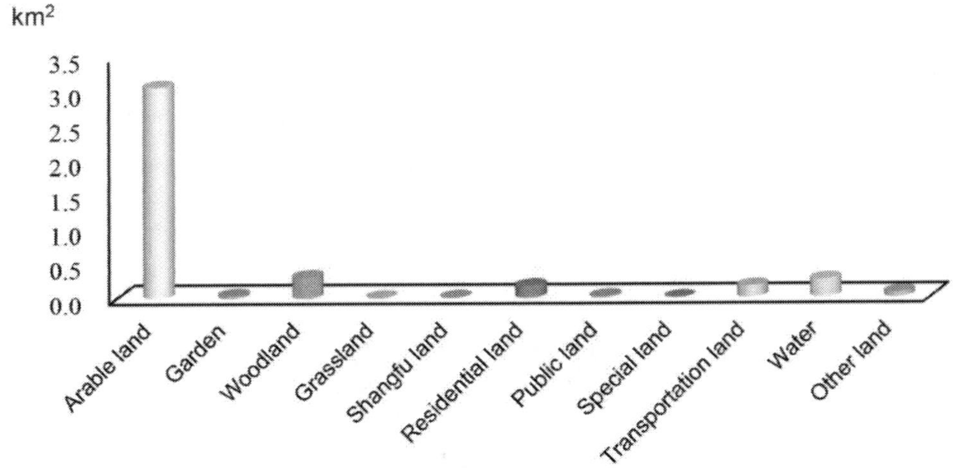

Fig. 1. Distribution map of land use status

3 Characteristics of Eco- Geological Conditions

3.1 Topography and landform conditions

Strawberry Plant Area is located in the valley area of the Wuliehe River Basin in Longhua County, with an elevation ranging between 561m and 841m. There are high mountains on both sides and low river valley in the middle. The topography is undulating, and typical river valley landscapes are developed (Figure 2 left). It is an important concentration area of agriculture, ecology and life (Figure 2 right).

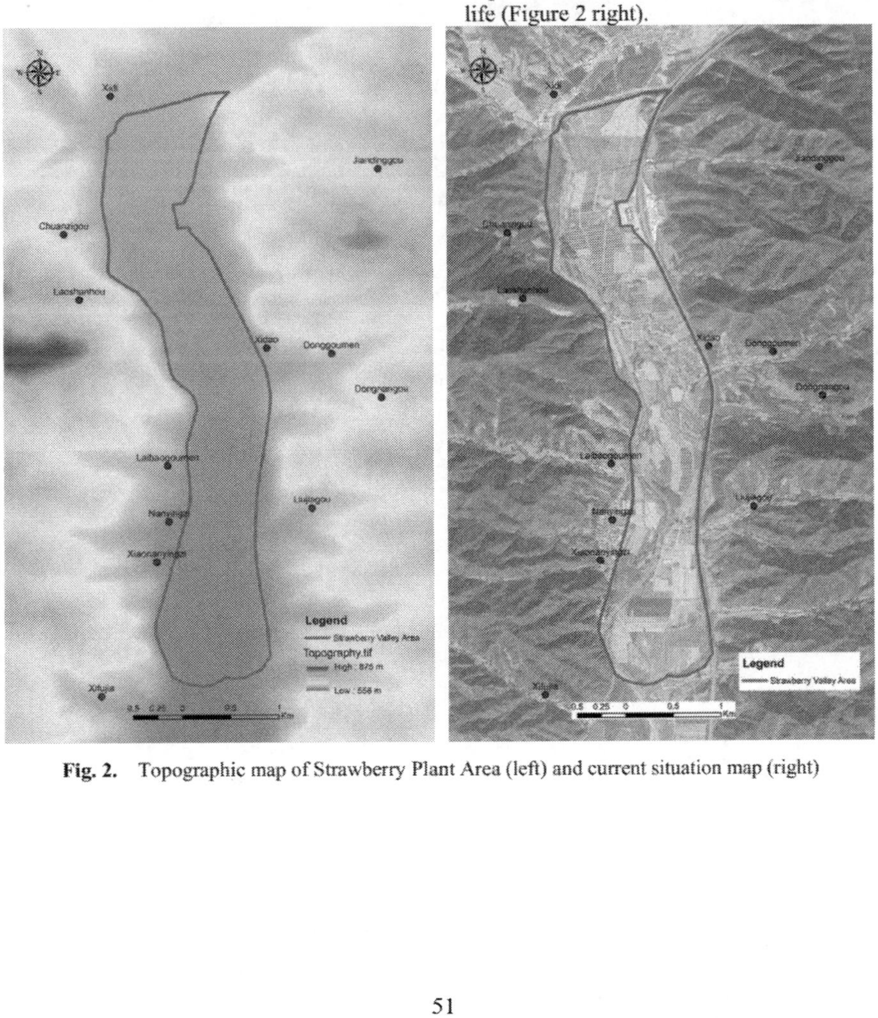

Fig. 2. Topographic map of Strawberry Plant Area (left) and current situation map (right)

3.2 Eco-geological conditions

The main types of geological formation in this area are loose accumulation of Quaternary, including aeolian loess, fluvial sand and gravel [4,5]. The bedrock is mainly composed of late Jurassic (J3G) and Late Triassic (T3Q) medium-fine-coarse biotite monzonitic granites, Neoarchean (Ar3Sgn) potash feldspathic gneiss (the original rock is monzonitic granite), and Paleoproterozoic (Pt1Dηδο) quartz monzodiorite (Fig.3).

4 Analysis and discussion

4.1 Comprehensive evaluation of land quality

According to Land Quality Geochemical Evaluation Standards (DZ/T 0295-2016) [6], in the Strawberry Plant Area, the evaluation grade of land quality is fourth-class, covering an area of 2,148,280.88 m2, accounting for 53% of the total area （fig.4a）. The evaluation grade of soil environment is mainly third-class, with a risk of slight pollution (fig. 4b). The evaluation grade of soil nutrient is mainly third-class, covering an area of 3,542108.09 m2, accounting for 87.3% (fig.4c). The evaluation results of these soil quality are mainly due to the poor soil conditions, with the texture dominated by gravel soil, and nutrients easily leached in partial area. The statistics of various comprehensive soil indexes are shown in Table 1.

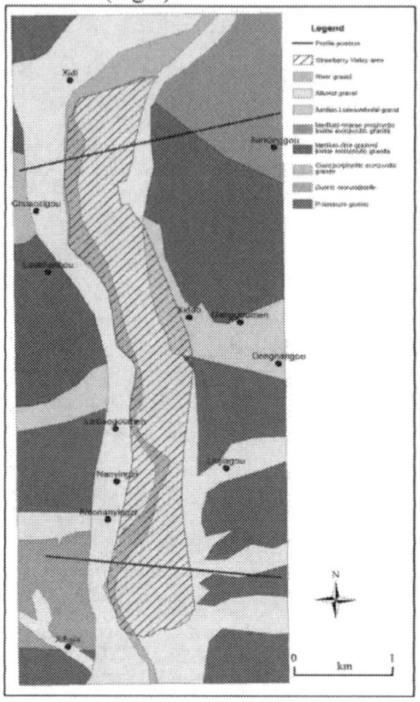

Fig. 3. Eco-geological map of Strawberry Plant Area

Table 1. Statistical Table of Comprehensive Soil Indexes in Strawberry Plant Area

Level	Soil comprehensive quality		Soil nutrient		Soil environment	
	Area(km2)	Percentage(%)	Area(km2)	Percentage(%)	Area(km2)	Percentage(%)
First	0	0	0.02	0.5	0	0
Second	0	0	0.18	4.5	0.47	11.6
Third	0.47	11.6	3.54	87.3	2.15	53.0
Fourth	2.15	53.0	0.31	7.8	1.44	35.5
Fifth	1.44	35.5	0	0	0	0

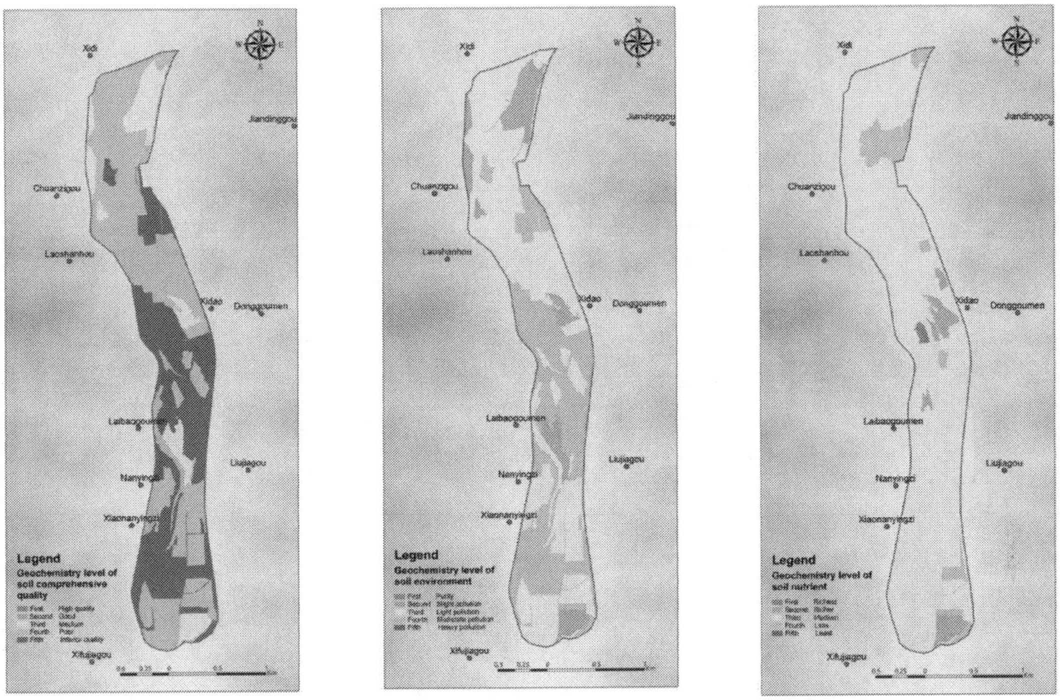

a) Soil comprehensive quality b) Soil environment c) Soil nutrient

Fig. 4. Grade map of soil comprehensive quality, environment, nutrient in Strawberry Plant Area

4.2 Comprehensive Geochemical Status of Soil Nutrients

The Strawberry Plant Area is obviously deficient in N, less P and rich in K [7]. The evaluation grade of N element is mainly fourth-class (Fig.5a), with a content of between 0.75 and 1.0 mg/kg, and covering an area of 1,881,54.56 m2, accounting for 46.4%. The evaluation grade of P element is mainly third-class (Fig.5b), with a content of between 0.6 and 0.8 mg/kg, and covering an area of 3,461,169.42 m2, accounting for 85.3%. There is mainly rich in K element, with the first-class area of 3,485,987.69 m2, accounting for 85.9%, and the rest as second-class. The evaluation grade of SOM is mainly fourth-class (Fig.5c), with a content between 10 mg/kg and 20 mg/kg, and with an area of 3,727,788.35 m2, accounting for 91.9%. The statistics of individual indexes of soil nutrients are shown in Table 2.

Table 2. Statistical Table of Individual Indexes of Soil Nutrients in Strawberry Plant Area

Level	N		P		K		SOM	
	Area (km2)	Percentage (%)	Area (km2)	Percentage (%)	Area (km2)	Percentage (%)	Area (km2)	Percentage (%)
First	0	0	0.02	0.5	3.49	85.9	0	0
Second	0.04	0.8	0.26	6.4	0.57	14.1	0	0
Third	0.54	13.3	3.46	85.3	0	0	0.05	1.3
Fourth	1.88	46.4	0.29	7.1	0	0	3.73	91.9
Fifth	1.60	39.4	0.03	0.7	0	0	0.28	6.8

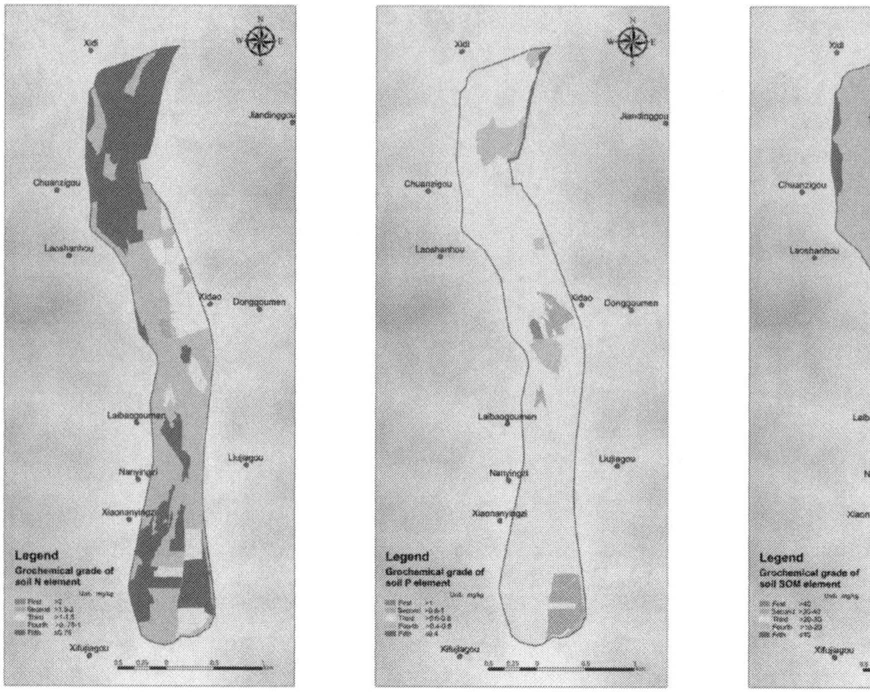

a) N b) P c) SOM

Fig. 5. Geochemical grade map of soil N, P, SOM in Strawberry Plant Area

4.3 Soil single element geochemical grade status

The soil in this area is rich in trace beneficial elements such as molybdenum (Mo, Fig.6a), manganese (Mn, Fig.6b), zinc (Zn, Fig.6c), ferrum (Fe, Fig.6d), magnesium (Mg, Fig.6e), and calcium (Ca, Fig.6f). The soil trace elements reach the medium and rich level, namely second-class and above. Only the first-class of Mo element accounts for more than 50%, and that of other elements account for less than 5%. The statistics of indexes of single soil nutrient elements are shown in Table 3.

The evaluation grade of selenium (Se, Fig.6g), boron (B, Fig.6h), germanium (Ge, Fig.6i) and other elements in the soil are mainly fourth or fifth class, and the Se content is between 0.071 mg/kg and 0.19 mg/kg, which is lower than the selenium-rich land standard (with a content of more than 0.4 mg/kg) [6].

Table 3. Statistical Table of Indexes of Single Soil Nutrient Elements in Strawberry Plant Area

Level	Se		B		Ge	
	Area(km2)	Percentage(%)	Area(km2)	Percentage(%)	Area(km2)	Percentage(%)
First	0	0	0	0	0	0
Second	0	0	0	0	0	0
Third	33008.7	0.8	0	0	236545.5	5.8
Fourth	1713731.5	42.2	543558.1	13.4	3474926.4	85.7
Fifth	2310182.9	56.9	3513365.1	86.6	345451.1	8.5
Level	Mo		Mn		Zn	
	Area(km2)	Percentage(%)	Area(km2)	Percentage(%)	Area(km2)	Percentage(%)
First	2121960.4	52.3	0.06	1.4	0.17	4.1
Second	1638584.8	40.4	1.53	37.7	3.26	80.5
Third	169459	4.2	2.29	56.3	0.53	13.0
Fourth	58531.6	1.4	0.16	3.8	0.07	1.7
Fifth	68387.2	1.7	0.03	0.7	0.03	0.7
Level	Fe		Mg		Ca	
	Area(km2)	Percentage(%)	Area(km2)	Percentage(%)	Area(km2)	Percentage(%)

https://doi.org/10.1051/e3sconf/202235001009

First	0.06	1.4	0	0	0	0
Second	0.21	5.3	0.09	2.2	0.07	1.7
Third	2.49	61.3	2.99	73.6	3.96	97.6
Fourth	1.27	31.2	0.95	23.5	0.03	0.7
Fifth	0.03	0.7	0.03	0.7	0	0

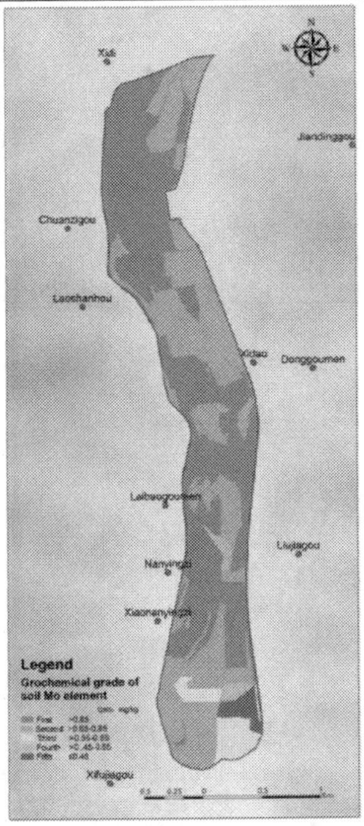

a) Mo

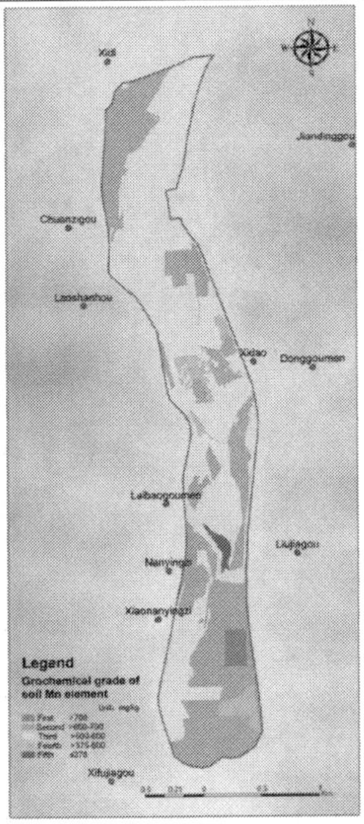

b) Mn

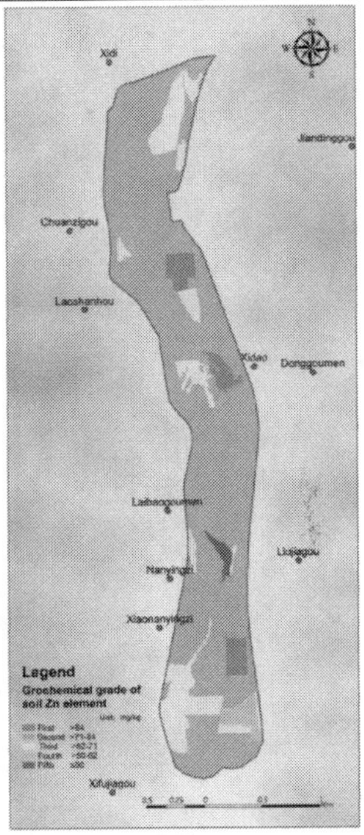

c) Zn

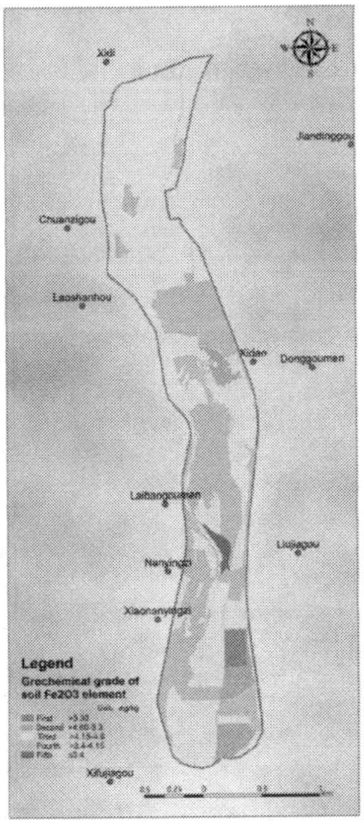

d) Fe

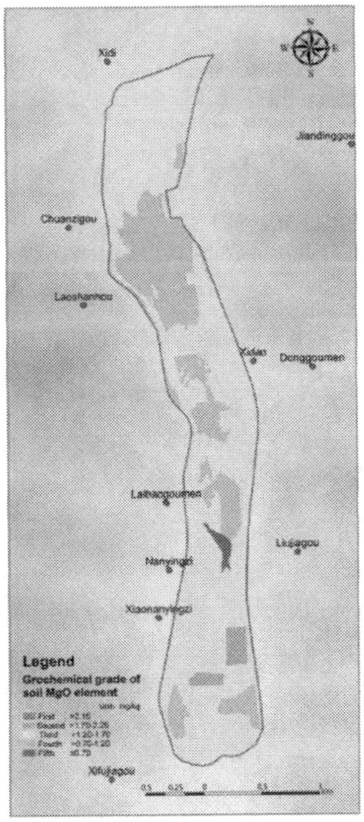

e) Mg

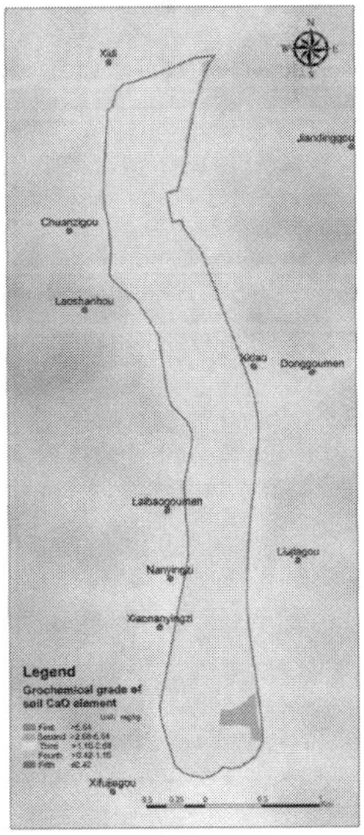

f) Ca

https://doi.org/10.1051/e3sconf/202235001009

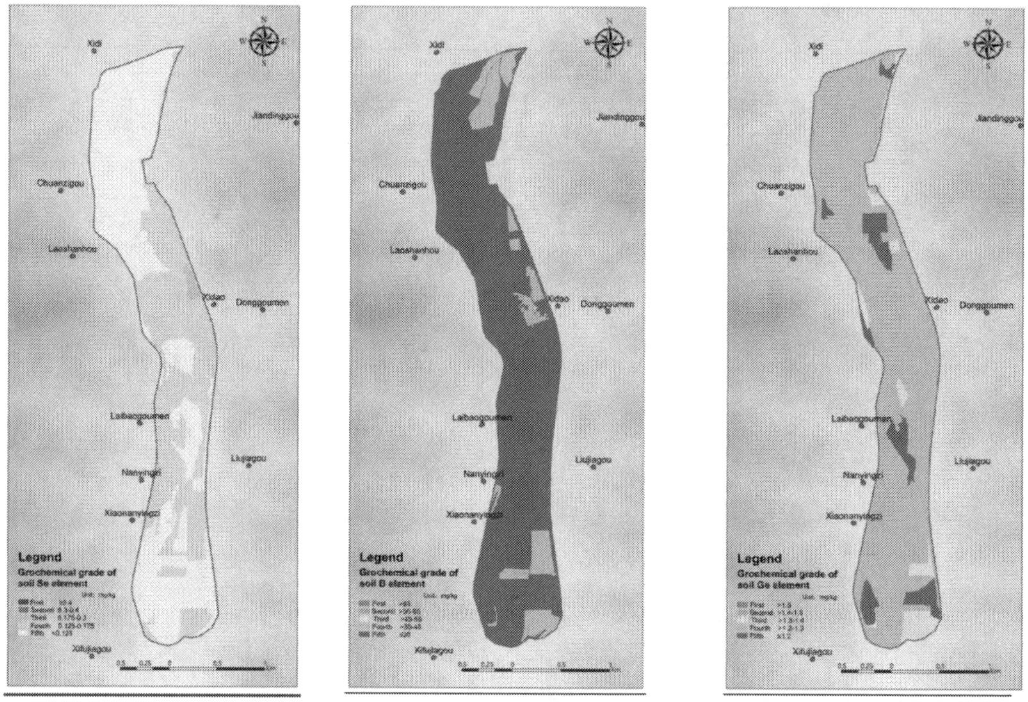

g) Se h) B i) Ge

Fig. 6. Geochemical grade map of soil nutrient elements in Strawberry Plant Area

5 Conclusions

1) Strawberries are plants that love K and P. The soils in the survey area generally have high potassium content, but the evaluation grade of effective potassium belongs to the third-class, and the content of phosphorus and nitrogen is uneven. It is recommended to adjust the formula fertilization structure according to the survey results. According to the needs of different elements for strawberry growth, it is recommended to use P+K fertilizer together. On the other hand, strawberry growth requires B and S elements, but these elements are in a lack status in this area. It is recommended that trace elements such as B and S should be fertilized in a targeted manner according to the characteristics of the soil background elements, while reducing the supply of rich elements in the soil such as Zn, to avoid excessive heavy metal elements or accumulated pollution caused by artificial fertilization.

2) Comparing the land quality between the southern area and the northern area, the absorption of trace elements in the soil during strawberry growth will lead to soil depletion. It is recommended to use the soil cultivation method in the strawberry planting area and use the crop rotation method for soil restoration in 3-5 years. Or according to the analysis of the characteristics of soil texture and nutrient content, most of the soil in the non-strawberry planting area is located on the dual structure of the river, and the soil is mainly a mixture of loamy soil and residual layer, with good nutrient texture, which can be used as a high-quality land resource concentration area for replacement of guest soil.

3) There is no heavy metal pollution in the land in the survey area, and the comprehensive evaluation results of the soil environment have no first-class (clean) and fifth-class (severely polluted). The evaluation grade of soil environment is mainly third class, accounting for more than 50%, which is greatly affected by the pH value. Therefore, it must be to pay attention on control of soil pH during the process of planting strawberry to maintain the balance of soil pH.

4) The content of Se, B, Ge and other elements in the soil in the survey area is lower than the enrichment standard. The Se content is between 0.076 mg/kg and 0.193 mg/kg, which is lower than the standard of selenium-enriched land (with a content of more than 0.4mg/kg), and the evaluation grade is fourth class (accounting for 45.6%) or fifth class (accounting for 52.8%). However, the evaluation grade as first class of the trace elements such as Fe, Zn, and Mn (rich grade) account for less than 5%, and only Mo element as rich grade accounts for more than 50%. Thus, the background content of beneficial elements in the soil is not high.

5) The floodplain in Strawberry Plant Area is generally less than 1m in soil layer that can be beneficial to farming. The lower part is a gravel layer, which has a poor ability of retaining water and fertilizer. On the other hand, the groundwater in floodplain is relatively shallow, and man-made fertilization will cause the nitrogen and phosphorus in the river to exceed the standard. Therefore, it is recommended to limit the floodplain area to the development scope of facility agriculture and set it as a buffer zone in the basin, that is, an ecological protection zone.

References

1. YH Li, BG Li, XC Liu, F Gao, BS Shi. Studies on assimilation and utilization of the minieral elements with Yuan Mei preventive drug in replanted shed strawberry fields, Journal of Agricultural University of Hebei, 28(6): 58-63. (2005)
2. HJ Lei, K Feng, ZH Zhang, HW Pan, M Zang. Growth and quality of potted strawberry under aerated drip irrigation in the three typical soils in Henan Province, Journal of Drainage and Irrigation Machinery Engineering, 35(2): 158-164. (2017)
3. LQ Zhang, SS Xu, LL Song, CY Wu, K Duan, QH Gao. Environmental quality analysis and evaluation of heavy metal content in soil of strawberry main production area in Shanghai, Acta Agriculturae Shanghai, 32(6): 113-117. (2016)
4. YL An, LQ Wan, X Li, ZQ Yin, XF Wei, ZX He, FC Jia. Spatial structure and distribution characteristics of heavy metals in the soil in Chengde, Hydrogeology & Engineering Geology, 47(6): 119-131. (2020)
5. HY Sun, XF Wei, FW Gan, H Wang, ZX He, FC Jia, J Zhang. Determination of Heavy Metal Geochemical Baseline Values and Its Accumulation in Soils of the Luanhe River Basin, Chengde, Environmental Science, 40(8): 3753-3763. (2019)
6. Specification of land quality geochemical assessment (DZ/T 0295-2016), Ministry of Land and Resources of the People's Republic of China, 2016.6.12.
7. CZ Hang, D Liang. Spatial Variability of Soil Available Potassium in Cultivated Land in Chengde City, Journal of Chengde Petroleum College, 19(2): 48-53. (2017)

A Multi-step Prediction Method of Urban Air Quality Index Based on Meteorological Factors Analysis

Yu Zhang[1], *Mingxiang* Yang[2,*], *Fengyu* Yang[1], and *Ningpeng* Dong[2]

[1]School of Software, Nanchang Hangkong University, Nanchang 330063, China
[2]China Institute of Water Resources and Hydropower Research, Beijing 100038, China

Abstract. With the development of science and technology, Industry, transportation and other industries used to discharge a large number of pollutants into the atmosphere, which results in air pollution. When air pollution become serious, it will do great harm to human health. High-precision Air Quality Index(AQI) prediction is as important as weather prediction. People could arrange traveling and their life according to the highly precise prediction results, so as to better protect their own health. Considering a lot of complex factors, we choose several potential meteorological factors and historical data to precisely forecast AQI. The principal component analysis (PCA) is introduced in our method to conduct dimension reduction on nine meteorological factors, in order to reduce noise of data and the complexity of the model calculation, which improves the accuracy of AQI prediction as a result. Then the data of meteorological factors after PCA and historical AQI are input into the multi-step prediction model based on LSSVM to train and refine it. Finally, we set up the experiment with data of meteorological factors and AQI. Experimental results show that the method proposed in this paper has better prediction accuracy over classical ARIMA method and has better generalization than ARIMA method as well.

1 Introduction

In recent years, both domestic and foreign countries have paid great attention to air pollution and air quality. With the continuous development of industries such as industry and transportation, the amount of pollution source emissions has gradually increased. Automobile emissions, industrial emissions and residential emissions have become normal, making the normal composition of air occur. Change, which in turn produces air pollution. In 2012, my country officially used AQI (Air Quality Index, AQI) to quantitatively describe air quality. The better the air quality, the smaller the AQI; conversely, the worse the air quality, the greater the AQI.

According to the National Air Quality Status in 2020 issued by the Ministry of Environmental Protection, although the concentration of pollutants has declined by 2020, heavy air pollution has still occurred on a large scale, and the annual average AQI in some cities is as high as 110. The AQI is not only a collection of local evaluation indicators, but also affected by many other factors. Among them, meteorological factors will have a direct or indirect impact on the AQI index, such as pollution sources in surrounding cities on air pollution. Through air circulation, it will directly affect the local AQI value. AQI is a parameter that directly reflects the air quality, so it is necessary to predict the city's AQI with high accuracy, and the multi-step prediction method of AQI can effectively avoid the cumulative error of the system in the single-step prediction, and the prediction effect is better. In this way, people can arrange travel matters in advance based on the AQI forecast results and can take precautionary measures in advance to reduce or even eliminate the harm of air pollution to the human.

Domestic and foreign experts have been studying the impact of meteorological factors on air quality. Research has shown that air quality is closely related to meteorological factors. Sheng Chen and Guangyuan Kan [1] used the one-year data of the six pollutants time series of air pollution and the Air Quality Index (AQI) as model inputs to analyze the impact of air pollutants on the changes in AQI.

More and more scholars take meteorological factors as the main consideration to predict the air quality index. For example: Xin Gao and Xuan Wang [2] analyze and forecast based on PEK and air quality index (AQI) based on machine learning. Jian Xue and Yan Xu [3] use the time series of the O-U model to improve the accuracy of predicting AQI changes. Qunli Wu and Huaxing Lin [4] proposed an optimal hybrid model that combines the advantages of secondary decomposition (SD), artificial intelligence methods and optimization algorithms for AQI prediction. Cyuan-Heng Luo and Hsuan Yang [5] proposed a prediction method called adaptive iterative forecasting (AIF), which can predict the PM2.5 value in the next few hours based on the trend of historical data. In 2018, Yi Lin and Long Zhao [6] proposed a new air quality prediction algorithm based on cloud model granulation (CMG). The cloud model is extracted for each particle from the original data space to the feature space, which can simplify the uncertainty Time series modeling process. In the same year, Yuzhe Yang and Zijie Zheng [7] proposed a Gaussian plume model based on neural network (GPM-NN) to physically characterize

* Corresponding author: yangmx@iwhr.com

the particle diffusion in the air and an adaptive monitoring algorithm, constructing an accurate AQI mapping can provide more High AQI prediction accuracy. The distribution of pollutant concentration is affected by factors such as historical pollution data and weather conditions. Ghaemi Z and Alimohammadi A[8] use the perspective that the distribution of pollutant concentration is affected by factors such as historical pollution data and weather conditions, and design an online algorithm based on LaSVM. Yes, the speed has increased significantly. In 2019, Xinghan Xu and Weijie Ren [9] proposed the SLI-ESN model in order to accurately predict the PM2.5 time series in monitoring and predicting the air quality index (AQI), which improved the impact of historical evolution on the current state. The mRMR feature selection method is introduced to reduce redundant information and improve efficiency. In 2021, Wang Z, Li J [10] believe that accurate forecast and early warning of air pollution episodes is crucial for the air pollution control. Therefore, this article selects meteorological factors and historical AQI as the influencing factors of AQI to predict AQI. AQI multi-step forecasting is the same as weather forecasting, multi-step forecasting has also become a key research object. In 2016, Yang, Xiaoping et al. [11] integrated natural factors, humanistic factors, and self-evolution factors into a vector autoregressive model to predict Beijing air quality in multiple steps. The experimental results show that the model is robust and capable. It can predict haze conditions more accurately and is sensitive to sudden changes in air quality.

Based on the current research status, although there are methods to predict AQI based on meteorological factors, there are few studies used for multi-step prediction, and there are not many analytical studies based on meteorological factors. This article will analyze in detail the impact of 9 meteorological factors on AQI. And based on the analysis results, multi-step prediction of AQI is carried out.

2 Materials and Methods

2.1 The predictive AQI method proposed in this paper

There are many meteorological factors that affect AQI, including wind speed, weather, temperature, humidity, etc. This paper first analyzes the correlation between various meteorological factors and AQI. Considering the mutual influence of various meteorological factors, the data may contain overlapping feature information. Using PCA to remove overlapping feature information can effectively reduce the computational complexity of the model and improve the performance of the model. In multi-step forecasting, in order to avoid accumulated errors, direct multi-step forecasting methods are used. Since LSSVM can overcome the shortcomings of long training time, randomness of training results and over-learning, and has been applied in many fields such as pattern recognition, the PCA-based LSSVM method is selected to predict AQI. The forecast scheme is shown in Figure 1.

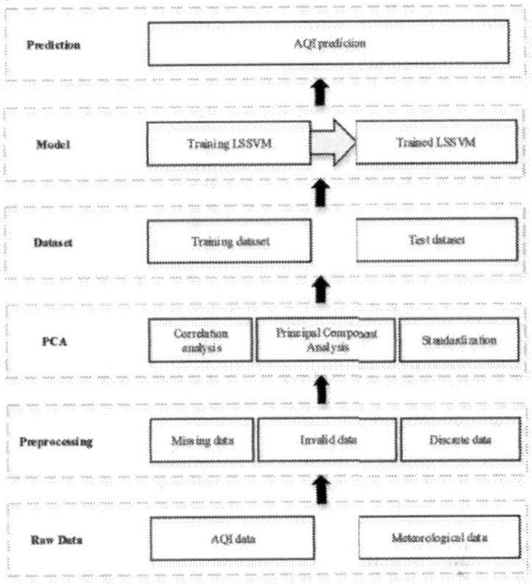

Fig. 1. General prediction process.

2.1.1 Preprocess the sample data

Some of the acquired sample data is missing, and there are some illogical data, which need to be preprocessed. The preprocessed data is $x^{'} = \{x_1^{'}, x_2^{'}, x_3^{'}, ..., x_m^{'}\}$.

2.1.2 Normalize $x^{'}$

Since the sample data is not the same dimension data, it needs to be standardized and converted to dimensionless data to perform correlation analysis and principal component analysis. The standardization method used in this paper is 0-1 standardization, and the standardization equation is:

$$x = \frac{x-\mu}{\sigma} \quad (1)$$

μ is the mean and σ is the standard deviation. Standardize x^' to convert the sample data into $x^{''}=\{x_1^{''}, x_2^{''}, x_3^{''}, ..., x_m^{''}\}$.

2.1.3 Correlation analysis of AQI and meteorological factor data

Perform correlation analysis on meteorological factor data and AQI. If there is no correlation or weak correlation, the feature will be directly removed. Combine the data with strong correlation with AQI into a new matrix $Y=\{y_1, y_2, ...y_i,\}$ where I≤m-1 and the remaining data are combined into $Y^{'} = \{Y_1^{'}, ..., Y_{m-i}^{'}\}$.

2.1.4 Principal component analysis

Principal component analysis is performed on the sample data set $Y = \{Y_1, Y_2, ...Y_i\}$, and the correlation matrix $Y^{'}$ of Y is obtained by eigenvalue $\lambda = [\lambda_1, \lambda_2, ..., \lambda_i]$ and the contribution rate of the eigenvalue. $c = [c_1, c_2, ..., c_i]$. Extract the components with a cumulative contribution rate of 85% as the main component; according to the eigenvalue λ and the component α eigenvalue, the corresponding eigenvector$d = [d_1, d_2, ..., d_k]$ is obtained. The dimensionality reduction data y of the sample data set Y after principal component analysis can be obtained from the sample data set Y and $D^T = [d_1, d_2, ..., d_k]$.

$$y = [y_1, ..., y_k] = D^T \quad (2)$$

2.1.5 Train the LSSVM model

Combine the matrix y obtained by principal component analysis and the historical AQI of the previous α day into the input data of the training set x^{train} ,the target AQI data is used as the output of the LSSVM training set to train the LSSVM model.

2.1.6 AQI Forecast

Use the trained LSSVM model to predict the target AQI of the test set.

3 Results & Discussion

3.1 Data interpretation

This paper selects three cities of Nanchang, Beijing and Nanjing as the research objects to predict its AQI. The data obtained is from 2018.12.21-2020.06.08, a total of 2703 items, including meteorological factor data and AQI data; the meteorological factor data includes average humidity, rainfall rate, maximum temperature, rainfall, sunny rate, average wind speed, minimum temperature There are nine types of data including, wind direction and average air pressure. The 2703 pieces of data are divided into training data for training models and test data for testing models. 90% are training data and 10% are test data. Some of the obtained data is missing, and some of the data is unreasonable. Preprocessing is needed to prevent it from having a greater impact on the prediction results.

The missing values of the eight types of meteorological factors: average humidity, rainfall rate, maximum temperature, sunny rate, average wind speed, minimum temperature, wind direction, and average pressure, and missing values of AQI are directly taken as the average value of the previous day and the next day. The replacement value of the missing value; for the processing of the missing value of precipitation, all the data consistent with the rainfall rate and sunny rate of the day are selected, and the average rainfall of these data is used as the replacement value of the missing value.

Regarding the wind direction factor, the processing method adopted in this paper is to directly digitize the wind direction. The unsustainable wind direction is digitized as 0, the southeast wind is digitized as 1, the northeast wind is digitized as 2, the southwest wind is digitized as 3, and the northwest wind is digitized as 4.

3.2 Correlation analysis

Analyze the correlation between AQI and meteorological factors and the correlation between meteorological factors in the selected three cities, using standard T test for analysis, and the analysis results are shown in Table 1.

Table 1. Correlation coefficient between AQI and meteorological factors.

Meteorological factors	Nanchang	Beijing	Nanjing
Humidity	-0.311	-0.244	-0.238
Rain percentage	-0.271	-0.280	-0.311
Maximum temperature	-0.236	-0.283	-0.306
Rainfall	-0.158	-0.118	-0.180
Sunny percentage	0.252	0.229	0.261
Wind speed	-0.309	-0.205	-0.225
Minimum temperature	-0.361	-0.376	-0.415
Atmospheric pressure	0.369	0.164	0.115
Wind direction	-0.071**	-0.174	-0.099

(Note: No asterisk indicates significant correlation p≤0.01 ** indicates significant correlation p≤0.05)

It can be seen from Table 1 that the selected nine types of meteorological factors are all significantly related to AQI; the correlation between the nine types of meteorological factors and AQI in the three cities is consistent, and the air pressure and weather rate are positively correlated with AQI, that is, as the air pressure rises The high AQI also increases, and the other eight types of meteorological factors are negatively correlated with the AQI, that is, the AQI will decrease with the increase of the value of these eight types of meteorological factors. Therefore, it is feasible to use meteorological factors as the influencing factors of AQI to predict AQI.

3.3 Principal component analysis

Principal component analysis is divided into two steps. First, KMO and Chi-square test are used to verify the feasibility of principal component analysis in the city, and then principal component analysis is performed.

Table 2. KMO and chi-square test

Factor analysis	Nanchang	Beijing	Nanjing
KMO test	0.715	0.602	0.569
Bartlett test	6071.24	4322.83	4341.36

It can be seen from Table 2 that the KMO statistic values of the three cities are 0.715, 0.602, 0.569, which are significantly greater than 0.5; the approximate chi-square values are 6071, 4322, and 4341, which are far greater than the critical value of 18.3; and the significance is less than 0.05 These all further verify that the principal component analysis of meteorological data can be carried out.

The principal component analysis method is used to conduct principal component analysis on the experimental data of the three cities of Nanchang, Beijing, and Nanjing. The results of the analysis are shown in Table 3.

Table 3. Principal component analysis results of three cities

Component	Nanchang		Beijing		Nanjing	
	Variance %	Cum. Variance %	Variance %	Cum. Variance %	Variance %	Cum. Variance %
1	32.34	32.34	30.35	30.35	30.73	30.73
2	29.14	61.48	23.95	54.304	22.41	53.14
3	11.67	73.15	11.98	66.29	12.82	65.96
4	10.07	83.22	10.41	76.70	10.60	76.56
5	9.39	92.61	9.57	86.262	8.92	85.48
6	4.16	96.77	7.03	93.290	8.44	93.91
7	1.94	98.70	4.27	97.56	3.94	97.85
8	0.91	99.61	2.05	99.61	1.72	99.57
9	0.39	100.00	0.39	100.00	0.43	100.00

Table 3 shows the principal component analysis results of the three cities. The leftmost column is the 9 components arranged in ascending order of variance. The data of each city consists of 2 columns. The first column "Variance%" is the variance. It also shows the contribution rate of the component, and the second column is the cumulative contribution rate. The first column of each city is the corresponding variance. From Figure 3-2, we can see that the second column is nine eigenvalues arranged by size.

According to the principle of the principal component analysis method, the main components can be selected according to the cumulative contribution rate, and several principal components with a cumulative contribution rate greater than 85% are taken to replace all the components to reduce the dimension of calculation. It can be seen from Table 3 that the cumulative contribution rates of the first five components of the three cities are 92.61%, 86.26%, and 85.48%, which are all greater than 85%. Therefore, the first five components can be selected as the principal components for AQI prediction.

The PCA method performs dimensionality reduction and denoising processing on the meteorological factor data of the three cities of Nanchang, Beijing, and Nanjing. The meteorological factor data of the three cities are all reduced from 9-dimensional data to 5-dimensional data. While removing redundant information, it can greatly Simplify the training and prediction process of LSSVM and improve the training speed.

3.4 LSSVM model based on PCA

This paper selects the three-day historical AQI and the meteorological factors of the previous day in the three cities of Nanchang, Beijing, and Nanjing as input data, and the AQI of the day as the output data to verify the feasibility of the PCA method. The prediction results are shown in Table 4.

Table 4. Comparison of LSSVM prediction results with or without PCA in three cities

	Nanchang		Beijing		Nanjing	
	MAPE	RMSE	MAPE	RMSE	MAPE	RMSE
Without PCA	0.400	24.362	0.391	**24.542**	0.391	25.663
With PCA	**0.395**	**21.182**	**0.342**	26.693	**0.271**	**23.070**

It can be seen from Table 4 that the prediction results of the LSSVM method with principal component analysis in the AQI prediction results of Nanchang and Nanjing are slightly smaller than the prediction results of the LSSVM method without principal component analysis in both MAPE and RMSE; there is no principal component in the prediction results of Beijing AQI The RMSE is small in the predicted results of the analysis, but the MAPE is small in the predicted results of the principal component analysis.

Because the RBF kernel function is suitable for large samples as well as small sample data, suitable for high-dimensional space and low-dimensional space, as well as the advantages of low parameters and low computational complexity, the RBF kernel function is selected as the LSSVM model. The kernel function of, then the regression equation of LSSVM is:

$$f(x) = \sum_{i=1}^{n} \alpha_i exp\left(-\frac{||x-x_i||^2}{2\sigma^2}\right) + b \quad (3)$$

σ is the parameter in the RBF kernel function.

3.4.1 Parameter setting

Because in all the method models in this paper, there are a total of days of historical AQI, and the penalty parameter C in the LSSVM model needs to be determined in the kernel function parameter σ.

(1) Determination of historical AQI days

The most recent historical AQI has the greatest impact on the current AQI, and the longer the time, the smaller the impact until the impact on the current AQI is negligible. Therefore, this paper selects the days with the best effect in 1-4 days according to the experimental results. The experimental analysis is carried out on three cities of Nanchang, Nanjing, and Beijing. The historical weather factors of the previous day are selected as the influencing factors, and the number of days of the historical AQI is from 1-4 to predict the current AQI.

Table 5. The impact of historical AQI days in the three cities a

Nanchang Beijing Nanjing

	Nanchang		Beijing		Nanjing	
	MAPE	RMSE	MAPE	RMSE	MAPE	RMSE
1	0.390	21.133	0.260	22.059	0.265	22.902
2	0.383	20.676	0.266	22.176	0.268	22.831
3	0.395	21.182	0.301	27.372	0.271	23.070
4	0.407	21.510	0.302	25.427	0.275	23.313

It can be seen from the table that among the prediction results of the three cities with different historical AQI days as the influencing factor, the MAPE and RMSE of the Nanchang prediction result are the smallest; the MAPE and RMSE of the Beijing prediction result are both the smallest when a=2, but a= The MAPE and RMSE at 2 o'clock are very close to the values at a=2; the MAPE of Nanjing prediction results is the smallest at a=1, the RMSE at a=2 is the smallest, and the MAPE at a=2 is the value at a=1 very close. In summary: when a=2, the overall prediction effect is the best, so a=2.
(2) Determination of the parameters in the penalty parameter C and the kernel function σ

In this paper, the grid search method is used to optimize C and σ, that is, to try all possible C and σ, and then use cross-validation to verify, and find the C and σ with the highest accuracy. Cross-validation is an exhaustive search method. Although it is time-consuming compared to other methods, the complexity is not too high when there are fewer parameters, and the grid search method can find the global optimal solution, so this paper selects cross Verification to verify the parameters.

3.4.2 Analysis of multi-step forecast results

The prediction results of the three cities of Nanchang, Beijing and Nanjing based on the PCA+LSSVM method are shown in Table 6, Table 7, and Table 8, respectively.

Table 6. Error analysis of Nanchang AQI multi-step prediction

Days	MAPE	MAPE error increased by %	RMSE	RMSE error increased by %
1	0.384		21.182	
2	0.490	27.6%	24.641	16.3%
3	0.589	20.2%	27.637	12.2%
4	0.604	2.5%	28.824	4.3%

It can be seen from Table 6 that as the number of forecast days increases, the accuracy rate gradually decreases. It is predicted that the MAPE and RMSE of the AQI on the second and third days will increase more than the previous day, and the MAPE will increase by 27.6% and 20% respectively; the RMSE will increase by 16.3% and 12.2% compared with the previous day. Although MAPE and RMSE have increased a lot, MAPE and RMSE themselves are relatively small.

Table 7. Error analysis of multi-step prediction of Beijing AQI

Days	MAPE	MAPE error increased by %	RMSE	RMSE error increased by %
1	0.267		22.536	
2	0.288	7.9%	24.894	10.5%
3	0.297	3.1%	25.279	1.5%
4	0.317	6.7%	26.279	4.0%

It can be seen from Table 7 that as the number of forecast days increases, the accuracy rate gradually decreases. It is predicted that the MAPE and RMSE of the second day will increase more than the previous day. The MAPE will increase by 7.9% and the RMSE will increase by 10.5%.

Table 8. Error analysis of Nanjing AQI multi-step prediction

Days	MAPE	MAPE error increased by %	RMSE	RMSE error increased by %
1	0.285		24.094	
2	0.300	5.3%	24.000	-0.4%
3	0.328	9.3%	25.731	7.2%
4	0.338	3.0%	26.341	2.4%

It can be seen from the table that as the number of forecast days increases, the accuracy rate gradually decreases. It is predicted that the MAPE and RMSE of the AQI on the second day will increase more than the previous day, and the MAPE will increase by 79.3% compared with the previous day; the RMSE will increase by 7.2% compared with the previous day. Since the MAPE and RMSE of the first day's prediction results are small, the accuracy rate decreases with the increase of the number of days is small, so the overall prediction effect is better.

3.4.3 Comparison results with ARIMA method

ARIMA is a classic time series multi-step forecasting method. On the experimental data of three cities, the results of the algorithm in this paper and the ARIMA algorithm are compared. The results of MAPE are shown in Table 9, and the results of RMSE are shown in Table 10.

Table 9. MAPE between ARIMA and Ours

	Nanchang			Beijing			Nanjing		
days	ARIMA	Ours	Improve by %	ARIMA	Ours	Improve by %	ARIMA	Ours	Improve by %

1	0.474	**0.384**	-20.0	0.305	**0.267**	-12.5	0.311	**0.285**	-8.4
2	0.638	**0.490**	-23.2	0.346	**0.288**	-16.8	0.346	**0.300**	-13.3
3	0.679	**0.589**	-13.3	0.354	**0.297**	-16.1	0.366	**0.328**	-10.4
4	0.691	**0.604**	-12.6	0.362	**0.317**	-12.4	0.357	**0.338**	-5.3

Table 10. RMSE between ARIMA and Ours

days	**ARIMA**	**Ours**	**Improve by %**	**ARIMA**	**Ours**	**Improve by %**	**ARIMA**	**Ours**	**Improve by %**
1	23.34	**21.18**	-9.2	24.63	**24.09**	-2.2	25.83	**25.11**	-2.8
2	30.38	**24.64**	-18.9	27.03	**24.00**	-11.2	28.05	**23.82**	-15.1
3	31.42	**27.64**	-12.0	27.46	**25.73**	-6.3	28.25	**25.95**	-8.1
4	31.89	**28.82**	-9.6	27.08	**26.34**	-2.7	27.00	**24.66**	-8.7

It can be seen from Table 9 that the method proposed in this paper predicts that the MAPE of AQI for 1-4 days in Nanchang, Beijing, and Nanjing is less than the MAPE predicted by the ARIMA method, and the error reduction ranges from 5.3% to 23.2%. Most of the MAPE errors are reduced. Not less than 12%. Overall, the forecast results have been greatly improved.

It can be seen from Table 10 that the method proposed in this paper predicts that the RMSE of the 1-4 days AQI of the three cities is less than that of the ARIMA method. The error reduction range ranges from 2.2% to 18.9%, and the average error is about 10%. The overall prediction result is relatively good. Big improvement. However, the improvement of RMSE relative to MAPE is not stable enough. The reason is that MAPE is a relative error, and RMSE is an absolute error. The value of AQI is inherently high and low, and the accuracy of prediction is more reasonable to be measured by MAPE. In general, the PCA+LSSVM method proposed in this paper is more accurate than the ARIMA method in predicting results.

4 Conclusions

This paper focuses on analyzing the impact of meteorological factors on AQI, combining meteorological factors and historical AQI as the influencing factors of AQI, and establishing a model to predict AQI. The main conclusions are as follows:

(1) Consider the influence of multiple meteorological factors on AQI. Since multiple meteorological factors influence each other, there is likely to be overlapping feature information. Analyzing the correlation between various meteorological factors and AQI, the results show that all the nine meteorological factors selected in this paper have a significant relationship with AQI, and most of the meteorological factors also have a significant correlation relationship. The influence of AQI is feasible, and the principal component analysis of meteorological factors is also feasible.

(2) Using PCA to reduce the dimensionality and denoising of meteorological factor data will not have a negative impact on the results of LSSVM model predicting AQI. On the contrary, due to its denoising characteristics, it can improve the accuracy of LSSVM prediction to a certain extent.

(3) AQI prediction method based on PCA and LSSVM. The experimental results show that the prediction effect is best when the historical AQI as an impact factor is 2 days.

(4) AQI multi-step forecasting method based on PCA and LSSVM, as the number of forecast days increases, the forecast error will gradually increase, and the error increase will also decrease. Compared with the ARIMA method, the prediction results of the multi-step prediction method based on PCA and LSSVM are smaller in MAPE and RMSE. The method proposed in this paper has better prediction effect and more applicability than the ARIMA method.

This work is supported by the National Natural Science Foundation of China (U1865102) and Beijing Science and Technology Plan Project (Z201100001820022).

References

1. S. Chen, G. Kan, J. Li, et al, Investigating china's urban air quality using big data, information theory, and machine learning. Polish Journal of Environmental Studies, **27(2)**, 565-578 (2018)
2. X. Gao, X. Wang, H. Sh, Multifractal cascade analysis on the nature of air pollutants concentration time series over china. Aerosol and Air Quality Research, **19(9)**. (2019)
3. J. Xue, Y. Xu, L. Zhao, et al, Air pollution option pricing model based on AQI. Atmospheric Pollution Research, **10(3)** 665-674. (2019)
4. Q. Wu, H. Lin, A novel optimal-hybrid model for daily air quality index prediction considering air pollutant factors. Science of The Total Environment, **683** 808-821. (2019)
5. C. Luo, H. Yang, L.P. Huang, et al. A Fast PM2.5 Forecast Approach Based on Time-Series Data Analysis, Regression and Regularization. 2018 Conference on Technologies and Applications of Artificial Intelligence (TAAI). (2018)
6. Y. Lin, L. Zhao, H. Li and Y. Sun, Air quality forecasting based on cloud model granulation. Eurasip Journal on Wireless Communications & Networking, **2018(1)** 106. (2018)
7. Y. Yang, Z. Zheng, K. Bian, L. Song, H. Zhu, Realtime profiling of fine-grained air quality index distribution using uav sensing. IEEE Internet of Things Journal, **(99)** 1-1 (2017)
8. Z. Ghaemi, A Alimohammadi, M Farnaghi. Lasvm-based big data learning system for dynamic prediction of air pollution in tehran. Environmental Monitoring and Assessment: An International Journal. (2018).
9. X. Xu, W. Ren, Prediction of Air Pollution Concentration Based on mRMR and Echo State Network. Applied Sciences, **9(9)** 1811 (2019)

https://doi.org/10.1051/e3sconf/202235001010

10. Z. Wang, J. Li, Establishment and operational application of china national air quality forecast and early warning device. (2021)
11. X. Yang, Z. Zhang, Z. Zhang, L. Sun, C. Xu, L. Yu, A long-term prediction model of beijing haze episodes using time series analysis. Computational Intelligence & Neuroscience, 2016, 17. (2016)

Research on Key Pollutant Reduction Program of Environmental control in a Small River Basin of Guangxi

Haoyang Song[1], *Junchao Li*[1], *Jinyuan* Jiang[1], *Jiao* Li[2], *and Wei Tan* [1,*]

[1]Chinese Research Academy of Environmental Sciences, Environment Engineering Design Center, Beijing 100012, China
[2]Technical Centre for Soil, Agriculture and Rural Ecology and Environment, Ministry of Ecology and Environment, Beijing 100012, China

Abstract. In order to strengthen the prevention and control of water pollution in rural rivers, it is necessary to analyze the pollution causes in detail, calculate the pollutant reduction amount and put forward the comprehensive treatment measures. Aiming at the small basin of a county in the southwest Guangxi, this paper carried out the comprehensive improvement of water environment, including specifically investigation the water pollution degree of the basin, calculating present pollution load and water environment capacity, and proposing pollutant reduction targets in accordance with the requirements for water quality targets. The basin has poor water quality background and is seriously polluted by rural living pollution sources, livestock and poultry breeding pollution sources and planting pollution sources, in order to solve these problems, the water quality assurance project, rural living sewage governance project and large-scale livestock and poultry breeding pollution governance project are put forward. Through project quantity accounting and pollutant reduction calculation, the water quality target can be achieved.

1 Introduction

Rural rivers are important natural resources for rural development, and have extremely important ecological functions as an important part of large and medium-sized rivers. However, with the rapid development of rural society and economy, the ecological environment of small basin in rural areas deteriorates sharply, resulting in a number of pollutant water quality indicators exceed the standard, and directly affect the quality of life of people on both sides of the river. It is urgent to strengthen the prevention and control of water pollution in rural rivers. For polluted rivers, pollutant reduction should be calculated according to the current pollution load and water environment capacity to provide a reference for integrated governance measures in small basins [1-2].

It is necessary to calculate the water environment capacity and the target reduction amount in order to formulate a more targeted implementation plan of comprehensive water environment improvement in small basin. At present, there are many studies on the calculation of pollution load reduction. Taking the comprehensive improvement of water environment in the river basin as an example, Huang comprehensively considered the estimation method of water capacity and the control of discharge reduction target, and carried out the calculation research on water environment capacity and the target reduction amount using one-dimensional calculation mode [3]. According to the small basin river system situation and characteristics of different river water quality, Zhang et al calculated the water environment capacity through basin segmentation, carried out calibration and validation of hydrological model parameters based on the measured data, proposed the method for calculating water environment capacity of river segments in small basin area, and determined the pollutant reductions, so as to formulate key project measures to reduce pollutants, and ensure that the water quality of the whole line is up to standard [4]. In this study, the pollution status and pollution load of a small rural basin in southwest Guangxi were calculated and analyzed, the water environment capacity was calculated according to the water quality target, the target reduction amount of pollutants was determined, pollutant reduction measures were proposed and the target accessibility analysis was carried out to ensure the river water quality target. This study will not only provide a reference for the calculation of water environment capacity and pollutant reduction targets in river basin, and provide technical support for the comprehensive remediation of water environment in small river basin, but also is of great significance to solve the problems of river water environment and eliminate the inferior class V water bodies in rivers.

2 Materials and methods

* Corresponding author: weitan_2012@163.com

2.1 Study area

Taking a small river basin in southeastern Guangxi province as the study area, it includes four rivers (hereinafter referred to as Four Rivers): YQ River, LW River, DB River and BML River. Four Rivers have a basin area of 303 km², a river length of 68.3 km, which flows through 29 administrative villages in 3 towns.

2.2. Calculation method of the amount of pollutant source into the river

(1) Township living sewage

Township living sewage treatment project is point source pollution discharge, whose pollutant discharge calculation formula is:

$$W_i = C_i \times Q \times 10^{-3} \quad (1)$$

Where, W_i is the pollutant discharge of township sewage treatment plant (kg/m), C_i is the discharge concentration of pollutants from the township sewage treatment plant (mg/L), and Q is the discharge of township sewage treatment plant (m3/d).

(2) Rural living sewage

Based on the number of rural populations in the basin, the output of living sewage in the basin is calculated by using the pollution coefficient method [4-6].

$$W_{living} = \alpha_i \times N_{agriculture} \times 10^{-3} \quad (2)$$

Where, W_{living} is the discharge of rural living pollutants (kg/d), N agriculture is the number of rural population (person), and α_i is the rural living sewage discharge coefficient (g/person·day) .

(3) Non-point source of living garbage

$$W_i = \gamma_i \times M \quad (3)$$

Where, W_i is the discharge of living waste pollutants i (kg/d), and γ_i is the conversion coefficient of pollutant discharge in garbage.

(4) Livestock and poultry breeding pollution sources

$$W_{Livestock} = \alpha_2 \times N_{pig} \times 10^{-3} \quad (4)$$

Where, $W_{Livestock}$ is the discharge of livestock and poultry breeding pollutants, N_{pig} is the number of converted active pigs (head), and α_2 is the sewage discharge coefficient of livestock and poultry (g/head · day).

(5) Planting

Agricultural non-point source pollution includes farmland surface runoff pollution and farmland solid waste pollution.

Farmland surface runoff pollution discharge calculation formula is:

$$W_{farmland\ surface\ runoff} = A_{farmland} \times \varphi_i / 1000 \quad (5)$$

Where, farmland surface runoff pollution discharge is the product of farmland area (A farmland) and pollutant loss coefficient (φ_i) within the basin [5-6].

Farmland solid waste pollutants are mainly plant residues from agricultural production:

$$W_{Farmland\ solid\ waste} = 30\% \times A_{farmland} \times \varepsilon \times \alpha_3 \quad (6)$$

Where, $A_{farmland}$ is the farmland area in the basin. ε is the solid waste coefficient from farmland. α_3 is the main pollutants content of farmland solid waste non-point source (α_3).

(6) Soil erosion pollution

The calculation formula of soil erosion pollution source discharge is:

$$W_{soil\ erosion} = 25\% \times A_{total} \times \mu \times \alpha_4 \quad (7)$$

Where, A_{total} is the total area of the basin (km2). μ is the soil erosion modulus. α_4 is the content of pollutants in sediment.

The formula for calculating the amount of pollutants inflow into the river is:

$$W = \sum (W_i \times \beta_i) \quad (8)$$

Where, W is the amount of pollutants inflow into the river (kg/d), i is the type of pollution source, W_i is the discharge of pollutants from all pollution sources (kg/d) and β_i is the river inflow coefficient of all pollution sources [7-8].

2.3 Pollution capacity calculation method

The environmental capacity calculation formula is:

$$W = 365 \times 0.0864 \times Q_h \times [C_s \times exp(K \times L / 86400u) - C_0] \quad (9)$$

Where, W is the environmental capacity, t/a, Q_h is the river flow during dry season, m3/s, C_s is the standard concentration of river water quality control, mg/L, K is the comprehensive degradation coefficient of pollutants, 1/d, L is river spacing, m, u is the river velocity, m/s and C_0 is the river background concentration, mg/L.

The pollutant reductions can be calculated as follows:

$$X = P - W \quad (10)$$

Where, X is the pollutant reduction target (t/a), P is the amount of pollutants into the river (t/a) and W is the environmental capacity (t/ a).

3 Results & Discussion

3.1 Water environment quality

According to the analysis of water quality monitoring data of the Four rivers in 2019, the water quality of the four rivers in 2019 was inferior class V, which all did not meet the water environment function requirements (class V standard), and the main exceeding factors were

 https://doi.org/10.1051/e3sconf/202235001011

ammonia nitrogen and total phosphorus. In YQ River, LW River, DB River and BML River, the average annual exceeding multiple of ammonia nitrogen was 15.75, 3.09 and 2.93 times, respectively (the average annual exceeding multiple of ammonia nitrogen in BML River reached the standard), and the average annual exceeding multiple of total phosphorus was 2.86, 3.45, 2.96 and 1.4 times, respectively.

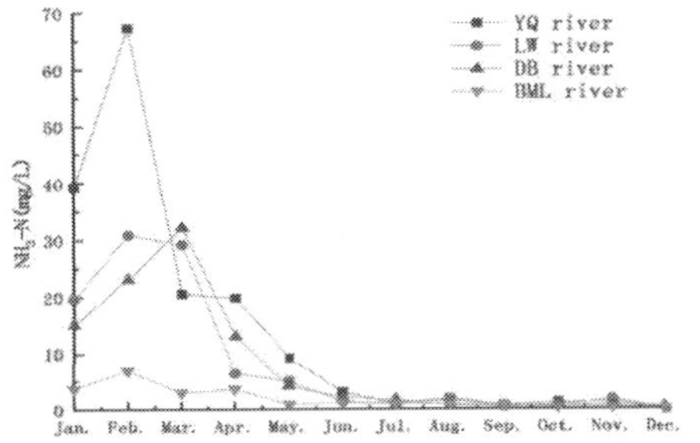

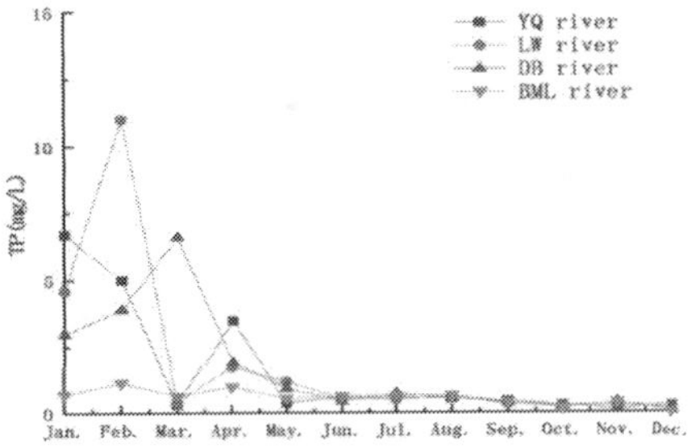

Fig. 1. Concentrations of ammonia nitrogen (left) and total phosphorus (right) of Four Rivers in 2019

3.2 Pollutants calculation results

The discharge load of pollutants is calculated based on the statistical data of 2019. The results are shown in Table 1. The discharge of COD, NH_3-N, TN and TP are 8827.21 t/a, 259.72 t/a, 916.17 t/a and 189.54 t/a, respectively. Generally speaking, livestock and poultry pollution sources are the main pollution sources.

The main reasons that livestock and poultry breeding has become the main source of pollution are first, the proportion of large-scale breeding is low, where in the first quarter of 2020, the current amount of B county is 413,000, while the amount of large-scale breeding is only 81,000. Second, the aquaculture methods in breeding farms below large scale are backward, where traditional aquaculture accounts for a large proportion and environmental pollution governance level is backward. Third, as a traditional industry in county, breeding supports the lives of local people. Whereas there are a large number of breeding farms below large scale, it is difficult to regulate the discharge of breeding pollution, and the pollution of livestock and poultry has not formed a special supervision and management system, thus livestock and poultry breeding pollution has become one of the main factors affecting the regional water environment quality in B County.

Soil erosion pollutant discharge is large, among which COD discharge is 2022t/a, ammonia nitrogen discharge is 259.72t/a, total nitrogen discharge is 916.17t/a, and total phosphorus discharge is 189.54t/a.

Table 1. Statistics of pollution source discharges in Four Rivers' basin in 2019

Pollution sources		COD (t/a)	Ammonia nitrogen (t/a)	Total nitrogen (t/a)	Total phosphorus (t/a)
Living sewage	Rural living sewage	547.45	54.30	121.47	17.35
	Township living sewage	26.18	2.62	7.85	0.26
Living garbage	Rural living garbage	77.39	1.55	2.32	3.87
	Township living garbage	5.15	0.10	0.15	0.26
Livestock and poultry breeding	Livestock and poultry breeding below large scale	2452.21	39.07	151.27	18.80
	Large-scale livestock and poultry breeding	3114.84	83.76	114.32	16.56
Planting industry	Fertilizer loss pollution	265.46	11.68	113.30	2.46
	Farmland solid waste	316.54	66.64	102.18	28.88
Soil erosion	Soil erosion	2022.00	0.00	303.30	101.10
Totally		8827.21	259.72	916.17	189.54

 https://doi.org/10.1051/e3sconf/202235001011

According to formula (1)-(8), the total amount of COD pollutants into the river is shown in Fig.2, among which the pollutants discharge from large-scale livestock and poultry breeding accounts for the highest proportion of 44.42%. The total amount of ammonia nitrogen pollutants into the river is 246.88t/a, where the pollution sources from large-scale livestock and poultry breeding accounts for the highest proportion of 30.87%. Again, the total amount of total phosphorus discharge is 91.83t/a, where the pollution sources from planting industry accounts for 35.92%.

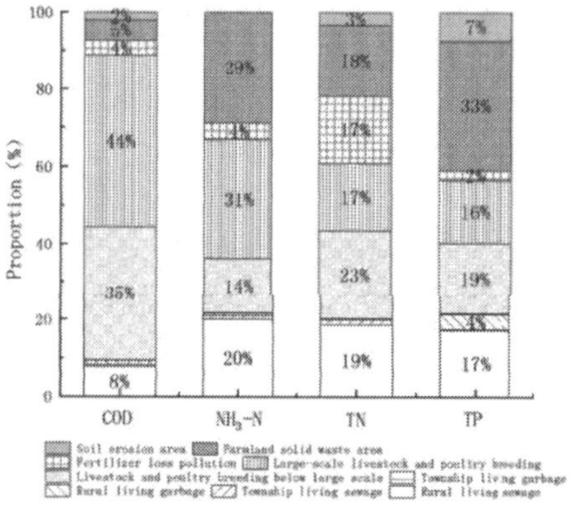

Fig. 2. Proportion of main pollution sources into the river in Four Rivers' basin

3.3. Analysis of pollutant reduction targets

The river flows in each section are measured data in dry season, which are 0.34, 0.48, 0.64 and 0.17 m^3/s respectively. According to Formula (9), the environmental capacities of COD, ammonia nitrogen and total phosphorus are 474.64, -549.33 and -86.15 t/a respectively. The reduction amount can be calculated from the difference between river inflow amount and environmental capacity, the results are shown in Fig.3.

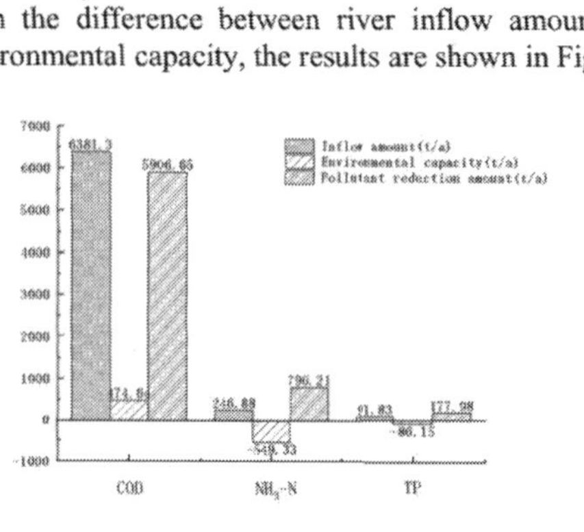

Fig. 3 The pollutant reduction in the basin

3.4 Project reduction program

In order to improve pollution control requirements, the following reduction principles shall prevail, the key pollution sources are first reduced, pollutants in the control unit must be reduced, key sources are emphatically reduced, and livestock and poultry breeding source is the focus to reduce. The Pollutant reductions of comprehensive treatment projects in four rivers' basin was shown in Tab.3.

3.4.1 Water quality guarantee project

Through pollution load analysis, it can be seen that the internal pollution and planting pollution in the four rivers account for a relatively large proportion. The wetland is constructed to reduce non-point source runoff of rural life, non-point source sediment of farmland agriculture, dissolved nitrogen and phosphorus and other pollutants along the river, and purify the water quality of runoff into the river. The ecological buffer zone has an obvious effect on the removal of nitrogen and phosphorus in agricultural production, and can intercept and absorb 40%-60% of the nutrients absorbed in soil particles. Using artificial floating island, planting willow herb, calamus, reed and other aquatic plants, through the absorption and attachment function of plant roots, the concentration of nitrogen and phosphorus pollutants in the water would be reduced, so as to achieve the effect of purifying water. The serious sedimentation of bottom mud in water body has a certain impact on water body and is also detrimental to industrial and agricultural water consumption. Therefore, it is necessary to carry out dredging treatment project.

3.4.2 Rural living sewage treatment project

At present, there are 484 natural villages in the region that have not carried out rural living sewage treatment project. Through field investigation, it is found that the rural living sewage construction land is insufficient, and the villages are scattered and complex. The living connection rate is insufficient and the living connection is not standardized. In order to solve the above problems, the rural living sewage treatment project needs to reconstruct 326 ecological regulation and storage ponds including the use of waste ponds,, and 101 functional wetlands. As to supporting pipe network projects, it constructs new 17km residential pipe network, 31km ecological ditch, 95km culvert repair, 24 hydrolysis acidification tank and contact oxidation tank.

3.4.3 Pollution control project for livestock and poultry breeding below large scale

As to the pollution control project for livestock and poultry breeding below large scale, a total of 1,723 livestock and poultry pollution treatment sites are contained and along with the construction of aerobic composting sites and livestock and poultry wastewater treatment sites in the 'septic tank + oxidation pond' model, the dry manure removal process is promoted in a coordinated manner. Above all, with cross-border four rivers' section water quality requirements (surface water V standard) as the goal, the comprehensive basin treatment projects of cross-border four rivers. Eventually, after the implementation of all projects, the reduction amount of COD, ammonia nitrogen and total phosphorus is 6434.81 t/a, 635.75 t/a and 368.91t/a respectively, which are all higher than the pollutant reduction target of 5906.65 t/a, 796.21 t/a and 177.98t/a, reaching the overall reduction target. This project is of great significance to solve the problems of river water environment and eliminate the inferior class V water body.

Table 2. Pollutant reductions of comprehensive treatment projects in four rivers' basin

No.	Name	Content	Pollutant reductions（t/a）		
1 Water quality guarantee project			COD	NH_3-N	TP
1.1	Artificial wetland	A total of 75,700 m^2 wetland is constructed.	2399.14	299.71	96.26
1.2	Ecological buffer zone	A total of 58.8km ecological farmland buffer zone is constructed.			
1.3	Artificial floating island	A total of 12,830 m^2 artificial floating is constructed.			
1.4	Internal source governance engineering	A total of 11,020m^3 mud is removed and a total of 354,086m^3 aquaculture waste water is treated.			
2 Rural living sewage treatment project			262.26	43.18	30.48
2.1	Pipe network project	A total of 17km residential pipe network, 31km ecological ditch, and 95km culvert repair, are constructed.			
2.2	Living sewage governance	A total of 326 ecological regulation and storage ponds, covering an area of 428 mu, and 101 functional wetlands, covering an area of 406 mu are constructed.			
3 Pollution control project for livestock and poultry breeding below large scale			3773.41	292.86	242.17
3.1	Aerobic fertilizer composting site	A total of 30,860m^3 of manure storage tank for livestock and poultry breeding, 115,725m^3 of compost tank and 7,715m^2 of fertilizer warehouse are constructed.			
3.2	Oxidation pond	16457.6m^2 oxidation pond is constructed.			
Totally			6434.81	635.75	368.91

4 Conclusion

(1) The ammonia nitrogen and total phosphorus of Four Rivers is seriously over the standard and the overall water quality is line with the surface water class V standard. The change of water quality is seriously affected by seasons and residents' activities.
(2) According to the analysis of river inflow, large-scale livestock and poultry breeding and rural living sewage are the main pollution sources.
(3) Through the implementation of rural sewage treatment projects, livestock and poultry pollution prevention and control projects and other project measures, the pollutants pollution of inflow into the river can be reduced and the overall reduction target can be reached.

Acknowledgments

Supported by the National Key Research and Development Program of China (NO.2018YFD1100505).

References

1. H. Le, Study on measurement method of water environment capacity and target reduction: a case study of comprehensive water environment regulation in Liangtan River Basin of Chongqing. J. Three Gorges ecological environment monitoring 1-14 (2021)
2. R. Xu, R. Zhao, Y. Huang, X. Song, Y. Xie, Study on water pollution control in Jiuzhou river basin based on TMDL model. J. Environmental science and management, 45(07):72-76. (2020)
3. L. Zhang and Y. Liu, W Sun. Bobai. Estimation of water environmental capacity and pollutant reduction allocation in Caoqiao River watershed of Taihu lake basin. J. Journal of lake sciences, **21(04)** 502-508. (2009)
4. Ministry of Housing and Urban-Rural Development, PRC. Technical Guide of Guangxi Rural Living Sewage Treatment (Trail). http://mohurd.gov.cn/wjfb/201101/t20110107_201895.html (2010)

5. Z. Song, S. Cui, Q. Fu,Q. Gao, K. Li, S. Gao, Y. Zhao, C. Jia, Preliminary estimation and source analysis of pollutants into river in Harbin City, J. Journal of irrigation and drainage, **39(03)** 134-144. (2020)
6. Y. Qin, J. Xiao, L. Xu, W. Zhao, Estimation and application of non-point source pollutant discharge coefficient in typical area of Minjiang River. J. Sichuan Environment, **40(04)** 138-144 (2021)
7. Institute of Environment and Sustainable Development in Agriculture. Nanjing Institute of Environmental Sciences. Ministry of Environmental Protection. Manual of pollutant Discharge Coefficient and Production-Discharge Coefficient of livestock and poultry breeding of the first National Census of Pollution Sources. (2009)
8. D. Yang, Study on improving water quality in watershed. J. Chemical Engineering and Equipment, (07) 61-264 (2020)

Influence Factors on Removal of 2,4-DCP by Sulfided Nanoscale Zerovalent Iron

Yajun Li[1], *Yongxiang* Zhang [1,*], and *Jinhao* Wang[1]

[1]Beijing University of Technology, Faculty of Architecture, Civil And Transportation Engineering,100124, 100 Pingleyuan, Chaoyang District, Beijing, China

Abstract. 2,4-Dichlorophenol (2,4-DCP) has been extensively applied for chemical and pharmaceutical production, resulting in severe environmental pollution. In this paper, the sulfided nanometer zero-valent iron (S-nZVI) was synthesized and applied to remove 2,4-DCP. The experimental tests displayed that when the sulfur iron mole ratio was 0.129, the elimination rate for 2.4-DCP was 91.9%, and the removal rate declined when the sulfur-to-iron proportion increased. As the initial concentration of 2, 4-DCP improved from 10 to 40mgL^{-1}, the elimination rate of 2,4-DCP declined from 92.6% to 65.3%. The elimination effect of S-nZVI on 2,4-DCP increased with rising temperature. The removal rate of 2,4-DCP varied under various pH conditions. The removal efficiencies were reduced from 75.5% to 48.8% when the initial pH ranged from 5 to 3. When pH is 11, the removal rate is 97.9%. Kinetics of degradation reaction of 2,4-DCP under different conditions were conducted. The process for removing 2,4-DCP was in accord with the pseudo-first-order kinetics model. The initial pH and sulfur mole ratio played a decisive role, which determined the removing rate of 2,4-DCP. The findings can guide more efficient S-nZVI reactivity towards the target contaminants in water remediation.

1 Introduction

2,4-dichlorophenol (2,4-DCP) drew much attention because of its high toxicity and non-biodegradability, causing considerable harm to humans and the environment[1,2]. Therefore, it is necessary to develop remedial technology. Methods for eliminating 2,4-DCP contaminants include the microorganism method and chemical oxidation methods[3,4]. However, the different solutions to the problem have some disadvantages, such as toxic intermediate products and the requirement of microbial domestication. Hence, the development of an efficient chemical reduction technology is desired.
Nowadays, nanoscale zero-valent iron particles (nZVI) have become a better repair material for removing organic pollutants [5,6]. However, there are still many technical challenges in water remediation with the implementation of nZVI. The main issue restricting broad-scale nZVI use is that nZVI particles can react with water to form iron oxide layers outside the nZVI particles and inhibit the reaction between nZVI and contaminants[7,8]. Increasing its selectance for objective contaminants rather than water is necessary to increase efficiency. Sulfidation is an emerging method to stabilize nano zero-valent iron particles. Recently, nanometer zero-valent iron sulfide (S-nZVI) has attracted extensive attention from researchers due to its favourable properties in water remediation[9]. Researchers found that sulfur can form sulfide layers clinging to nano zero-valent iron, avoid contact with oxygen and decrease the passivation of the materials, to result in the decrease of reactivity[10]. In addition, the hydrophobicity and the electron transfer efficiency of sulfide layers were superior to oxide layers. Thus the removal efficiency of pollutants is higher[11].

In this work, S-nZVIs were synthesized and used to remove 2,4-DCP. The influences of reaction factors (temperature, initial concentration, dosage, and sulfur iron mole ratios) on the treatment efficiency of 2,4-DCP pollutants were studied. According to the experimental results, the sulfur mole ratio significantly influences the 2,4-DCP removal.

2 Materials and Methods

2.1 Material

2,4-dichlorophenol ($C_6H_4Cl_2O$), o-chlorophenol (C_6H_5ClO), p-chlorophenol (C_6H_5ClO), and phenol (C_6H_6O) was acquired from Shanghai Maclin Biochemical Technology Co., LTD (Shanghai, China). The experimental medicines were obtained from Tianjin Fuchen Chemical Factory (Tianjin, China) included potassium borohydride (KBH_4), hydrochloric acid (HCl), ferrous sulfate ($FeSO_4{\cdot}7H_2O$), sodium dithionite ($Na_2S_2O_4$), and sodium hydroxide (NaOH). All chemicals were analytical grade. Nitrogen (N_2) was purchased from Beijing Shunchi East ring dry ice

* Corresponding author: yxzhang@bjut.edu.cn

operation Center (Beijing, China). All Ultrapure water was used to prepare the chemical solutions.

2.2 Preparation of Sulfided nanometer zero-valent iron

S-nZVI particles were prepared based on the one-step method. Firstly, the solution of $FeSO_4 \cdot 7H_2O$ was formed by adding $FeSO_4 \cdot 7H_2O$ into three-necked flasks. Then, the mixed $NaBH_4$ and $Na_2S_2O_4$ solutions were slowly poured into $FeSO_4 \cdot 7H_2O$ solutions, constantly stirring, where the ratio of BH_4^-/Fe^{2+} was 2.2. After reduction, it continues stirring for 30 minutes. The entire process referred to above was carried out under the N_2 atmosphere. The resulting SNZVI suspensions were extracted through vacuum filtration, and the solid was collected and cleaned with deoxygenated ultrapure water three times. The as-prepared S-nZVI particles were vacuum-dried before use.

2.3 Batch experiments

S-nZVI was evaluated using batch-type experiments of 2,4-DCP degradation at different sulfur iron mole ratios (0.026, 0.064, 0.129, 0.257), pH values (3, 5, 7, 9, 11), initial concentration of 2,4-DCP (10, 20, 30, 40 mg/L) and reaction temperature (15, 25, 30, 40℃). 0.4 g S-nZVI was dispersed in 200 mL 2,4-DCP solutions(20mg/L), located in a constant temperature shaking table (200 rpm) at 25±1 ℃. At a given time interval, a 5ml sample was extracted with an injection syringe and filtered by a 0.45 um filter membrane before use. All experiments were carried out in triplicate. The concentration of 2,4-DCP was analyzed by high-performance liquid chromatography (HPLC, LC-2030) equipped with a C18 column, and the UV detector wavelength was 284 nm.

3 Results & Discussion

3.1 Effect of Sulfur mole ratio

Fig 1(a) depicts the removal efficiency at different sulfur molar ratios. The results showed that when the sulfur molar ratio was 0.129, the highest removal rate was 91.9%. However, the removal rate decreased with the sulfur molar ratio increased to 0.257. Table (1) displays the pseudo-first-order rate constant for 2,4-DCP removal versus S-nZVI sulfur molar ratio. At the low sulfur molar ratio, as the sulfur molar ratio increase to 0.129, the K_{obs} is up to 0.02339 h^{-1}, which indicates that the removal rate continues to accelerate. When the sulfur molar ratio is more than 0.257, the K_{obs} decrease when the sulfur molar ratio is low because the surface of nZVI is not completely coated by the sulfide layer. Exposed nZVI could directly react with 2, 4-DCP. When the sulfur molar ratio was 0.064, the surface of nZVI was entirely covered with sulfide layer; the sulfur content was too low for removal efficiency to slow. With the increase of sulfur content, the removal efficiency and reaction duration are the best. The removal efficiency decreases when the sulfur molar ratio is more than 0.129. This is because too much FeS and FeS_2 on the surface of nZVI could plug the active surface sites, thus decreasing the results. Furthermore, the higher bandgap (0.95eV) could slow down electron transfer[12].

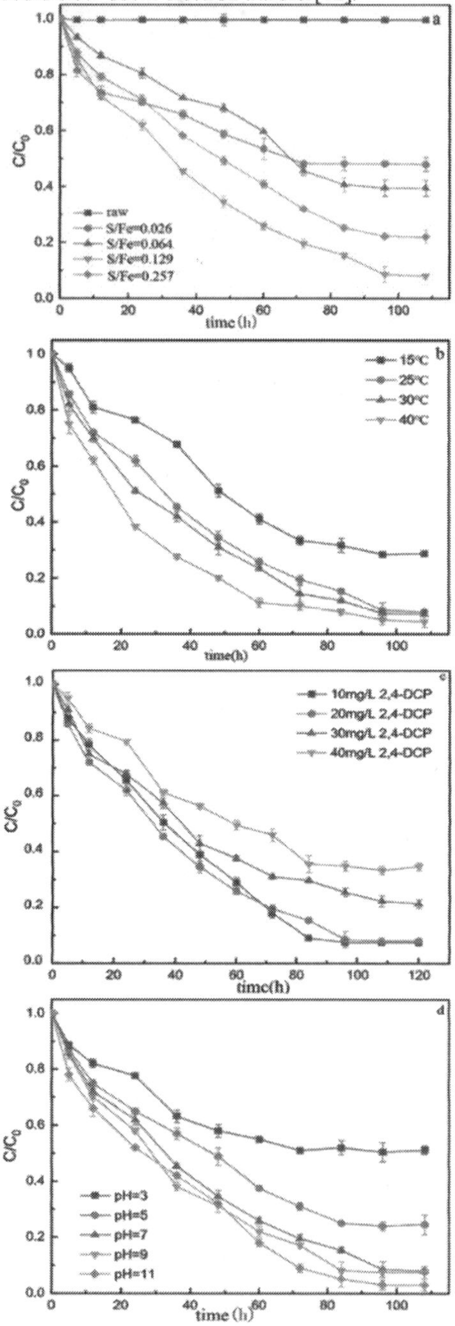

Fig.1 (a) The removal effect of 2,4-DCP at different sulfur molar ratios, **(b)** The removal effect of 2,4-DCP at different reaction temperatures, **(c)** The removal rate of 2,4-DCP at different initial concentrations, **(d)** The removal effect of 2,4-DCP at different initial pH.

3.2 Effect of reaction temperatures

As shown in fig1 (b), the results indicated that the reaction rate remarkably increased with the reaction temperatures build-up. The degradation efficiencies were only up to 71.2% when initial reaction temperatures of 15 ℃. With the reaction temperatures ranging from 15 to 40 ℃, the removal efficiencies are up to 95.5%. As shown in Table 2, the observed rate constants were 0.01627–0.03217 h^{-1}, respectively, at reaction temperatures 15-40 ℃. This effect was ascribed to the fact the number of activated molecules increases with the reaction temperatures increase, the mass transfer velocity of 2,4-DCP increases, thus accelerating the reaction process. However, low temperature inhibits the reaction process.

Table 1. The fitting parameters at different ferrite molar ratios

ferrite molar ratios	reaction rate constant k_{obs} (h^{-1})	R^2
0.026	0.0937	0.94553
0.064	0.0986	0.98984
0.129	0.2339	0.99487
0.257	0.1568	0.99774

Table 2. The fitting parameters at different reaction temperatures

reaction temperatures	reaction rate constant k_{obs} (h^{-1})	R^2
15	0.01627	0.98157
25	0.02339	0.99487
30	0.02591	0.99321
40	0.03217	0.97928

3.3 Effect of the initial concentration of 2,4-DCP

Fig 1(c) studied the relationship between the initial concentration of 2,4-DCP and removal efficiency that with the 2,4-DCP concentration in the initial solution of 10, 20, 30, and 40 mg L^{-1}, the removal efficiency of 2,4-DCP were 92.5%,91.9%, 78.6%, 65.3%, respectively. With an increase in the initial concentration of 2,4-DCP, the equilibrium time is prolonged. As shown in table 3, k_{obs} decreased with an initial concentration of pollutants. In theory, a high concentration of 2,4-DCP can increase the probability of practical collision between 2,4-DCP and nanoparticles. However, the dosage of S-nZVI was constant when it increased the concentration, and the reactive sites were limited, decreasing the degradation rate.

Table 3. The fitting parameters at different original concentrations of 2,4-DCP

initial concentrations of 2,4-DCP	reaction rate constant k_{obs} (h^{-1})	R^2
10	0.02493	0.97878
20	0.02339	0.99487
30	0.0153	0.99243
40	0.01153	0.99447

3.4 Effect of initial pH

Fig2 showed the zeta potentials of nZVI and S-nZVI at various pH values. The zeta potentials value of nZVI and S-nZVI first decreases and increases as the pH increases. The isoelectric point of nZVI(pH_{ZPC}=6.40) is less than S-nZVI (pH_{ZPC}=7.23), which also confirmed that sulfur could change the surface properties of nZVI. Furthermore, S-nZVI is more stable at high pH combined with Zeta potential measurement results. As shown in Fig 1 (d), the pH value significantly affects the elimination process of 2,4-DCP by S-nZVI. The removal efficiencies were reduced from 75.5% to 48.8% when the initial pH ranged from 5 to 3. At low pH, the corrosion rate of Fe^0 and a large amount of hydrogen are produced to decrease the removal rate because of the dissolution of the sulfide layer.

On the contrary, the degradation efficiencies were up to 97.9% on the alkaline condition at an initial pH of 11. This result is consistent with previously reported sulfided nanoscale zero-valent iron[13]. Table 4 indicated that the pseudo-first-order model had a good correlation(R^2=0.95). Alkaline conditions are favourable for the reaction, and removal efficiency for S-nZVI is highest (97.9 %) at pH = 11.

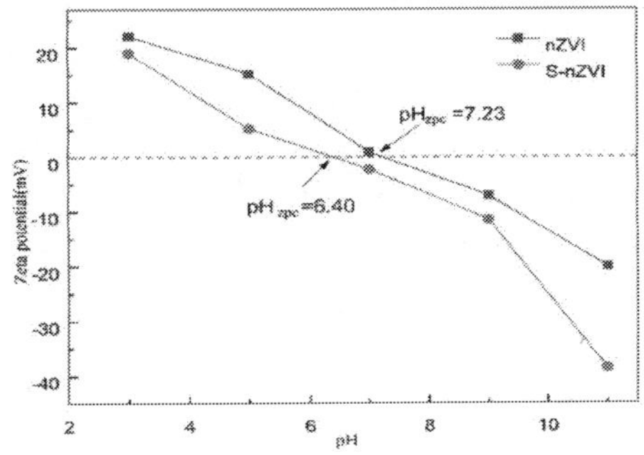

Fig.2 The Zeta potential plots of nZVI and S-nZVI

Table 4. The fitting parameters at different different initial pH

initial concentrations of 2,4-DCP	reaction rate constant k_{obs} (h^{-1})	R^2
3	0.00873	0.95661
5	0.01584	0.98657
7	0.02339	0.99487
9	0.02676	0.98323
11	0.03291	0.95921

4 Conclusions

This study investigated the effect of diverse experimental conditions for removing 2,4-DCP. The initial pH and sulfur mole ratio played a significant role, which determined the removal rate of 2,4-DCP. The findings of the experiment can guide more efficient S-nZVI reactivity towards the objective contaminants in water remediation.

Acknowledgements

This paper was funded by China's National key research and development program (No.2016YFC0401404).

References

1. L. Gan, B. Li, M. Guo, X. Weng, T. Wang, Z. Chen, Mechanism for removing 2,4-dichlorophenol via adsorption and Fenton-like oxidation using iron-based nanoparticles. J. Chemosphere, **206** 168-174. (2018)
2. W. Junjie, F. Xin, L. Yu, H. Jinqiang, Z. Na, L. Zhifeng, L. Yajing, Y. Chuangxin, L. Wanqi, Effect of mesoporous silica molecular sieve coating on nZVI for 2,4-DCP degradation: morphology and mechanism during the reaction. J. Chem. Eng. Process., **135** 68-81.(2018)
3. N. Yan, M. An, J. Chu, L. Cao, G. Zhu, W. Wu, L. Wang, Y. Zhang, B.E. Rittmann, More rapid dechlorination of 2,4-dichlorophenol using acclimated bacteria. J.Bioresour. Technol, 2021. **326** p. 124738. (2021)
4. C. Wu, L. Zhou, Y. Zhou, C. Zhou, S. Xia, B.E. Rittmann. Dechlorination of 2,4-dichlorophenol in a hydrogen-based membrane palladium-film reactor: Performance, mechanisms, and model development. J.Water Res. **188** 116465. (2021)
5. K. Hye-Jin, L. Megan, N. Bhanuphong, D.T. Robert, V.L. Gregory, Effect of emplaced nZVI mass and groundwater velocity on PCE dechlorination and hydrogen evolution in water-saturated sand. J. Hazard. Mater. J, **322** 136-144. (2017)
6. Y. Shih, K. Hsia, C. Chen, C. Chen, C. Dong. Characteristics of trichloroethene (TCE) dechlorination in seawater over a granulated zero-valent iron. J. Chemosphere, **216** 40-47 (2019)
7. X. Wang, S. Cong, P. Wang, J. Ma, H. Liu, P. Ning. Novel green micelles Pluronic F-127 coating performance on nano zero-valent iron: Enhanced reactivity and innovative kinetics. J.Sep. Purif. Technol, **174** 174-182. (2017)
8. L. Jing, F. Mingjie, L. Miao, L. Xiang, Cr (VI) removal from groundwater using double surfactant-modified nanoscale zero-valent iron (nZVI): Effects of materials in different status. J.Sci. Total Environ., **717** 137112. (2020)
9. E.J. Kim, J.H. Kim, A.M. Azad, Y.S. Chang. Facile Synthesis and Characterization of Fe/FeS Nanoparticles for Environmental Applications. J.ACS Appl Mater Interfaces, 2011. 5: 1457-62. (2011)
10. Y. Su, A.S. Adeleye, A.A. Keller, Y. Huang, C. Dai, X Zhou, Y Zhang. Magnetic sulfide-modified nanoscale zero-valent iron (S-nZVI) for dissolved metal ion removal. J.Water Res. 74: 47-57. (2015)
11. J. Li, X. Zhang, Y. Sun, L. Liang, B. Pan, W. Zhang, X. Guan, Advances in Sulfidation of Zerovalent Iron for Water Decontamination. J. Environ. Sci. Technol., **51** 13533-13544 (2017)
12. Z. Cao, X. Liu, J. Xu, J. Zhang, Y. Yang, J. Zhou, X. Xu, G.V. Lowr. Removal of Antibiotic Florfenicol by Sulfide-Modified Nanoscale zero-valent Iron. J. Environ. Sci. Technol, **51** 11269-11277. (2017)
13. E.J. Kim, J.H. Kim, A.M. Azad, Y.S. Chang. Remediation of Trichloroethylene by FeS-Coated Iron Nanoparticles in Simulated and Real Groundwater: Effects of Water Chemistry. J. Ind. Eng. Chem. Res. **52** 9343-9350. (2011)

Ae-Based Failure Warning Mechanism During Cemented Paste Backfilling: Experimental Investigation

Chunde Ma[1,2], *Zihe* Wang[1*], *Jiaqing* Xu[1], *Guanshuang* Tan[1], and *Zhihai* Lv[1]

[1]School of Resources and Safety Engineering, Central South University, Changsha 410083, Hunan, China
[2]Center for Advanced Studies, Central South University, Changsha 410083, Hunan, China

Abstract. To research the AE warning mechanism of cemented paste backfilling (CPB), the uniaxial compression test with AE monitoring was carried out on CPB. The effects of different moisture content, cement proportion, and loading rate on AE characteristics of CPB were analyzed. And a damage model based on AE is established. The results show that the increase of water content, gypsum content, and loading rate can enhance the ductility of CPB and lengthen the AE warning area. However, the increase of moisture content will lead to the delay of AE signal peak, which is bad for the early warning. With the increase of loading rate, the failure mode of specimens gradually changes from splitting failure to shear failure. The cracks produced under different failure modes can correspond to the peaks of AE signals. The damage model based on AE shows that the damage process of CPB samples can be divided into three stages, linear-exponential-linear. The experimental results can further research the optimal conditions for ae monitoring of backfill, and provide a theoretical basis for ae monitoring and warning of backfill.

1 Introduction

In China, a large amount of waste rock generated during underground mining accumulates on the surface, which not only occupies land space but also releases harmful elements into the air and water [1]. Filling mining technology is a burgeoning mining technology which uses solid offal such as tailings, construction waste, and abandoned rocks to carry out crushing, screen grading, and then cement and fill mined-out areas [2-4]. It can be used to reduce ore dilution and decrease mining cycles to improve production efficiency [5-6], which brings great benefits to the economy, safety, and environmental protection. In the last few decades, different filling mining techniques had been developed and applied to mine goaf management [7-8]. However, it was found in the engineering practice that the filling mining technology mentioned above has some disadvantages, such as high cost and insufficient utilization of the wastes in the mining process. As a consequence, cementing paste filling has been widely used because of its ability to reduce the cost of waste rock storage and treatment and its environmentally friendly feature [9-10].

As a porous medium consisting mainly of cementing materials, waste rock aggregate, and pores [11], the AE characteristic of CPB is closely related to the choice and the ratio of cementing material, and the aggregate particle size grading, etc. [12-14]. Many types of researches indicated that the loading rate has a significant effect on the properties of materials. Different materials such as rock, concrete, and backfill exhibit different behaviors under different loading rates [15-17].

AE monitoring technology, as one of the important development directions of safety detection and non-contact prediction methods, can detect individual fracture events and provide a basis for the mechanical property analysis of solid materials [18-20]. Many scholars have applied AE technology to study the mechanical properties of solid materials [21-22]. Establishing reasonable failure criteria and using AE signal characteristics as the failure criterion of materials in nondestructive testing is of great importance for the research of the AE characteristics of CPB samples. However, the relationship between the AE characteristics and failure modes of CPB samples and the relationship between AE signals and crack failure surface have been barely researched by scholars.

In this paper, a uniaxial compression test on cemented red sandstone was carried out. Combined with AE detection data, the effects of different moisture content, loading rates, and the ratio of cementing materials on AE characteristic of CPB were discussed. The failure modes of CPB samples under different loading rates and the corresponding relationship between the AE ringing counts and the crack failure surface was revealed. The experimental results have important guiding significance for monitoring AE signals of CPB failure signs and analyzing failure mechanism of CPB in the process of filling mining.

* Corresponding author: csuwangzihe@163.com

2 Materials and Experimental Equipment

2.1 Experimental Materials

2.1.1 The aggregate of CPB samples

As a kind of common solid waste material produced by coal mining, red sandstone can be used as the aggregate of CPB. In this experiment, red sandstone with a natural density of 2.48g/cm^3 and an average strength of 45MPa was used as the material for CPB sample preparation. First, complete red sandstone blocks were crushed into small pieces, followed by further mechanical crushing of the larger ones. Then the grading screen was used to classify the particles, and the gravel particles with a diameter range of 0~2mm, 2~5mm, 5~8mm, 8~10mm, and 10~12mm were uniformly selected as the test samples. According to Talbot Index theory:

$$\frac{M_i}{M}=\left(\frac{d_i}{d_m}\right)^n \quad (1)$$

where n is the Talbot index, Mi is the mass of aggregate whose diameter is less than d_i, and M is the total mass of aggregate. D_m is the maximum diameter of red sandstone material particles, which is 12mm in this experiment. In this paper, a gradation scheme with a Talbot index of 0.5 was used for sample preparation. According to Formula 1, the mass and percentage of each particle size are shown in Table 1.

Table 1. The particle size distribution of the sample

Weight of each particle size (g)				
0~2mm	2~5mm	5~8mm	8~10mm	10~12mm
114.24	66.36	47.88	27.16	24.36

As one of the important factors studied in this paper, cementing material has a great influence on the stress-strain properties and AE characteristics of the CPB sample. In the preparation of the cementing sample, cement and gypsum are often mixed in different proportions as the cementing material of the sample In this paper, 42.5 grade Portland cement and hemihydrate gypsum are used as cementing materials.

2.1.2 Preparation of Specimens

In this paper, the total mass of the cementing materials was set as 100g, and CPB samples were prepared with three ratios of cement to gypsum: 3:1,1:0 and 1:1. Homogeneous slurry consisting of water and cementing materials was adequately mixed and stirred with crushing red sandstone particles. Poured the mixture into a cylindrical mold with a diameter of 50mm and a height of 100mm, and vibrated the bubbles out of the slurry. After initial solidification, the samples were removed from the mold and cured for 28 days in a constant temperature curing box. The two ends were smoothed out after being removed from the curing box, and the test samples were obtained.

As for the samples with different moisture content, the DHG-9035A drying oven was used to dry the samples with a cement/gypsum ratio of 3:1 at 105°C for 24 hours to get the drying samples. After taking out from the drying box, the samples were carried out sealing preservation immediately to ensure the moisture content as the lowest state. The saturated samples were obtained after 24 hours soaking. The moisture content in dry state, natural state and saturated state were measured as 0%, 3.01% and 10.20% respectively. Some samples are shown in Fig. 1.

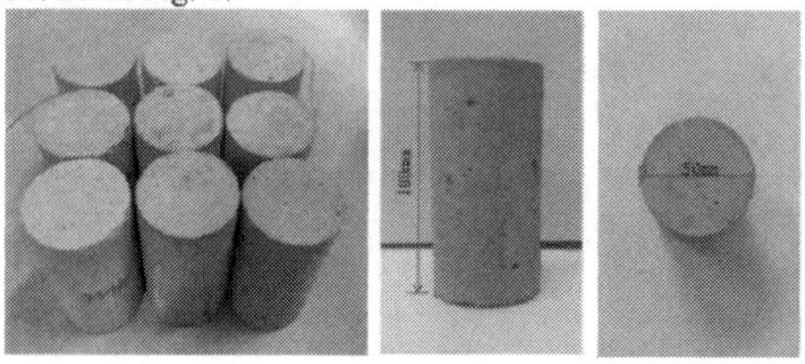

Fig. 1. CPB samples

2.2 Experimental Equipment

WHY-200/10 microcomputer-controlled automatic experimental machine as shown in Fig.2(a) was used for uniaxial compression test on CPB, whose load accuracy can be controlled within 1%, displacement accuracy can be controlled within 0.5%, and the maximum load range is 300kN.

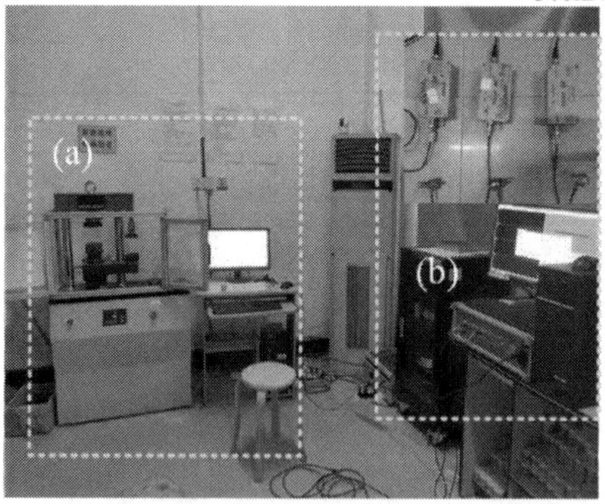

Fig. 2. The loading and AE monitoring integrated acquisition system, (a) WHY-200/10 microcomputer-controlled automatic experimental machine (b) AE monitoring system

The AE monitoring system is shown in Fig.2(b). Set the sampling frequency as 2.5MHz and threshold value as 40dB.The AE sensors were attached equally spaced to the sample. Vaseline was applied between the probe and the sample as a coupling agent to eliminate the influence of head load on AE signals. AE signals were transmitted into the AE acquisition system through the preamplifier, and AE data is formed after noise reduction and screening by the instrument.

2.3 Experimental Methods

CPB test schemes with different cementing material ratios, water content and loading rate were set to study the effects of cementing materials, moisture content and loading rate on the mechanical properties and AE characteristics of CPB samples. The specific parameters of the test scheme are presented in Table 2.

Table 2. Experimental scheme of uniaxial compression

Sample	Mass of cementing material (g)		Loading rate	Moisture content
	Cement	Gypsum	(N/s)	(%)
CPB-1				0
CPB-2	75	25	150	3.01
CPB-3				10.20
CPB-4	100	0		
CPB-5	75	25	150	3.01
CPB-6	50	50		
CPB-7			100	
CPB-8	75	25	150	3.01
CPB-9			200	

Certain discrete factors in the mechanical properties of CPB have resulted in the manufacturing operation. To ensure the universality of the test results, three groups of repeated tests were carried out for each sample, and the mechanical parameters were averaged.

3 Results and discussion

3.1 AE characteristic of CPB

3.1.1 AE ringing count characteristics of CPB

It is generally accepted that parameters such as AE ringing count are generated by crack propagation and rock failure. The information on crack propagation in the process of specimen compression can be obtained by AE monitoring. However, there are some problems in the process of tests, such as coupling of sensor and sample, a slight difference of frequency response of the different sensor, and signal attenuation caused by different propagation distance. By comparing the signals received by the sensors with equidistant Settings, AE ringing counts were collected and obtained. After calibration, the time-load-AE ringing count diagrams are shown in Fig.3-5. The results show that during the loading process, the AE ringing count will go through three stages:

(I) Quiet period. At the initial loading stage, tiny cracks and elastic deformation appeared in the sample. AE ringing count is rarely generated, which is in the quiet period of acoustic emission.

(II) Rapid development stage，after the elastic stage, plastic deformation occurs on the samples, cracks generate and expand rapidly until failure. At this time, AE signals appear frequently, the ringing count increases sharply, and the peak value of the AE ringing count appears.

(III) Stable period, in the post-peak phase, the AE ringing count less than the peak value is continuously generated.

3.1.2 Effect of moisture content on AE characteristic

As can be seen from Fig.3, with the increase of water content, the level of AE ring-counts decreases rapidly, but the time range of AE ring-counts increases greatly.

The AE ringing counts generated at yield phase of samples in dry condition were mostly above 500, while the average number of samples in natural condition was around 80 and of samples in saturation condition was between 0 and 50. This was mainly due to the fact that the internal cementation structure of the specimen was damaged after being immersed in water. The sample was softened by water and showed a creeping tendency, and the severity of the internal damage has lessened.

The relative warning area of the sample, which is the ratio between the rapid development period and the whole loading process, obviously increased. In the dry state, no obvious AE ringing counts appear until about 110s, while in the natural state and saturated state samples, obvious AE ringing counts began to appear around the 70s and 20s.

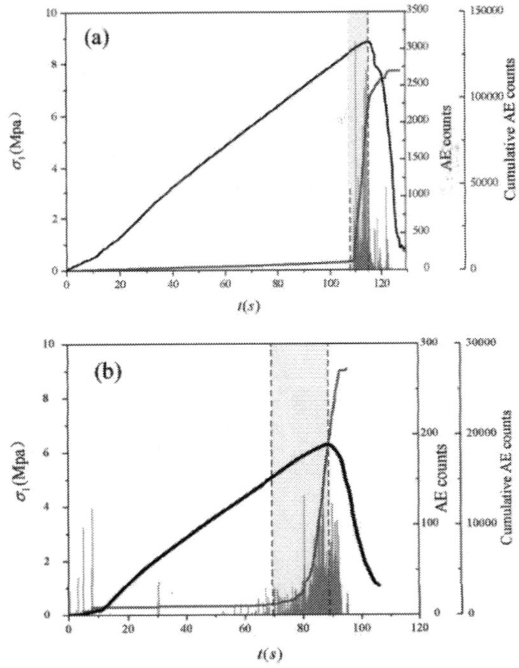

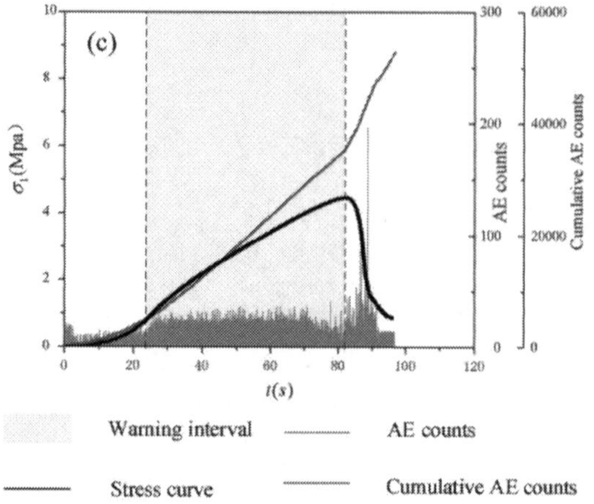

Fig. 3. Evolution of accumulative count of CPB samples with different moisture contents: (a)dry state (b) natural state (c) saturated state

From the time of the AE signal distribution, samples in dry condition had been in a long AE quiet period since the elastic stage. The cracks expanded rapidly at the approach of failure. At yield phase, high-level AE signals suddenly appeared, followed by a peak of the ringing count. The distribution of AE ringing counts was concentrated, and the sample had a tendency of sudden failure. The AE signals of samples in the natural state were more concentrated in the late elastic stage and in yield phase. There were only a few AE signals generated by fracture compaction in the compaction stage, and the maximum value of the AE ringing count appeared near the peak. In saturated samples, new micro-cracks were always generated and expanded after the crack compaction process, and AE signals were ubiquitous in the whole loading process. The maximum value of the AE ringing count appeared after the peak, and there were several similar peaks of the AE ringing count near the maximum point. No sudden increase in AE ringing counts occurred, and the slope of the cumulative ringing count curve had little change.

According to the above, moisture content has a great influence on the type of rock damage. With the increase of moisture content, the CPB specimen experienced significant softening and strength reduction, and the failure type changed from brittle failure to ductile failure. The ability of samples to absorb elastic energy is weakened, releasing energy gradually. Meanwhile, with the increase of moisture content, the peak value of AE ringing counts lags obviously, which greatly weakens the predictive effect of acoustic signals on backfill failure.

3.1.3 Effect of cementing material on AE characteristic

The AE characteristics of CPB samples with different ratios of cementing materials can be seen in Fig.4. With the increase of the proportion of gypsum, the AE ring-counts produced by the development of cracks appeared earlier, and the cumulative AE ring-counts increased more slowly before the peak, the relative warning area becomes significantly longer.

For samples made entirely of cement, the AE ring count became higher and more intensive around the 270s, and the sample reached the peak strength after about 10s. For the sample cement and gypsum ratio of 1:1, the AE ringing counts started to increase significantly at about 72s, and the sample reached the peak strength after about 15s. However, for the sample with cement and gypsum ratio of 1:3, the AE ringing counts began to increase at about 48s, and the sample reached the peak strength after about 18s.

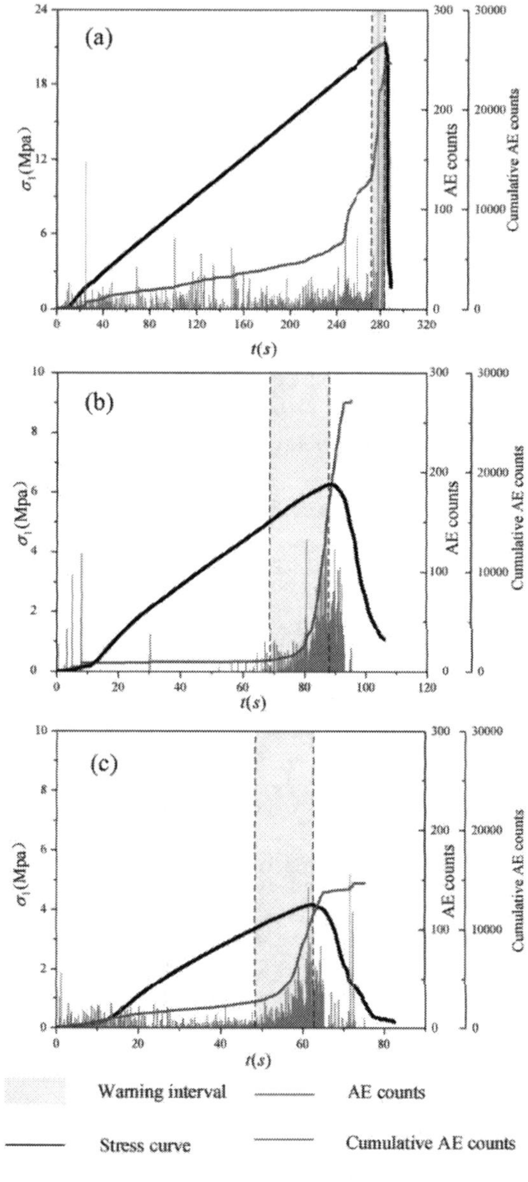

Fig. 4. Evolution of accumulative count of CPB samples with different cementing material ratio: (a)C/G=1:0 (b) C/G=3:1 (c) C/G=1:1

The samples with a cement/gypsum ratio of 1:0 presented higher AE ringing counts than those with cement/gypsum ratios of 3:1 and 1:1 in the compaction and elastic stage, and the damage was more severe during loading. The yield phase of the stress-strain curve was relatively short, and brittle failure of the sample occurred without obvious premonition. However, in the samples that used gypsum to replace cement with equal mass, the AE ringing counts in the compaction and elastic stages were at a low level. The strength of the weak bond inside the sample was significantly lower than that of the sample with a cement/gypsum ratio of 1:0. There was an obvious transition period of AE ringing counts in yield phase, and the crack propagation before failure had an obvious omen.

It indicated that gypsum can increase the ductility of samples, making it possible to predict the occurrence of failure by picking up AE disaster signals. In engineering applications, how to set the proportion of gypsum added to make CPB samples meet both strength requirements and economic needs has become a focus of attention.

3.1.4 Effect of loading rate on AE characteristic

Fig.5 shows the CPB AE ring counting curve at different loading rates. The maximum level of AE ring count has not changed significantly and remains at about 100. With the increase of loading rate, the relative warning area becomes longer, and the peak value of AE ringing count lags obviously relative to the peak intensity. The peak value of the AE ringing count of the sample of 100N/s occurred before the peak intensity, and the peak value of the AE ringing count of the sample of 150N/s occurred from before to after the peak intensity, while the peak value of the AE ringing count of the sample of 200N/s appeared 10s after the peak intensity.

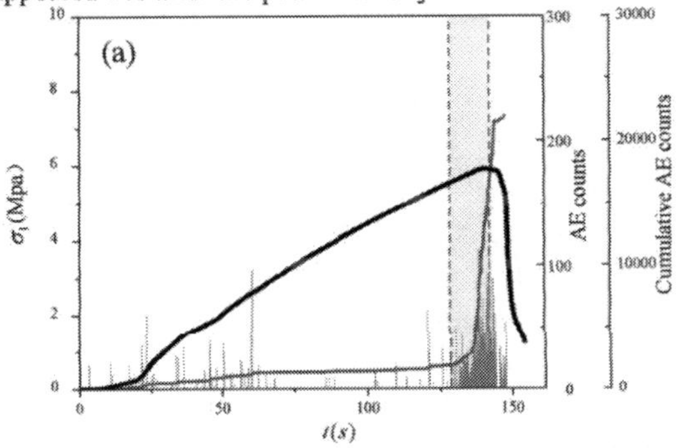

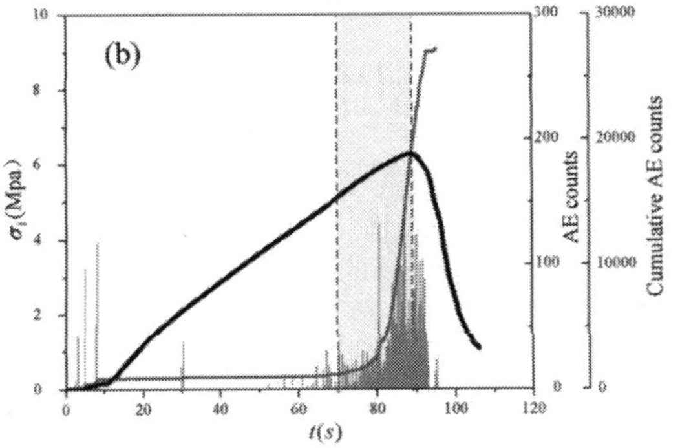

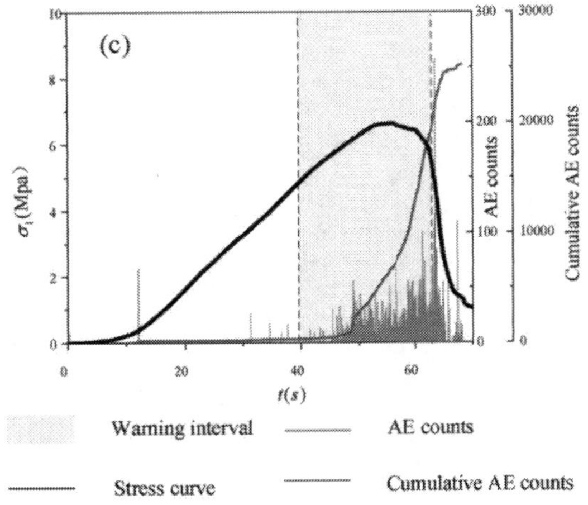

Fig. 5. Evolution of accumulative count of CPB samples with different loading rates: (a) 100N/s (b) 150N/s (c) 200N/s

The loading rate has a great influence on the AE characteristics of CPB samples, as is shown in Fig.6. In the early stage of loading, the properties of samples with different loading rates were not very different. After the yield stage, CPB samples with different loading rates showed different AE characteristics. Under the loading rate of 100N/s, yield phase of the stress-strain curve of samples was very short, and the failure mode was inclined to brittle failure. The cumulative AE ringing counts increased within a short time before reaching the peak strength, and then the ringing counts decreased rapidly. Several ringing counts peaks appeared around the failure of the samples. Yield phase of the total stress-strain curve of the sample with a loading rate of 150N/s became longer, low-level AE ringing counts frequently occurred before reaching the peak strength. The AE ringing counts without fluctuating significantly indicated that there were small cracks but no main failure surface generated inside samples. When the peak strength was reached, the peak value of AE ringing counts appeared with the main failure surface generated. For the samples with a loading rate of 200N/s, the failure mode changed from brittle failure to ductile failure. AE ringing counts appeared continuously and frequently in a long stage with multiple peaks formed. After the peak strength, the strength of the CPB sample did not decrease rapidly, some deformation still occurred under the condition of maintaining high-stress level, and a large number of AE signals could still be acquired. Subsequently, the maximum AE ringing count appeared, and the stress dropped rapidly, indicating that the sample was destroyed and lost its strength.

The concentrated generation of AE ringing counts in a certain stage can often indicate the process of a failure surface from crack propagation to crack destruction, which is represented as a peak and valley in the AE ringing count diagram. The level of the AE ringing count tended to rise, and dropped rapidly after the occurrence of a maximum, which means that a stress surface cracked and failed under the function of stress, and the stress distribution inside the sample changed.

Fig.6 shows the partially enlarged drawing of AE ringing counts and the failure surface distribution during sample failure. The sample with a loading rate of 100N/s showed splitting failure, with only one major failure surface corresponding to a peak in the AE ringing count diagram. And the rest were fine minor split cracks. The sample with a loading rate of 150N/s was characterized by both splitting failure and shear failure. There were some local shear cracks and two through cracks on the sample, one of which had the characteristics of shear failure. The characteristics of shear failure of the sample with a loading rate of 200N/s became more apparent. Except for a through-crack, a shear surface was inclined to segment the sample, and several cracks intersected the shear surface, forming several failure surfaces with local coincidence. It can be clearly seen from Fig.7(c) that the AE ringing counts of samples with a high loading rate presents multiple discontinuous dense areas with multiple peaks, which corresponds to the formation of multiple local shear cracks and major failure surfaces. Combined with the figures, it can be verified that in the sample with a loading rate of 200N/s, local failure occurred first, resulting in a local shear surface that affected the strength of the sample but did not cause complete failure of the sample. With the subsequent loading, the local shear surfaces were expanded and merged into a penetrating shear plane, and the sample was destroyed.

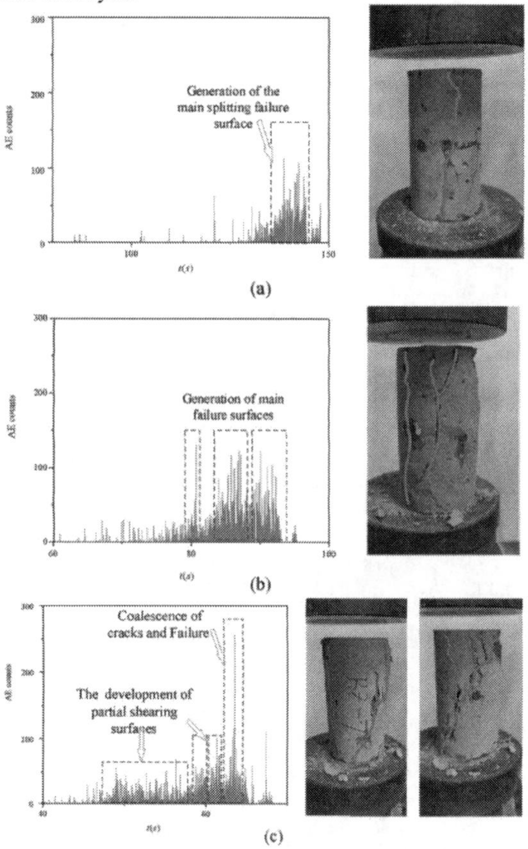

Fig. 6. Local magnified view of AE ringing count and corresponding failure surface distribution during sample failure: (a) 100N/s (b) 150N/s (c) 200N/s

3.2 Damage model based on AE ringing count

The AE signal generated during the loading process of the CPB sample can represent the expansion, connection, and coalescence of cracks inside the sample, which will cause internal damage of rock, resulting in rock deterioration and failure. Compared with AE energy and impact, AE ringing count can reflect the development of microscopic cracks better [44]. In this paper, AE ringing counts were used to represent the internal damage evolution law of CPB samples during the loading process, which can accurately warn the failure and instability of CPB through non-destructive monitoring.

The damage variable D is introduced

$$D = \frac{A_d}{A} \tag{2}$$

where A_d is the total area of micro defects in the compression section, and A is the total area of the compression surface under initial nondestructive conditions.

The AE ringing count C_w when unit area element damaged is

$$C_w = \frac{C_0}{A} \tag{3}$$

where C0 is the cumulative AE ring count when the whole section A of nondestructive material is completely destroyed.

At this point, the damage variable D when the damaged area reaches Ad can be expressed as

$$D = \frac{C_d}{C_0} \tag{4}$$

where Cd is the cumulative AE ringing count when the damaged area reaches A_d.

However, during the experiment, when the sample reached the peak strength, the sample section did not completely fail. Therefore, the damage critical value Du was introduced to modify the damage variable as

$$D = D_u \cdot \frac{C_d}{C_0} \tag{5}$$

For the sake of calculation, Du is usually evaluated as

$$D_u = 1 - \frac{\sigma_p}{\sigma_c} \tag{6}$$

where σ_p is the residual strength of the sample, and σ_c is the peak strength of the sample.

Therefore, the relationship between sample damage and time-based AE ringing count is shown in Fig.8. It can be seen from the figure that the time-damage curve has obvious characteristics of periodic changes, showing a linear - exponential - linear change trend basically. (I) In the initial stage, corresponding to the crack compaction stage in the specimen compression process, the damage value is close to 0. (II) In the stable development stage, corresponding to the elastic deformation stage, the damage shows a linear change trend. (III) In the rapid development stage, corresponding to the failure stage of the sample, the damage value presents an exponential upward trend. (IV) In the later stage of failure, the damage value changes linearly again and gradually approaches the maximum damage value D_{max} at the corresponding strain softening and residual deformation stages.

By piecewise function analysis of time-damage curve, the fitting relationship between damage value D of CPB sample and loading time T under different conditions can be obtained as follows:

$$D = \begin{pmatrix} c_1 t + c_2, & t < t_a \\ c_3 \exp(-t / c_4) + c_5, t_a < t < t_b \\ c_6 t + c_7, & t_b < t < t_c \end{pmatrix} \tag{7}$$

Table 3. Piecewise fitting parameters

	c_1	c_2	c_3	c_4	c_5	c_6	c_7	t_a	t_b	R^2
CPB-1	2.35E-4	0.00126	-3.87E24	1.90201	0.52684	0.05051	-5.23443	109	111.87	0.959
CPB-2.5.8	3.81E-4	0.01001	1.51E-11	-3.63176	0.03396	0.05311	-4.12946	65.48	88.09	0.995
CPB-3	0.0074	-0.0855	3.87E-6	-8.10168	0.4299	0.01408	-0.55788	77.11	89.47	0.996
CPB-4	9.36E-4	-0.0060	-2.83E7	13.092	0.4679	0.03626	-9.3766	243	270.67	0.948
CPB-6	0.0029	0.01966	0.00122	-9.64423	-0.0931	3.50E-4	0.90096	51.91	64.88	0.993
CPB-7	6.81E-4	-0.00259	0.00134	-19.6533	-1.2666	0.01256	-1.03367	135.96	144	0.995
CPB-9	7.21E-4	-0.0106	1.11E-4	-7.2733	-0.03223	0.0118	-0.02618	45	64.26	0.997

After piecewise fitting with Origin, the fitting results and parameters are shown in Table 3

When the time-damage curve reached the second stage, a large amount of AE signals began to appear and the damage value began to rise rapidly. Therefore, the damage value reaching the second stage can be judged as the precursor of peak failure of the sample, the phase from ta to tb in the curve can be regarded as a relative warning area [23].

4 Conclusion

In this paper, a uniaxial compression test was carried out on the CPB sample with a Talbot index of 0.5, and the influences of different loading rates, moisture content, and cementing material ratio on the AE characteristic of CPB were analyzed. The failure mechanism of CPB samples were discussed based on their AE characteristics. The conclusions are as follows.

(I) The ductility of CPB samples can be changed by loading rate, moisture content, and ratio of cementing materials. With the increase of loading rate, moisture content, and gypsum quality in the cementing material, the ductility of CPB increases, while the strength decreases. The corresponding AE signal is the warning area before the failure occurs, which shows that the AE ringing times gradually increase, and the cumulative AE ringing times curve begins to show an obvious upward trend. Some AE signals are generated after the peak strength.

(II) CPB sample without gypsum has higher strength and hardness, and sudden and severe damage occurs when its strength reaches a peak value. And there is no obvious tendency for the AE ringing counts to increase before failure. The addition of gypsum will reduce the strength and peak strain of the CPB sample, but it can increase the ductility of the CPB sample and significantly increase the AE ringing counts level before the failure. Therefore, CPB can be an early warning by adjustment of cement ratio, and the failure of backfill can be dealt timely with AE signal monitoring.

(III) The moisture content will change the mechanical properties and AE signal distribution of the CPB sample. With the increase of moisture content, the strength of CPB samples and the average level of AE ringing counts decrease, meanwhile, AE ringing counts are more widely distributed, and the peak value of AE ringing counts will lag. Therefore, during the filling process, the moisture content of the filling area should be controlled to ensure the strength of CPB and the timeliness of AE monitoring.

(IV) The failure mode of CPB varies with the loading rate. With the increase of loading rate, the splitting failure of single crack gradually changes to shear failure of multiple cracks

(V) The damage model based on AE indicates that the damage value of the sample presents a linear-exponential - linear change, in which the stage of exponential change is a relative warning area, which can be used as a basis for judging the impending failure of CPB in AE monitoring.

This work was funded by the National Natural Science Foundation of China (52074352).

References

1. Z. Bian, X. Miao, S. Lei, et al. The Challenges of Reusing Mining and Mineral-Processing Wastes. Science, **337(6095)** 702-703, (2012)
2. X.X. Miao, J.X. Zhang, M.M. Feng. Waste-filling in fully-mechanized coal mining and its application. Journal of China University of Mining and Technology, **18(4)** 479-482, (2008)
3. S. Cao, W. Song. Effect of filling area time on the mechanical strength and ultrasonic properties of

cemented coarse tailing backfill. International Journal of Mineral Processing, **166**(62-68), (2017)

4. Z. Lu, M. Cai. Disposal Methods on Solid Wastes from Mines in Transition from Open-Pit to Underground Mining. Procedia Environmental Sciences, **16**(715-721), (2012)
5. D.R. Tesarik, J B Seymour, T R Yanske. Long-term stability of a backfilled room-and-pillar test section at the Buick Mine, Missouri, USA. International Journal of Rock Mechanics and Mining Sciences, **46(7)**: 1182-1196, (2009)
6. M.L. Bloss. An operational perspective of mine backfill//Y Potvin, T Grice. Eleventh International Symposium on Mining with Backfill. Perth; Australian Centre for Geomechanics. 15-30 (2014)
7. V.I. Cheskidov, A.V. Reznik. Features of Hydraulic Fill Formation in Mining Water-Bearing Lignite Deposit. Journal of Mining Science, **55(2)**: 273-279, (2019)
8. W.Y. Lv, Z.H. Zhang. Research on the Belt Loader Waste Rock Gravity Backfill Mining Technology. Applied Mechanics and Materials, **253-255**(253-255), (2012)
9. X. Li, J. Du, L. Gao, et al. Immobilization of phosphogypsum for cemented paste backfill and its environmental effect. Journal of Cleaner Production, **156**(137-146), (2017)
10. L. Chen, X. Xu, J. Wu, et al. Characteristics Variation of Tailings Using Cemented Paste Backfill Technique. Water, Air, & Soil Pollution, **225(5)**: 1974, (2014)
11. P. Jongpradist, S. Youwai, C. Jaturapitakkul. Effective Void Ratio for Assessing the Mechanical Properties of Cement-Clay Admixtures at High Water Content. Journal of Geotechnical and Geoenvironmental Engineering, **137(6)**: 621-627, (2011)
12. J. Wu, M. Feng, G. Han, et al. Experimental Investigation on Mechanical Properties of Cemented Paste Backfill under Different Gradations of Aggregate Particles and Types and Contents of Cementing Materials. Advances in Materials Science and Engineering, **2019**: 11, (2019)
13. Z.X. Liu, X.B, Li. Study on Fractal Gradation of Tailings and Knowledge Bank of Its Cementing Strength. Chinese Journal of Rock Mechanics and Engineering, **(10)**: 1789-1793, (2005)
14. J. Wu, M. Feng, Z. Chen, et al. Particle Size Distribution Effects on the Strength Characteristic of Cemented Paste Backfill. Minerals, **8(8)**: 322, (2018)
15. E. Yilmaz. Sustainability and Tailings Management in The Mining Industry: Paste Technology. Mugla Journal of Science and Technology, **1**: (2018)
16. X. Zhang, G. Ruiz, A.M. Elazim. Loading rate effect on crack velocities in steel fiber-reinforced concrete. International Journal of Impact Engineering, **76** 60-66, (2015)
17. A.L. Rosa, R.C. Yu, G. Ruiz, et al. A loading rate dependent cohesive model for concrete fracture. Engineering Fracture Mechanics **82** 195-208, (2012).
18. X.Q. Fan, S.W. Hu, J. Lu, et al. AE properties of concrete on dynamic tensile test. Construction and Building Materials, **114** 66-75, (2016).
19. J.W. Archer, M.R. Dobbs, A. Aydin, et al. Measurement and correlation of AEs and pressure stimulated voltages in rock using an electric potential sensor. International Journal of Rock Mechanics and Mining Sciences, **89** 26-33, (2016).
20. X. Yang, X.R. Liu, K. Zhao, et al. Application of AE time series in the study of the damage process of tailing cemented backfill under uniaxial compression. Electronic Journal of Geotechnical Engineering, **18** I 1739-1756, (2013)
21. X. Liu, K. Liu, K. Zhao, et al. Strain rate effects and AE characteristics of tailing cement backfill under uniaxial compression. Electronic Journal of Geotechnical Engineering, **18** R 3919-3932, (2013)
22. A.Y. Cao, G.C. Guang, L.M. Dou, et al. Damage evolution law based on AE of sandy mudstone under different uniaxial loading rate. Journal of Mining & Safety Engineering, **32(06)**: 923-928+35, (2015)
23. J.X. Ren, S. Jing, C.K. Liu. Experimental Study on Deformation and Failure Laws and AE Characteristics of Weak Cemented Sandstone. Coal Technology, **38(06)**: 1-4, (2019)

Research on Synthesis of Feco@N-C and Activation Persulfate

Jun Chen[1,*], *Shangqing* Xia[1], *Longchi* Zhao[1], *Sisi* Cao[1], *Cheng* Zong[1], and *Jie* Jin[1]

[1]Hefei University, School of Biology, Food and Environment, Anhui Key Laboratory of Sewage Purification and Eco-restoration Materials, 230601, Hefei, Anhui, China

Abstract. Based on the characteristics of persulfate (PS) activation that can quickly and effectively remove organic pollutants in water, a new catalytic material of carbon nitride supported Fe-Co diatom (FeCo@N-C) was prepared for PS activation to degrade orange II (OII). The XRD pattern showed that eight characteristic diffraction peaks, including $CoFe_2O_4$ diffraction absorption peaks. The FT-IR inferred that C-N/C=N, C=O, and O-H functional groups may exist. SEM observed the surface morphology of the synthetic material, and Fe and Co can be successfully loaded on the synthetic material. The activation effect of FeCo@N-C activator on PS was significantly improved, and the decoloration removal rate reached 81.66% at 20 min, 57% higher than the PS alone, FeCo@N-C activator had a good pH and temperature adaptability. FeCo@N-C activator could promoted the production of free radicals which can degrade OII efficiently.

1 Introduction

Advanced Oxidation Process (AOPs) can directly mineralize high concentration organic wastewater with poor biodegradability and high relative molecular weight or improve the biochemical properties of pollutants through oxidation, and can completely mineralize or decompose most organic compounds, which has a good application prospect [1-2]. Orange II (OII) is a typical anionic azo dye. Due to its advantages of brilliant color, stable coloring and low price, OII is widely used in the dyeing of various textiles [3]. However, OII aqueous solution has a high salt content, which will have a serious impact on the activities of aquatic organisms when release into water. Therefore, OII was selected as the degradation target. The degradation of OII was carried out by the activation of persulfate (PS) and the main active oxide was $SO_4^{\cdot-}$. Compared with AOPs with (·OH) as the main active substance, the activation process of AOPs base on PS is suitable for a wide pH range, still has a higher active site under alkaline conditions, and the half-life is longer than that of single activation, which is conducive to the long-term degradation of the target pollutant OII [4-5]. Transition metals can be activated efficiently at ambient temperature, such as Ag, Cu, Fe, Co, Mn and other transition metal ions. In addition to the transition metal for PS activation, there are also non-transition metal materials for PS activation, and the most used is carbon materials. Carbon material itself uses natural advantages, such as large specific surface area, large pores, low economic cost, no worries about secondary pollution, and good stability [6]. We can neutralize two catalytic pathways, so that transition metal ions can be loaded onto carbon-based materials. Two or more active groups could improve the activation efficiency of PS. Based on the characteristics of PS activation that can quickly and effectively remove organic pollutants in water [7], a new catalytic material carbon nitride supported Fe-Co diatom (FeCo@N-C) catalyst was prepared for PS activation, the synergistic catalysis of Fe and Co had also been reported [8-9]. The coprecipitation pyrolysis method was used to prepare FeCo@N-C material, nitrogen doping graphene also exhibited excellent catalytic performance because doping N could endow the high binding energy with PS, which enhances electrontransfer from graphene to PS [10-12], and during this process XRD was used. FT-IR, SEM, and other means could intuitively observe the surface morphology of the material.

2 Materials preparation and methods

2.1 Experimental reagents and instruments

Iron nitrate nonahydrate ($Fe(NO_3)_3 \cdot 9H_2O$) were purchased from Sinopharm Chemical Reagent Co.LTD. Cobalt-nitrate hexahydrate ($Co(NO_3)_2 \cdot 6H_2O$) were purchased from Tianjin Bodi Chemical Co., LTD.absolute methanol(CH_3OH), hydrochloric acid,sodium hydroxidewere purchased from Xilong Scientific Co., Ltd.PS ($K_2S_2O_8$) were purchased from Unixide Initiator Shanghai Co., LTD. Carbamide ($CO(NH_2)_2$) was purchased from Tianjin Damao Chemical Reagent Factory. Tropeolin($C_{16}H_{11}N_2NaO_4S$), melamine ($C_3H_6N_6$) were purchased from Shanghai Maclean Biochemical Technology Co., LTD. The water used in the whole study was deionized. All above chemicals and reagents were of analytically pure without further purification.

* Corresponding author: chenjun@hfuu.edu.cn

2.2 Synthesis of FeCo@N-C

The preparation of FeCo@N-C was completed by precipitation and carbonization, and the specific steps were as follows.

2.2.1 Preparation of precursor

0.202 g of $Fe(NO_3)_3{\cdot}9H_2O$, 0.145 g of $Co(NO_3)_2{\cdot}6H_2O$ and 3.003 g of urea put them into an agate mortar (during the synthesis process, the total amount of iron nitrate 9 hydrate and cobalt nitrate hexahydrate was 1 mmol, and the ratio of metal salt to urea was 30-100. In this experiment, the production ratio was 100), and 10 mL anhydrous ethanol was added to fully grind for 30min to obtain reddish-brown solute. The solid obtained was dried at 60 ℃ for 6 h in an air blast drying oven, which was the precursor [13].

2.2.2 Preparation of catalyst

Before pouring the obtained precursor into the crucible, it was best to grind it into uniform powder, and then set the heating program (3 ℃/min-5 ℃/min) to heat up to 800 ℃ for high-temperature carbonization for 2 h, and then naturally cool to room temperature to took out black solid, namely FeCo@N-C [14]. concentration was respectively arranged. Using pure water as a reference, a quartz cuvette with the optical path of 10mm was used to measure its absorbance at a wavelength of 484 nm, and the standard curve was prepared.

2.3 Experimental steps for degrading dye wastewater

Took 10 mL of OII dye waste from the standard reserve solution into a 100 mL volumetric flask and dilute it with pure water to the scale line to obtain 100 mg/L OII dye waste. Repeat this step three times to obtain three 100 mL 100 mg/L OII dye waste. Took three 150 mL conical bottles. Pour the required reagents and materials into the conical flask from the volumetric flask, and oscillate in a 25 ℃ constant temperature oscillator (100 rpm) for 5 min, 10 min, 15 min, 20 min, 25 min, and 30 min. Then, used a 5 mL disposable syringe to absorb 3 mL of dye waste liquid and filter it with a 0.45 um disposable needle filter. A small amount of anhydrous methanol was added to stop the reaction. Then the absorbance was measured at wavelength 484 nm [15], and the concentration was calculated by using the standard curve, and then the removal rate was calculated. The decoloration efficiency was calculated by the following formula:

$$W = (C_0 - C_e) / C_0 \times 100\% \quad (1)$$

Where W was decoloration efficiency
C_0 was the initial concentration of O II dye waste liquid (mg/L)
C_e was the instantaneous concentration of O II dye waste liquid (mg/L)

2.4 Analytical methods

TD-3500 X-ray diffractometer (Dandong Tongda) was employed to radiate the sample to obtain their X-ray diffraction (XRD) patterns. The functional groups on the surface of the sample was analyzed by Fourier transform infrared spectrophotometer (FT-IR, Nicolet 6700, America), and the spectra was recorded in the range of 4000 to 1000 cm^{-1}. The SEM spectra of the sample was observed by SU8010 cold field emission scanning electron microscope (Thermo Fisher Scientific, USA).

3 Results and discussions

3.1 characterization and physicochemical characteristics analysis of FeCo@N-C

To determine the main components of the synthetic material, the material was characterized by XRD. Jade software was used to retrieve the physical phase of the material, and then the data were derived and processed to obtain the graph shown in Fig. 1 (a). The figure above was the XRD pattern of FeCo@N-C in the range of 2θ 20-80°. It can intuitively see the characteristic diffraction peaks of $CoFe_2O_4$ at 2θ=30.1°, 35.5°, 43.1°, 57.0°, and 62.8°, according to the standard JCPDS (card # 22-1086). indicating that Fe and Co were successfully combined on N-C.

The absorption spectrum after data processing was shown in Fig. 1 (b). In the infrared spectrum, the wavelength or wave-number position of the absorbed light will form absorption peaks. According to this characteristic, some functional groups in the synthetic material can be viewed. C=O at 1735 cm^{-1} and possible O-H at 2930 cm^{-1}.

To intuitively observe the synthetic materials and prove that iron and cobalt were successfully loaded on the surface of carbon nitride, SEM was used to observe FeCo@N-C at different multiples. The SEM spectrum of the sample was shown in Fig. 1 (c). The overall shape of carbon nitride material was flake with microspores, it can be observed that there were many Fe and Co granular objects attached to N-C material, which were spherical.

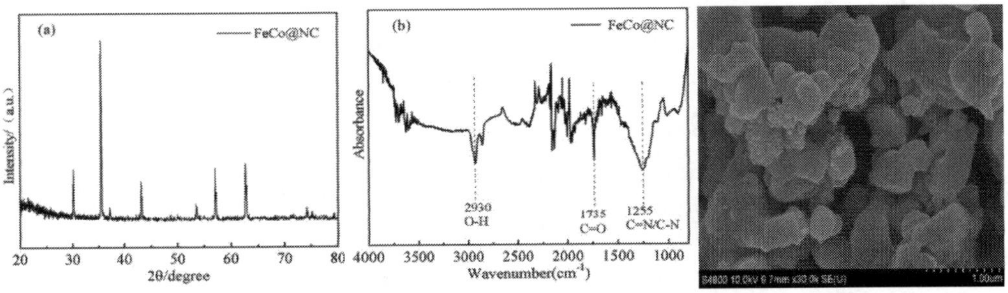

Fig. 1. (a) XRD patterns, (b) FT-IR spectra, (c) SEM spectra of FeCo@N-C.

3.2 Study on the degradation efficiency and influencing factors of OII by FeCo@N-C activation PS

3.2.1 Study on degradation efficiency of OII by FeCo@N-C activation PS

Under the initial experimental conditions of the same temperature, pH and initial concentration of OII, took 3 150 mL conical bottles, took 100 mL100 mg/L OII dye waste liquid, added 0.05 g FeCo@N-C, 2 g PS, 0.05 g FeCo@N-C and 2 g PS, respectively. Oscillate in a 25℃ thermostatic oscillator (100 rpm), the absorbance was measured, and the concentration was calculated using the standard curve, and then the removal rate was calculated. The experimental results were shown in Fig. 2.

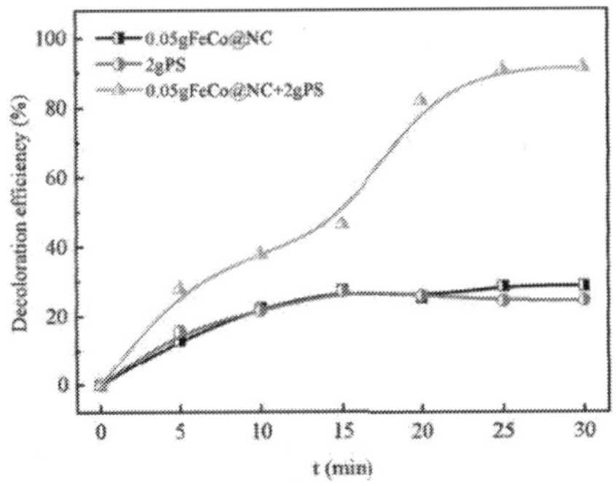

Fig. 2. Effect of catalyst on degradation of OII

As can be seen from Fig. 2, when FeCo@N-C prepared in the experiment was put in alone, it could not remove OII well, and the removal rate was 28.41% after 30 min. When PS was added alone, the reaction was slow due to the low activity of PS itself, and the removal rate was only 24.25% at 30 min. However, when PS was activated by the catalyst, the discoloration removal rate reached 81.66% at 20 min, showed a good catalytic effect. The decolorization efficiency of PS activated by Fe and Co alone on carbon nitride was 50.32% and 63.67% in 20 min, respectively, which was lower than that activated by FeCo@N-C.

3.2.2 FeCo@N-C Influence of dosage

Under the premise of the same initial concentration of OII, pH, temperature, and PS dosage, took 3 150 mL conical bottles, took 100 mL100 mg/L OII dye waste liquid, add 2 g PS, and press FeCo@N-C of 0.5 g/L, 1.0 g/L, and 1.5 g/L, respectively. Oscillate in a 25℃ thermostatic oscillator (100 rpm), the absorbance was measured, and the concentration was calculated using the standard curve, and then the removal rate was calculated. At the same time, the influence of FeCo@N-C dosage on catalytic PS was studied. The experimental results were shown in Fig. 3.

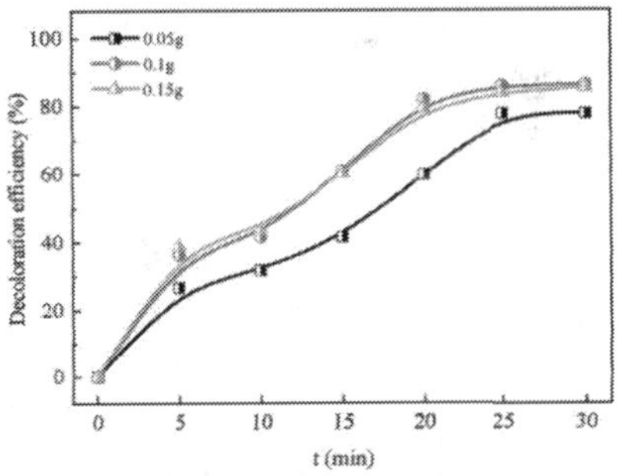

Fig. 3. Effect of FeCo@N-C dosage on degradation of OII

According to Fig. 3, the addition amount of FeCo@N-C to catalytic potassium PS was slightly lower in the case of 0.5 g/L, while there was no significant difference in case of 1 g/L and 1.5 g/L. However, compared with 0.5 g/L, the experimental results showed that the instantaneous removal rate of OII could be accelerated with the increase of the additional amount, indicated that the loaded bimetallic atom sites on the material were relatively uniform. Its catalytic efficiency could achieve the optimal effect at 1 g/L and finally tends to balance.

3.2.3 Effect of initial pH of reaction on the degradation of OII

OII was a weak acid substance, and the initial pH of 100 mg/L OII simulated dye waste liquid prepared was 5.56. Take 4 150 mL conical bottles, and take 100 mL100 mg/L OII dye waste liquid. The pH was adjusted to 3, 5, 7, and 10 with 0.1 mol/L HCl and NaOH, then 0.05 g sludge-based activated carbon and 2 g PS was added. Oscillate in a 25℃ thermostatic oscillator (100 rpm) and took samples every 5 min for 6 times in total, the absorbance was measured. Use the standard curve to calculate the concentration, and then calculate the removal rate. The effect of the initial pH of solution on catalytic PS was analyzed. The experimental results were shown in Fig. 4.

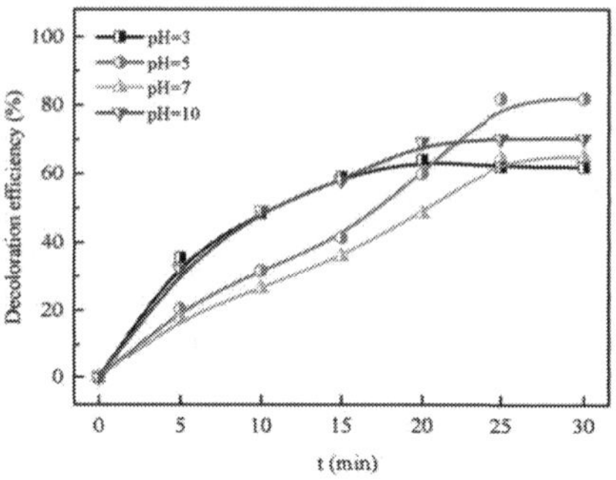

Fig. 4. Effect of different pH on degradation of OII

The pH value of the reaction solution has an important influence on the types of free radicals generated in advanced oxidation technology and the degradation of organic pollutants. After adding PS to the conical flask, the pH value was adjusted to 3, 5, 7, 10. Fig. 4 showed that OII had a better degradation effect under weak acid conditions. The final removal rate of 3, 7, and 10 fluctuates little, indicating that the material was adaptable to a wide range of pH and the higher the pH was, the more negative charge on the surface of FeCO@N-C. OII was an anionic material, resulting in electrostatic inhibition, so the pH of 10 was weaker than that of 5. When the pH of the reaction solution was 3, the degradation rate of OII was slower than that of pH 5, which may be due to the formation of hydrogen bond between H^+ and O-O bond in PS under strongly acidic conditions, which led to the reduction of PS activation efficiency as well as the degradation rate [16].

3.2.4 Influence of temperature on degradation of OII

Took three 150 mL conical bottles, added 100 mL of 100 mg/L OII dye waste liquid, then added 0.5 gFeCo@N-C and 2 g PS, in a thermostatic oscillator (100 rpm), the oscillations were carried out at 10℃, 25℃ and 40℃, respectively, and samples were taken every 5min for a total of 6 times, the absorbance was measured. The concentration was calculated using the standard curve, and the removal rate was calculated. The result was shown in Fig. 5.

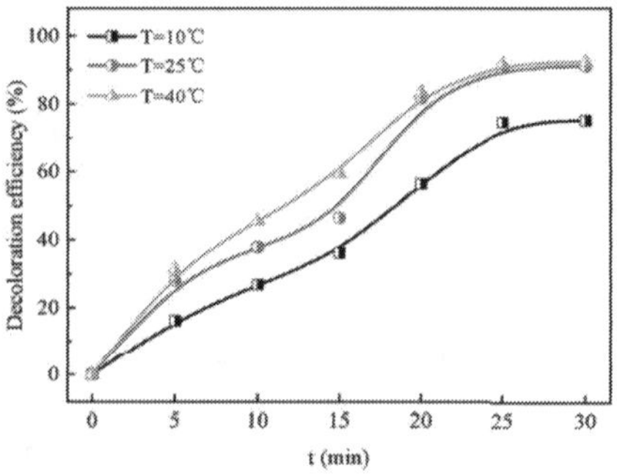

Fig. 5. Effect of temperature on degradation of OII

Fig. 5 illustrated the: reaction to 10 min, the removal rate increased with the increase of temperature was small, the removal rate of the OII small-amplitude increases with the increase of the temperature quickly, this may be due to the increase of the temperature, thus speeding up the intermolecular Brownian motion, accelerating the oxide free radical generation, so as to speed up the degradation rate [17]. However, there was no significant difference between 25 ℃ and 40 ℃ (the removal rate reached 81.66% and 83.45% respectively at 25 min), indicating that temperature had little overall influence on the experimental effect [18]. Considering practical factors, 25 ℃ was the optimal temperature for this experiment.

3.3 Mechanism and kinetics of degradation of OII by FeCo@N-C activation PS

3.3.1 Kinetic simulation of degradation of OII by FeCo@N-C activated PS

The reaction kinetics of the chemical reaction in the degradation process of OII was analyzed and the reaction rate constant was obtained. The kinetic analysis of the degradation of OII by activated PS was a complex process. The quasi-first-order kinetic equation and pseudo-second-order kinetic equation were used to linearly fit the dosage of different FeCo@N-C materials. The following was the dynamic model.
Quasi-first-order dynamics model was:

$$-\ln(C_0 / C) = K_1 t \quad (2)$$

Pseudo-second-order dynamics model was:

$$(1 / C - 1 / C_0) = K_2 t \quad (3)$$

Where C was the instantaneous concentration of OII at time;
C_0 was the initial concentration of OII;
K_1 quasi-first-order reaction rate constant (min^{-1});
K_2 was pseudo-second-order reaction rate constant (min^{-1});
t was degradation time;

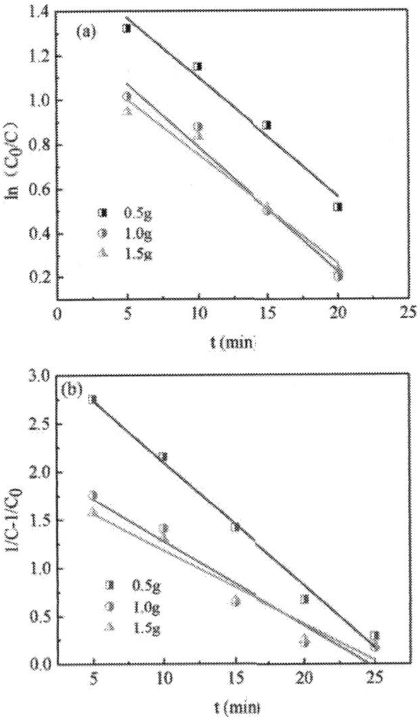

Fig. 6. (a) Quasi-first-order kinetic curves of degradation of OII with different FeCo@N-C dosage, (b) Pseudo-second-order kinetic curves of OII degradation with different FeCo@N-C dosage.

As can be seen from Fig. 6 (a) and 6 (b), degradation processes of OII under different FeCo@N-C dosage were in line with quasi-first-order dynamic model and pseudo-second-order dynamic model. When FeCo@N-C dosage was 0.5 g, 1.0 g and 1.5 g, respectively,the corresponding first-order reaction rate constant (K_2) and second-order reaction rate constant (K_2) were -0.05383, -0.05658 and -0.04949, respectively. -0.12818, -0.08749, -0.07624. It can be seen that as the dosage increases, the slope increases, which means the reaction rate accelerates. According to the analysis, this was because the increase of material dosage, the removal of OII reaction system has more active sites, more active groups, the activation of PS produced more free radicals, which led to a faster degradation rate.

3.3.2 Mechanism of activation of PS by FeCo@N-C

Took three 100 mL 100 mg/L OII dye waste liquid into conical bottles, added 0.5 g FeCo@N-C and 2 g PS into three conical bottles, and add 40mL methanol (MeOH) and 40mL tert-butanol (TBA) to two of them respectively to quench reactive oxygen species (ROS). Oscillate in a 25℃ thermostatic oscillator (100 rpm), the absorbance was measured, and the concentration was calculated using the standard curve, and then the removal rate was calculated.

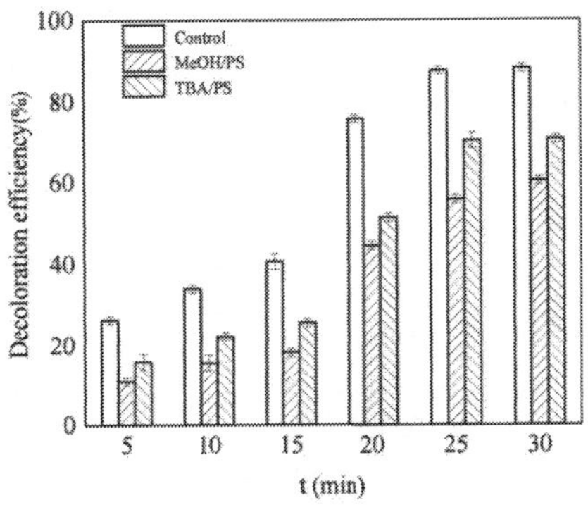

Fig. 7. Influence of MeOH and TBA quenching

Under the same temperature and pH, MeOH and TBA, two quenching agents, were selected to identify ROS involved in degradation of OII. As shown in the Fig. 7, when MeOH was added to quench $SO_4{\cdot}^-$, ·OH and $^1O_2$ the degradation efficiency of OII was significantly reduced, indicating that $SO_4{\cdot}^-$, ·OH and $^1O_2$ played a role, but after the addition of TBA, it only decreased by about 10%. TBA alcohol mainly quenches ·OH, but the effect was not obvious, indicating that the degradation effect of $SO_4{\cdot}^-$, $^1O_2$ on OII was greater than that of ·OH.

The activation mechanism was mainly divided into the following aspects, the responsible ROS for OII degradation in PS/FeCo@N-C system were $SO_4{\cdot}^-$, ·OH and $^1O_2$. The reaction between PS ion and water produced $SO_4{\cdot}^-$, ·OH and $^1O_2$. The prepared catalyst contained Fe and Co ions, which could activate metal PS to produce $SO_4{\cdot}^-$, and the material of non-metallic catalyst itself has a certain activation performance to PS. The mechanism was expressed in the following chemical equation in order:

$$S_2O_8^{2-} + 2H_2O \rightarrow 2SO_4^{2-} + SO_4{\cdot}^- + O_2^- + 4H^+ \quad (4)$$

$$SO_4{\cdot}^- + H_2O \rightarrow SO_4^{2-} + H^+ + {\cdot}OH \quad (5)$$

$$S_2O_8^{2-} + e- \rightarrow {\cdot}SO_4^- + SO_4^{2-} \quad (6)$$

4 Conclusion

FeCo@N-C was synthesized to activate PS and decompose OII., and the degradation rate of 81.66% can be achieved in 20 min. The experimental results showed that this material had a certain effect on PS activation. The dosage of FeCo@N-C for degradation of OII accorded with the quasi-first order kinetic model and pseudo second order kinetic model. The reaction mechanism was the reaction of PS ions with water to produced $SO_4{\cdot}^-$, ·OH and $^1O_2$. $SO_4{\cdot}^-$plays a major role in activation. The prepared catalyst contained Fe and Co ions, which could activate metal PS to produced $SO_4{\cdot}^-$. N-doped carbon shells could provided a large number of active sites for PS activation, the material of non-metallic catalyst itself had certain activation performance to PS. FeCo@N-C activated PS to produced free radicals that rapidly destroyed the chromophile groups of OII. This led to the fading of OII

This research was supported by the National Key R&D Program of China (2020YFC1908601, 2020YFC1908602). Hefei University graduate innovation and entrepreneurship program. Science Development Fund Project of Hefei University (22040521004).

References

1. M. Rayaroth, C. Aravindakumar, N. Shah, et al. Advanced oxidation processes (AOPs) based wastewater treatment-unexpected nitration side reactions-a serious environmental issue: A review. J. Chemical Engineering Journal: 133002 (2021).
2. Y. Deng, R. Zhao. Advanced Oxidation Processes (AOPs) in Wastewater Treatment. J. Current Pollution Reports, **1(3)**:167-176 (2015).
3. Y. Liu, K. Yan, S. Xue, A review on dyeing wastewater treatment. J. Textile Dyeing and Finishing Journal, **36(07)**:8-12 (2014).
4. X. Gu, Y. Wang, Z. Miao, et al. Degradation of trichloroethylene in aqueous solution by persulfate activated with Fe(III)–EDDS complex. J. Research on Chemical Intermediates, 43(1):1-13 (2017).
5. D. Ge, Y. Zhu, G. Li, et al. Identifying the key sludge properties characteristics in Fe2+-activated persulfate conditioning for dewaterability amelioration and engineering implementation. J. Journal of environmental management, 296: 113204 (2021).
6. Y. Gao, Q. Wang, G. Ji, et al. Degradation of antibiotic pollutants by persulfate activated with various carbon materials. J. Chemical Engineering Journal: 132387 (2021).
7. Q. Xu. The Degradation of Methyl Orange by SodiumPersulfate with Co-loaded Activated Carbon as Catalyst. J. Chemical Engineering & Equipment, **(11)**:11-15 (2016).
8. X. Gu, Y. Ji, J. Tian, et al. Combined MOF derivation and fluorination imparted efficient synergism of Fe-Co fluoride for oxygen evolution reaction. J. Chemical Engineering Journal, **427**: 131576 (2022).
9. G. Huang, C. Wang, C. Yang, et al. Degradation of bisphenol A by peroxymonosulfate catalytically activated with $Mn_{1.8}Fe_{1.2}O_4$ nanospheres: synergism between Mn and Fe. J. Environmental science & technology, **51(21)**: 12611-12618 (2017).
10. X. Duan, Z. Ao, H. Sun, et al. Nitrogen-doped graphene for generation and evolution of reactive radicals by metal-free catalysis. J. ACS applied materials & interfaces, **7(7)**: 4169-4178 (2015).
11. P. Duan, T. Ma, Y. Yue, et al. Fe/Mn nanoparticles encapsulated in nitrogen-doped carbon nanotubes as a peroxymonosulfate activator for acetamiprid degradation. J. Environmental Science: Nano, **6(6)**: 1799-1811 (2019).
12. H. Sun, X. Peng, S. Zhang, et al. Activation of peroxymonosulfate by nitrogen-functionalized sludge carbon for efficient degradation of organic pollutants in water. J. Bioresource Technology:244-251 (2017).
13. W. Nan, Y. Lei, Q. Wang, et al. Facile synthesis of FeCo@N-C core–shell nanospheres supported on graphene as an efficient bifunctional oxygen electrocatalyst. J. Nano Research, **10(7)**:2332-2343 (2017).
14. P. Duan, Y. Qi, S. Feng, et al. Enhanced degradation of clothianidin in peroxymonosulfate/catalyst system via core-shell FeMn@N-C and phosphate surrounding. J. Applied Catalysis B: Environmental, **267**: 118717 (2020).
15. B. Zhu, H. Cheng, J. Ma, et al. Bi_2MoO_6 microspheres for the degradation of orange II by heterogeneous activation of persulfate under visible light. J. Materials Letters, **261**: 127099 (2020).
16. G. Wang, S. Chen, X. Quan, et al. Enhanced activation of peroxymonosulfate by nitrogen doped porous carbon for effective removal of organic pollutants. J. Carbon, **15**:730-739 (2017).
17. C. Cai, J. Liu, Z. Zhang, et al. Visible light enhanced heterogeneous photo-degradation of Orange II by zinc ferrit ($ZnFe_2O_4$) catalyst with the assistance of persulfate. J. Separation and Purification Technology, **165**:42-52 (2016).
18. Y. Kuang, J. Du, R. Zhou, et al. Calcium alginate encapsulated Ni/Fe nanoparticles beads for simultaneous removal of Cu (II) and monochloeobenzene. J. Journal of Colloid and Interface Science, **447**:85-9 (2015)

Study on Human Thermal Comfort in Asymmetric Radiant Heat Environment in Large Space

Rong Hu[1], *Haolin* Wang[1], and *Yueming* Li[1,*]

[1]Guilin University of Electronic Technology, School of architecture and traffic engineering, Guilin 541004, China

Abstract. In order to study indoor temperature distribution in a large space and thermal comfort of human, an environmental test and questionnaires were carried out at an airport in Xi'an in summer. The results show that the temperature distribution in the space is asymmetric. The room temperature increases from cooling floor to 0.1m high, but then it has little change within 1.5m high. However, the temperature fluctuation in horizontal is significantly influenced by internal equipment, lighting, elevator and etc. According to the environmental parameters and subjective voting, 55% of the PMV values are greater than 0.5, and more than half of the thermal feelings near the heat source and glazing curtain wall are "warmer". Properly increasing the wind speed of air vent could effectively improve the environmental comfort.

1 Introduction

As a large space, the airport needs to carry a large number of people. With the continuous growth of the number of domestic terminals, how to make most passengers satisfied with the indoor environment of the terminal is the focus of some current research. Geng et al. [1] Conducted environmental measurement and questionnaire survey on the indoor environmental quality of eight major airport terminals in China. It is found that there is a high correlation between subjective satisfaction and objective environmental measurement. Based on the analysis of Kano model, it is found that "thermal comfort" and "space layout" are the basic factors affecting the overall satisfaction of passengers. and the research on thermal comfort of three airports in Brazil by Jacqueline elhage Ramis et.al [2]. is to find out the main causes of passenger complaints through on-site temperature measurement. Kotopouleas et al. [3,4] A field measured was conducted in three airport terminals in the UK, and a questionnaire survey was conducted carried out among 3087 personnel. The average sensitivity of staff to temperature change is 1.6 times that of passengers. Passengers and staff demonstrated a high level of thermal tolerance and acceptable temperature range, with an average of 6.1 °C in summer and 6.7 °C in winter. Jia et al. [5] Through the field study of several terminals, it is concluded that the passenger can withstand the temperature change in a wide range, and the adaptability to indoor thermal environment tends to be same as the residence time increased.

A radiant cooling/heating systems was first attempt to be applied in Terminal 3 of Xi'an airport in China, and it has been put into operation in 2012. The indoor thermal environment controlled by the cooling floor should attract the attention of many researchers and engineers. With the consideration, an environment test and questionnaires were carried out in summer. The key environmental parameters that have significnt impacts on comfort level can be determined through analyzing the measured data.

2 Methodology

The terminal adopts independent temperature and humidity control technology in some large spaces. Radiant heating/cooling floors are responsible for partial sensible load, and a displacement ventilation system mainly treat fresh air and room latent load. In this test, relevant parameters of thermal environment are tested in a certain area of terminal 3 of Xi'an Xianyang International Airport. In order to observe the impacts of internal source and glazing curtain wall on indoor thermal environment, the testing region (Fig. 1) is divided into three parts. The first part is close to the heat sources such as shops and elevators, and the air vent is placed not far away. The second part is the area without any heat source. The third part is near to glazing curtain wall.

(a) Real scene of testing region.

* Corresponding author: liyuemingff@163.com

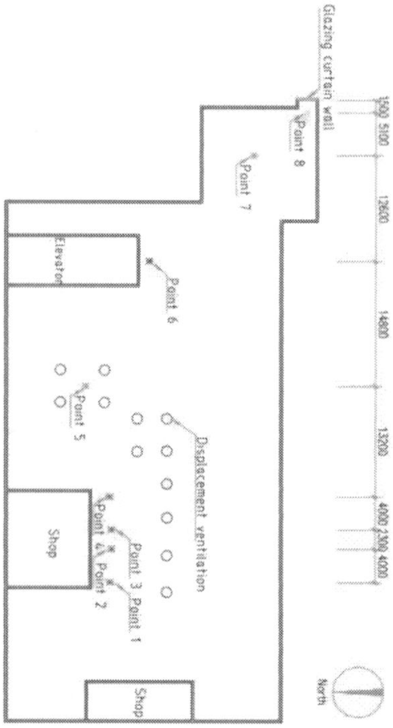

(b) Layout of testing region and measuring points.

Fig. 1. Testing region in T3 terminal.

PMV index is generally used to analyze the environment. The thermal comfort index PMV (predicted mean vote) reflects the deviation degree of human thermal balance. It is a comprehensive comfort index prposed by Professor P.O. Fanger [6] , which represents the feeling of most people in the same environment. These evaluation criteria are summarized into the temperature index to obtain the thermal comfort equation.

$$\begin{aligned} PMV &= [0.303exp(-0.036M)+0.0275] \\ &\times\{M-W-3.05[5.733-0.007(M-W)-P_a] \\ &-0.42[(M-W-58.2)-0.0173M(5.867-P_a) \\ &-0.0014M(34-t_a)-3.96\times10^{-8}f_{cl} \\ &\times[(t_{cl}+273)^4-(t_{mrt}+273)^4]-f_{cl}h_c(t_{cl}-t_a)\} \end{aligned} \quad (1)$$

$$PPD=100-95*exp-(0.03353*PMV^4+0.2179*PMV^2) \quad (2)$$

Where,

t_{cl}: Outer surface temperature of clothes, ℃, from reference, [6]

t_a: Air temperature around human body, ℃, testing value.

P_a: Partial pressure of water vapor around human body, kPa, from reference. [6]

h_c: Convective heat transfer coefficient W / (m^2 · K), from reference. [6]

f_{cl}: Area coefficient of clothing, clo， from reference. [6]

RH: Relative Humidity, %, testing value.

v: Speed, m/s, testing value.

MRT: Mean radiant temperature, ℃, calculating value by equation

$$T_m=[T_g^{\ 4}+\frac{h_c}{\sigma_e}(T_g\text{-}T_a)] \quad (3)$$

T_g: Globe temperature, ℃, testing value.

MET: Metabolic equivalent, met, calculation value.

Therefore, the testing parameters and corresponding instruments list in Table. 1 and Figure.2 . The RC-4HC and AZ-8778 is placed at each measuring point. To observe the vertical temperature gradient changes, the instruments are respectively placed 0.1m, 0.6m, 1.1m and 1.5m off the ground.In the sitting position, 0.1m corresponds to the foot, 0.6m corresponds to the waist and knee, and 1.1m corresponds to the neck. When standing, 0.1m corresponds to the foot, 0.6m corresponds to the waist, 1.1m corresponds to the knee and 1.5m corresponds to the neck. Floor temperature is measured with a thermometer gun and averaged by multiple readings. Use AR866A to measure the wind speed at 1.1m for each measuring point.

Table 1. Measured range and accuracy of the instruments.

Instrument	parameter	Accuracy
Temperature and humidity self recorder RC-4HC	*Ta*(℃) RH(%)	±0.5℃
Black Ball Temperature RecorderAZ-8778	*Tg*(℃)	±0.6℃
Anemometer AR866A	*v*(m/s)	0.01m/s
Thermometric F566-2	*Ta*(℃)(0m)	±1%

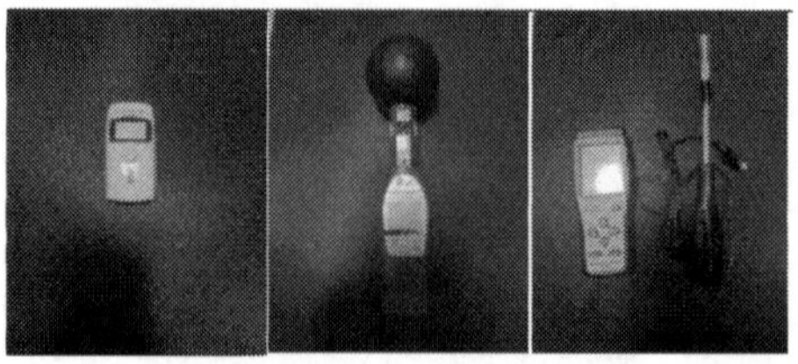

Fig. 2. Instrument diagram of Jingchuang RH-4HC, AZ-8778 and AR866a

TSV is used as an index to evaluate human thermal sensation in the environment, The subjective investigation contents of human psychological thermal response include: local thermal sensation and thermal comfort of forehead, back, arm, hand and lower leg, overall thermal sensation and thermal comfort, etc. ASHRAE 7 continuous scale [7] is used to evaluate local

 https://doi.org/10.1051/e3sconf/202235001015

and overall thermal sensation, and intermittent scale is used to evaluate local and overall thermal comfort.

2.1 Environmental parameter analysis

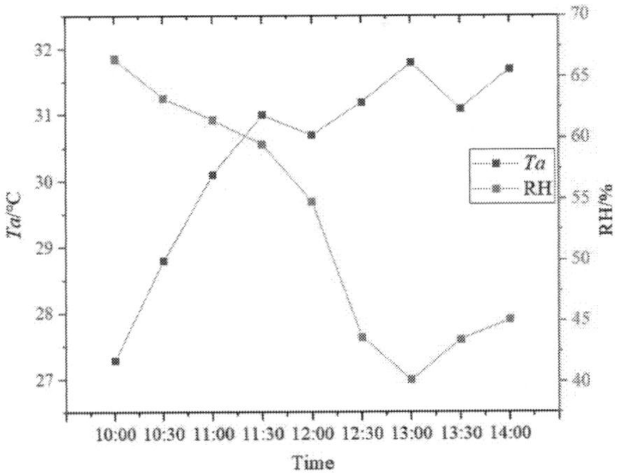

Fig. 3. Weather conditions during testing period

The test time is 10:00-14:00, and outdoor air temperature and relative humidity duiring the testing period are shown in Figure. 3. The outdoor temperature increases with time, while the relative humidity decreases.

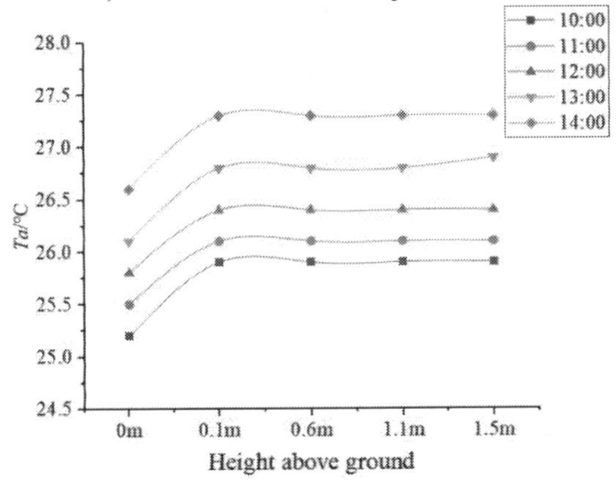

Fig. 4. Vertical Temperature Distribution at Point 1.

Fig. 4 illustrates that the air temperature at the vertical height at point 1 increases with time. The room temperature increases obviously from cooling floor to 0.1m high, but then it has little change with the height. The temperature variation law of other measuring points is similar to that of point 1. The reason is that the radiation cooling floor works terminal 3.Radiant cooling floor exchanges heat with indoor air by means of radiant and convective heat transfer. The cool air from displacement air vent (1m high) encounters other heat sources such as people, the temperature rises, and the air floats up. By inquiring and investigating the thermal sensation of the four parts of the personnel, there is no obvious discomfort with the difference in vertical temperature. This result has a similar distribution law with the previous test results of Zhao [7] et al. In the vertical height of the terminal.

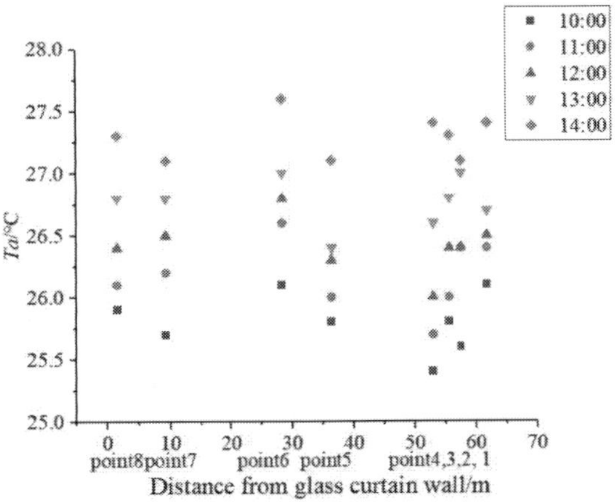

Fig. 5. Air temperature at each measuring point 1.1m from the ground.

Fig. 6. Thermal imaging of glazing curtain wall at point 8 at 13:00

Since the temperature variation can be ignored in the range of 0.1m and 1.5m high, the temperatures at 1.1m height are used to present the temperature variation in horizontal direction.

temperatures at 1.1m height of the eight points are shown in the figure. 5. The air temperature of all measuring points increases with time. The peak temperature value of each testing point occurred at 14:00 p.m. Overall, the temperatures at points 5 and 7 are relatively low because there is no heat source surrounding and they are close to air vents. There is a charging pile next to point 1, the heat emitted by the charging pile increases the air temperature. Point 6 is close to the escalator. Point 8 is close to the glazing curtain wall. As Figure 5 illustrates, the temperature of the curtain wall is between 28-30.1 ℃, The heat source emits heat to the space through radiation and convective heat transfer (mainly radiation). The heat dissipation of elevator has the greatest impact on the indoor, followed by the heat dissipation of store lights and charging piles. The temperature of point 1 is generally higher than that of points 2-4, because point 1 is at the charging pile, more and more people use chargers. The temperature near the curtain wall is obvioulу affected by the outdoor air temperature.

Inputting the measured environmental parameters *Ta*, RH, *v* (The whole process is in a static wind state, and the wind speed is recorded as 0.01m/s), The metabolic equivalent can be approximately regarded as 1met, and

https://doi.org/10.1051/e3sconf/202235001015

the thermal resistance of the clothing worn by the subjects can be approximately regarded as 0.6clo, which is a typical summer clothing thermal resistance.and *Tg* into the PMV to obtain the PMV value of each point (The height is 1.1m from the ground):

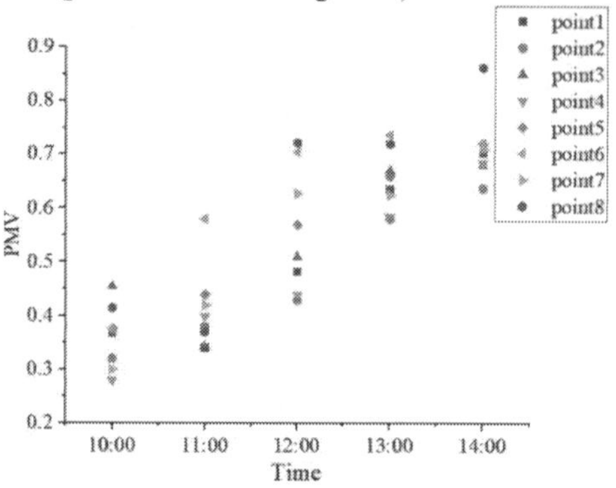

(a) PMV value

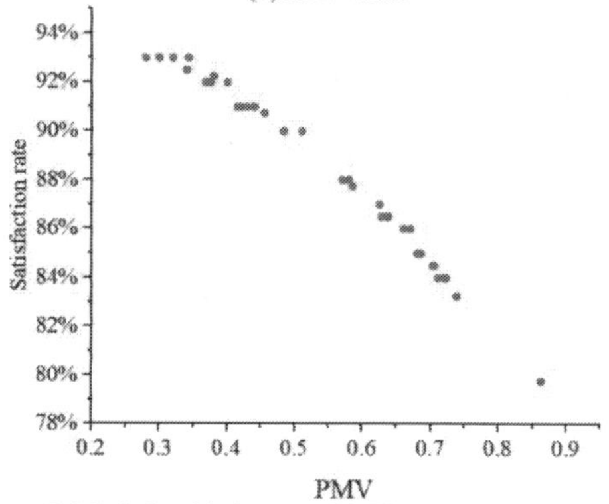

(**b.**) Relationship between PMV and satisfaction

Fig. 7. PMV value and satisfaction

The PMV value is shown in Figure. 7-a, and 55% is greater than 0.5 during testing period. According to the thermal comfort requirements of ASHRAE standard 55-2017, PMV greater than 0.5 is relatively hot.From the numerical point of view, after 11:00, the PMV value at point 6 and point 8 is the highest among all points, which is basically caused by the high air temperature and high black ball temperature at these points, implying that the heat from elevator and glazing curtain have obvious impacts on comfort level.

Figure.7-b shows the relationship between PMV and satisfaction. Satisfaction is calculated by PMV-PPD model. It is a predicted value, which reflects people's evaluation of the current environment. As the value of PMV increases, the satisfaction of people to the environment decreases.

2.2 Subjective questionnaire analysis

Reference [8] suggested that 100 subjects were randomly selected in an experiment, The situation of testers is shown in the table. 2. The tester is in a sitting or standing position. The sedentary time of the passengers is more than 15 minutes.

Table 2. Testers Situation.

Tester Information		Number
Gender	Male	56
	Female	44
Identity	Passenger	12
	Staff	88
Age	<18	4
	19-30	25
	31-45	32
	45-60	27
	>60	12

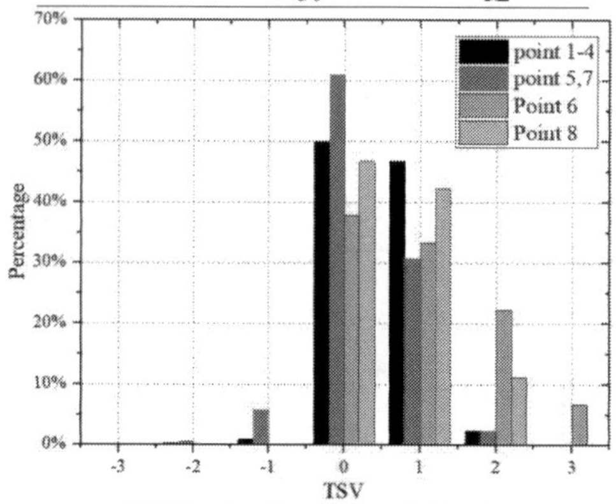

(a) TSV value distribution of all points

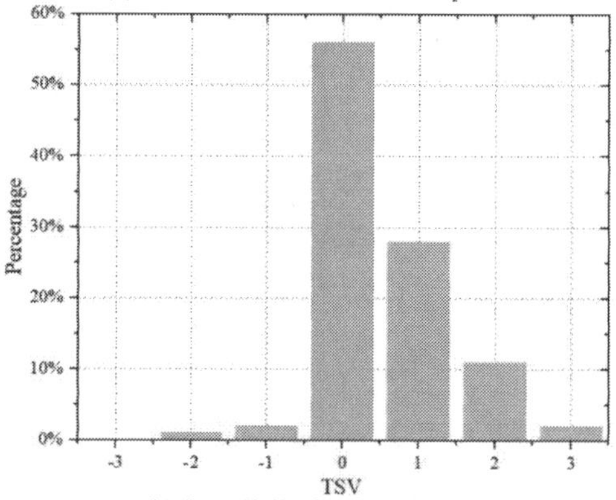

(b) Overall distribution of TSV

Fig. 8. TSV value distribution of all points

Figure. 8 shows the distribution of TSV values and overall values of all points. The environment of points 5 and 7 is considered "neutral" , more than 50%.People at points 1-4 and 6 of the heat source and point 8 of the glazing curtain wall feel "hot" more than people without heat source. Compared with the PMV value in the figure 8, TSV can also reflect the same thermal feeling of different positions in the space, and more than 80% interviewees are satisfy the thermal environment.

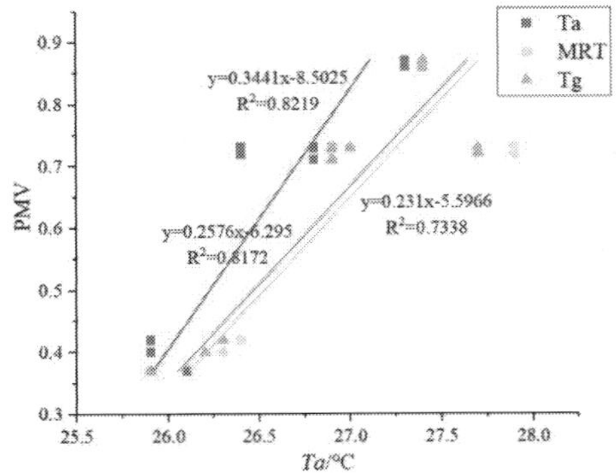

Fig. 9. Linear regression diagram of different temperature parameters and PMV

Figure. 9 shows the relationship between PMV values and *Ta*, MRT and *Tg*. It can be seen that the air temperature in the environment is most closely related to PMV. Appropriately reducing the air dry bulb temperature or increasing the wind speed could improve the environment comfort.

3 Conclusion

This paper tests the environmental parameters and subjective questionnaire survey of a space in the terminal, and obtains the following conclusions.

1.The temperature in space is asymmetric. 55% of the PMV values are greater than 0.5, and many values are greater than 0.5. Most people feel "hot" near to the heat source and glazing curtain wall, hoping that the environment will be "cold".

2.According to the fitting image and curve of PMV(Fig.9), The air temperature has the greatest impact on environment comfort level.Reducing the air temperature could significantly improve the environment.

3.During the test, the wind speed in the environment is in a static state, which is 0.01m/s. Increasing the air volume of displacement ventilation should significantly improve the environmental situation. It can also consider adding vents under heat sources or seats.

The research is supported by Guangxi Natural Science Foundation (number : 2018GXNSFBA050022)

References

1. Y. Geng, J. Yu, B. Lin, Z. Wang, Y. Huang, Impact of individual IEQ factors on passengers' overall satisfaction in Chinese airport terminals, Build. Environ. **112** 241–249 (2017)
2. J.E. Ramis, The impact of thermal comfort in the perceived level of service and energy costs of three Brazilian airports. Journal of Transport Literature., **7** 192-206 (2013)
3. A. Kotopouleas, M. Nikolopoulou, Thermal comfort conditions in airport terminals: indoor or transition spaces? Build. Environ. **99** 184–199 (2016)
4. A. Kotopouleas, M. Nikolopoulou, Evaluation of comfort conditions in airport terminal buildings, Build. Environ. **130** 162–178 (2018)
5. X Jia, B Cao. Field studies on thermal comfort of passengers in airport terminals and high-speed railway stations in summer. Build. Environ. **206** 108319 (2021)
6. P.O. Fanger, Thermal comfort Malabar F L: Robert E Krieger Publishing Company, (1982)
7. K Zhao, X Liu. On-site measured performance of a radiant floor cooling/heating system in Xi'an Xianyang International Airport. Solar Energy. **180** 274-286 (2014)
8. ISO Standard 7730. Moderate thermal environments-determination of the PMV and PPD indices and specification of the conditions for thermal comfort. Switzerland: International Standards Organization.

Adsorption Performance of Glyphosate on Modified Shell Powder/Ce-N-Tio₂

Wei Zhang[1,3], *Zhe* Liang[2,*], *Hai* Lin[4], *Jinkai* Shu[1,3], and *Aihe* Wang[1,3]

[1]Hunan City University, College of Municipal and Geomatics Engineering, 413000 Yiyang, China
[2]Shenyang Jianzhu University, College of Municipal and Environmental Engineering, 110168 Shenyang, China
[3]Hunan City University, Hunan Rural Drinking Water Quality Safety Engineering Technology Research Center, 413000 Yiyang, China
[4]Yiyang Commodity Quality Supervision and Inspection Institute, 413000 Yiyang, China

Abstract. Glyphosate is a typical dissolved organic phosphorus that can cause adverse effects on plants, animals, and humans. Modified shell powder/Ce-N-TiO_2 was synthesized by high-temperature doping, and the prepared modified shell powder/Ce-N-TiO_2 was characterized and used response surface methodology to find out the optimum condition for adsorption removal rate. The zeta characterization indicates that the doping of Ce shifts the isotropic potential of the material to the right. The FT-IR and XPS characterization indicated that glyphosate was successfully adsorbed, and the two-two interactions of pH, contact time and dosing amount had a more obvious effect on adsorption by response surface analysis. The process of glyphosate adsorption by modified shell powder/Ce-N-TiO_2 was consistent with the quasi-secondary kinetic model and Langmuir model, and the maximum adsorption amount was 66.98 mg/g. The above results may be useful for the study of other systems of glyphosate wastewater treatment.

1 Introduction

Glyphosate is a broad spectrum, post emergency and non-selective herbicide, widely used for post-emergence weed control [1-3]. Considering that come into contact with glyphosate is toxicologically harmful and can cause to serious health problems, such as cancer, liver and tissue rupture, endoderm destruction, lymphoma, etc [4-5]. So it is essential to choose a reliable application for removing glyphosate.

Many of the available treatment technologies share several drawbacks in respect of their applicability, mainly in view of the expansion of the scale [6-10]. So on, adsorption exist as an operative correction strategy to remove glyphosate pollutants from the aqueous phase, in existence some advantages, such as low cost and simple operation [11-13].

In this study, TiO_2 was modified by heat treatment and doping, and glyphosate was used as the target pollutant. The surface structure, functional groups and bonding energy changes were investigated before and after adsorption by means of scanning electron microscopy, infrared spectroscopy and XPS characterization, and its adsorption properties on glyphosate were explored by adsorption kinetics and thermodynamics.

2 Materials and methods

2.1 Instruments and materials

Experimental instrument: UV-4802S ultraviolet visible spectrophotometer; KSY-14-16 muffle furnace; GJ-2 sealed sample pulverizer; XSB-88 top impact vibrating screen machine; LLS-20-L ultrapure water machine; YC-S30 constant temperature water bath shaker; SJ-6 pH meter for laboratory; 0.22 μM water syringe filter.

Experimental materials: Waste oyster shell; Glyphosate (≥95%) was purchased from Shanghai Shifeng Biotechnology Co,Ltd; Cerium nitrate hexahydrate, purchased from Beijing mairuida Technology Co,Ltd; Urea is analytical pure and purchased from Sinopharm Chemical Reagent Co,Ltd; Tetrabutyl titanate (98.0%) purchased in Shanghai Aladdin Biochemical Technology Co,Ltd; Isopropanol, purchased from Shanghai Aladdin Biochemical Technology Co,Ltd; Sodium hydroxide is analytical pure and purchased from Tianjin Hengxing chemical reagent manufacturing Co,Ltd; Potassium bromide is analytical pure and purchased in Sinopharm Chemical Reagent Co,Ltd; Sodium nitrite is analytical pure and purchased at Shanghai Aladdin Biochemical Technology Co,Ltd; Nitric acid is analytical pure and bought in Hengyang Kaixin Chemical Reagent Co,Ltd. The experimental water is ultrapure water.

2.2 Experimental method

Preparation of modified shell powder/Ce-N-TiO_2 Taken a 100ml beaker and added 60ml isopropanol into it, then

* Corresponding author: 249713349@qq.com

added 0.173g prepared shell powder (calcination temperature 900 °C, sieving particle size 120 mesh, starch dosage 15%, calcination time 2.5h and alkali treatment) was optimized by single factor orthogonal experiment. Shaken with ultrasonic for 10min, then dropped 15ml tetrabutyl titanate under magnetic stirring, and stirred uniformly for 20min to formed solution A. Measured 20ml isopropanol and 2.4ml deionized water, then mixed with 1.313g cerium nitrate hexahydrate and 0.655g urea, and the pH was adjusted to 3 with nitric acid to formed solution B. Slowly dropped solution B into solution a, stirred at a uniform speed for 2h to form sol. Aged at room temperature for 12h, dried and ground in a 65 °C blast drying oven, and calcine in a 450 °C muffle furnace for 4h to obtain modified shell powder/Ce-N-TiO_2.

Adsorption kinetics experiment. The glyphosate solution with 100ml initial concentration of 1000mg/L was poured into the 250ml cone bottle, adjusted the initial pH to 4.0±0.2. Added 0.5g modified shell powder/Ce-N-TiO_2 mixed, shaken for 7h at 25°C and 160r/min. Sampled every 0.5h and measured the absorbance, calculated the adsorption amount. Each experiment was conducted in parallel three times. In addition, the quasi first-order kinetic model is shown in equation (1) and the quasi second-order kinetic model is shown in equation (2) to fit the data.

$$ln(q_e - q_t) = lnq_e - k_1 t \quad (1)$$

$$\frac{t}{q_t} = \frac{1}{k_2 \cdot q_e^{2}} + \frac{1}{q_e} \cdot t \quad (2)$$

in which qe and qt are the equilibrium adsorption capacity of glyphosate and the adsorption at time t (mg/g), k_1 is the equilibrium constant of the quasi-first-order kinetic model (min^{-1}), k_2 is the equilibrium constant of the quasi-second-order kinetic model [g/(mg·min)].

Adsorption thermodynamics experiment. Added separately 100ml glyphosate solution with initial concentration of 1000, 900, 800, 700, 600 and 500mg/L into several 250ml conical flasks. Adjusted the initial pH to 4.0±0.2. Added 0.5g modified shell powder/Ce-N-TiO_2 into it, oscillated for 6h at 15, 25 and 35°C and 160r/min. Measured the change of glyphosate concentration in the solution and calculated the adsorption capacity. Each temperature gradient was parallel three times. Langmuir equation (3) and Freundlich equation (4) are used for linear fitting.

$$\frac{C_e}{q_e} = \frac{1}{b \cdot q_{max}} + \frac{1}{q_{max}} C_e \quad (3)$$

$$1nq_e = 1nK + \frac{1}{n} \cdot 1nC_e \quad (4)$$

in which b is Langmuir adsorption equilibrium constant (L/mg), q_{max} is the maximum adsorption amount (mg/g), C_e is the adsorption equilibrium concentration (mg/L), K is Freundlich adsorption equilibrium constant.

2.3 Determination method of glyphosate concentration and calculation method of adsorption capacity

The concentration of glyphosate was determined by nitrite ultraviolet spectrophotometry.

Formula (5) is used as the evaluation index of glyphosate adsorption effect of modified shell powder/Ce-N-TiO_2.

$$q_e = \frac{(C_0 - C_e)V}{m} \quad (5)$$

3 Results and discussion

3.1 Characterization analysis

3.1.1 Surface morphology analysis

The surface morphology of modified shell powder/Ce-N-TiO_2 before and after adsorption was analyzed by scanning electron microscope. The results are shown in Figure 1 and Figure 2.

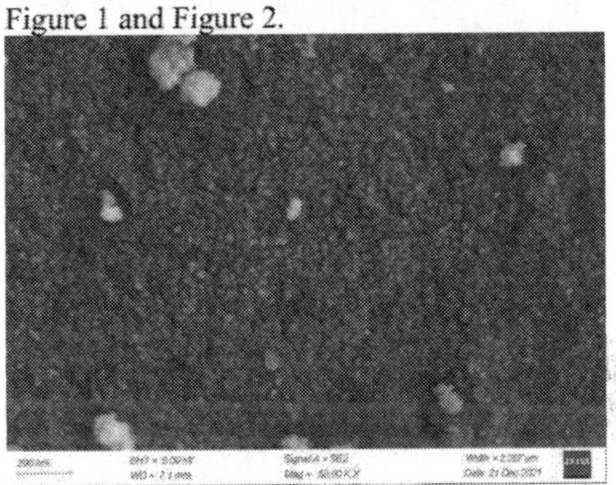

Fig. 1. SEM image before adsorption.

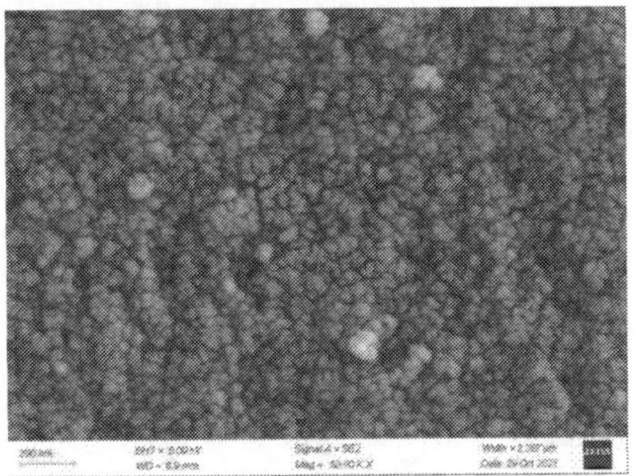

Fig. 2. SEM image after adsorption.

It can be seen from SEM Figure 1 and Figure 2 before and after adsorption. Before adsorption, the surface of the material is blocky, and after adsorption, there is obvious aggregation on the surface, and the surface is dense honeycomb.

3.1.2 Zeta potential analysis

The zeta potential change curves before and after

glyphosate adsorption by modified shell powder/Ce-N-TiO_2 are shown in Figure 3.

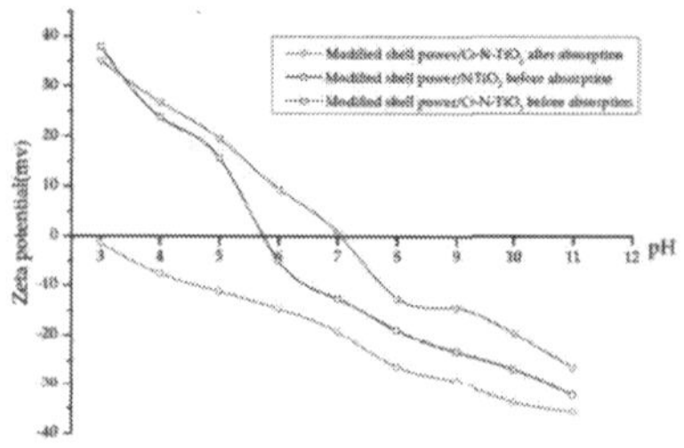

Fig. 3. Zeta potential at different pH conditions.

As shown in Figure 3, the isoelectric point of modified shell powder/N-TiO_2 is 5.7, and that of modified shell powder/Ce-N-TiO_2 is 7. The isoelectric point of modified shell powder/Ce-N-TiO_2 shifts left after glyphosate adsorption. The isoelectric point of metal oxides is determined by the deprotonation and protonation of hydroxyl groups on their surfaces. The leftward shift of modified shell powder/Ce-N-TiO_2 isoelectric point after glyphosate adsorption indicates that glyphosate forms negatively charged inner layer complexes with the surface of modified shell powder/Ce-N-TiO_2.

3.1.3 Surface functional group analysis

The functional group changes of the modified shell/Ce-N-TiO_2 before and after adsorption were analyzed by infrared spectroscopy. The results show that in Figure 4.

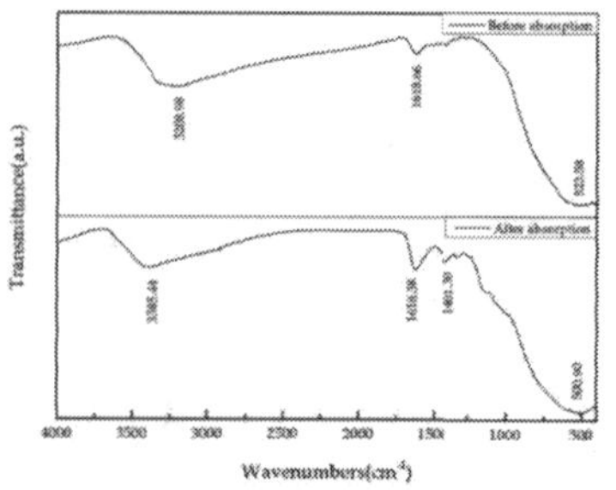

Fig. 4. FT-IR spectra before and after adsorption.

As shown in Figure 4, a symmetric stretching vibration of the carboxylate was observed at 1401.30 cm^{-1} after the adsorption of glyphosate, indicating the adsorption of glyphosate by it [14].

3.1.4 XPS analysis

The modified shell powder/Ce-N-TiO_2 before and after adsorption were analyzed by XPS. The results show that in Figure 5.

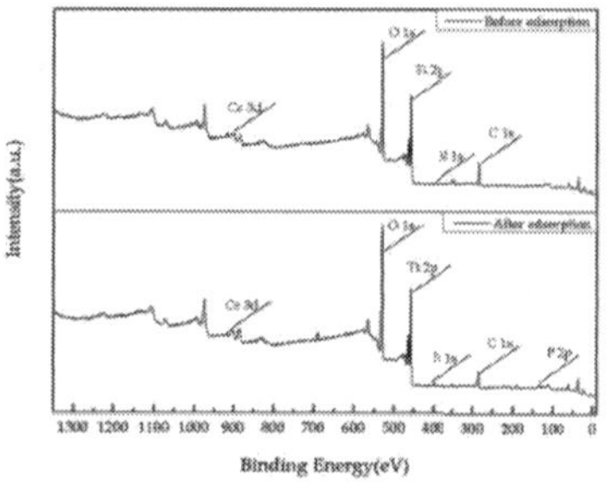

Fig. 5. XPS graphs before and after adsorption.

As shown in Figure 5 after the adsorption of glyphosate, element P appeared, indicating that glyphosate was successfully adsorbed by modified shell powder/Ce-N-TiO_2.

3.2 Response surface analysis

In this experiment, three factors were selected for investigation: pH, contact time and dosing amount, and response value: removal rate of glyphosate. The response surface analysis of three factors and three levels was carried out by using BOX-Behnken design principle, and 17 experimental groups were obtained to analyze the interaction between the influencing factors and obtain the optimization equation in the reaction process. The experimental design factor levels are shown in the Table 1 , and the results of experimental runs are shown in the Table 2.

Table 1. Experimental design factor level table.

Level	Factor		
	A pH	B Contact time (h)	C Dosing amount (g/L)
-1	3	4	2
0	4	5	5
1	5	6	8

Table 2. Experimental design and results.

Number	pH	Contact time (h)	Dosing amount (g/L)	Experimental value removal rate
4	5	6	5	61.46
11	4	4	8	62.67
8	5	5	8	63.25
13	4	5	5	61.33

12	4	6	8	65.59
14	4	5	5	61.33
17	4	5	5	61.33
15	4	5	5	61.33
16	4	5	5	61.33
1	3	4	5	57.45
5	3	5	2	55.57
3	3	6	5	64.17
2	5	4	5	56.18
6	5	5	2	55.11
9	4	4	2	56.92
7	3	5	8	63.03
10	4	6	2	61.39

The data in Table 2 were analyzed using Design Expert 10 to obtain a quadratic polynomial regression equation between the response values and each influencing factor. $Y=16.09167+16.65917*A+0.072083*B+1.62958*C-0.36000*A*B+0.056667*A*C-0.12917*B*C-1.95875*A^2+0.44375*B^2-0.014583*C^2$

The actual and predicted values of this polynomial model are compared in Figure 6 and Figure 7.

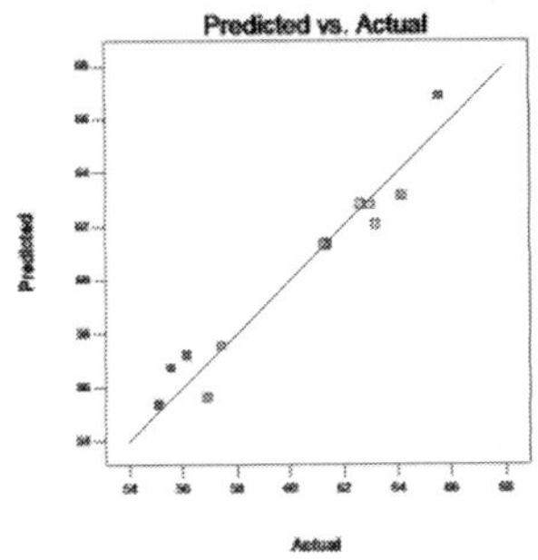

Fig. 6. Distribution of true and predicted values of glyphosate removal rate.

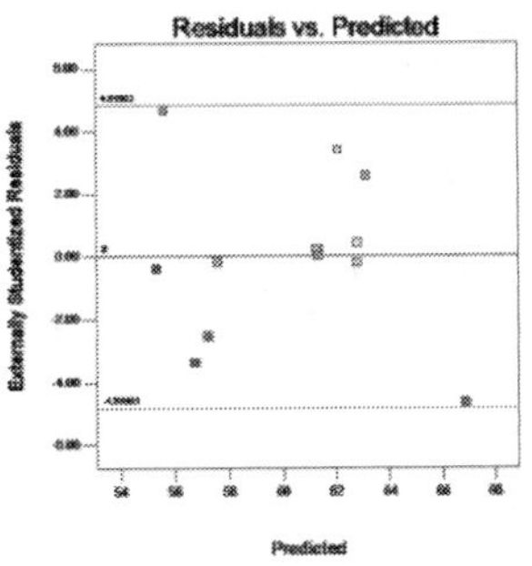

Fig. 7. Distribution of true and predicted values of glyphosate removal rate.

Table 3. Variance analysis results of response surface regression model.

Source	Sum of Squares	df	Mean Square	F Value	P-value	
Model	148.90	9	16.54	13.79	0.0011	Significant
pH	1.45	1	1.45	1.21	0.3077	
Time	34.46	1	34.46	28.73	0.0011	
Amount	41.85	1	41.85	34.89	0.0006	
AB	0.52	1	0.52	0.43	0.5319	
AC	0.12	1	0.12	0.096	0.7652	
BC	0.60	1	0.60	0.50	0.5020	
A^2	16.15	1	16.15	13.47	0.0080	
B^2	0.83	1	0.83	0.69	0.4332	
C^2	0.073	1	0.073	0.060	0.8128	
Residual	8.40	7	1.20			
Lack of Fit	8.40	3	2.80			
Pure Error	0.000	4	0.000			
Cor	157.29	1				

 https://doi.org/10.1051/e3sconf/202235001016

Total	6

As can be seen from Table 3, the P-value of this model is less than 0.0500, which is a good fit and significant. The fitted correlation coefficient R^2=0.9466 and adjusted coefficient of determination R_{adj}^2=0.8780, this regression equation has credibility and can respond to the true value.

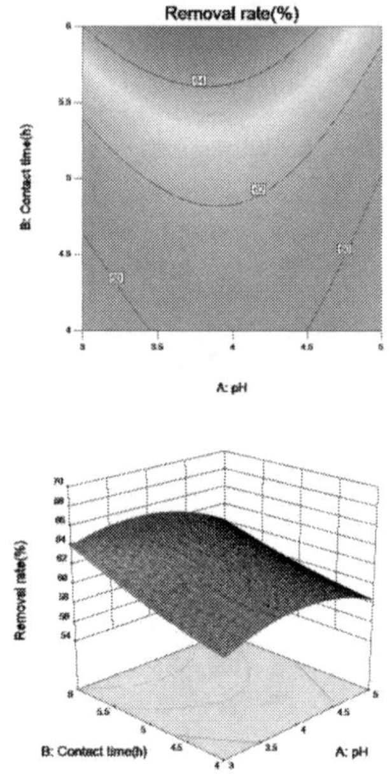

Fig. 8. Interaction of pH and contact time.

Figure 8 shows that the surfaces are relatively flat, indicating that the interaction between these two influencing factors does not have a significant effect on the removal rate. Figure 9 shows the effect of the interaction between pH and dosing amount, and it can be seen that the interaction between the two is significant.

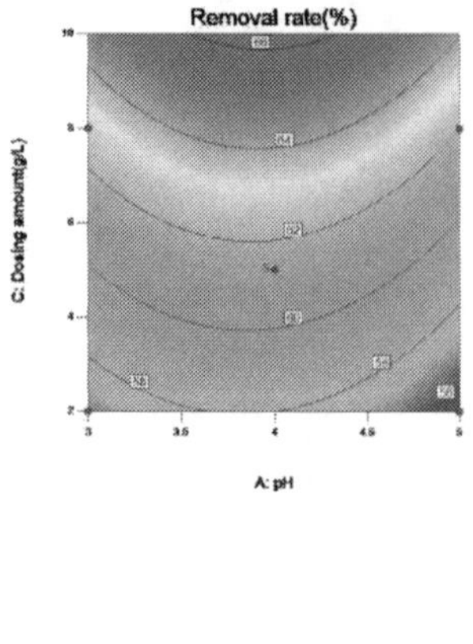

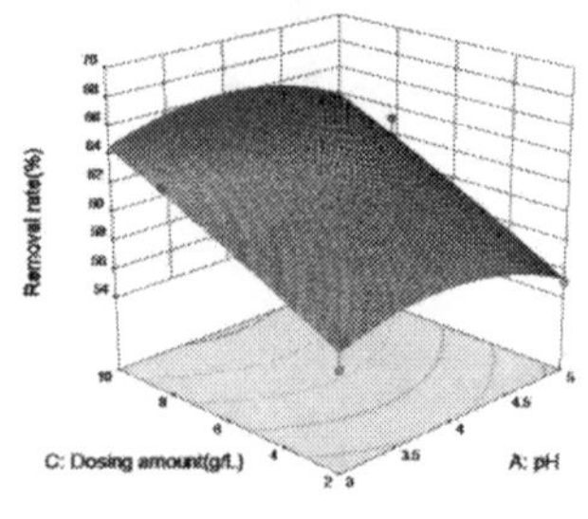

Fig. 9. Interaction of pH and dosing amount.

Figure 10 shows the effect of the interaction between contact time and dosing amount, and it can be seen that the interaction is significant.

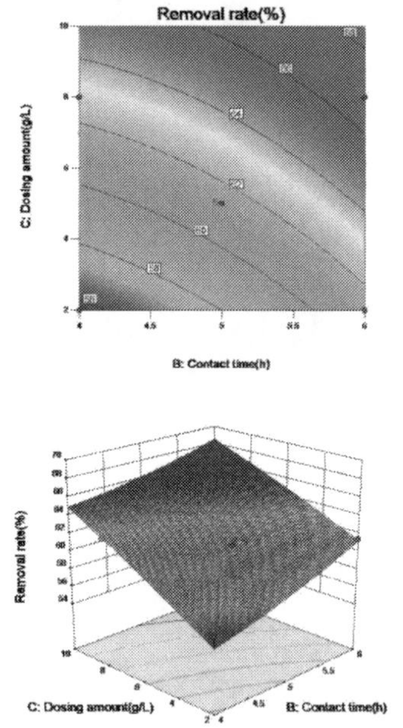

Fig. 10. Interaction of contact time and dosing amount.

According to the response surface analysis above, the effect of three factors on the removal rate of modified shell powder/Ce-N-TiO_2 on glyphosate containing wastewater, and the interaction between two factors were the same as the regression data analysis. This response surface model is reliable for optimization analysis and prediction.

3.3 Adsorption of glyphosate on modified shell powder/Ce-N-TiO2

3.3.1 Adsorption kinetics

Under the conditions of initial concentration of 1000 mg/L, pH = 4.0±0.2, temperature of 25℃ and rotating speed of 160 r/min, the effects of different contact time on the removal of glyphosate from modified shell powder/Ce-N-TiO_2 were investigated. The results are found in the Figure 11.

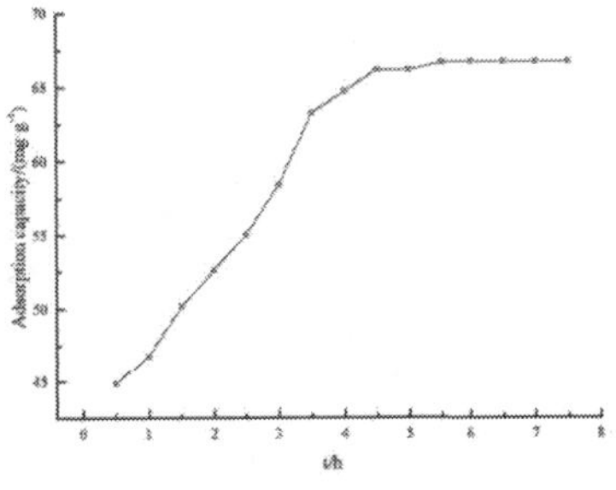

Fig. 11. Effect of adsorption time on the removal of glyphosate by modified shell powder/Ce-N-TiO_2.

As seen in Figure 11 that the adsorption capacity gradually increased in the first 4.5h and reached the highest value of 66.15 mg/g at 5h. The quasi-first-order and quasi-second-order kinetic models are used to fit the data in the Figure 12 and Figure 13.

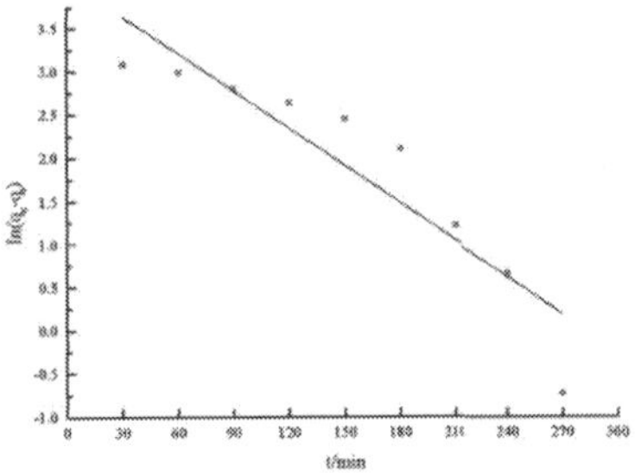

Fig. 12. Quasi-level kinetic model fitting curve.

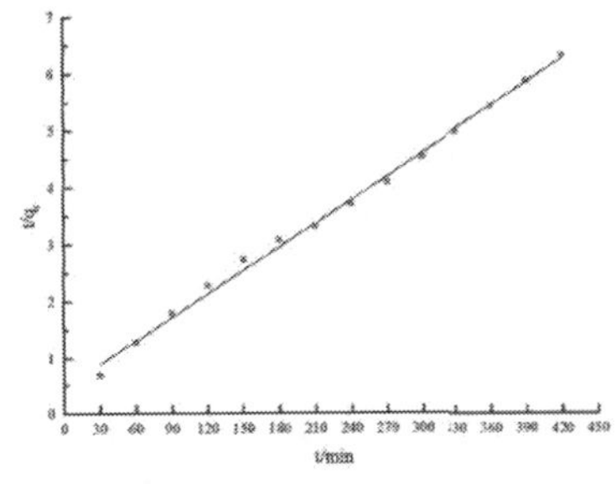

Fig. 13. Quasi-secondary kinetic model fitting curve.

Table 4. Kinetic fitting parameters for the adsorption of glyphosate by modified shell powder/Ce-N-TiO_2.

q_e	Quasi-level kinetic model			Quasi-secondary kinetic model		
	q_{cal}	k_1	R^2	q_{cal}	k_2	R^2
66.63	58.67	0.0144	0.827	72.94	0.0004	0.995

It can be seen from the Figure 10 , Figure 13 and Table 4 that the adsorption process of modified shell powder/Ce-N-TiO_2 is better matched with the quasi-second-order kinetic model, and the adsorption process is mainly chemical adsorption.

3.3.2 Adsorption thermodynamics

Under the conditions of pH = 4.0±0.2, rotating speed of 160r/min and contact time of 6h, the adsorption isotherms of glyphosate at 15, 25 and 35°C were carried out respectively. Results are presented in the Figure 14.

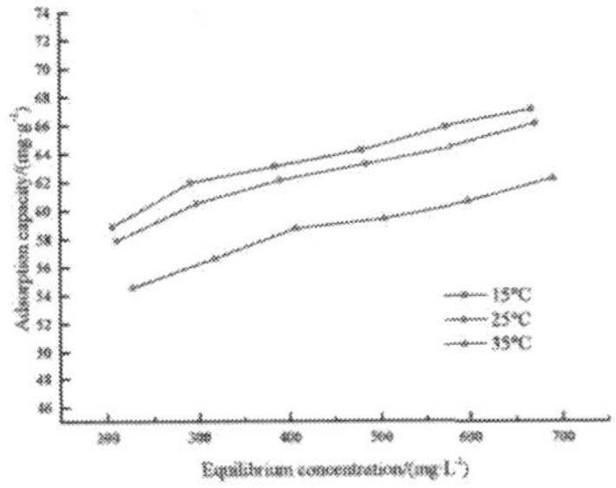

Fig. 14. Isotherm of glyphosate adsorption by modified shell powder/Ce-N-TiO_2.

As can be seen in Figure 14 with the increase of temperature, the adsorption counts of glyphosate by modified shell powder/Ce-N-TiO_2 decreases slightly, indicating that low temperature is conducive to adsorption, which is an exothermic reaction process. The maximum adsorption capacity was 66.98 mg/g at 15℃. Langmuir and Freundlich adsorption isotherms were used for fitting, and the results are shown in the Figure 15 and Figure 16.

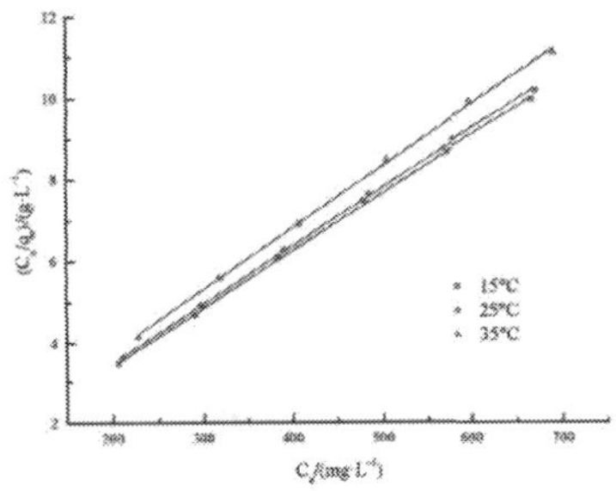

Fig. 15. Langmuir adsorption isotherm

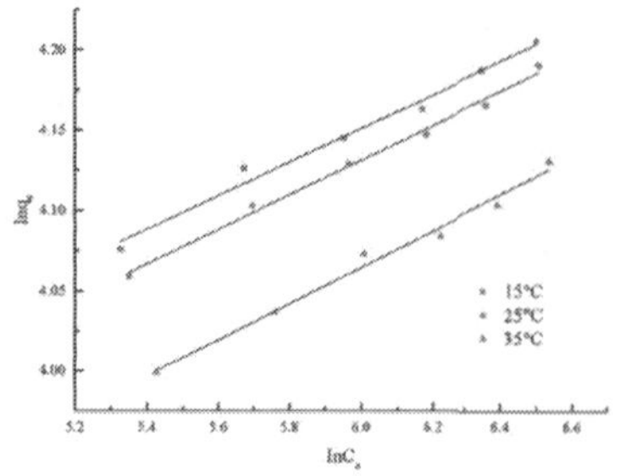

fitting.

Fig. 16. Freundlich adsorption isotherm fitting.

Table 5. Adsorption isotherm fitting parameters for glyphosate adsorption by modified shell powder/Ce-N-TiO_2.

T(℃)	Langmuir parameters			Freundlich parameters		
	q_{cal}	b	R^2	1/n	lnK	R^2
15	71.07	0.0222	0.999	0.1043	3.525	0.983
25	70.13	0.0209	0.999	0.1072	3.488	0.990
35	66.31	0.0189	0.999	0.1125	3.390	0.983

It can be seen from the Figure 15 , Figure 16 and Table 5 that the Langmuir model has a higher degree of fitting and is more suitable to describe the adsorption of glyphosate on modified shell powder/Ce-N-TiO_2. It shows that the adsorption process can be explained by single-layer coverage and uniform distribution.

4 Conclusion

(1)The successful adsorption of glyphosate by modified shell powder/Ce-N-TiO_2 and the formation of negatively charged inner layer complexes on the surface of modified shell powder/Ce-N-TiO_2 were demonstrated by SEM, zeta potential, FT-IR and XPS characterization.

(2)The effects of pH, contact time and dosing amount on the removal rate were analyzed by response surface analysis, which indicated that the interaction of two-two factors had significant effects on the adsorption and removal of glyphosate.

(3)Modified shell powder/Ce-N-TiO_2 adsorption of glyphosate was in accordance with the quasi-secondary kinetic model and Langmuir model, and the maximum adsorption amount was 66.98 mg/g.

This paper is one of the phase results of the National Natural Science Foundation of China (No: 42071122), Hunan University Student Innovation and Entrepreneurship Training Program (S202111527011, S202011527003).

References

1. D. Feng, A. Soric, O. Boutin, Treatment technologies and degradation pathways of glyphosate: A critical review. Sci Total Environ: 140559 (2020)
2. Peillex C, Pelletier M. The impact and toxicity of glyphosate and glyphosate-based herbicides on health and immunity. J Immunotoxicol, **17**:163-174 (2020).
3. J. Joensson, R Camm, T. Hall Removal and degradation of glyphosate in water treatment: a review. Journal of Water Supply: Research and Technology—AQUA, **62**:395-408 (2013).
4. Ingaramo P, Alarcón R, Muñoz-De-Toro M, Luque E.H. Are glyphosate and glyphosate-based herbicides endocrine disruptors that alter female fertility? Mol Cell Endocrinol, **518**:110934 (2020).
5. J.O. Ighalo, O.J. Ajala,A.G. Adeniyi , E.O. Babatunde, M.A. Ajala Ecotoxicology of glyphosate and recent advances in its mitigation by adsorption. Environ Sci Pollut R, **28**:1-14 (2020).
6. L. Cao, D. Ma, Z. Zhou, C. Xu, C. Cao, P. Zhao, Q. Huang Efficient photocatalytic degradation of herbicide glyphosate in water by magnetically separable and recyclable BiOBr/Fe3O4 nanocomposites under visible light irradiation. Chem Eng J, **368**:212-222 (2019).
7. R. Danial, S. Sobri, L.C. Abdullah, M. N. Mobarekeh FTIR, CHNS and XRD analyses define mechanism of glyphosate herbicide removal by electrocoagulation. Chemosphere, **233**:559-569 (2019).
8. S. Firdous, S. Iqbal, S. Anwar Optimization and modeling of glyphosate biodegradation by a novel Comamonas odontotermitis P2 through response surface methodology. Pedosphere, 30:618-627 (2020).
9. H. Rubí-Juárez, S. Cotillas, C. Sáez, P. Ca Izares, Barrera-Díaz C, Rodrigo M.A. (2016) Removal of herbicide glyphosate by conductive-diamond electrochemical oxidation. Applied Catalysis B Environmental, **188**:305-312.
10. M. Xie, Z. Liu, Y. Xu Removal of glyphosate in neutralization liquor from the glycine-dimethylphosphit process by nanofiltration. Journal of Hazardous Materials, **181**:975-980 (2010).
11. J.C. Diel, D.S.P. Franco, I.D.S. Nunes, H.A. Pereira, Moreira K.S, de L. Burgo T.A, et al. Carbon nanotubes impregnated with metallic nanoparticles and their application as an adsorbent for the glyphosate removal in an aqueous matrix. Journal of Environmental Chemical Engineering, **9**:105178 (2021).
12. J.M. Salman, F.M. Abid Preparation of mesoporous activated carbon from palm-date pits: optimization study on removal of bentazon, carbofuran, and 2, 4-D using response surface methodology. Water Sci Technol, **68**:1503-1511 (2013).
13. G.L. Dotto, G. Mckay Current scenario and challenges in adsorption for water treatment. Journal

of Environmental Chemical Engineering, **8**:103988 (2020).

14. Y. Huang, Z. Li, K. Yao, C Chen, C. Deng, Y Fang, et al. Suppressing toxic intermediates during photocatalytic degradation of glyphosate by controlling adsorption modes. Applied Catalysis B: Environmental, **299**:120671 (2021).

E3S Web of Conferences **350**, 01017 (2022) https://doi.org/10.1051/e3sconf/202235001017
EREGCE 2022

Seasonal and Spatial Variation of Nitrogen Distribution for Leaves of *Phyllostachys Pubescens* and Its Response to Precipitation Exclusion

Yong-hui Cao[1*], *Ben-zhi* Zhou[1], and *Xiao-ming* Wang[1]

[1]Research Institute of Subtropical Forestry, Chinese Academy of Forestry; Qianjiangyuan Forest Ecosystem Research Station, National Forestry and Grassland Administration, 311400 Hangzhou,Zhejiang, China

Abstract. The leaf nitrogen content in different canopy position for different ages of individuals for *Phyllostachys pubescens* was studied systematically with the season change in the experiment of precipitation exclusion in the bamboo field. The results showed that under different treatments (natural growth and precipitation exclusion treatment), the leaf nitrogen content of bamboo varied with seasons and in the decreasing order of spring, autumn, summer and winter. The change of leaf nitrogen content with bamboo age was also affected by the growing season. Precipitation exclusion treatment changed the seasonal and age difference of leaf nitrogen content in bamboo. In spring, the age differences in leaf nitrogen content of bamboo were significant ($p<0.05$). The leaf nitrogen content of bamboo varied with the height of the canopy, and the vertical distribution gradient was affected by the growing season. It increased with the increase of canopy height in summer and autumn, which also proved the applicability of the "light-nitrogen" hypothesis to the bamboo. Under two treatments, the leaf nitrogen content was linearly positively correlated with bamboo age. It changed with the change of soil water content in different growing seasons under treatments. Under precipitation exclusion, it was more sensitive to soil water content. This study confirmed the existence of the vertical distribution gradient of nitrogen in the leaves of *P. pubescens* and its variation with the growing season.

1 Introduction

Nitrogen is an indispensable and important nutrient element for plant growth, and an important limiting factor for the productivity of forest ecosystems. Nitrogen is an important component of the photosynthetic system and enzymes necessary for photosynthesis. The 30-40% of nitrogen in leaves participates in the carboxylation reaction, and the proportion of leaf nitrogen allocated to the carboxylation system determines the final photosynthetic efficiency. It is undoubtedly an important guarantee for the increase of carboxylation rate and net photosynthetic rate [1-2].

Photosynthesis of leaves within canopy under different light conditions will help to improve carbon harvesting and growth adaptability and resistance of the whole plant [3-4]. Due to the light gradient distribution in the canopy and the need to utilize resources effectively for plants, the plant leaf traits also change significantly in the canopy. Compared with leaves in shaded environment in canopy, leaves under higher light conditions have larger leaf mass per area, nitrogen content and photosynthetic capacity per leaf area [5-6], which are explained as the optimal distribution theory of plant resources, especially nitrogen resources in the canopy layers [7]. The gradient distribution of nitrogen content in canopy leaves is one of the basic characteristics of tree canopy. With the increase of atmospheric CO_2 concentration and the change of rainfall distribution pattern, the effect of nitrogen gradient pattern within canopy on leaf photosynthesis and the improvement of plant productivity are more important [6].

Moso bamboo (*Phyllostachys pubescens*) is an important forest resource in southern China with the advantages of fast growth, early maturity, high yield, wide range of uses and high income. Today, the bamboo forest is an increasing carbon sink. However, under the global climate change, the frequency and intensity of extreme climate (weather) events are increasing, and extreme drought events have posed unprecedented challenges to ecosystem structure and function. As an important condition for plant growth, water is closely related to the productivity of *P. pubescens*. Therefore, drought has an important potential impact on the dynamic change of the nitrogen gradient within the canopy of *P. pubescens* and its productivity. *P. pubescens* is rich in research content [8-11]. However, as an important growth trait, the temporal and spatial distribution of nitrogen in *P. pubescens* leaves, the vertical distribution gradient in canopy and its changes with drought have not been reported. In this paper, the field rain interception and drought experiment of *P. pubescens* was carried out to study the dynamic changes of nitrogen content in leaves of different ages of *P. pubescens* in different seasons and canopy layers. The response to drought stress was used to confirm the existence of a vertical gradient of nitrogen distribution in

* Corresponding author: fjcyh77@sina.com

P. pubescens leaves and its response to drought with seasons.

2 Materials and methods

2.1 Study area

The research was carried out at the Miaoshanwu auxiliary station of the Qianjiangyuan Forest Ecosystem Positioning Research Station of the State Forestry and Grassland Administration, located in Fuyang District, Hangzhou city (119°56′～120°02′ E, 30°03′～30°06 ′N). The terrain of this area belongs to the remnants of the Tianmu Mountains in the low hilly area of western Zhejiang, and belongs to the mid-subtropical monsoon climate zone. The annual average temperature is 16.1℃ (the highest temperature is 40.2℃, the lowest temperature is -14.4℃), and the annual average precipitation is 1441.9 mm. The soil is acidic red soil.

2.2 Materials and sample plot settings

The bamboo forest in the experimental area was planted in the 1960s and is a natural regeneration forest under extensive management. The slope of the sample plot is about 20° and due south, and the altitude is 169 m. Three standard plots of 20m × 20m were established, and the diameter at breast height (DBH) and height of all bamboo plants in the plots were measured. The bamboo forest density was 3 875 plants•hm-2, the canopy density was 0.95, the DBH of bamboo ranged from 4.0 to 13.6 cm, the average DBH was 9.8 cm, and the average bamboo height was 13.2 m. There are almost no shrubs and herbs under the forest, with an average thickness of about 2.0 cm of litter. The mountain is split once every 2 years, no fertilization, no ploughing, only spring bamboo shoots are harvested.

2.3 Rain interception drought experiment

In late July 2014, 2 plots of 10m × 10m were selected respectively within each of the 3 standard plots of 20m × 20m in the moso bamboo forest for the control and the rain interception drought experiment plots respectively, and so there are 3 natural growth (control) plots and 3 rain interception drought treatment plots. The sparse shrubs in the rain-intercepting drought plot were removed, and the canopy method was used to simulate the rain interception. Specific methods: In the rain interception drought plot, in order to remove the edge effect and ensure that the dry area is 10m × 10m, a rain interception greenhouse (with an area of 11m × 11m) of PVC waterproof sheet is built at a height of 1.5m above the ground to intercept the rain. Use glue to glue the gap between the plastic sheet and the bamboo. In order to ensure that the slope, topography and stand conditions are as consistent as possible with the control plot, one side of the greenhouse is parallel to the contour line. A trench with a depth of about 50cm and a width of about 20cm was excavated around the sample plot, and a 50cm-deep iron sheet was buried along the trench. The inside of the trench was also covered with a plastic film to prevent the infiltration of water from the side and to allow better drainage. For the control plot, only similar trenches were dug around, and no other treatment was performed.

2.4 Nitrogen content determination

After 2 years of growth under two treatment experiments, in clear and cloudless weather in the middle months of each month in March (spring), August (summer), October (autumn) and December (winter) of 2016, and March of the following year, samples were selected for continuous determination of leaf nitrogen content, leaf water content, soil water content and other indicators. In each plot, 2 Moso bamboos of different ages with the similar DBH were selected. With the help of the ladders temporarily built in the field, the sunny branches in the upper, middle and lower canopy of Moso bamboo with different age were hooked by using a sickle, and all the leaves were quickly picked and brought back to the laboratory for the determination of nutrient content and water content of the leaves. The 30 leaves were taken from each treatment (respectively for age and canopy) for nitrogen content determination as three replicates, and the average value was taken at last.

Determination of leaf nitrogen content: The remaining leaf samples were dried in an oven at 102℃ for 2h and at 65℃ for 24h and ground through a 100-mesh sieve for the determination of leaf nitrogen content. Nitrogen content of leaves was determined by Foss Kjeldahl analyzer, and the digested liquid of the test sample was digested with concentrated sulfuric acid-hydrogen peroxide solution for the determination of total nitrogen (*N*mass).

2.5 Soil moisture content determination

During the experiment, soil samples were taken synchronously from the field in the plots to measure soil water content. Four soil profiles were collected in the rain interception plot and the control plot as replicates, respectively, and the sampling depth was 0-20 cm, the soil was packed in an aluminum box and brought back, and 20g of fresh soil was weighed with an electronic balance (0.01g) and dried at 80℃ to constant weight. The calculation formula is: soil water content = [(fresh weight - dry weight)/fresh weight] × 100%. In this experiment, there were seasonal differences in soil water content at a depth of 0-20 cm in the control and the rain interception plots. The soil water content in the control plot was the highest in winter and spring (average of 29.82% and 30.15%, respectively), followed by summer (The average soil water content was 27.12% in autumn (October), and it was the lowest in autumn (23.42% on average). The seasonal differences of soil water content in the rain interception plots were not significant (the average soil water content was in the range of 4.44%-4.92%). The soil water content in the control plots of moso bamboo in each growing season was significantly higher than that in the rain interception plots.

2.6 Data analysis

The differences of nitrogen content in leaves of different ages among seasons and canopy layers were processed by Spss10.0 and Excel 2007, respectively, and each parameter was expressed as mean ± SE.

3 Results

3.1 Response of seasonal variation of leaf nitrogen content to drought

The research showed that the nitrogen content of leaves under natural growth (control treatment) was 23.24±2.24g·kg-1, which was slightly lower than that of leaves under rain interception drought treatment (24.39±1.14g·kg-1). However, the difference between the two treatments was not significant (P>0.05) (Fig. 1(a)). There were seasonal differences in the nitrogen content of leaves under different treatments, but they were all insignificant (P>0.05). The nitrogen content of leaves under the two treatments all varied with seasons as spring>autumn>summer>winter (Fig. 1(b)). In the same season, there was no significant difference in nitrogen content among different treatments (P>0.05). It can be seen that the drought did not significantly change the nitrogen content of bamboo leaves and its seasonal variation trend.

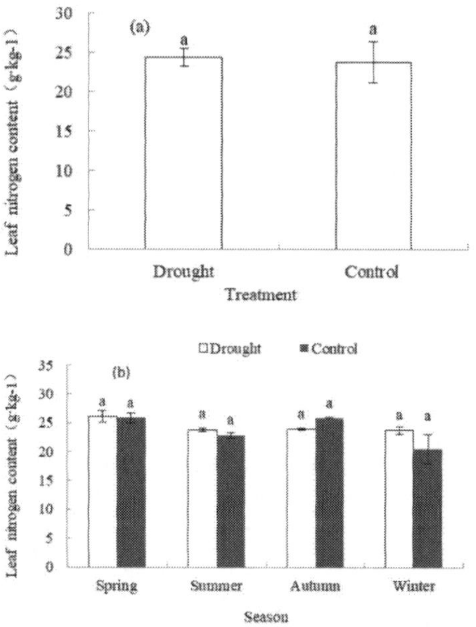

Fig. 1. The seasonal change of leaf nitrogen content for different ages of *P. pubescens* between control and drought stress (Note: Different lowercases in the same treatment denote significant differences at p<0.05 among the seasons and different uppercases in the same season denote significant differences at p<0.05 among the treatments. The same below.)

3.2 Response of leaf nitrogen content for different age of bamboo to drought

The variation of leaf nitrogen content with bamboo age was also affected by the growing season (Fig. 2). Under natural growth, the leaf nitrogen content of Moso bamboo with different bamboo ages in spring showed a trend of three-degree bamboo > two-degree bamboo > one-degree bamboo, that is, the leaf nitrogen content increased with the increase of bamboo age. The change of leaf nitrogen content in summer was just the opposite and it decreased with the increase of bamboo age. In autumn and winter, it showed "first decrease and then increase". Under the rain interception, the nitrogen content of leaves increased with the increase of bamboo age in spring and summer, and the opposite was in winter, and it decreased with the increase of bamboo age. In autumn, it showed the changing trend of the second degree > the third degree > the first-degree of bamboo.

Compared with the natural growth, the interception of rain changed the seasonal and age differences of nitrogen content in leaves for different age of Moso bamboo. The analysis of variance showed that in spring, the difference of bamboo age in leaf nitrogen content was significant (P<0.05), but there was no significant difference between treatments (P>0.05).

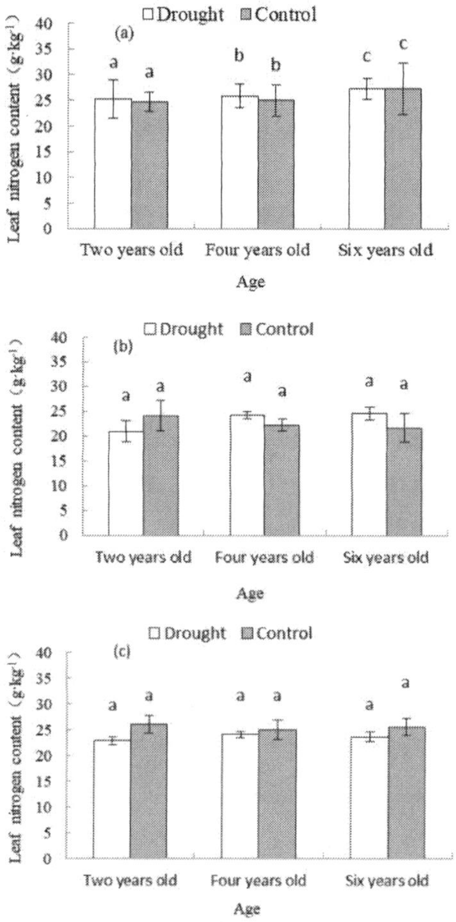

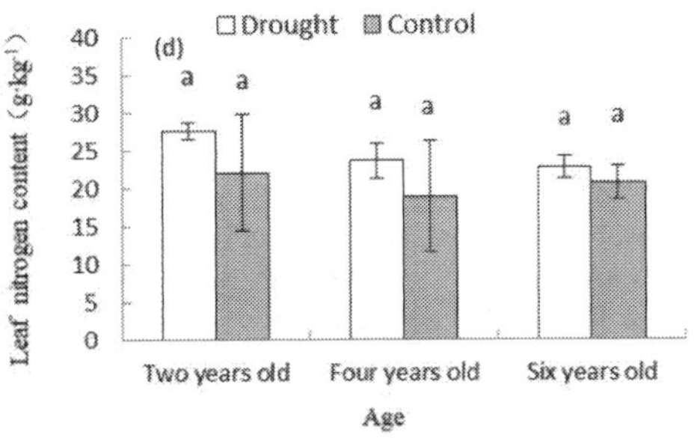

Fig. 2. Seasonal variation of leaf nitrogen content for different ages of *P. pubescens* under different treatments (Note: Figure (a), (b), (c) and (d) denotes spring, summer, autumn and winter, respectively.)

3.3 Response of leaf nitrogen content variance within canopy height to drought

The result showed that the nitrogen content of leaves changed with the canopy height for moso bamboo, and the gradient distribution characteristics were affected by the growing season (Fig. 3). Under natural growth, it increases with the canopy height in summer and autumn, while that of leaves in spring and winter tends to decrease with the increase of canopy height. Under the drought, that of leaves in summer was the highest in the upper part of the canopy, followed by the lower part. In spring and autumn, it decreased with the increase of canopy height, and was the highest in the middle of the canopy and the smallest in the upper part in winter. It can be seen that under natural growth and rain interception, the nitrogen content of leaves varies with the canopy height in spring as the upper part of the canopy < middle part < lower part, and the upper part of the canopy is the highest in summer.

The study found that the difference in the canopy gradient variation pattern of nitrogen content in Moso bamboo leaves between treatments was also affected by the growing season (Fig. 3). Except in autumn, it is lower under natural growth than that under drought in other seasons, indicating that the rain interception increases the nitrogen content of bamboo leaves in each canopy layer.

Analysis of variance showed that in spring, it was significantly different among canopies (P=0.008<0.01), however, in the same canopy location it was not significant between the control and the rain interception treatment (P>0.05). In autumn, there was significant differences between treatments (P<0.05 and P<0.01, respectively).

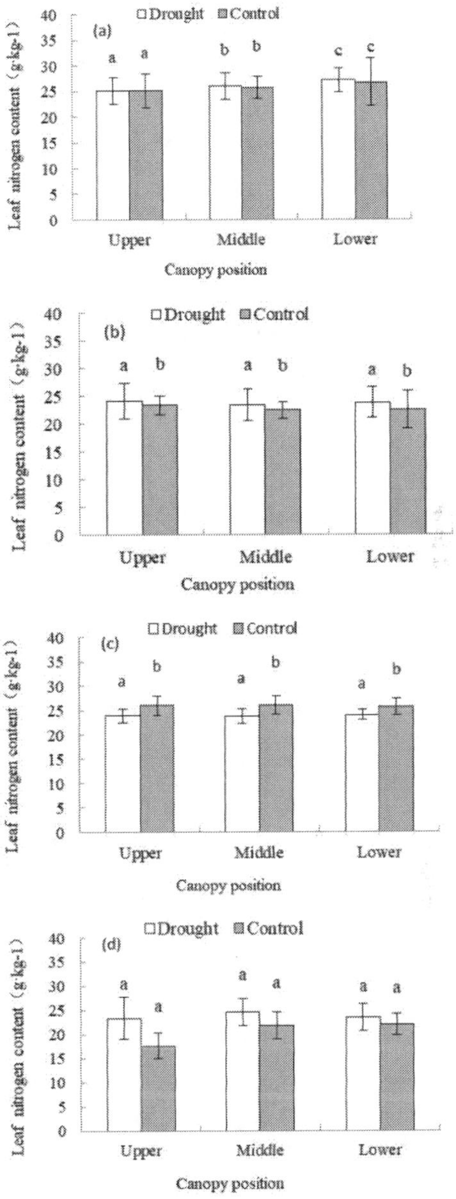

Fig. 3. Canopy variation of leaf nitrogen content for different ages of *P. pubescens* under different treatments (Note: Figure (a), (b), (c) and (d) denotes spring, summer, autumn and winter, respectively.)

3.4 Correlation of leaf nitrogen content with bamboo age, soil and leaf water content, and maximum net photosynthetic rate

It was linearly correlated with the increase of bamboo age under two treatments ($R2$=0.055 and $R2$=0.313, Figure 4), but not significant (P>0.05).

The nitrogen content of leaves changed with the change of soil water content in each growing season (Fig. 5). Under natural growth, it decreased with the increase of soil water content, but the correlation was not significant ($R2$=0.174). After the rain interception, there was a significant linear positive correlation with soil water content, indicating that the nitrogen content of bamboo leaves was more sensitive to the soil water content after the rain interception. Under natural growth, there was a small linear positive correlation between nitrogen content and leaf water content in *P. pubescens*, and a weak negative correlation existed under rain interception (Fig. 6).

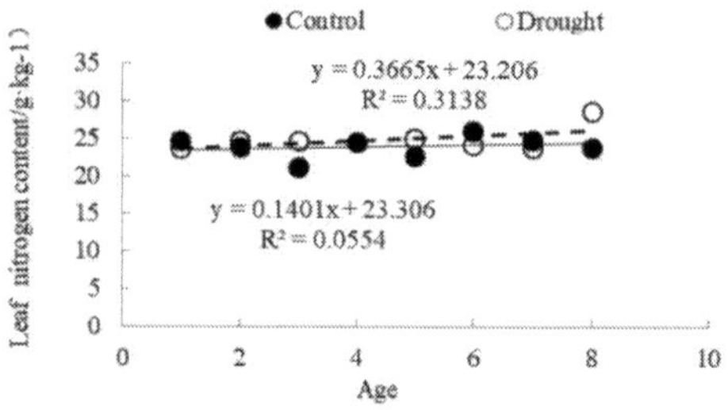

Fig.4. The correlation between leaf nitrogen content and age of *P. pubescens* under control and drought stress

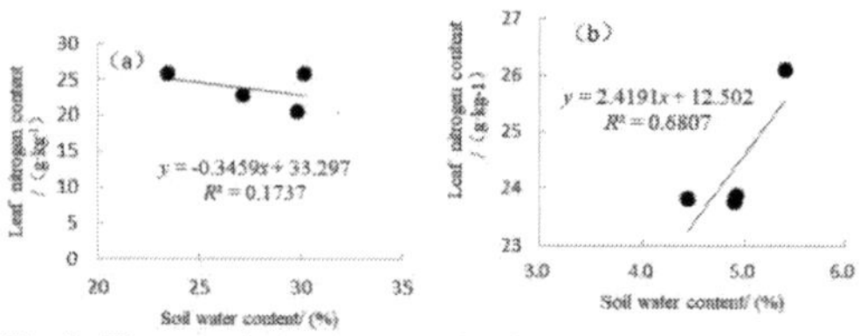

Fig.5. The correlation between leaf nitrogen content and soil water content for different growth stage of *P. pubescens* under control (a) and drought stress (b)

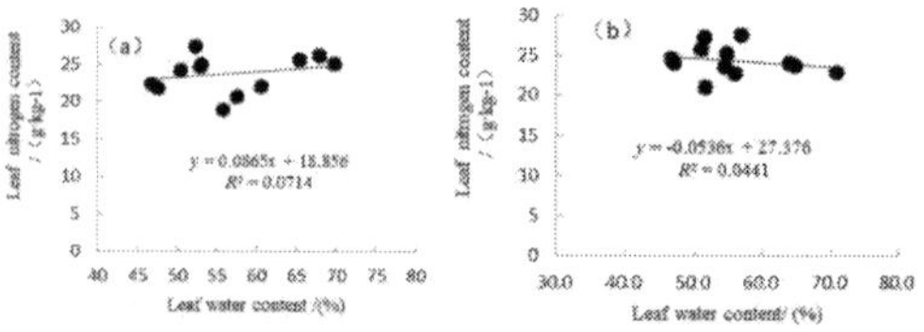

Fig.6. The correlation between leaf nitrogen content and leaf water content for different growth stage of *P. pubescens* under control (a) and drought stress (b)

4 Discussion

Nitrogen is one of the main limiting factors for plant growth [12]. The vertical gradient distribution of nitrogen content in canopy leaves is one of the important and common characteristics of tree canopy [13-14]. The maximum photosynthetic rate of trees is closely related to leaf nitrogen content and light radiation [2,15], the change of vertical distribution gradient of leaf nitrogen is regarded as a plastic response of plants to optimize nitrogen use during carbon assimilation [16]. Based on the basic physiological relationship between light radiation and nitrogen, the case studies of most scholars showed that when the leaves of trees in higher light contain higher nitrogen content than the leaves in shade, the photosynthesis of tree will reach the maximum, which is the "light-nitrogen" hypothesis [4,17-18].

There are many factors that affect the nitrogen gradient distribution in the canopy. There is a redistribution of nitrogen in canopy leaves at different growth stages of plants [6,18-19]. Therefore, at different growth stages of plants, the distribution of nitrogen in leaves determines the dynamic gradient distribution of nitrogen and its impact on canopy photosynthetic carbon fixation [19]. However, the effects of drought on the distribution of nitrogen gradient in the canopy are less studied.

At present, more studies have focused on traditional cultivation management techniques on bamboo forests [20-22], and there are recent reports on the limiting factors for the maximum height growth of bamboo plants [23]. Systematic research has also been carried out on its canopy photosynthesis [8,24-25]. However, there is still a lack of understanding of the distribution pattern of nitrogen in the canopy of *P. pubescens* and its correlation with soil water content and leaf water content, especially in the south of china where seasonal drought is becoming more and more obvious under global climate change.

Nitrogen is an important factor limiting the growth of *P. pubescens* [20,26-28]. Some scholars believed that different productivity levels affect the nitrogen distribution ratio of various organs of *P. pubescens*, and has a greater impact on the nitrogen distribution ratio of leaves and bamboo branches [29]. There are seasonal changes in the nitrogen content of bamboo leaves [20]. This study also found that the nitrogen content of leaves under two treatments varied with seasons as spring>autumn>summer>winter. In this study, the change of nitrogen content in leaves with bamboo age was affected by the growing season. Under natural growth, that of leaves increases with the increase of bamboo age in spring, but decreases with the increase of bamboo age in summer, which may be related to the consumption of nutrients by bamboo shoots in spring.

Scholars believed that the effect of canopy position on leaf nitrogen content during tree growth is staged [13]. The upper leaves of the canopy are in a high radiation environment and are prone to water shortage, which will affect the absorption, utilization and redistribution of nitrogen by plant leaves [14]. There is a significant linear correlation between the nitrogen content of manchurian ash leaves and the vertical height of the canopy, and it increases linearly with the increase of the vertical height of the canopy [14]. In this study, under natural growth, the nitrogen content of *P. pubescens* leaves basically increased with the canopy height in summer and autumn, which is consistent with the research results that the photosynthesis of *P. pubescens* in the upper part of the canopy is greater than that of the leaves in the lower part of the canopy[11] , and it indicated that the distribution pattern of nitrogen content in Moso bamboo leaves along the canopy gradient in summer and autumn is conducive to the demand for nitrogen in leaf photosynthesis, which also confirms the applicability of the "light-nitrogen" hypothesis to Moso bamboo in this growing season.

Effective leaf water supply can maintain stomatal opening to the maximum extent [30], thereby promoting photosynthesis. In this study, the nitrogen content of *P. pubescens* leaves decreased with the increase of canopy height in spring and autumn under drought. It can be seen that the rain interception has changed the canopy distribution pattern of nitrogen in Moso bamboo leaves to a certain extent. The canopy difference of nitrogen content becomes smaller. After rain interception, the response of leaf nitrogen content to soil water content is more sensitive, and it tends to increase with the increase of soil water content, and it is consistent with the result of Gao et al. (2019) that soil water content and leaf nitrogen are positively correlated in the process of vegetation restoration on the Loess Plateau [31]. Under natural growth, it is less affected by soil water content. The rain interception made the nitrogen content of leaves slightly higher than that of the naturally grown ones, which is in agreement with Fang Xuan et al. (2019) who believed that the leaf nitrogen of Chinese fir is more sensitive to water, and appropriate water deficit can increase the nitrogen concentration in the leaves [32]. A study on the relationship between leaf water content and photosynthesis at different leaf positions in the three-leaf stage of summer maize under continuous drought showed that leaf photosynthesis is closely related to its water content [33]. The results of this study provide a theoretical basis for sustainable management techniques based on nitrogen distribution in bamboo leaves under the background of climate change.

Acknowledgments

This study is funded by the project of Basic Public Welfare Research Program Project of Zhejiang Province(LGF20C160001).

References

1. Q.H. Feng, R.M. Chen, Z.M. Shi, et al. Response of Rumex dentatus foliar nitrogen and its allocation to altitudinal gradients along Balang Mountain, Sichuan, China. Chin. J. Plant Ecol., **37(7)** 591-600 (2013)
2. S. Yan, L. Zhang, Y.S. Jing, et al. Variations in the relationship between maximum leaf carboxylation rate and leaf nitrogen concentration. Chin. J. Plant Ecol., **38(6)** 640-652 (2014)
3. A. Boonman, N.P.R. Anten, T.A. Dueck, et al. Functional significance of shade induced leaf senescence: an experimental test using transgenic tobacco [J]. Am. Nat., **168(5)** 597–607 (2006)
4. R.E. McMurtrie, R.C. Dewar, Leaf-trait variation explained by the hypothesis that plants maximize their canopy carbon export over the lifespan of leaves. Tree Physiol., **31(9)** 1007–1023 (2011)
5. U. Niinemets, F. Valladares, Photosynthetic acclimation to simultaneous and interacting environmental stresses along natural light gradients: optimality and constraints Plant Biol., **6(3)** 254–268 (2004)
6. S.V. Archontoulis, J. Vosa, X. Yina, et al. Temporal dynamics of light and nitrogen vertical distributions in canopies of sunflower, kenaf and cynara Field Crops Res., **122(3)** 186-198 (2011)
7. M.S. Peltoniemi, R.A. Duursma, B.E. Medlyn, Co-optimal distribution of leaf nitrogen and hydraulic conductance in plant canopies Tree Physiol., **32(5)** 510-519 (2012)
8. G.S. Wen, L.Y. Zhang, R.M. Zhang, et al. Temporal and spatial dynamics of carbon fixation by Moso bamboo (Phyllostachys pubescens) in subtropical China Bot. Rev., **77(3)** 271-277 (2011)
9. Y.Q. Ying, J. Guo, J.F. Wei, et al. Effects of drought stress on physiological characteristics of Phyllostachys edulis seedlings Chin. J. Plant Ecol., **30(2)** 262-266 (2011)
10. J.L. Yuan, G.S. Wen, M.R. Zhang, et al. Water potential with Phyllostachys edulis in its fast-growth period J. Zhejiang A&F University, **32(5)** 722-728 (2015)
11. Y.H. Cao, B.Z. Zhou, X.M. Wang, et al. Effects of Canopy Height on Photosynthetic Physiology Characteristics of Phyllostachys pubescens Leaves Acta bota. boreali-occidentalia sin, **36 (11)** 2256-2266 (2016)
12. Z.M. Shi, J.C. Tang, R.M. Chen, et al. A review of nitrogen allocation in leaves and factors in its effects Acta ecol. Sin., **35(18)** 5909-5919 (2015)
13. X.B. Chen , S.J. Han, Z.H. Zhang, et al. Nutrient dynamics in Quercus mongolica leaves at different canopy positions Chin. J. Applied Ecol., **22(9)** 2272-2278 (2011)
14. J.X. Tian, L.P. Wei, N.P. He, et al. Vertical variation of leaf functional traits in temperate forest canopies in China Acta ecol. Sin., **38(23)** 8383~8391 (2018)
15. J.C. Tang, Z.M. Shi, D. Luo, et al. Photosynthetic nitrogen-use efficiency of Manglietia glauca seedling leaves under different shading levels Acta ecol. Sin., **(22)** 122-131 (2017)
16. S. Delagrange, Light- and seasonal-induced plasticity in leaf morphology, N partitioning and photosynthetic capacity of two temperate deciduous species Environ. & Exp. Bot.,**70(1)** 1-10 (2011)
17. C. Field, Allocating leaf nitrogen for the maximization of carbon gain: leaf age as a control on the allocation program Oecologia, **56(2-3)** 341–347 (1983)
18. T. Hirose, M.J.A. Werger, Maximizing daily canopy photosynthesis with respect to the leaf nitrogen allocation pattern in the canopy Oecologia, 72(4):520-526 (1987).
19. L. Hallik, Ü. Niinemets, O. Kull, Photosynthetic acclimation to light in woody and herbaceous species: a comparison of leaf structure, pigment

content and chlorophyll fluorescence characteristics measured in the field Plant Biol., **14(1)** 88-99 (2012)

20. X.P. Gu, X.L. Wu, Y.D. Wang, Study on diagnosis of nitrogen nutrition in Phyllostachys pubescens J. Zhejiang For. Sci. Techn., **24(2)** 1-4 (2004)
21. P.J. Gao, Y.H. Qiu, Z.Q. Zhou, et al. Productivity and photosynthetic ability of Phyllostachys edulis with nitrogen fertilization J. Zhejiang A&F University, **31(5)** 697-703 (2014)
22. G.L. Liu, S.H. Fan, B.H. Guo, et al. The Carbon, Nitrogen and Phosphorus Contents of Phyllostachys edulis with Different Ages Chin. J. Trop. Crops, **37(2)** 279-285 (2016)
23. K.F. Cao, S.J. Yang, Y.J. Zhang, et al. The maximum height of grasses is determined by roots Ecol. Lett., **15(7)** 1–7 (2012)
24. H. Komatsu, Y. Onozawa, T. Kume, et al. Canopy conductance for a Moso bamboo (Phyllostachys pubescens) forest in western Japan. Agric. For. Meteorol., 156 111-120.doi:10.1016/j.agrformet.2012.01.004 (2012)
25. Y.H. Cao, B.Z. Zhou, X.M. Wang, et al. The photosynthetic characteristics of Moso bamboo (Phyllostachys pubescens) for different canopy leaves Adv. Mat. Res., **726-731** 4274-4279 (2013)
26. M.Y. Du, S.H. Fan, G.L. Liu, et al. Stoichiometric characteristics of carbon, nitrogen and phosphorus in Phyllostachys edulis forests of China. Chin. J. Plant Ecol., **40(8)** 760-774 (2016)
27. S.M. Chen, T.T. Hu, L.H. Luo, et al. Rapid estimation of leaf nitrogen content in apple-trees based on canopy hyperspectral reflectance using multivariate methods. Infrared Phys. Techn.,111, 103542–. doi:10.1016/j.infrared.2020.103542 (2020)
28. D.L. Jiang, B.L. Yang, X.L. Cheng, et al. The stoichiometry of leaf nitrogen and phosphorus resorption in plantation forests. For.Ecol.Manag., 483:118743. doi:10.1016/j.foreco.2020.118743. (20 20)
29. B.H. Guo, G.L. Liu, S.H. Fan, et al. Distribution patterns and stoichiometry characteristics of C, N, P in Phyllostachys edulis forests of different productivity levels[J]. Scientia Silvae Sinicae, **50(6)** 1-9 (2014)
30. S. Manzoni, G. Vico, A. Porporato, et al. Biological constraints on water transport in the soil-plant-atmosphere system Adv.Water Resour., **51** 292-304 (2013)
31. D.X. Gao, W. Zhang, C.J. Ren, et al. Ecological stoichiometry characteristics of soil and leaves during the recovery process of typical vegetation on the Loess Plateau Acta ecol. Sin., **39(10)** 3622-3630 (2019)
32. X. Fang, J. Wang, B. Wang, et al. Effects of simulated soil warming and precipitation exclusion on N and P metabolisms in Cunninghamia lanceolate[J]. Acta ecol. Sin., **39(10)** 3526-3536 (2019)
33. F. Wang, Q.J. He, G.S. Zhou, (2019) Leaf water content at different positions and its relationship with photosynthesis when consecutive drought treatments are applied to summer maize from the 3-leaf stage[J]. Acta ecol. Sin., 39(1):254~264.

The Removal of Cd^{2+}, Cu^{2+} and Pb^{2+} from Aqueous Solution with Palygorskite: The Effects of Humic Acid

Sichun Ding[1,2], *Yuting* Tang[1,2,*], *Xiaobin* Chen[1,2], *Xiaoqian* Ma[1,2], and *Tao* Huang[1,2]

[1]School of Electric Power, South China University of Technology, 510640, Guangzhou, China.
[2]Guangdong Province Key Laboratory of Efficient and Clean Energy Utilization, 510640 Guangzhou, China.

Abstract. To investigate the impact of Humic acid (HA) on Palygorskite (PAL) as the adsorbent for heavy metal ions, this study was carried out to solve the existing dispute in the effects of HA on PAL sorption capacity. The sorption properties of PAL and HA-loaded PAL (HA/PAL) for Cd^{2+}, Cu^{2+}, Pb^{2+} in batch experiments were evaluated. PAL and HA/PAL effects on metal ion elimination were examined under various test settings (time, dosage and starting concentration) were compared. The sorption efficiency of PAL was significantly higher than that of HA/PAL. The N_2 sorption-desorption and FTIR results showed that HA successfully adhered to the PAL surface without changing its crystal structure but reducing its specific surface area. Cd^{2+}, Cu^{2+}, and Pb^{2+} for PAL and Cd^{2+} and Cu^{2+} for HA/PAL were modelled by pseudo-first-order kinetics; the pseudo-second-order rate conformed to Cu^{2+} for PAL and Pb^{2+} for HA/PAL. the Langmuir linear fitting was consistent with Cd^{2+}, Cu^{2+} and Pb^{2+} for PAL and HA/PAL. Pb^{2+} for PAL and Cd^{2+}, Pb^{2+} for HA/PAL were consistent with the Freundlich isotherm. The results proved that HA was not conducive to improving the adsorption performance of PAL.

1 Introduction

During the production process of certain industries, all kinds of heavy metals are let out into the nature in the form of sewage. Due to their nonbiodegradability and bioaccumulation, once the quantity of heavy metal ions in the aquatic environment surpasses the permissible level, heavy metal ions inflict irreparable damage to human health and the ecological balance [1]. Palygorskite (PAL) has excellent salt tolerance and a rapid hydration rate [2]. Two aspects give PAL good sorption properties. Humic acid (HA) is important in modulating heavy metal geochemical behavior and processes [3]. HA has a large number of functional groups, leading to many different metal ion binding sites [3]. Humus soil in aged refuse provides us with a large source of HA, and the resource utilization of humus soil is of great significance for waste treatment. It is a problem worth studying whether the advantages of the both can be combined through the way of load. However, the majority of scholars have not reached a consensus on whether HA is beneficial to PAL sorption of heavy metal ions. Li et al. found that HA-PAL has good sorption of Ni^{2+} in wastewater [4]. In contrast, Rouhaninezhad et al. found that the introduction of HA into PAL could reduce the removal rate of Cr^{6+} in solution [5]. It provides the original intention of this paper. As a result, further research on the implications of HA on the PAL sorption of various heavy metal is required.

2 Substances and Techniques

2.1 Substances

Guangzhou Qianhui Co., LTD. in China supplied the PAL. Therefore, it has been widely used in the environmental protection field. Tianjin Guangfu Institute of Fine Chemicals in China supplied the humic acid. Shanghai Macklin Co., LTD. in China, provided the $Cd(NO_3)_2{\cdot}4H_2O$. $Cu(NO_3)_2{\cdot}3H_2O$ was provided by Tianjing Damao Chemical Reagent Factory, China. $Pb(NO_3)_2$, NaOH, and HCl were supplied by Guangdong Guangshi Reagent Technology Co., Ltd., China. All the above reagents were guaranteed reagents. The water used for the whole experiment was deionized water.

2.2 Preparation

In order to improve the purity of the adsorbent, PAL was first purified with 3 mol/L HCl. After that, it was cleaned with deionized water to neutralize it before being dried for use. Next, 0.5 g HA was weighed accurately and placed in a beaker with 500 ml of water. Then, 0.1 mol/L NaOH was dribbled continuously into the solution until HA was completely dissolved. Following filtration, the filtrate was put in a beaker and the pH was adjusted to 3. Afterwards, at constant temperature of 60 ℃, 5 g purified PAL and HA solution were fully stirred in a magnetic agitator for 6 h and allowed to sit for 12 h. Next, the mixture was rinsed with deionized water and filtered multiple times until the supernatant contained no HA. The HA-loaded PAL was crushed to a 200-mesh powder

* Corresponding author: tangyuting_43@163.com

after drying at 105 °C continuous temperature for 8 hours. Finally, the HA-loaded PAL powder was stored in a dry environment. The preparation steps of the adsorbent referenced Jiang et al [6].

2.3 Experiment method

Batch sorption tests were performed with various heavy metal ions (Cd^{2+}, Cu^{2+}, Pb^{2+}), reaction periods (10, 30, 60, 120, 180 min), PAL or HA/PAL doses (0.05, 0.1, 0.2, 0.3 g), and heavy metal starting concentrations (40, 60, 80 mg/L). The metal ions solution is kept constant at a volume of 20 ml. Measure all the Cd^{2+}, Cu^{2+} and Pb^{2+} original pH values at 5.4. The temperature is kept constant at 35 ℃. at a specific level of concentration, pour the heavy metals solution into 50 ml capped plastic centrifuge tubes. The solution received a particular amount of adsorbent. The mixture oscillated in an isothermal water bath shaker for a certain time at a rotation speed of 200 rpm. The supernatants were collected after centrifuging the suspensions for 2 minutes at 4000 rpm. Use an inductively coupled plasma emission spectrometer (ICP-OES) to determine. The heavy metals residual concentration. Finally, the HA/PAL heavy metal ion sorption capacity (q) and sorption efficiency (η) could be calculated.

$$q = \frac{(C_0 - C_e) \times V}{m} \quad (1)$$

$$\eta = \frac{C_0 - C_e}{C_0} \times 100 \quad (2)$$

where C_0 is the heavy metal starting concentration, and C_e is the heavy metal equilibrium concentration (mg/L); V is the solution volume (L); and m is the mass of adsorbent (g).

2.4 Characterization analyses

Using the Fourier transform infrared spectroscopy (FTIR) was to characterize the adsorbent functional groups. Using the Brunauer–Emmett–Teller (BET)was to determine the textural qualities of the samples. The aforesaid characterisation was carried out at the SCUT Analysis and Test Center and Key Laboratory of Efficient And Clean Energy Utilization of Guangdong Province.

2.5 Sorption kinetics

The sorption rate of the pseudo-first-order (PFO) kinetic equation and the quantity of adsorbent pores are relative, whereas the sorption rate and the square of the number of holes are dependent in the pseudo-second-order (PSO) kinetic equation [7]. the PFO and PSO kinetic equation are pursuing:

$$q_t = q_e\left(1 - e^{-K_1 t}\right) \quad (3)$$

$$\frac{1}{q_t} = \frac{1}{K_2 q_e^2 t} + \frac{1}{q_e} \quad (4)$$

where at equilibrium and time t (min), the heavy metal removal amount is q_e (mg/g), and q_t (mg/g), respectively; and K_1 (min^{-1}) is the Eqs. (3) rate constants; and K_2 (g/(mg•min)) is the Eqs. (4) rate constants.

2.6 Sorption isotherms

The goal of the isotherm was to explore at the sorption patterns of adsorbents and heavy metals, as well as the distribution of heavy metal ions after the sorbent-sorbate system achieved equilibrium, sorption isotherms were employed. A heterogeneous adsorbent surface with heterogeneous energy active sites is postulated in the Freundlich [7]. The Langmuir isotherm, on the other hand, implies that the active sorption sites are equitably spread and that the surface of the adsorbent is monomolecular [7]. Eqs. (5) and (6) express the Freundlich and Langmuir isotherm equations, respectively.

$$q_e = K_F C_e^{\frac{1}{n}} \quad (5)$$

$$q_e = \frac{K_L q_m C_e}{1 + K_L C_e} \quad (6)$$

where q_e (mg/g) is the heavy metal equilibrium removal quantity; C_e (mg/L) is the concentration in balance; K_F (mg$^{(1-n)}$/(L^{-n} • g)) is the sorption constant of Freundlich; and n is the inhomogeneity factor of particle surface. The Langmuir theoretical maximum sorption quantity is q_m (mg/g), while the Langmuir bonding term connected with sorption strength is K_L (L/mg).

3 Discussions and analyses

3.1 Effects of time on heavy metal sorption

The sorption rates of Cd^{2+}, Cu^{2+}, and Pb^{2+} for PAL and HA/PAL rose over time and eventually reached equilibrium at 120 minutes (Fig. 1). The sorption of Cd^{2+}, Cu^{2+} and Pb^{2+} for PAL and HA/PAL went through three stages, and the patterns were similar. In the first stage (0 - 10 minutes), The sorption rate was extraordinarily quick, and three heavy metal ions were completely absorbed. In the next stage (Cd^{2+}, Cu^{2+}: 10 - 120 minutes; Pb^{2+}: 10 - 30 min), the sorption rate slowed down, and the sorption process gradually tended to equilibrium. In the last step (Cd^{2+}, Cu^{2+}: after 120 minutes; Pb^{2+}: after 30 minutes), the sorption reached equilibrium. After adsorption equilibrium, Cd^{2+}, Cu^{2+}, and Pb^{2+} removal rates for PAL were 87 %, 91 %, and 99 %, respectively, and Cd^{2+}, Cu^{2+}, and Pb^{2+} removal rates for HA/PAL were 80 %, 56 %, and 98 %, respectively. Compared with PAL, HA/PAL has a slightly lower instantaneous adsorption rates. The quick response during the early stage of adsorption might be attributed to the immediate interchange of heavy metals with cations at the PAL and HA/PAL particles exterior planar and edge locations. After the rapid reaction, Cd^{2+}, Cu^{2+} and Pb^{2+} may slowly diffuse into the exchange sites of PAL and HA/PAL. Therefore, 120 minutes was determined as the optimal contact time.

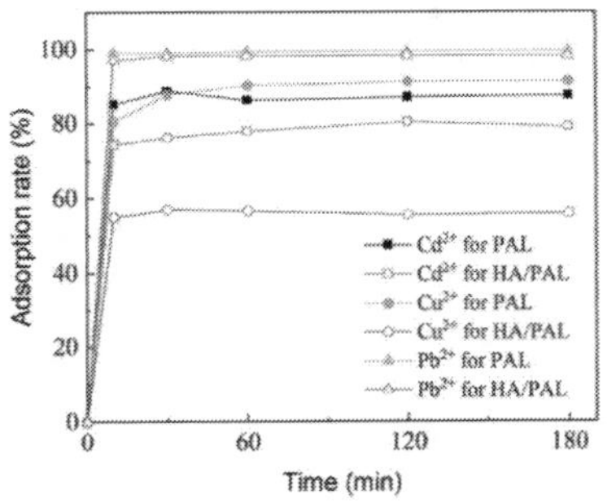

Fig. 1. Effects of time on heavy metal sorption in PAL and HA/PAL. (pH = 5.4, temperature = 35 ℃, dosage = 0.2 g, volume = 20 ml, starting concentration = 80 mg/L)

3.2 Effects of dosage on heavy metal sorption

The sorption rates increased with increasing adsorbent dosage (Fig. 2 (a)). This was due to the fact that increasing the dose supplied additional sorption sites for heavy metal ions [8]. When the PAL dose exceeded 0.05 g, the sorption rate of Pb^{2+} did not increase. The sorption rate of Pb^{2+} for HA/PAL reached the highest when the dosage was 0.2 g. After adding more than 0.2 g HA/PAL, the sorption rate of Cd^{2+} and Cu^{2+} for PAL did not change much, and the sorption rate of Cd^{2+} for HA/PAL did not change much. When the HA/PAL dosage reached 0.3 g, it still increased significantly. The sorption rate of most heavy metal ions did not increase obviously after 0.2g dosage. After comprehensive consideration, 0.2 g was considered the best additive amount for PAL and HA/PAL.

The sorption capacity decreased with increasing dose, contrary to the changing trend of the sorption rate (Fig. 2 (b)). This rule was widely used as part of the heavy metal sorption investigation. The accepted explanation was that increasing the adsorbent dose caused heavy metal ions to compete with the adsorbent, the agglomeration of adsorbent particles, and the lengthening of the diffusion route [8]. This is because adsorbent particle collisions can produce particle polymerization, which reduces adsorption surface area while increasing diffusion route length. Meanwhile, the increase of the dose increased the probability of particle collision and aggregation. Therefore, when most Cd^{2+}, Cu^{2+} and Pb^{2+} were adsorbed by PAL and HA/PAL, the number of heavy metal ions bound to the adsorption site did not increase with increasing dosage.

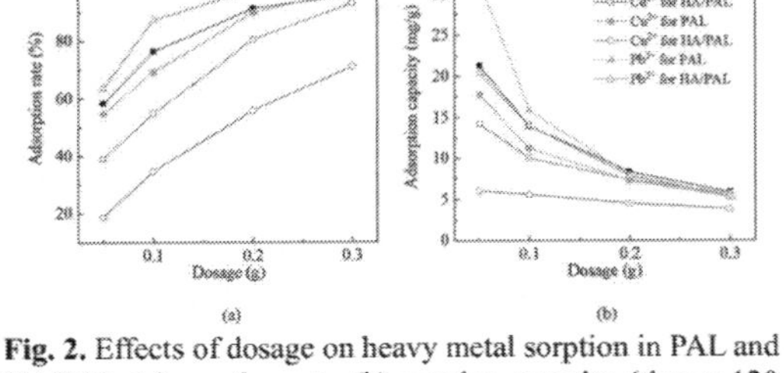

Fig. 2. Effects of dosage on heavy metal sorption in PAL and HA/PAL: (a) sorption rate: (b) sorption capacity. (time = 120 min, pH = 5.4, temperature = 35 ℃, volume = 20 ml, starting concentration = 80 mg/L)

3.3 Effects of starting concentration on heavy metal sorption

The sorption rates decreased as the starting concentration increased (Fig. 3 (a)). The fundamental explanation for this is that while the dosage was fixed, the adsorbent afforded restricted sorption sites [9], leading to a result that the remaining ion quantities heightened along with the heavy metal starting concentrations grew. The starting concentration had a decreasing influence on the rise in heavy metal ion adsorption capability. Therefore, the sorption rates decreased with increasing starting concentration. However, the change in starting concentration had almost no influence on the PAL and HA/PAL sorption of Pb^{2+}, indicating that these two adsorbents provided sufficient sorption sites for Pb^{2+} in the vicinity of 40-80 mg/L. In addition, the increase in starting concentration did not significantly decrease the sorption rates of PAL to Cd^{2+} and Cu^{2+}, indicating that PAL had relatively sufficient sorption sites for Cd^{2+} and Cu^{2+}. However, the sorption rates of HA/PAL for Cd^{2+} and Cu^{2+} decreased significantly, indicating that the addition of HA reduced the sorption sites of Cd^{2+} and Cu^{2+} in PAL.

The sorption capacity increased with increasing starting concentration (Fig. 3 (b)). Because the adsorbent effective sorption sites could meet the number of heavy metals. When the starting concentration was low, the concentration of the solution has a considerable impact on the capacity for adsorption. Furthermore, when the starting concentration grew, so did the driving power of heavy metal ion migration to sorption sites. Therefore, unit mass adsorbent could adsorb more heavy metal ion. However, when the starting concentration hit a certain threshold, the sorption site attained saturation and the sorption capacity peaked, as seen from the HA/PAL sorption capacity curve of Cu^{2+}. In contrast, other sorption capacity curves in Fig. 3 (b) show an upward trend, proving that the sorption sites were not saturated, and the sorption capacity of PAL and HA/PAL did not reach the limit.

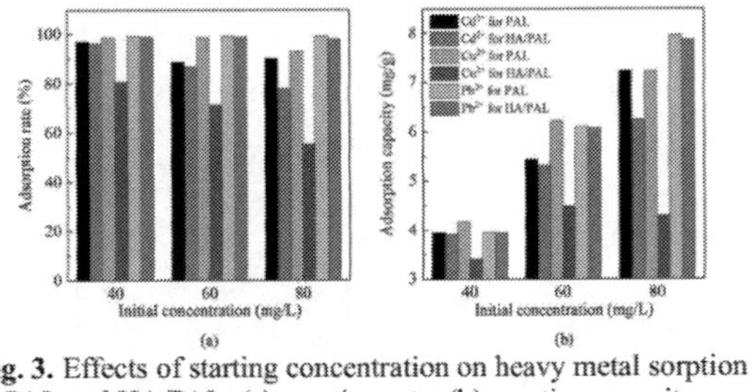

Fig. 3. Effects of starting concentration on heavy metal sorption in PAL and HA/PAL: (a) sorption rate; (b) sorption capacity. (time = 120 min, pH = 5.4, temperature = 35 ℃, volume = 20 ml, dosage = 0.2 g

3.4 BET of PAL and HA/PAL

The N_2 sorption-desorption isotherm results indicated

that PAL and HA/PAL also possessed a H3-type hysteresis loop and typical IV(a)-type isotherm (Fig. 4), in accordance with the IUPAC classification [10]. In the 0.8-1 relative pressure range, PAL and HA/PAL hysteresis loops indicated that many mesopores were evenly distributed in the material. In the whole range of relative pressures, the N_2 sorption capacity of PAL was not improved by HA loading. The BET used the static volume method for testing. when PAL was loaded with HA, the pore diameter and volume, specific surface area were all decreased. (Table 1), which was not conducive to PAL physical sorption. HA penetrated the large pores of PAL, forming smaller holes (Fig. 5), leading to a result that active sites and driving forces decrease.

Table 1. BET test results of PAL and HA/PAL

	PAL	HA/PAL
BET surface area (m^2/g)	60.1947	46.9859
Pore volume (cm^3/g)	0.1047	0.0501
Sorption average pore diameter (Å)	69.5532	42.6476

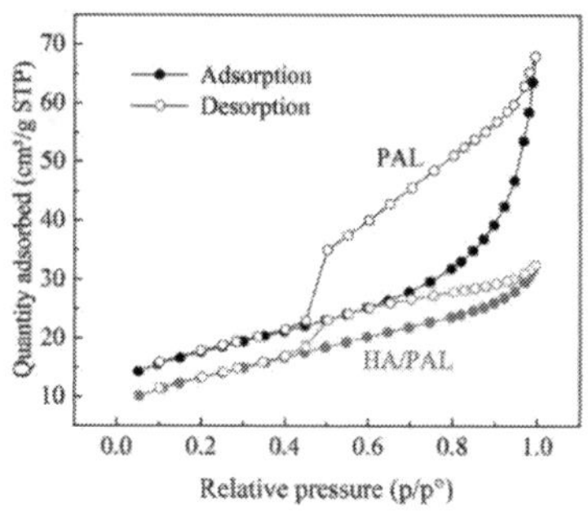

Fig. 4. PAL and HA/PAL N_2 sorption-desorption isotherms

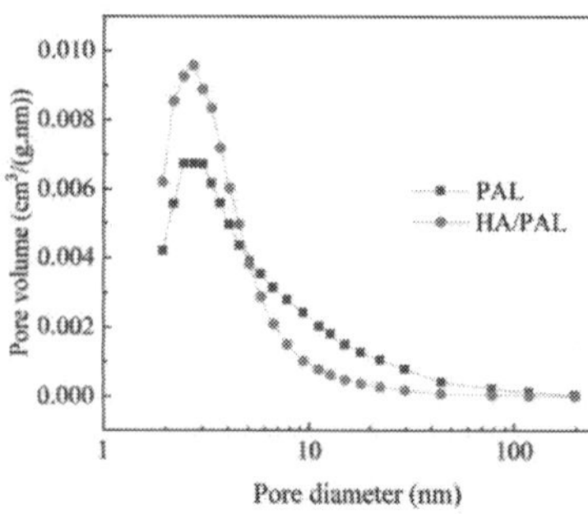

Fig. 5. Pore volume - pore diameter curve for PAL and HA/PAL

3.5 The FTIR of PAL and HA/PAL

Fig. 6 is the FTIR of PAL and HA/PAL. PAL contains distinct peaks at 3624, 3435, 913, and 836 cm^{-1} that are linked to R-OH stretching oscillations [11]. The HA/PAL characteristic peaks are at 3625, 3443, 922, and 839 cm^{-1}, and they barely changed slightly, showing that the HA load did not modify the structure of PAL, but the decrease in characteristic peak intensity suggested that the HA load decreased the amount of PAL functional groups, consistent with the BET study results.

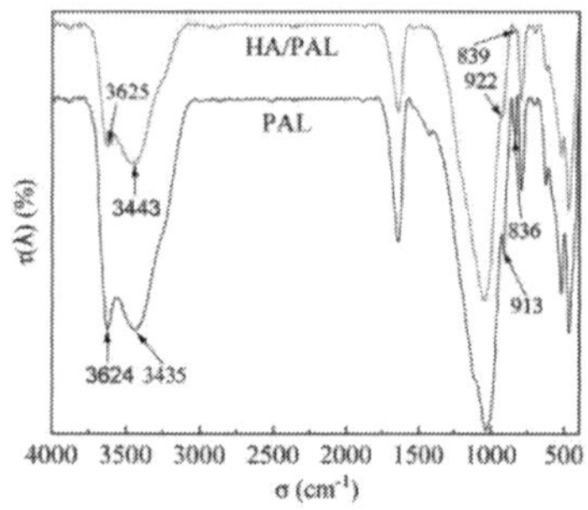

Fig. 6. FTIR of PAL and HA/PAL

3.6 Sorption kinetic results

According to Table 2, the best model for Cd^{2+}, Cu^{2+}, and Pb^{2+} sorption on PAL and Cd^{2+}, Cu^{2+} sorption on HA/PAL was the PFO model (R^2 = 0.9995, 0.9987, 0.9999, 0.9982, 0.9995, respectively). In addition, Cu^{2+} sorption on PAL and Pb^{2+} sorption on HA/PAL proceeded the PSO rate expression (R^2 = 0.9979, 0.9259, respectively). Table 2 also shows that HA significantly reduced the sorption rate of PAL, indicating that PAL reached sorption equilibrium earlier than HA/PAL.

Table 2. Cd^{2+}, Cu^{2+}, and Pb^{2+} sorption kinetics model fitting parameters on PAL and HA/PAL

Adsorbent	Heavy metal ion	PFO			PSO		
		q_e (mg/g)	K_1 (min-1)	R^2	q_e (mg/g)	K_2 (g/(mg·min))	R^2
PAL	Cd^{2+}	7.81	0.37	0.9995	7.83	0.52	0.3772
	Cu^{2+}	7.58	0.26	0.9987	7.76	0.09	0.9979
	Pb^{2+}	8.17	2.43	0.9999	8.17	2.39	0.6328
HA/PAL	Cd^{2+}	7.00	0.30	0.9982	7.09	0.19	0.8238
	Cu^{2+}	4.70	0.05	0.9995	4.73	0.99	0.3075
	Pb^{2+}	8.06	0.05	0.5524	8.08	1.03	0.9259

3.7 Sorption isotherm results

The isothermal sorption line was studied at 35 ℃, the dosage of adsorbent was 0.2 g, the solution's starting concentration was 80 mg/L, and its volume was 20 mL, and the pH value was 5.4. In accordance with Table 3, the sorption of Cd^{2+}, Cu^{2+}, and Pb^{2+} on PAL and HA/PAL was reported to be compatible with the the Langmuir, suggesting the monolayer nature of sorption [12]. By confining the sorption to a monolayer, Langmuir fitting revealed a substantial interaction between Cd^{2+}, Cu^{2+}, and Pb^{2+} and adsorbents. The sorption of Pb^{2+} on PAL and the sorption of Cd^{2+}, Pb^{2+} on HA/PAL were consistent with the linear fitting of the Freundlich isotherm. *1/n* was a heterogeneous factor, with *n* values of 1.19, 5.29 and 2.19, respectively, indicating that the sorption effect was good [13]. The sorption of Pb^{2+} on PAL and the sorption of Cd^{2+}, Pb^{2+} on HA/PAL conformed to the characteristics of multilayer sorption [14].

Table 3. Sorption isotherms model fitting parameters of Cd^{2+}, Cu^{2+} and Pb^{2+} on PAL and HA/PAL

Adsorbent	Heavy metal ion	Freundlich			Langmuir		
		n	K_f	R^2	q_m	K_L	R^2
PAL	Cd^{2+}	3.85	3.73	0.8265	7.35	0.83	0.9063
	Cu^{2+}	5.34	5.41	0.6673	7.65	3.05	0.9977
	Pb^{2+}	1.19	11.30	0.9953	36.45	0.42	0.9974
HA/PAL	Cd^{2+}	5.29	3.61	0.9991	6.67	0.70	0.9956
	Cu^{2+}	6.01	2.51	0.6740	4.61	0.51	0.9911
	Pb^{2+}	2.19	6.36	0.9162	11.70	1.29	0.9522

4 Conclusion

The characterisation and study of these two adsorbents revealed that the crystal structure of PAL was not altered by HA, but its sorption sites were lowered. The sorption characteristics of adsorbents are still mainly physical sorption. It is generally known, HA primarily binds heavy metal ions by complexation, which is a chemical reaction. Therefore, although the inclusion of HA increased the amount of functional groups in PAL, it also competed for heavy metal ion adsorption sites. Obviously, this competition dominated the whole adsorption process. The ability of PAL to remove heavy metals was not improved by HA loading. Overall, the findings of this investigation clearly support the contrary viewpoint. HA does not give play to the advantages of its functional groups, and Therefore, the inclusion of HA in PAL sorption of heavy metals should be avoided.

Acknowledgments

The Guangzhou Science and Technology Innovation Development Special Fund [202002030105] and the Guangdong Province Natural Science Foundation of China [2114050000829] sponsored this research..

References

1. G. Tian, W. Wang, L. Zong, A. Wang (2017). MgO/palygorskite adsorbent derived from natural Mg-rich brine and palygorskite for high-efficient removal of Cd(II) and Zn(II) ions. Journal of Environmental Chemical Engineering **5**: 1027-1036.
2. S.A. Al-Hammadi, A.A. Al-Absi, O.A. Bin-Dahman, T.A. Saleh (2018). Poly(trimesoyl chloride-melamine) grafted on palygorskite for simultaneous ultra-trace removal of methylene blue and toxic metals. J Environ Manage **226**: 358-364.
3. W. Shi, C. Lu, J. He, H. En, M. Gao, B. Zhao, B. Zhou, H. Zhou, H. Liu, Y. Zhang (2018). Nature differences of humic acids fractions induced by extracted sequence as explanatory factors for binding characteristics of heavy metals. Ecotoxicology And Environmental Safety **154**: 59-68.
4. J.J. Li, Y.W. Zheng, X. Feng, C. Lv, X.T. Liu, Y.P. Zhao, L. Chen(2019). Adsorption removal of Ni(II) and phenol from aqueous solution by modified attapulgite and its composite hydrogel. Environ Technol **42**(15): 2413-2427.
5. A.A. Rouhaninezhad, S. Hojati, M.N. Masir (2020). Adsorption of Cr (VI) onto micro- and nanoparticles of palygorskite in aqueous solutions: Effects of pH and humic acid. Ecotoxicol Environ Saf **206**: 111247.
6. H. Jiang, W. Zhang, S. Zhou, Y. Duan (2015). Adsorption properties and mechanism of U(VI) onto humic acid /attapulgite composites. Chinese Journal of Environmental Engineering **9**: 705-710.
7. M. Niu, G. Li, L. Cao, X. Wang, W. Wang (2020). Preparation of sulphate aluminate cement amended bentonite and its use in heavy metal adsorption. Journal of Cleaner Production **256**: 120700.
8. Y.S. Chang, P.I. Au, N.M. Mubarak, M. Khalid, P. Jagadish, R. Walvekar, E.C. Abdullah (2020). Adsorption of Cu(II) and Ni(II) ions from wastewater onto bentonite and bentonite/GO composite. Environ Sci Pollut Res Int **27**: 33270-33296.
9. E. Mahmoudi, S. Azizkhani, A.W. Mohammad, L.Y. Ng, A. Benamor, W.L. Ang, M. Ba-Abbad (2020). Simultaneous removal of Congo red and cadmium(II) from aqueous solutions using graphene oxide-silica composite as a multifunctional adsorbent. J Environ Sci (China) **98**: 151-160.
10. Q. Chen, R. Zhu, L. Deng, L. Ma, Q. He, J. Du, H. Fu, J. Zhang, A. Wang (2019). One-pot synthesis of novel hierarchically porous and hydrophobic Si/SiOx composite from natural palygorskite for benzene adsorption. Chemical Engineering Journal **378**: 122131.
11. W. Yan, D. Liu, D. Tan, P. Yuan, M. Chen (2012). FTIR spectroscopy study of the structure changes of palygorskite under heating. Spectrochim Acta A Mol Biomol Spectrosc **97**: 1052-1057.
12. C.V. Lazaratou, D. Panagiotaras, G. Panagopoulos, M. Pospíšil, D. Papoulis (2020). Ca treated Palygorskite and Halloysite clay minerals for Ferrous Iron (Fe+2) removal from water systems. Environmental Technology & Innovation **19**: 100961.
13. A.P. Nazar de Souza, Y.E. Licea, M.V. Colaço, J.D. Senra, N.M.F. Carvalho (2021). Green iron oxides/amino-functionalized MCM-41 composites as adsorbent for anionic azo dye: kinetic and isotherm studies. Journal of Environmental Chemical Engineering **9**(2): 105062.
14. S. Radoor, J. Karayil, A. Jayakumar, J. Parameswaranpillai, S. Siengchin (2021). Efficient removal of methyl orange from aqueous solution using mesoporous ZSM-5 zeolite: Synthesis, kinetics and isotherm studies. Colloids and Surfaces A: Physicochemical and Engineering Aspects **611**: 125852.

Experimental Study on the Preparation of Improved Soil Substrate from Coal Gangue

Qianyi Liu[1], *Yingte* Li[1], *Linlin* Du[1], *Liu* Ma[1], and *Lianman* Xu[1,*]

[1] Liaoning University, School of Environment, 110036, No.66, Chongshan Middle Road, Huanggu District, Shenyang, Liaoning Province, China

Abstract. With the continuous development of China's industry, the disorderly discharge and accumulation of coal gangue in mining areas have become a key ecological problem, which not only occupies a large amount of land, but also causes various pollution phenomena. Therefore, the comprehensive utilization of coal gangue has become a research hotspot in the field of environmental governance. In this paper, by mixing coal gangue and fly ash with straw in different ratios to prepare compound soil substrate and growing lettuce, the effects of different ratios of the three wastes and coal gangue particle sizes on crop growth were analyzed by measuring cotyledon area, plant height and other indexes. The experimental results showed that the best proportion of soil matrix improvement was 42.9% coal gangue, 7.1% fly ash, 10% straw and 40% soil, where the coal gangue particle size proportion was 20% fine particles, 20% medium particles and 60% coarse particles. This experiment mixed coal gangue with typical solid wastes, which will, to a certain extent, solve the problem of solid waste dumping and resource scarcity in China and provide a scientific reference for the comprehensive utilization of the three wastes and the repair of soil in mining areas.

1 Introduction

In the process of the continuous promotion of the national coal market policy, the coal consumption of China, the largest coal producer and consumer in the world, keeps increasing, reaching 3.68 billion tons in 2018, accounting for 50.2% of the world coal consumption [1]. Coal gangue is an associated resource produced in the process of coal production and processing. According to statistics, coal gangue storage occupies an area of about 106hm2 in China, and nearly 4 billion tons of the storage makes coal gangue one of the most important solid waste in China [2]. Due to the wide distribution area, high probability of environmental pollution and long duration [3], the open-air stacking of coal gangue will not only cause soil, air and groundwater pollution, but also induce landslides and debris flows, thus causing adverse effects on the natural landscape and ecological environment of the mining area.

At present, coal gangue treatment technology is mainly divided into ground and underground treatment technology. The research and utilization of coal gangue ground treatment technology mainly focuses on coal gangue power generation, road paving, production of building materials, production of chemical raw materials and agricultural application [4]. Shen, H.S., He, W.H. [5] et al. used coal gangue instead of concrete coarse aggregates to carry out the compressive strength test of coal gangue-based concrete and underground hardening tests, which verified the good effect of coal gangue-based concrete in underground hardening projects. Gu, L.Y. [6] et al. prepared porous ceramics with good compressive strength by using Huainan gangue and fly ash as raw materials, soda as sintering agent and sawdust as the pore-forming agent. Jin B. [7], et al. used sludge, gangue and shale as materials to prepare sintered brick, and the results showed that the strength of sintered bricks could reach to grade MU15 with 10% sludge, 40% gangue and 50% shale. Wang, Y.K. [8], et al. used coal gangue and bauxite to prepare Mullite-based ceramsite and prepare Fe/C/Mullite-based ceramics composite absorbing material by wet chemical synthesis technology. Ling, C.Y., Zhang, R. [9] et al. designed the experimental process of preparing polymeric aluminum ferric silicate PEASS from coal gangue, and discovered that the PFASS flocculant from coal gangue had the best effect on oil-field sewage after comparing with other flocculants. So, it is significant to adopt reasonable green and low-carbon technology to recycle solid waste, reduce ecological environmental pollution and achieve sustainable development. However, the processing technology of coal gangue is often limited by the input cost and the benefits generated, and some utilization directions have practical technical bottlenecks [10]. Therefore, developing a green and low-carbon comprehensive treatment method of coal gangue with low cost, simple operation, good effect has excellent economic and social benefit.

* Corresponding author: xulianman@lnu.edu.cn

2 Materials and Methods

2.1 Experimental procedure

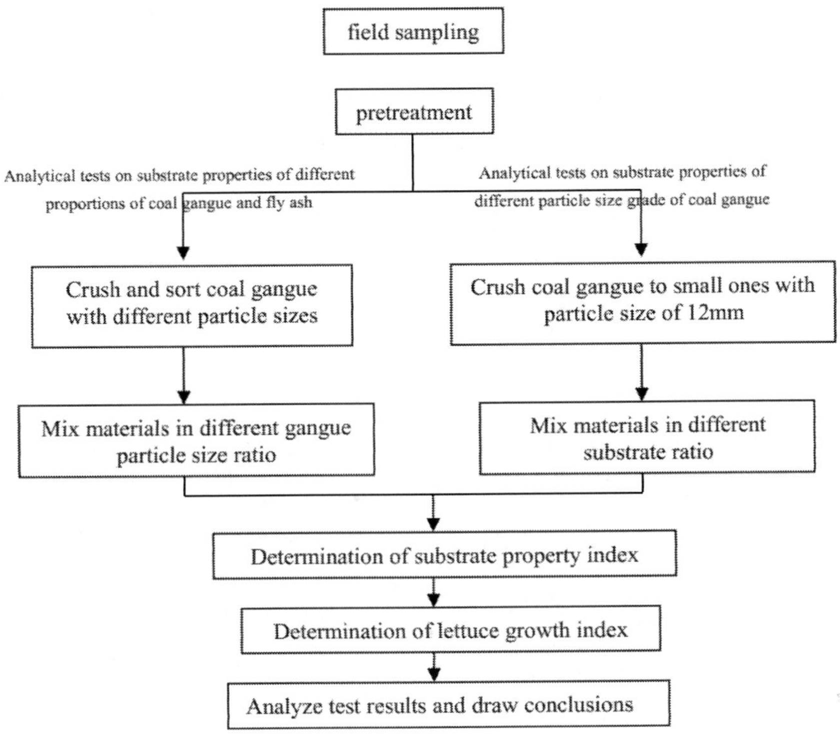

2.2 Experimental materials

The experiment used gangue, fly ash, straw and soil as the base materials to prepare the composite substrate. The coal gangue was taken from Xiaonan Mine coal gangue mountain of Tiefa Coal Group in Diaobingshan, Liaoning, and the fly ash was taken from Datang Power Plant of Tiefa Group; the corn straw and soil were taken from the farmland of Liuhuangdi Village in Diaobingshan, Liaoning near the coal gangue mountain.

American Dasusheng Lettuce seeds were selected as the test crop. The American Dasusheng lettuce grows fast, with a fertility period of about 45 days, high adaption and strong cold-endurance, and it can be produced annually, making it an ideal variety for protected cultivation and open-field planting in spring and autumn.

2.3 Experimental design

Simulated potted plants were chosen in the experiment. In order to avoid the problems of the low temperature and insufficient light when planting seedlings in laboratory, the experimental site was set in Gaoli greenhouse in Yuhong District, Shenyang.

A pot of 330 mm in length, 230 mm in width and 100 mm in depth was used for the simulation. Lettuce seeds were first placed in a gauze-laid Petri dish, and then placed in a biochemical and thermostatic incubator at 20°C for hydroponic germination after adding a small amount of ultrapure water. After 2 days, 20 germinated seeds were evenly sown in pots containing seven different substrates, and three replicate groups were set for each substrate, covered with about 1-2 cm thickness of substrate and watered thoroughly.

In this experiment, the basic physicochemical properties of the raw material of the substrate were firstly analyzed by conventional agrochemical analysis, and the heavy metal contents in coal gangue and fly ash were measured. The results are showed in Table1 below. The results showed that the heavy metal contents of the raw materials used were in accordance with the agricultural land requirements

Table 1. Basic physicochemical properties of substrate materials and heavy metal content

Raw materials	Coal gangue	Fly ash	Corn straw	soil
pH	8.60	9.42	5.12	8.16
Soil moisture（%）	4.30	0.06	10.62	31.85
Organic matter（g/kg）	71.31	24.24	153.22	69.18
Total N（mg/kg）	759.34	677.81	1408.11	830.91
Alkali-hydrolyzed N（mg/kg）	50.97	25.42	94.80	63.04
Total P（mg/kg）	1787.75	1020.76	7661.05	1322.51
Available P（mg/kg）	13.09	90.48	331.83	78.11
Total K（mg/kg）	630.80	938.96	1072.80	708.24
Available K（mg/kg）	193.97	110.55	3070.85	187.59
Hg（mg/kg）	0.92	0.12		
As（mg/kg）	0.06	0.11		
Pb（mg/kg）	46.68	5.85		
Cd（mg/kg）	0.58	0.21		
Cr（mg/kg）	54.16	30.14		

2.3.1 Analytical tests on substrate properties of different proportions of coal gangue and fly ash

Crush large coal gangue to small ones with particle size of 12mm, crush corn straw to 1mm-5mm, sift soil through a 10 mesh sieve and remove the plant residues, stones and other debris contained in the raw material. Mix the gangue without particle gradation and fly ash in the ratio of 2:1, 3:1, 4:1, 6;1, 8:1, 1:0 with a fixed amount of straw (10%) and soil (40%) to prepare mixed substrates, and record the six different substrates as R1-R6 (where R1 is 33.3% gangue, 16.7% fly ash; R2 is 37.5% gangue, 12.5% fly ash; R3 for 40.0% coal gangue, 10.0% fly ash; R4 for 42.9% coal gangue, 7.1% fly ash; R5 for 44.4% coal gangue, 5.6% fly ash; R6 for 50% coal gangue, 0% fly ash). And set up the farmland soil near the gangue mountain for the control group CK.

2.3.2 Analytical tests on substrate properties of different particle size grade of coal gangue

The coal gangue in the study area is predominantly blocky with particle size larger than 6mm, and the largest particle size is even up to 50-70mm. So planting in such a soil structure is bound to be unsuccessful. Thus it must be crushed first, and then sorted if necessary in response to the requirement of the soil substrate structure where crops grow.

Crush and sort coal gangue by using sieves of different apertures. Then set gangue with particle size less than 3mm as fine particles, 3-6mm as medium particles, greater than 6mm as coarse particles. Fine particles, medium particles, coarse particles of gangue is used as topsoil in the ratio of 8:1:1, 1:1:1, 2:1:2, 2:2:1, 1:3:1, 1:1:3, just to ensure that the total weight of the pots remain unchanged. Record the six different substrates as M1-M6 (where M1 is 80.0% fine particles, 10.0% medium particles, 10.0% coarse particles; M2 is 33.3% fine particles, 33.3% medium particles, 33.3% coarse particles; M3 is 40.0% fine particles, 20.0% medium particles, 40.0% coarse particles; M4 is 40.0% fine particles, 40.0% medium particles, 20.0% coarse particles; M5 is 20.0% fine particles, 60.0% medium particles, 20.0% coarse particles; M6 is 20.0% fine particles, 20.0% medium particles, 60.0% coarse particles). Well mix gangue, fly ash, straw (in the ratio of 4:1:1) and 1kg of quantitative soil as soil substrate, and then conduct potted plant simulation tests. And set up unsorted gangue substrate for the control group CM.

2.4 Analysis method

2.4.1 Determination of substrate property index

Soil physicochemical indicators are related to important properties such as soil nutrients, buffering and water and fertility retention, thus determining soil fertility and soil productivity, So the determination of several physicochemical indicators is important for exploring soil production performance. Take the mixed substrate in pots respectively before and after planting lettuce and put it in two sealed bags. Then label and well store for the determination of pH, organic matter, alkali-hydrolyzed nitrogen, available phosphorus and available potassium in the mixed substrates. The pH was determined directly by pH meter, organic matter by potassium dichromate volumetric method and external heating, alkali-hydrolyzed nitrogen by alkali-hydrolyzed reduction diffusing method, available phosphorus by sodium bicarbonate extraction and molybdenum-antimony resistance colorimetry, and available potassium by ammonium acetate extraction with flame photometer method.

Contents of gangue, fly ash, straw and soil in the mixed substrate were quantitative except for the particle size of gangue, and the nutrient content of the substrate and other chemical characteristics only had slight changes, but the composition of the soil particle size would make a greater difference on the texture and structure and other physical characteristics, thus affecting the growth of crops. Therefore, analytical tests on substrate properties of different particle size grade of coal gangue focused on the determination of physical properties, such as soil bulk density, porosity, and moisture, which reflected the properties of each substrate of different particle grade. In this experiment, soil bulk density was determined by cutting ring method, and soil porosity and moisture were measured based on soil bulk density.

2.4.2 Determination of lettuce growth index

The number of days required for lettuce cotyledons to emerge, for full lettuce seedlings, and the emergence rate of lettuce were recorded during the first two weeks after planting. Select three lettuce plants of each pot as samples. Count the number of leaves per plant. Measure the height and stem thickness (bottom root of the stem) of the plant. Select five leaves of each sample and measure the leaf area with a leaf area meter. Then measure the fresh and dry weights of each lettuce sample with an electronic balance. Finally, the samples were sterilized in an oven at 105°C for 5 min, dried at a constant temperature of 80°C to a constant weight, and weighed with an electronic balance for the dry weight, and take the average value.

3 Results & discussion

3.1 Analysis results of the substrate properties of different ratios of coal gangue and fly ash

3.1.1 The results of the determination of the physicochemical properties of the substrates

The results of the determination of the physicochemical properties of the substrates are shown in Table 2 below.

Table 2. Substrate property indexes in different ratios of coal gangue and fly ash

Group number	pH	Organic matter (g/kg)	Alkali-hydrolyzed N (mg/kg)	Available P (mg/kg)	Available K (mg/kg)
R1	8.18±0.14	66.75±0.81	65.90±1.05	89.29±0.98	420.53±3.22
R2	8.02±0.25	69.71±0.69	67.42±0.93	86.30±1.57	426.06±4.35
R3	7.88±0.16	72.84±0.62	68.38±0.75	84.48±1.34	431.31±3.90
R4	7.67±0.09	75.97±0.58	72.41±0.83	81.35±1.09	438.94±5.86
R5	7.43±0.19	76.09±0.74	73.94±0.76	78.26±0.87	443.25±3.94
R6	7.26±0.20	78.65±0.53	76.15±0.54	76.97±0.96	450.08±6.01
CK	8.16±0.11	69.18±1.03	63.04±0.66	78.11±1.31	187.59±2.06

The results showed that in the experiment of soil substrate with different ratios of coal gangue and fly ash, with the increase of the proportion of coal gangue, the pH value of the substrate was gradually decreasing, which was due to the reduction of the content of alkaline material-fly ash; compared with the second national soil census grading standards, soil acidity and alkalinity gradually tended to neutral, the pH of R1-R4 were between 7.5-8.5, which was weakly alkaline, the PH of the substrate in R5 and R6 groups was less than 7.5 and neutral.

For organic matter content, in addition to R1 group whose content was less than the control group, the rest of the groups were greater than the control group content, and with the proportion of coal gangue increasing, soil organic matter content also gradually increased, and R6 group content reached a maximum of 78.65g/kg.

Alkali-hydrolyzed nitrogen and available potassium in the substrate increased with the gangue content. But compared the alkaline-hydrolyzed nitrogen content of the substrate and the farmland soil near the gangue mountain, the amount didn't increase much; Conversely, the available potassium content could reach more than twice that of the control soil group. And the available phosphorus in the substrate decreased while the gangue content increased. According to the second national soil survey and relevant standards, the main nutrient content

of the substrates with different gangue and fly ash different ratio mechanism all reached the first standard (organic matter > 40g/kg, available nitrogen > 0.2mg/kg, available phosphorus > 40mg/kg, available potassium > 200mg/kg). Overall, the contents of organic matter, alkaline-hydrolyzed nitrogen, available phosphorus and available potassium in the compounding substrate were significantly higher than the control group, except for individual groups, which reached the fertility level of fertile soil. Moreover, the value of compounded organic matter reduced on an alkaline basis, which was beneficial to plant growth, so it was completely feasible to grow crops with compounded substrates.

3.1.2 The results of the determination of the growth indexes and biomass of lettuce

The results of the determination of the growth indexes and biomass of lettuce at 40 d in different ratios of coal gangue and fly ash are showed in Figure 1 below.

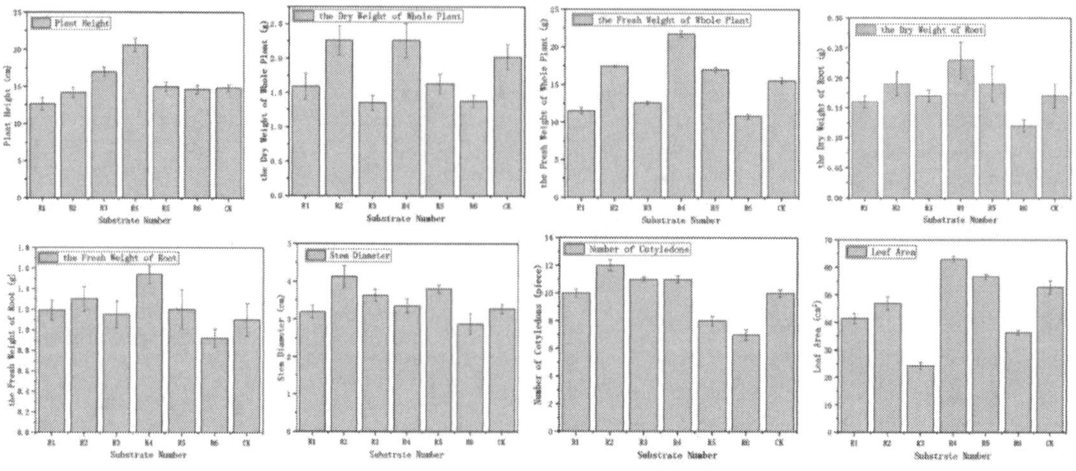

Fig. 1. Growth indexes and biomass of lettuce at 40 d in different ratios of coal gangue and fly ash

By measuring the plant height, stem thickness, leaf area and number of cotyledons of lettuce planted with different content of gangue, it could be seen that with the increase of the proportion of gangue, the stem thickness and leaf area showed two peaks, respectively at R2 and R4(the stem thickness peaked at 4.13cm at R2 and 3.80cm at R4; the leaf area peaked at 46.90cm2 at R2 and 62.90cm2 at R4), and exceeded the control group; the plant height showed one peak at R4 group, reaching 20.60 cm, and exceeded the control group; the number of cotyledons reached a maximum of 12 pieces at R2, and R2, R3, as well as R4 all exceeded the control group. The combined analysis showed that R2 and R4 were more favorable for lettuce growth. To draw further conclusions, parameters such as fresh weight and dry weight of the plants need to be analyzed.

From the results of the determination of the fresh and dry weights of the whole lettuce plant, we could draw conclusions that the fresh and dry weights of the whole plant and roots showed two peaks at R2 and R4(the fresh weight of the whole plant peaked at 17.43g at R2 and 21.74g at R4; the dry weight of the whole plant peaked at 2.26g at R2 and R4,; the root fresh weight peaked for the first time at 1.3g at R2 and reached a maximum of 1.54g at R4; the root dry weight peaked for the first time at 0.19g at R2 and reached a maximum of 0.23g at R4) with the increase of gangue content in the substrate, and R2 and R4 were higher than the control group; for the dry weights of the whole lettuce plant and roots, the values were lower than the control group except for R2 and R4; the growth indexes of R4 group were all the maximum, which indicated that the substrate ratio of R4 group can be best for the lettuce growth.

3.2 Analysis results of the substrate properties of different ratios of coal gangue particle sizes

3.2.1 The results of the determination of the physicochemical properties of the substrates

The physical properties such as substrate capacity, porosity and water holding capacity were measured in this group of tests. The soil capacity and water holding capacity were determined by cutting ring method, and the soil porosity was measured according to the soil capacity. The results are shown in the Table3 below.

Table 3. substrate property indexes in different ratios of gangue particle size

Group number	Soil bulk density (g/cm3)	Field water-holding capacity (%)	Total porosity (%)	Water-holding porosity (%)	Aeration porosity (%)	Air-water ratio
M1	1.46±0.03	32.51±0.96	38.44±0.93	57.91±1.46	18.31±0.81	0.31±0.02
M2	1.28±0.02	24.90±0.73	44.85±1.09	45.06±0.99	43.43±1.19	0.97±0.06
M3	1.21±0.05	26.25±0.83	47.47±1.52	36.30±1.94	42.52±0.98	1.17±0.05
M4	1.30±0.03	27.52±0.87	42.60±0.80	44.21±1.51	39.35±1.37	0.89±0.04
M5	1.23±0.01	23.94±0.91	46.21±1.62	43.15±1.24	44.86±1.41	1.04±0.03
M6	1.15±0.06	22.82±1.14	49.34±1.43	40.95±1.03	56.98±1.16	1.39±0.07
CM	0.97±0.02	22.03±0.95	64.90±0.97	22.48±0.62	58.06±1.26	2.58±0.04

The experimental results showed that the greater the particle size grade of gangue in the substrate, the smaller the bulk density, the smaller field water holding capacity, the larger the field porosity, the larger the air-water ratio, thus the better aeration, the poorer water holding capacity. Conversely, the smaller the particles, the larger the soil bulk density, the larger the field water holding capacity, the smaller the total porosity, the smaller the air-water ratio, thus the better water holding capacity, the poorer aeration.

3.2.2 The results of the determination of the growth indexes and biomass of lettuce

The results of the determination of the growth indexes and biomass of lettuce at 40 d in different ratios of coal gangue and fly ash are showed in Figure 2 below.

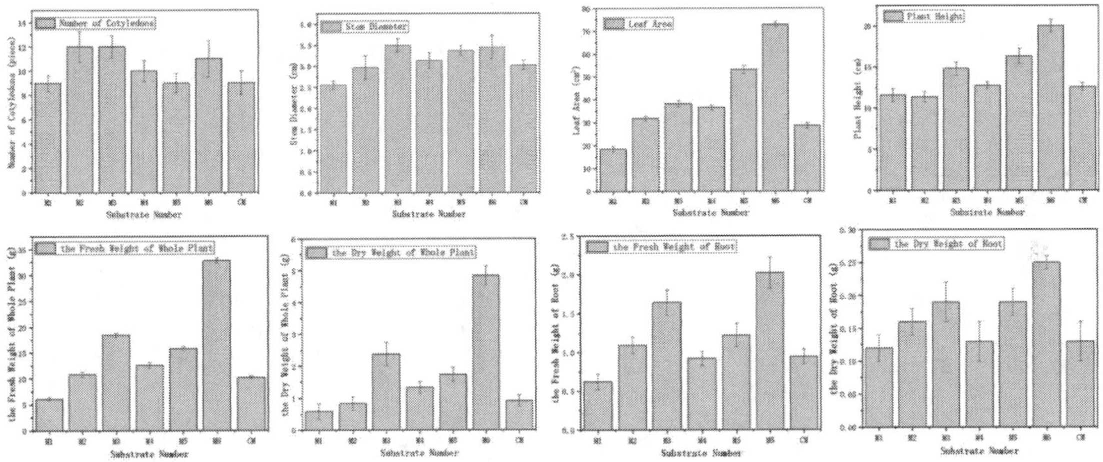

Fig. 2. Growth indexes and biomass of lettuce at 40 d in different particle size ratios

The plant height, stem thickness, leaf area and number of cotyledons of lettuce planted with different particle size compound substrates of coal gangue were measured, and it was found that with the increase of the proportion of coal gangue, the stem thickness, plant height and leaf area showed two peaks at M3 and M6 respectively(the stem thickness peaked at 3.5cm at M3 and 3.45cm at M6; the plant height peaked at14.80cm at M3 and 20.00cm at M6; the leaf area peaked at 38.10cm2 at M3 and 72.80cm2 at M6), and M3 and M6 were higher than the control group, and the plant height and leaf area values of M6 group were significantly higher than the other groups, 20.00 cm and 72.80 cm2 respectively. The number of cotyledons reached the maximum at M2, M3 and M6, and all of them exceeded the control group.

From the determination of the fresh and dry weights of the whole plant and roots, we could draw conclusions that the fresh and dry weights of the whole plant and roots showed two peaks at M3 and M6(the fresh weight of the whole plant peaked at 18.4g at M3 and 32.82g at M6; the dry weight of the whole plant reached a peak of 2.37g at M3 and a maximum of 4.84g at M6; the root fresh weight peaked for the first time at 1.64g at M3 and reached a maximum of 1.64g at M6; the root dry weight peaked for the first time at 0.19g at M3 and reached a maximum of 0.25g at M6), and both M3 and M6 are higher than the control group; the fresh and dry weights of the whole plant and roots were the minimum in the M1 group and the maximum in the M6 group. It indicates that the gangue particle size ratio of M6 group can be best for the growth of lettuce.

4 Conclusions

Lettuce was grown in soil substrates in different ratios of coal gangue, fly ash and in different particle size ratios. Then analyze and evaluate two aspects: the basic physicochemical properties of the substrate and the growth quality of lettuce. Finally select the best ratio for the preparation of improved soil substrate. The main conclusions are as follows:

(1) The alkali-hydrolyzed nitrogen and available potassium content in the substrate in different ratios of coal gangue and fly ash increased with the increase of the proportion of coal gangue, while pH and available phosphorus content decreased; the greater the particle size grade of gangue in the substrate, the smaller the bulk density, the smaller field water holding capacity, the larger the field porosity, the larger the air-water ratio, thus the better aeration, the poorer water holding capacity.

(2) It was feasible to improve the soil substrate of coal gangue mountain by compounding the gangue soil substrate from the different ratios of coal gangue and fly ash as well as the different particle size gradation of coal gangue in two aspects. Lettuce grew better in compounded substrates with the appropriate proportion of gangue and fly ash as well as the gangue particle size than the corresponding control group, and the substrate properties were also better. So the compounded substrate developed in this experiment could improve gangue soil substrate effectively to varying degrees, and was more conducive to lettuce growth.

(3) The test results showed that the best ratio of gangue and fly ash is 6:1; and the best ratio of gangue particle grade is 1:1:3(fine particles: medium particles: coarse particles). Thus we could draw a conclusion that the best ratio of the improved soil substrate of the mine gangue mountain is 42.9% gangue, 7.1% fly ash, 10% straw, 40% soil, where the gangue particle size ratio was 20% fine particles, 20% medium particles, 60% coarse particles .

Acknowledgments

This paper is one of the phase results of the Liaoning Provincial Science and Technology Plan Project "Research on Geological Disaster Prevention and Ecological Restoration Technology of Abandoned Tailings Ponds" (2020JH2/10300100) and the National Science and Technology Major Project for Water Pollution Control and Management "Key Technology Integration and Demonstration of Mine Ecological Degradation Area Restoration" (2015ZX07202-012).

References

1. Q. Miao, G. Meng, M. Chen, W.N. Ge, J.L. Qu, T.Y. Wang, Analysis and Guarantee of Coal Resources in China. ENERGY AND ENVIRONMENT, **02** 6-8+23 (2020)
2. C.Y. Yuan, H. Wang, Current Aituation and Prospect of Comprehensive Utilization of Coal Gangue in Qingyuan Area, Shanxi. COAL PROCESSING AND COMPREHENSIVE UTILIZATION, **02** 71-74 (2020)
3. J.H. Lei, Analysis of Pollution Hazards and Comprehensive Utilization of Coal Gangue. ENERGY AND ENERGY CONSERVATION, **04** 90-91+147 (2017)
4. N. Zhou, Y.N. Yao, W.J. Song, Z.W. He, G.H. Meng, Y. Liu, Present Situation and Prospect of Coal Gangue Treatment Technology. JOURNAL OF MINING AND SAFETY ENGINEERING, **37(01)** 136-146 (2020)
5. H.S. Shen, W.H. He, C.Z. Zhao, B.Z. Wu, Y.L. Lu Application Test of Coal Gangue-based Concrete in Underground Roadway Hardening Engineering. MINING SAFETY AND ENVIRONMENTAL PROTECTION, **48(06)** 65-71 (2021)
6. L.Y. Gu, The Investigation on Preparation of Porous Ceramic by Using Coal Gangue as the Raw Materials. MULTIPURPOSE UTILIZATION OF MINERAL RESOURCES, **05** 135-137 (2018)
7. B. Jin, L. Zhao, X. Wang, K.J., Zhao, J.W. Zhang, X.T. Zhang, L.S. Yang, Preparation of Sintered Brick with Sludge, Shale and Gangue. NON-METALLIC MINES, **44(05)** 39-41 (2021)
8. Y.K. Wang, B.S. Zhu, G.M. Li, L.P. Liang, Preparation of Fe/C/Mullite-based Ceramsite Composite Absorbing Materials by Recycling Solid Waste Coal Gangue. JOURNAL OF FUEL CHEMISTRY AND TECHNOLOGY, **49(02)** 238-246 (2021)
9. C.Y. Ling, R. Zhang, M. Niu, P.P. Bu, H.Z. Shi, X.Y. Li, Laboratory Experimental Study on Preparation of Flocculant for Water Treatment from Coal Gangue, PETROCHEMICAL INDUSTRY APPLICATION, **49(02)** 238-246 (2021)
10. D.Q. Deng, The Present Situation and Prospect of Coal Gangue Utilization in Southwesr China. CONSERVATION AND UTILIZATION OF MINERAL RESOURCES, **39(02)** 136-14 (2019)

Analysis of Rainstorm Flood Induced Flash Flood in Wende River Basin

Meihong Ma[1], *Yue* Zhou[1], *Shiyao* Bi[1], *Bingshun* He[2,*], *Qing* Li[2] and *Ting* Li[3]

[1]Tianjin Normal University, Geographical and Environmental Science Department, 300387, Tianjin, China
[2]China Institute of Water Resources and Hydropower Research, Research Center on Flood and Drought Disaster Reduction of the MWR, 100038, Beijing, China
[3]North China University of Science and Technology, College of Foreign Languages, 063210, Tangshan, Hebei, China

Abstract. Due to the joint influence of climate and topography, flash flood disasters in the Wende River Basin of Jilin Province occurred repeatedly, causing serious casualties and huge economic losses; in July 2017, three floods with large rainfall intensity and profound influence occurred in a row in the main stream of Wende River. Based on official measured data of the Wende River basin, three typical rainfall processes ("July 13", "July 20" and ","August 3 ") are selected to discuss the rainfall characteristics. The triggered characteristics, induced factors and the disaster mechanism of the basin are preliminarily explored from the aspects of rainfall weather system, temporal and spatial distribution of rainstorm and flood, geographical factors - to analyze the best countermeasures for flash flood prevention. The above will provide reference for the non-engineering measure project of flash flood prevention currently under construction.

1 Introduction

1.1 Background and significance

Flash flood disaster is a kind of flood disaster, which mainly refers to the steep flood caused by heavy rainfall in mountainous streams and ditches, as well as the induced river floods, landslides and mudflows [1]. Flood cause analysis is extremely important for the disaster study in small and medium-sized river basins. Song Qiumei, based on the rainfall and flood data of 1980-2012 in Beijing, found that the spatial and temporal distribution characteristics and intensity of rainstorm affect the flood pattern in this area [2]. Zhang Jintang and others analyzed the causes of rainstorm flood in Anhui section of the Yangtze River in 2016, and found that the cause of excessive flood on the south bank of Chaohu Lake was the uneven spatial distribution of rainfall [3].

The Wende River basin is located in the southeast of Yongji County, Jilin Province, and it is a primary tributary of the left bank of the Second Songhua River. The specific location of the Winter River basin in the study area is shown in Figure 1. The basin belongs to the northern temperate continental dry monsoon climate. It is hot and rainy in summer and cold and dry in winter. Affected by the weather system such as the North China cyclone, subtropical high rear shear and typhoon, the precipitation distribution in the river basin is not uniform, concentrated in the flood season. From July to August 2017, heavy rain hit the Wende River Basin, causing huge loss of life and wealth to Yongji County, 279800 people were affected, 31 people died or disappeared due to mountain torrents, and the total direct economic loss was 14 billion, among which "July 19-July 20" disasters and "August 2- August 3" disasters caused total direct economic losses of 3.645 billion yuan and 181 billion yuan respectively [4-7]. This paper selects the typical Wende River basin in Yongji County, Jilin Province as the research area. By analyzing the rainfall and flood characteristics of "July 13-July 14", "July 19-July 20", "August 2-August 3", it initially discusses the formation mechanism of flash flood disaster, in order to analyze the optimal flash flood defense countermeasures under the new defense situation, and accumulate experience in flood disaster analysis and flood control prevention and disaster reduction.

* Corresponding author: hebss@163.com

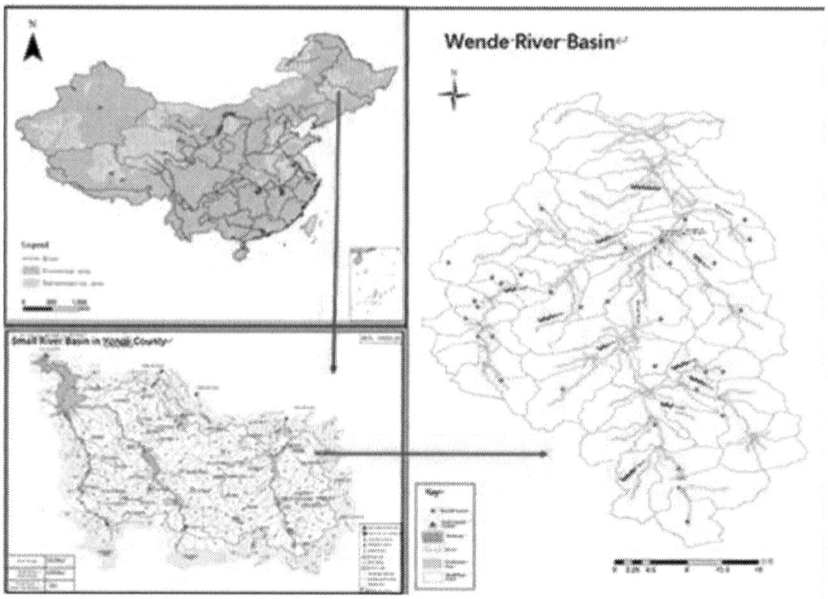

Fig. 1. Schematic Geographic Location of Wende River Basin, Yongji County, Jilin Province

2 Data introduction

The data are mainly obtained from the China Institute of Water Resources and Hydropower Sciences, mainly including DEM data, rainfall data, flood data in Wender River Basin, Yongji County, Jilin Province; both rainfall data and flood data are measured data. The rainfall data mainly includes the duration, rainfall, rainfall intensity and flood peak flow of the three rainfall events in Wende River Basin in 2017, which mainly come from the measured rainfall data of the hydrological station in front of the estuary and the measured rainfall data of 43 rainfall stations in the basin. At the same time, the data of 13 rainfall monitoring stations built by the hydrological department are also considered.

According to the division of precipitation data, the precipitation accumulation period is from 08:00 of the previous day to 08:00 of the current day. Based on the rainfall level adopting the meteorological definition of China, the daily rainfall is divided into three types: weak precipitation (0-25mm, light rain and moderate rain), general intensity precipitation (25-50mm, heavy rain), heavy precipitation (>50mm, rainstorm, heavy rainstorm and severe rainstorm); rainfall times is mainly defined by Taiwan RIT method, which means a rainfall starts when the rainfall is more than 4mm, and ends when the rainfall is less than 4mm for 6h.

3 Analysis of rainfall characteristic

In late July, the subtropical high stabilized and moved less. Jointly affected by the North China cyclone northward and the subtropical high [5-7], the Winter River basin suffered three flash floods from July 13 - July 14, July 19 - July 20, and August 3 - August 4, 2017. In "July 13", the maximum rainfall strength is 93mm, and the average rainfall is the highest rainfall in the three floods, up to 181.5mm. In "July 20", from 17:00 on 19th to 15:00 on 20th, there were two obvious rainfall processes: in the first round of rainfall, the rotation direction of the rainstorm centering along the watershed is clockwise; the new round of rainfall from 02:00 on the 20th has greater intensity and wider range. In "August 3", from 20:00 on August 2 to 21:00 on August 3, the third rainstorm with relatively small rainfall occurred in Yongji County, with average rainfall of 127.4mm. Specific precipitation of the three floods is shown in Table 1.

Table 1. Details of three floods

Rainfall time	Total rainfall duration (h)	Average Rainfall (mm)	Maximum Rainfall Strength (mm)
July 13	17	181.5	93
July 20	23	153	103
August 3	25	127.4	53.5

In terms of cumulative rainfall, the maximum rainfall is mainly concentrated in the northwest corner of the river basin, and the rainfall is relatively small in the southeast corner. The cumulative rainfall maximum increased from 100mm to 306.5mm. the largest accumulated rainfall in each period of rainfall is located in the upstream tributaries of Chundeng River (except Jinye Reservoir at 1 and 2 o'clock on the 20th), and those of the other periods are located near the Hongshiling Reservoir on Chundeng River. The southwest of the mouth basin is seriously affected by the rainstorm, while the main stream of Wende River is relatively less affected. The maximum 24-h accumulated rainfall of the three floods is shown in Figure 2.

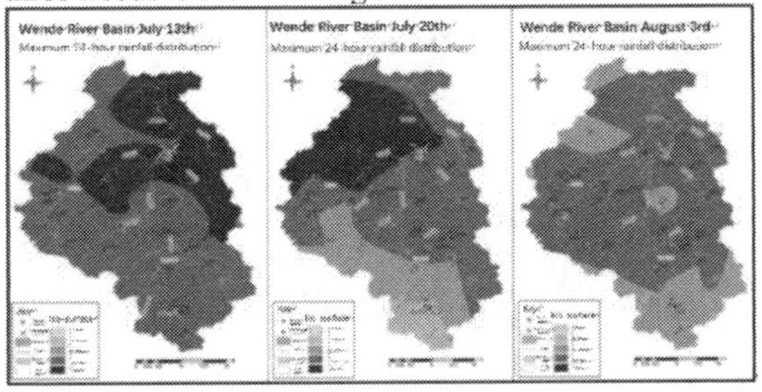

Fig. 2. Cumulative rainfall in three floods (maximum 24h)

4. Analysis of rainstorm and flood characteristics

4.1 Typical rainstorm and flood characteristics

4.1.1 "July 13" rainstorm and flood

Wende River in Yongji County is located in mountainous areas. Rising and falling rivers generally form fast yield and confluence, which often has the characteristics of fast formation, high intensity and short duration. On 13 July 2017, the average rainfall on the Winter River watershed reached 181.5mm. The water level of the front station began to rise at 17:00 on July 13th, and the water level began to overflow the top of the embankment at 21:25 on July 13th, reaching 260mm. At 0:00 on the 14th, there is a flood peak with an estimated discharge of 3350mm^3/s at the hydrological station in front of the mouth, with water level up to 228.05mm. The flood caused 90% of the town covered with water of 2-3 meters. The measured results of rainfall and flow on July 13 in Wende River Basin are shown in Figure 3. The entire rainfall center is near the outlet and the flood moves along the river from upstream to downstream.

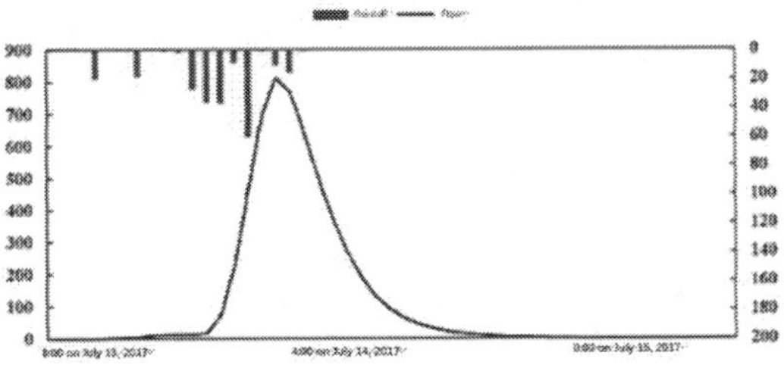

Fig. 3. Results of rainfall and flow on July 13 in Wende River

4.1.2 "July .20" rainstorm and flood

Basin

From 17:00 on July 19 to 15:00 on July 20, Yongji County was hit by the second heavy rain, with the average rainfall, maximum 1h rain intensity and maximum rainfall 153.0mm, 103mm/h, 339.5mm respectively. The rainstorm center is mainly distributed in the higher northwest of the basin and moves along the upstream of the river. The rainfall had two obvious processes about which the center of the first round turning clockwise around the Winter River basin and the second was more intensity and wider range; the station began to fall at 5:15 on 20th after the 226.8mm, peak of 2480mm^3 /s. The measured results of rainfall and flow of Wende River Basin July 20 are shown in Figure 4. The flood caused about 70 percent of the Yongji County city to drown, at a depth of 0.6m-2m.

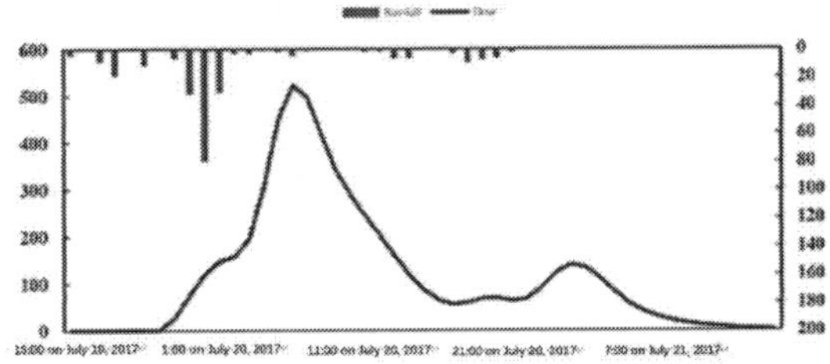

Fig. 4. Measured data of the rainfall and flow on July 20 in

4.1.3 "August 3" rainstorm and flood

Wende River Basin

From 20:00 on August 2 to 21:00 on August 3, the third rainstorm occurred in Yongji County, with the maximum average rainfall up to 127.4mm, 1h rain intensity up to 53.5mm, and the maximum rainfall reached to 180mm, but precipitation is smaller compared with the previous two. The rainstorm center is mainly distributed in the middle and lower reaches of the basin with relatively low terrain. The movement direction of the precipitation front is the same as that of the "July 13" and "July 20" floods, moving from the upstream to the downstream. At the front water level station in the Winter River basin, the water level rose from 01:00 on August 3. About seven hours later the water level reached the highest value of 224.98mm, 0.98mm higher than the warning water level (224mm), and the corresponding flow was 942mm^3/s. After that, the water level fell rapidly. At 14:00 on the 3rd, the water level fell to 221.36mm, 1.64mm lower than the warning water level.

4.2 Flood characteristics

At present, the warning water level of Wende river basin is 223mm and the dam crest elevation is 226mm; Characteristics of three rainstorm floods in 2017 are as follows:

In the "July 13" precipitation, the highest flood water level exceeds the dam crest elevation, and the rising flow is 8mm3/ s, peak flow is 3350mm^3/s. The highest peak level reaches 228.05mm, belonging to single peak flood. Compared with the July 28 flood in 2010 (peak flow of 3120mm^3/s), this flood peak flow was higher, becoming the maximum of the watershed. In the "July 20" precipitation, the highest flood level exceeds the dam crest elevation, with the 226.8mm. The peak flow is 2480mm^3/s and the rising flow is as high as 50mm^3/s. It belongs to multi peak flood, in which the peak modulus is 2.99. In the "August 3" precipitation, the highest flood level exceeds the dam crest elevation and warning water level. The peak flow is 943 mm^3/s, and the rising flow i stands in the middle of the three, which is 36mm^3/s, belonging to single peak flood.

4.3 Cause of flood investigation

The Wende River basin belongs to the continental dry-cold monsoon climate area in the northern temperate zone. The spatial and temporal distribution of precipitation is uneven, mostly concentrated in summer. The flood season rainfall can account for 70-80% of the annual rainfall, and the flood season mostly has southeast wind, and the warm and humid air flows northward, which results in heavy rain in the basin.

Main reasons: (1) affected by low altitude shear and warm and humidity flow, the precipitation intensity and range are large, the precipitation is concentrated, and the peak flow is heavy; (2) main rain experienced short rainfall time, the precipitation yield is large, and almost all of them produce runoff; (3) the superposition of flood peaks leads to large contribution of flood volume and rise of water level.

5 Preliminary disaster causing mechanism in drainage basin

5.1 Weather system triggering rainfall

The Wende River basin belongs to the northern temperate continental dry-cold monsoon climate zone. Under the joint influence of shear and warm and humid air flow, the subtropical high is stable and less dynamic, caused by the northward movement of the North China cyclone and the subtropical high. Due to the influence of high-altitude low vortex and shear at the back of subtropical high on the 20th-21st, the precipitation distribution on the 24th-25th was uneven, which means the low vortex system and shear precipitation are easy to cause severe rainstorm. At the same time, the rainfall front moves along the river direction, with the center of gravity moving from north to south.

5.2 Temporal and spatial distribution of rainstorm flood induced flash flood

5.2.1 Rainfall is the main cause

The temporal and spatial distribution of rainfall in Wende river basin is uneven and the rainfall is concentrated. Meanwhile, the rainfall front surface moves along the flow direction of the river. For example, the maximum rainfall is mainly concentrated in the northwest corner of the basin, while the relatively small rainfall is in the southeast corner. In addition, the areas with large precipitation, large precipitation intensity and the rainstorm center near the basin exit all induce the generation of flood.

5.2.2 Large flow

From 22:00 on the 13th to 0:00 on the 14th in the "July 13" flood, because of the large flow of the Bahu River, the water level of Wende River was raised, increasing the inundation loss suffered by the town in front of the river mouth.

5.2.3 The flood volume has a large contribution

Under the influence of rainstorm center distribution, the contribution ratio of each period and river reach to flood at different scales will be different. In "July 13" flood, rainfall impact is the same, and the flood contribution is mainly equivalent to the proportion of area. In the "July 19" flood, because the flood affected the spring reach, the flood volume contribution is twice that of the corresponding index.

5.3 Impact of geographical environment

The hydrological Station in front of the river mouth is located in the lower Wende River. Sijian River and Bahu River are tributaries of the Wende River. In the two floods of "July 20" and "August 3", the flood peaks of Wende River and Sijian River are superimposed in the section of Kouqian Town, while the flood of BAHU river is discharged, resulting in serious disasters. In the "July 13" flood, Wende River and Sijian River jointly superimposed flood peaks in Kouqian Town, and the disaster in kouqian town was further aggravated due to the large peak flow of BAHU river.

In addition, the terrain of Wende River basin is hilly, tilts from southeast to northwest, and the slope changes obviously; the main land use are woodland (97.5%), grassland (98%) and farmland (97.9%), and most of the surface is rough and exposed, the basin soil is mainly clay and pink loam, fine texture, resulting in slow infiltration rate. All of the above factors induced the occurrence of flash floods.

6 Analysis of countermeasures for optimizing flash flood prevention

According to the current situation that Wende River Basin often encounters heavy rainfall and concentrated rainfall, which is easy to induce flood disasters, this paper mainly puts forward the following suggestions:

(1) Strengthening the study on the rainfall and flow relationship of flash flood disasters. At present the early warning indicators are mainly rainfall or water level. We should strengthen the analysis of rainfall and runoff in the basin and the impact of flood on tributaries and tributaries, and study the transformation mechanism and causes of flood disasters

(2) Strengthening the construction of an intelligent

and automatic emergency defense system for falsh flood defense. Applying modern information technology such as artificial intelligence technology in pre-disaster, during-disaster and post-disaster defense emergency management; With the help of virtual scene to prevent flood, and clarify the relationship between multi factors that induce flash flood disaster, we should emphasize the importance of analyzing the main influencing factors, and building a passive to active flash flood disaster prevention system.

(3) Making overall planning for river treatment and building a high standard flood control system. In view of the large number and wide coverage of small and medium-sized rivers in China, we clarifying the key points of governance and formulating a comprehensive project plan for the governance of small and medium-sized rivers. At the same time, introducing foreign advanced control methods, such as adopting ecological methods to conserve water and soil and prevent water and soil loss.

(4) Improving the non-engineering flood control measures. Establishing and improving the flood control and rescue command system. The automatic measuring and reporting system should be used to timely and accurately grasp the rain and water conditions, preparing a reasonable long-term prediction scheme, and dispatching the water conditions step by step as planned. At the same time, research the flash flood risk system, and explore ways for appropriate flood defense and relief.

(5) Strengthening community disaster management and the dissemination of early warning information. Improving people's awareness of flood risk, and gradually improve the mass testing and prevention organization with Chinese characteristics. Using social software APP, combined with 5G technology to share disaster situations in real time.

7 Conclusion

This paper analyzes the process of three rainstorm floods in Wende River Basin in terms of rainfall characteristics, rainstorm causes and disaster factors to explore the causal mechanism of flood.

First, rainfall is the main factor; the three precipitations have large rainfall, large intensity, short duration and rainfall front moving along the river direction. Secondly, the flood peaks are superimposed and the contribution of the flood volume is large. During the "July 13" flood, the rainfall concentration periods of the three river sections are basically the same. At the same time, the proportion of flood volume contribution is large; Especially in the "July 19" precipitation, the proportion of flood contribution is twice that of the area. In addition, the geographical environment has also aggravated the disaster. The Bahu River is in the upper reaches and has a relatively large flow, leading to the rise of the Wende River water level, and the obvious slope, sparse vegetation and low soil infiltration rate all provide favorable conditions for the flood.

Based on the above situation, this paper proposes to further optimize flash flood prevention measures, such as comprehensive planning and control engineering, improving non-engineering flood control measures, strengthening group measurement and group prevention and advanced scientific measures, strengthening flood disaster prevention and providing reference for the ongoing flash flood disaster prevention work.

Acknowledgements

This work was supported by Deep learning-based research on the inducing factors and early warning of flash floods (42101086), Research on Key Technologies of Flash Flood Prevention Based on Multi-machine Learning Model (2019KJ086), Huaneng Lancang River Hydropower Co., Ltd.: Research on the Integrated Forecast System of Meteorology and Hydrology in the Lancang River Basin (Phase I) (JK2021/D02), Inspur Software Group Co., Ltd. entrusted project: Hainan Province Smart Water Network Information Platform (Phase I) Project(202109161126CGYRK03), National Key R&D Program of China(2021YFC3000202), China Three Gorges Corporation Scientific Research Project(0704183), Hainan Provincial Academician Innovation Platform Research Project (YSPTZX202142), State Key Laboratory of Hydrology, Water Resources and Hydraulic Engineering Science "The Belt and Road" Water and Sustainable Development Science and Technology Fund Key Project(021490311), and Independent exploration project of the State Key Laboratory of Water Cycle Simulation and Regulation in the Basin(SKL2020TS01). The authors wish to thank the anonymous reviewers for their comments and suggestions.

References

1. Y. Hong, P. Adhikari and J.J. Gourley, Flash flood. In Encyclopedia of Natural Hazards, Springer Netherlands, 324-325 (2013)
2. Q.M. Song, B. Zhu, D. Ma, etal, Analysis of rainstorm and flood in typical flash flood basin in Beijing, Hydrology, **037(003)**, 91-94,90 (2017)
3. J.T. Zhang, B.B. Li, Fang Hong, etal. Analysis of 2016 in Anhui section of Yangtze River Basin, Hydrology, **2017(37)**, 96 (2017)
4. Z.L. Xu and K.N. Yan, Study on the technical capability of flood control and flood fighting projects, Water conservancy and hydropower technology,(2017)
5. L.X. Chen, Y. Feng, M. Xue and X. Zhou, Analysis of 7 · 13 rainstorm and flood in Wende River[J].Northeast Water Resources and Hydropower Project, 2018 (2): 2-2.
6. P.L Qu, Y.L. Wang, L.Q. Han, J.X. Dong, M. Zhang and Z.H. Zhou, Analysis of storm flood in "2010 · 07" in Windriver River Basin, Hydrology, **2012 (02)**, 93-96 (2012)

7. F.C. Zhao, Analysis of 2017 flooding in Yongji County [J].Water Conservancy Planning and Design, **2018(9)**, 3-3 (2018)

Identify Parameters Hindering Renewable Hydrogen Production in France: Life Cycle Sensitivity and Uncertainty Analysis

Zongyue Zhang[1, 2, 3], *Mélanie* Douziech[2], *Paula* Perez-Lopez[2], *Qingrui* Wang[3, 4], and *Qing* Yang[3, 4, 5,*]

[1]China-EU Institute for Clean and Renewable Energy, Huazhong University of Science and Technology, 1037 Luoyu Road, Hongshan, 430074 Wuhan, Hubei, PR China
[2]Centre Observation, Impacts, Energie (O.I.E.), MINES ParisTech, PSL University, 1 Rue Claude Daunesse - CS 10207, 06904 Sophia Antipolis CEDEX, France
[3]State Key Laboratory of Coal Combustion, Huazhong University of Science and Technology, 1037 Luoyu Road, Hongshan, 430074 Wuhan, Hubei, PR China
[4]Department of New Energy Science and Technology, School of Energy and Power Engineering, Huazhong University of Science and Technology, 1037 Luoyu Road, Hongshan, 430074 Wuhan, Hubei, PR China
[5]John A. Paulson School of Engineering and Applied Sciences, Harvard University, 29 Oxford Street, 02138 Cambridge MA, USA

Abstract. Flourishing green hydrogen economy worldwide serves as a pillar for global energy transition and carbon-neutral targets. However, rare researches on the environmental impact of green hydrogen production have focused on national average resource availability and technology market share. Nor a detailed and holistic sensitivity and uncertainty analysis regarding both foreground and background parameters in the green hydrogen production life cycle could be found. To fill these gaps, we present this study as a comprehensive environmental impact investigation of renewable-electricity-based water electrolysis H_2 production in France, in terms of average resource availability and technology market share in France in 2019. Water electrolysis H_2 production with average French wind electricity through proton exchange membrane electrolysis unit was identified as the most environmentally-beneficial option among all pathways studied. By building the corresponding French national benchmark parametric life cycle model, the key parameters that determine the environmental impact of green hydrogen in France were revealed. Under current French context, environmental impact's uncertainty for solar-to-H_2 and wind-to-H_2 pathways is between 19%-26% and 8%-11%, respectively. Based on the unit process contribution investigation and further scenario analyses, we propose the guidelines and suggestions on improving life cycle renewability and sustainability to French green H_2 production.

1 Introduction

In the current global techno-economic context, fossil fuels, e.g., coal and natural gas, cover 80 % of global energy consumption[1]. The ever-increasing energy demand worldwide, especially the fossil fuel demand, becomes the primary driver of global warming[1], environment degradation[2, 3], and human disease[4, 5], pressuring the world's sustainable development[6]. Aiming for global decarbonization of energy services, hydrogen energy (H_2), a potentially green and renewable energy vector, gives a cross-boundary opportunity for the sustainable and renewable global energy landscape transition[7-10].

1.1 Hydrogen and hydrogen sector in France

As the lightest element in Mendeleev's periodic table, hydrogen is present from the first moment of the universe, taking around 75 % of the contemporary universe's total mass[11]. Colorless, odorless, and non-corrosive, hydrogen gas, i.e., the hydrogen molecule, is a suitable energy carrier on earth, with a much higher mass-energy density (120MJ/kg) than fossil fuels[11, 12].

At present, Europe currently consumes 339 TWh H_2 per year[13], of which around 90.6 % is fossil-based hydrogen[14]. In Europe 2018, only two hydrogen production plants (fossil-based) were equipped with carbon capture (CCS) facilities. For France, the country produced around 920,000 metric tons (Mt) H_2 in 2008[15], while the H_2 production capacity raised to 2088.8 Mt/day in 2019[16]. Besides, almost all merchant H_2 is produced from steam methane reforming and, to a lower extent, autothermal reforming and less commonly partial oxidation, both in Europe[7, 13] and France[15, 16]. Regarding water electrolysis H_2 based on low carbon or renewable electricity, their total installed capacity of electrolysis unit is around 58MW in EU27 2018[17], half of which is produced in Germany. The 58 MW electrolyzer generates less than 0.1 % of the total H_2 production in Europe.

One obvious thing is that the ambitious in hydrogen production sector decarbonization[18] is the crucial and inevitable step to fulfill the carbon emission reduction target and climate change goals announced by the French government successfully: 36 % CO_2 emission reduction by 2030 compared to 2005[19] in the National energy and climate plans of France; achieving a carbon-neutral

* Corresponding author: zy97science@hust.edu.cn

society by 2050[20]. Remarkably, the water electrolysis H_2 production based on renewable power and biomass sourced H_2 production draws the main focus of French society and government due to their renewable and sustainable potential, e.g., the negative-carbon effect of biomass sourced H_2 generation[21, 22].

The French government has established the "hydrogen deployment plan for the energy transition"[23, 24] to facilitate renewable & low carbon H_2 permeation. According to this plan, France will switch 20 % to 40 % fossil-based hydrogen nationwide consumed to renewable hydrogen electrolyzed from low carbon electricity or recovered from biomass by 2028[8]. The Multiannual Energy Program (la programmation pluriannuelle de l'énergie 2020, PPE 2020), released on 23rd April 2020 (for the period s of 2019 - 2023 and 2024 - 2028), provides a basket of policies adding financial support for hydrogen sector development[25]. Under the instruction of the PPE 2020, guarantees of origin scheme and Guarantees of Traceability, as has already been applied to the French biogas market, will also be developed[25]. French Ordinance No. 2021 - 167[26] defines the future context of the country's renewable & low-carbon hydrogen market. All these measures will finally standardize the co-financing models promoting the H_2 ecosystem deployment integrating different uses (mobility, industry, etc.) up to the local level in France. Besides, France also participated in the HyLaw project, IPCEI projects[27-30], etc., to identify, assess, and find prioritizing measures to the significant regulatory barriers, and thus to accelerate the development of a global hydrogen market.

Table 1. Critical advance in demonstration and commercialization of renewable and low-carbon H_2 in France.

Location and time	Responsible parties	Novelty and Significance
Vitry-le-François, Strasbourg 2019	Haffner Energy	Commercialization of the first highly efficient biomass sourced H_2 production recovering up to 70 % of the biomass energy with a relatively low price: € 6/kg[31].
Brittany, Normandy, and Loire 2020	Lhyfe	VHyGO project: Building a completely renewable H_2 ecosystem for west France region[32].
South region of Provence-Alpes-Côte d'Azur 2021	TotalEnergies, ENGIE	Masshylia project: Start design, develop, build, and operate France's currently largest renewable hydrogen production program[33].
Le Croisic 2022	Lhyfe, Chantiers de l'Atlantique	Start operation of first commercialized off-shore wind hydrogen production plant worldwide[34].

The nationwide favorable policies promotion has enabled France to become a thriving economy for sustainable hydrogen. Numerous robust actors in the fields of renewable & low-carbon hydrogen and fuel cells from both industries (Lhyfe[34, 35], ENGIE[36], TotalEnergies[33], etc.) and academia bring lots of new hydrogen production projects, and technologies come into being. Recent renewable & low-carbon H_2 production projects across Europe and France can be found from Fuel Cell and Hydrogen Joint Undertaking[37], International Energy Agency (IEA)[38], and EU Commission[39]. Several landmark project that marked critical advance in democratization and permeation of renewable & low-carbon H_2 for France is shown in Table 1 below.

1.2 Hydrogen production environmental impact

Serving as the basic for H_2 production sector decarbonization, identifying sustainable H_2 production pathways through environmental impact assessment of H_2 production has long been a hot spot for researchers. Life cycle assessment (LCA) methodology, which aims for assessing potential environmental impacts associated with all life cycle stages (e.g., extraction and processing of raw material, manufacture, distribution, and use, recycling or end-of-life treatment) of a commercial product, process, or service, is widely accepted and developed by researchers working on environmental impact assessment of H_2 production.

In terms of methodological innovation, Valente et al.[40-42] developed a harmonized approach to estimate the environmental impacts of different H_2 production technologies, currently focusing on three impact categories: global warming potential (GWP), cumulative energy demand (CED), and acidification potential (AP). Their harmonized LCA method has been applied to a vast of electrochemical[43], thermochemical[41], nuclear[44], and biological[45] H_2 production pathways. Their harmonization protocol to refine cradle-to-gate system boundaries suggests a functional unit of 1 kg industrial-grade hydrogen (purity >99 %) and compressed to 200 bars. To increase comparability of the results, they recommend using system expansion or economic allocation in the attributional LCA assessment when hydrogen is the main product or by-product, respectively[40].

However, for a region or country like France that aims to implement a highly sustainable H_2 sector, massive implementation for different renewable and low-carbon H_2 production technologies fed by various feedstocks is inevitable, along with continuous industry development. Current environmental impact assessment of H_2 production technologies rarely focuses on the average resource availability and technological market shares at the national or regional level. While such results could, however, provide the scientific basis to guide one nation's financing policy formulation[26], subsides determination[46], and tender process[47] to promote the renewable and low-carbon H_2 penetration.

Regarding specific technologies, embodied with huge sustainability potential, the environmental performance of water electrolysis H_2 production has been extensively studied[48-51]. The source of electricity used in the water electrolysis is the determining factor for the environmental impact of the H_2 produced[40, 49, 52, 53]. The majority of the research agrees that H_2 from wind, solar, and hydro electrolysis generally has a lower GWP, CED, and AP than the biomass sourced H_2 and much lower than the fossil-based H_2 without CCS[40-42, 49, 50]. In terms of GWP, CED, and AP, hydro electrolysis ranked as the best among water electrolysis H_2 pathways, followed by wind electrolysis, photovoltaic solar electrolysis, and thermal solar ranked as the worst[40, 42, 49, 50]. Nevertheless, some critical impact categories that are not supposed to be neglected, e.g., eutrophication potential human toxicity potential, ionizing radiation, photochemical Oxidation Potential, are seldomly discussed in the existing literature, which is the same for resources use mineral and metals, abiotic Depletion Potential[49].

Regarding the uncertainty (UA) and sensitivity analysis (SA), a detailed review of UA and SA in H_2 production LCA hasn't appeared yet. Most UA and SA for H_2 production focus on the techno-economic context[54-64], and only a few studies focus on the UA and SA of H_2 production on LCA. Furthermore, the little research available concentrates only on the UA and SA of foreground life cycle parameters related to H_2 production facilities and processes.

UA and SA towards the following foreground life cycle parameters have been carried out: reduction and recycling rate of electrolysis stack material[65]; electrolyzer efficiency[66]; energy mix applied during H_2 production[67], allocation strategy[68], and treatment of H_2 production by-product[67]; H_2 production catalyst & chemical consumption rate[69]; weighting factors among economic, social, environmental influence[70, 71]; pollutant emission ratio during facility installation[72]. However, regarding the LCA result, some upstream or downstream parameters may have much more influence than the foreground parameters related to H_2 production facilities[73]. According to the author's best knowledge, no research could be found that performs a detailed and holistic SA and UA regarding all the foreground and background life cycle parameters involved in renewable-electricity-based water electrolysis H_2 production life cycle. Only Yadav et al.[73] perform the SA and UA of several upstream parameters involved in solar electricity plant (e.g., PV plant performance ratio, PV plant efficiency, Global horizontal irradiation) in the life cycle CED and GWP analysis to solar-based high-temperature water electrolysis system.

Facing and looking forward to complementing the existing research gaps stated above, the main objective of this research is a comprehensive environmental impact investigation and comparison among renewable-electricity-based water electrolysis H_2 production pathways. This study highlights the following specific research objectives:

1) From the angle of average French resource availability and technology market share, build corresponding French national benchmark parametric life cycle model for wind electricity-based (wind-to-H_2) and solar electricity-based (solar-to-H_2) water electrolysis H_2 production pathways through proton exchange membrane electrolyzer (PEM) and alkaline electrolyzer (AE).

2) Calculate the multi-environmental impacts for all H_2 production pathways analyzed and identify the corresponding main contributors and hotspots, e.g., foreground and background activities and emissions, to 11 impact categories studied.

3) A detailed and holistic SA and UA regarding all the foreground and background life cycle parameters by propagating the uncertainties of each life cycle parameter into corresponding parametric life cycle models. And testing the importance of each parameter uncertainty in each impact categories' global uncertainty for all H_2 production pathways.

4) Based on further scenario variation and comparison, we put forward the guidelines and suggestions towards improving the life cycle renewability and sustainability to renewable-electricity-based water electrolysis H_2 production.

The following paper is structured as follows: Section 2 elaborates the life cycle assessment methodology applied by this research, where the goal and scope definition, hypothesis and method for life cycle inventory analysis; life cycle impact assessment; and life cycle result interpretation (methodology for holistic SA and UA included) for each H_2 production pathway analyzed were elaborated.

Section 3 systemically present and discuss the result obtained: environmental impact categories score hotspot unit process contribution; findings discovered from further scenario variation and comparison; as well as impact category score's sensitivity to life cycle parameters and corresponding uncertainty. Based on the research performed, Section 3 finished with the detailed guidelines and suggestions towards improving the life cycle sustainability of French green hydrogen production. Major conclusions of this research are extended into the Conclusion section.

2 Materials and Methods

2.1 LCA methodology

The 14000 series of environmental management standards of the International Organisation for Standardisation (ISO), in particular, ISO 14040[74] and ISO 14044[75], provide detailed and worldwide accepted procedures for conducting LCA studies.

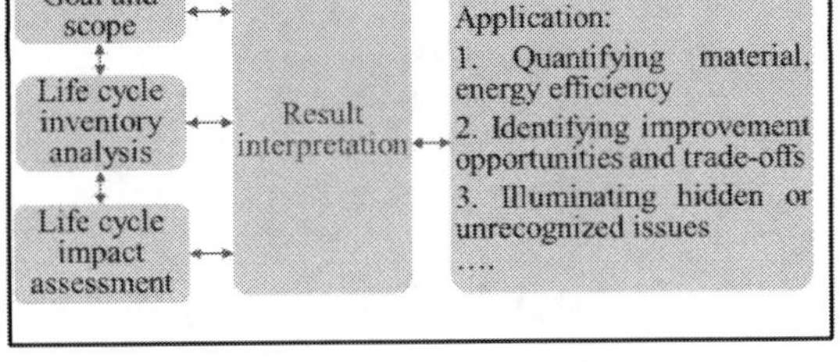

Fig. 1. General phases of life cycle assessment.

As pointed out by ISO, LCA can fulfill several goals: quantifying material and energy efficiency for a system; identifying improvement opportunities and trade-offs; highlighting hidden or unrecognized issues; promoting a wider communication about how to compare and improve highly complex industrial systems.

There are four general phases when conducting an LCA as shown in Fig. 1: Goal and scope definition; Life cycle Inventory analysis; Life cycle impact assessment (LCIA); Result interpretation.

2.2 Goal and scope definition

The goal of this work is to better understand the environmental sustainability of renewable-electricity-based water electrolysis H_2 production (RE-to-H_2) in France around 2019, from the angle of national average resource availability and technology market share, overcoming the existing limitation as stated above. The research result could serve as a basis and reference for governmental officials, policymakers, and companies, assisting in financing policy formulation to promote renewable & low-carbon H_2 penetration in France.

As mentioned in the Introduction section, the main tasks of this LCA research are, for each H_2 production pathway analyzed: 1) to build the corresponding French national benchmark parametric life cycle model, 2) to calculate the potential multi-environmental impacts, 3) to locate the primary hotspot of each impact categories studied, 4) to analyze the uncertainties and variability in the LCA results, thus becoming the first paper that performs a detailed and holistic SA and UA regarding both foreground and background life cycle parameters in RE-to-H_2 life cycle.

The functional unit of the research, common to all analyzed RE-to-H_2 pathways, is "1 MJ H_2 produced and being compressed to 200 bar". As indicated in Fig. 2 below, the life cycle boundaries of cradle-to-grave assessment include:

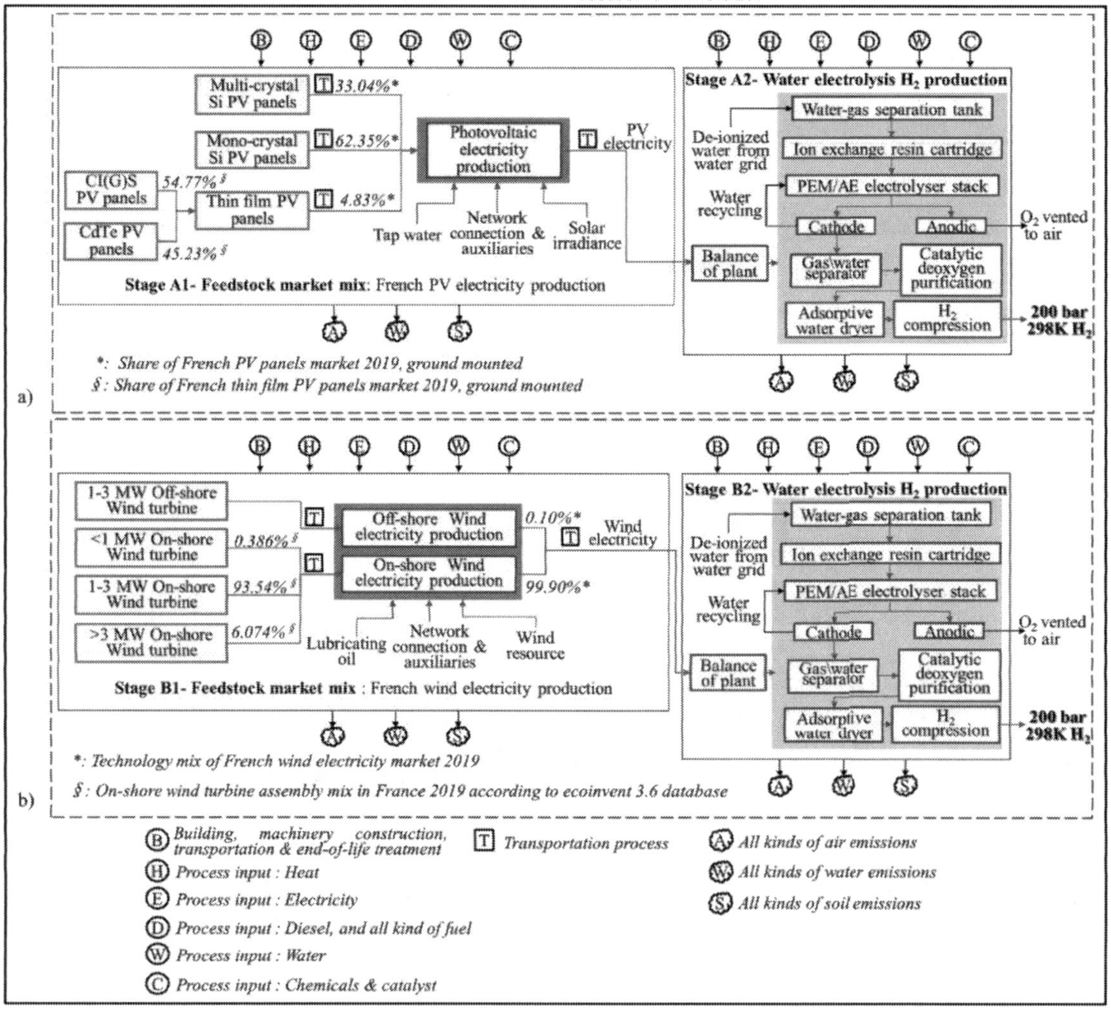

Fig. 2. Flowchart and life cycle boundaries of a) solar-to-H_2 pathways b) wind-to-H_2 pathways.

1) Photovoltaic solar electricity production market mix in France 2019, unified processes to each solar electricity production technologies clarified by ecoinvent v3.6 database, i.e., construction (material extraction,

transport, and end-of-life treatment) of PV panels and PV power plant network connection & auxiliaries; PV electricity production process.

2) mix of wind electricity production market in France 2019, unified processes to each wind electricity production technologies clarified by the ecoinvent v3.6 database, i.e., construction (material extraction, transport, and end-of-life treatment) of wind turbine assembly and wind power plant network connection & auxiliaries; wind electricity production process.

3) facility construction (material extraction, transport, and end-of-life treatment) of the water electrolysis unit (H_2 compression unit included).

4) water, energy & chemical consumption (related transport and end-of-life treatment included) during water electrolysis H_2 production, purification & compression process.

In each RE-to-H_2 pathway, the electricity generated by the PV panel or the wind farm will fulfill all the electricity demand of the water electrolysis unit (both the electrolyzer stack and all the auxiliaries included in the balance of plant, BOP), and the H_2 compressor.

Besides, in each pathway, the oxygen produced by the electrolysis unit is vented to the environment without a subsequent utilization and thus is considered environmental burden-free. Furthermore, after purification, the H_2 produced with a purity larger than 99.9 % is compressed to the standard H_2 transport pressure nowadays in Europe to ensure comparability to other studies.

The feedstock market mix, e.g., solar or wind electricity production market mix presented in Fig. 2 within the blue square, is embodied with high complexity under average resource availability and technology market share in France 2019. One obvious point is that the life cycle environmental impact for PV panels will vary based on PV panel types imported from different countries[76]. And according to different installation locations and applications, e.g., ground centralized PV panels, façade panels, and slanted or flatted mounted panels on the roof, the annual solar irradiation available on PV panels and the PV plant auxiliaries transportation distance will change. On the other hand, based on different technological details (e.g., rotor size, turbine capacity, tower type and size), more than 2000 wind turbine models are available and used in France. And the timely evolution of the technological specificities to PV and wind power plant components' production and installation will consume electricity under a very different electricity grid mix.

Thus, special attention should be paid when refining the life cycle boundaries and corresponding life cycle inventory (LCI) parameters. For the feedstock production market in each H_2 production pathway, this research follows the protocol and logic shown in Fig. 3 to define the life cycle boundaries, obtain corresponding LCI and parameters data. Thus, building the comparable French national benchmark parametric life cycle model best represents the reality, namely, France's average resource availability and technology market share.

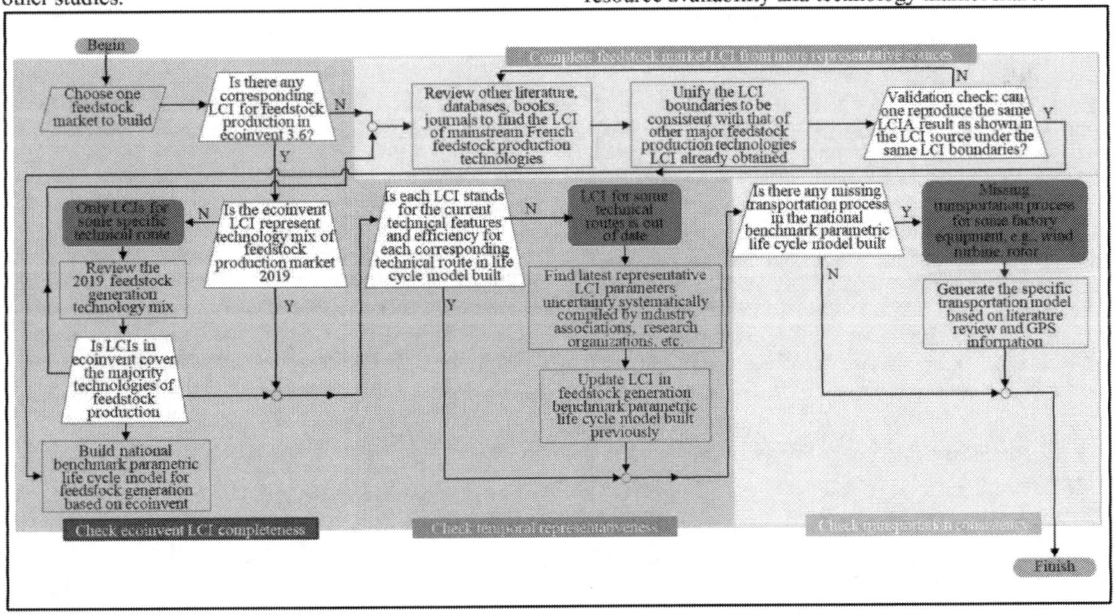

Fig. 3. Decision diagram for French national benchmark parametric LCA model harmonization regarding feedstock market mix.

As shown in Fig. 2 and Fig. 3, the French national benchmark parametric LCA models comprise the detailed modeling of the related transportation processes for each RE-to-H_2 pathway studied. In consist with methodology in the ecoinvent v3.6 database, all the transportation processes included in each RE-to-H_2 pathway are measured as metric ton*km (t*km hereafter).

2.3 Life cycle inventory

This section gives a detailed description of LCI for each RE-to-H_2 pathway studied, e.g., the data and hypothesis we collected and harmonized based on the new findings of the current H_2 production situation from various data sources and previous research. Corresponding French national benchmark parametric life cycle model, which integrated detailed LCI data of each pathway studied, is shown in Supplementary Materials. The ecoinvent database (version v3.6) “Allocation, cut-off by classification” is the primary source for the 2019 background LCI used in this research.

2.3.1 French solar electricity production market mix

2.3.1.1 PV plant construction: PV panels market mix in France

In this research, the solar electricity generation mix in mainland France 2019 mainly stands for the centralized photovoltaic electricity from on-ground mounted PV panels. Science large-scale captive H_2 production through centralized PV panels installed on the ground has long been drawing the main interest of the French government and companies[46, 77]. And due to data unavailability, the distributed & semi-centralized solar electricity generation is not included in the scope of this report. And it's hereby not included in the French national benchmark parametric LCA model. This research assumes a PV plant with 100 m^2 total area powered by the national level PV panel mix (one unit of French national benchmark PV plant).

As shown in Fig. 2 a), according to IEA's report on Task 12: PV Sustainability Activities[78-80], the open ground PV panels market mix in France 2019 is shared by three types of PV panels: multi-crystal Si PV panel (average, 62.35 %), mono-crystal Si PV panel (average, 33.04 %), and thin-film PV panel (average, 4.83 %). And the thin-film PV panel market of France in 2019 comprises mainly Cadmium telluride PV panels (CdTe, 54.77 %) and Copper indium gallium selenide PV panels (CIGS, 45.23 %). The PV panel production market has ushered in tremendous technological progress and efficiency improvement since 2005, namely, the time node in ecoinvent v3.6 database, as discussed in detail by Besseau[81] and the Methodology Guidelines for PV LCA released by IEA[82]. The latest LCI database for PV panels released by IEA[78] is used in this report to overcome the time hysteresis of the ecoinvent v3.6 database.

2.3.1.2 PV plant construction: network connection & auxiliaries

Following the protocol in Fig. 3, the network connection unit & auxiliaries apply the identical life cycle boundaries and LCI as defined in the ecoinvent v3.6 database, where the site preparation process, PV panel installation service, PV mounting system, and the DC/AC inverter, tap water consumption during PV plant installation and operation should be taken into consideration:

The DC\AC inverter used has a rated capacity of 500 kW. The LCI of the inverter, the PV panel installation service, and the tap water supplement are directly taken from the ecoinvent v3.6 database. The LCI of the PV mounting system used in this research is derived from IEA's report on Task 12: PV Sustainability Activities[78], overcoming time hysteresis of the ecoinvent v3.6 database as stated above. The requirement of the site preparation process (including mainly the diesel consumption for land preparation), PV panel installation service, tap water consumption during PV plant operation was recalculated from the data in the ecoinvent v3.6 database to match the installed capacity of the PV plant to satisfy the FU. The installed capacity of the inverter equals the PV model's installed capacity.

The amount of the PV mounting system, in m^2, equals the land area needed for the PV electricity plant, which is 100 m^2 as assumed. Equation (1) below calculates the 100 m^2 PV plant's installed capacity accessed in the solar-to-H_2 pathway.

$$P_{\text{PVplant}} = P_{\text{peak}} \times A_{\text{PVplant}} \times GCR \qquad (1)$$

Where,

P_{PVplant} : Nominal maximum power of the 100 m^2 French national benchmark PV plant.

P_{peak} : Nominal maximum capacity of the PV panels per m^2 under Standard Test Condition (STC), 115.06 Wp/m^2 on average[81].

A_{PVplant} : Land area for the PV plant, 100 m^2.

GCR : Ground cover ratio, which refers to the size of net PV modules, divided by the equivalent ground area of the PV power plant, under specific tilt and azimuth[83]. Besides PV panels, there should also be space dedicated to fire prevention and self-shading prevention in the PV plant. GCR is 36 % on average as calculated by Blondel Paul[84], under 30° tilt.

2.3.1.3 PV plant construction: transportation process

IEA's report on Task 12[78] includes detailed transportation requirements for each process involved in LCI of all kinds of PV panel production until the PV panel is transported from the production countries to European regional storage. In solar-to-H_2 pathway, to cover PV panel transportation from European regional storage to their installation site in mainland France, based on the transportation scenario of Romain Besseau[81], Celik et al.[85], and the road information on Google map, a two-stage land transportation process was modeled here:

All PV panels are first transported by train (700 km, average) and then by lorry (300km, average). The two-stage land transportation process was also applied to all the other components involved in the PV plant construction and operation, e.g., each component in the PV plant network connection unit & auxiliaries (PV panel mounting and installation system, DC/AC inverter, etc.). The mass of each component involved, e.g., PV panels mix, network connection unit & auxiliaries, can be found in their corresponding LCI as shown in supplementary materials.

2.3.1.4 PV plant operation: solar irradiance and PV electricity generation mix in France

The daily or yearly solar irradiance determines the PV electricity generation. The accumulative yearly solar irradiance (Wh/m^2) for 13 administrative regions of mainland France 2019 are taken from the CAMS radiation service dataset "AGATE" with McClear version 3 and CAMS radiation bias correction[86], as shown in the supplementary materials. To utilize the maximum annual solar irradiance potential, the tilt angle is fixed at 30 ° and azimuth at 180 ° for PV panels installed in the French national benchmark PV plant.

With France's yearly solar irradiance data, we can calculate the total electricity production during the whole PV power plant lifespan with Equation (2), as standardized by IEA[76].

$$Elec_{PV} = P_{PVplant} \times \frac{Irr}{I_{HTC}} \times PR \times Life_{PV} \times ALR \qquad (2)$$

Where,

$Elec_{PV}$: Accumulative electricity generation within the whole PV power plant lifespan, in kWh.

Irr : Annual solar irradiance at the PV power plant location under specific tilt and azimuth in $kWh/m^2/yr$. Maximum value[86] among 13 administrative regions in France 2019 as 1900.59 $kWh/m^2/yr$.

I_{HTC} : Solar irradiance under STC[87], which is 1 kW/m^2.

PR : Performance ratio (average, 0.8[78]), which indicates the efficiency of the PV power plant studied, namely the difference between the actual energy output and the theoretical energy output due to losses during PV electricity generation such as thermal and conduction losses[88].

$Life_{PV}$: PV power plant lifespan in years, 30 years on average[82], which equals PV panels lifespan.

ALR : Average loss ratio describes the PV panel's average efficiency loss among the system's lifespan. Based on the yearly loss ratio[89, 90], it should be calculated with Equation (3).

$$ALR = \frac{1}{Life_{PV}} \sum_{i=1}^{Life_{PV}} \frac{100 - YLR \times (i-1)}{100} = \frac{200 - YLR \times (Life_{PV} - 1)}{200} \qquad (3)$$

Where,

YLR : Yearly loss ratio (average, 0.7 %[82]), which indicates PV panel efficiency loss each year.

2.3.2 French wind electricity production market mix

2.3.2.1 Wind plant construction: wind turbine assembly market mix in France

To represent the national average wind electricity generation mix in mainland France 2019, in the wind-to-H_2 pathway, we assumed one unit of French national benchmark wind farm shared by the off-shore wind farm and onshore wind farm mix based on the reality summarized below and as shown in Fig. 2b).

French wind power observatory[91, 92], Offshore Wind in Europe 2020[93], etc., detailed illustrate the current wind electricity market (production, new installations, financing activity, etc.) in France. The national average wind electricity generation market in mainland France 2019 is distinguished between off-shore wind (average, 0.1 %) and onshore wind (average, 99.90 %) electricity production. One can calculate the specific ratio from France's annual total wind electricity production and the production amount by type. The annual accumulative wind electricity production in France 2019 is 34.1 TWh from both off-shore and onshore wind power[94]. And total off-shore wind electricity production can be calculated by Equation (4) below. Besides the off-shore wind electricity, other electricity comes from the onshore wind turbines.

Wind Power Database (28th July 2021 last version)[95] summarized the technical details for all wind power plants (farm location, turbine model and capacity, hub height, turbine number, etc.) in France. On average, the market share of onshore wind turbine assembly mix is 11.27 % (<1MW wind turbine): 82.16 % (1-3 MW wind turbine): 6.568 % (>3 MW wind turbine). Following the off-shore wind electricity market in France around 2019[93], in this research, the capacity of all French off-shore wind turbines fulls between 1-3MW. For both onshore and off-shore wind turbine assembly in each corresponding wind farm, this study took their LCI from the ecoinvent v3.6 database.

2.3.2.2 Wind plant construction: plant network connection & auxiliaries

Based on the capacity level of the wind turbine and onshore/off-shore features, the types and amount of network connection unit & auxiliaries needed, i.e., the DC\AC transformer, cables, tower, foundation, lubricating oil, will vary. In the wind-to-H_2 pathway, the LCI of network connection & auxiliaries to each corresponding wind turbine assemblies under different capacity levels are exported from the ecoinvent v3.6 database. We can directly find the amount needed for 1 kWh wind electricity production.

Table 2. Components in one unit of French national benchmark wind farm covered by the predefined two-stage transportation process.

Components	**Internal parts**
Rotor	Rotor blades, rotor hub, extender
Nacelle	Nacelle cover, gearbox, generator, shaft, yaw system, flanges, etc.
Tower	All kinds of steel, epoxy resin, and paint

Control electronics	All kinds of electronics in the nacelle
Network connection	Cables, transformer, sub-station with the circuit breaker, electricity meter, etc.

2.3.2.3 Wind plant construction: transportation process

Transportation of all components involved in wind plant construction and operation, e.g., wind plant network connection & auxiliaries, each type of wind turbine assembly, as shown in Table 2, from the production site to the installation site, are not covered by the ecoinvent database. Thus, following the logic of Fig. 3, to cover the transportation from the production site to the installation site, this research applied the same two-stage land transportation process as defined in Section 2.3.1 of each component involved, as shown in Table 2. For off-shore wind turbine assembly, this research applies overseas transportation to 10 km (on average). The mass of each component summarized in Table 2 can be found in their corresponding LCI shown in the supplementary materials.

2.3.2.4 Wind plant operation: wind electricity generation mix in France

In the wind-to-H_2 pathway, we can calculate the total electricity production during the whole lifespan of one unit of French national benchmark wind farm with Equation (4) below.

$$Elec_{wind} = Cf_{turbine} \times P_{Windplant} \times 8760 \times Life_{wind} \quad (4)$$

Where,

$Elec_{wind}$: Accumulative electricity generation within the whole wind farm lifespan, in kWh.

$Cf_{turbine}$: Capacity factor for off-shore or onshore wind turbine assembly, which equals the annual wind turbine net electricity output divided by turbines' maximum power capability (net energy output under turbine's rated power, 8760-hour uninterrupted annually working). On average, 24.50 % for onshore wind turbines[96] and 38 % for off-shore wind turbines[97, 98] in France 2019.

$Life_{wind}$: Wind farm lifespan in years, 20 years on average[81], which equals wind turbine lifespan.

$P_{Windplant}$: The total installed capacity of one unit 2019 French national benchmark wind farm, in kW, which can be calculated with Equation (5) below.

$$P_{Windplant} = R_{offshore} \times P_{offshore,1\text{-}3MW} + R_{onshore} \times \sum_i \left(R_{turbine,i} \times P_{onshore,i}\right) \quad (5)$$

Where,

$R_{offshore}$: Ratio of off-shore wind electricity to total wind electricity generated in France 2019, 0.1 % as calculated.

$P_{offshore,1\text{-}3MW}$: Representative capacity of off-shore wind turbine assembly under 1-3 MW capacity level in France 2019, which is 2 MW as calculated in ecoinvent 3.6 database.

$R_{onshore}$: Ratio of off-shore wind electricity to total wind electricity generated in France 2019, 99.90 % as calculated.

i : Different wind turbine capacity levels, e.g., <1 MW, 1-3 M, >3 MW.

$R_{turbine,i}$: Market share of different capacity level turbine assembly in French onshore wind turbine assembly mix 2019. For turbine assembly with a capacity level of <1MW, 1-3MW, >3MW, on average, the ratio is 11.27 %, 82.16 %, 6.57 %.

$P_{onshore,i}$: Representative capacity for each onshore turbine capacity level as calculated in ecoinvent 3.6 database. For turbine assembly with a capacity level of <1MW, 1-3MW, >3MW, it is 800KW, 2MW, 4.5MW, respectively.

2.3.3 Water electrolysis unit

2.3.3.1 Water electrolysis unit and electrolysis H_2 production

As shown in Fig. 2, this research compares two types of water electrolysis units: proton exchange membrane electrolyzer and alkaline electrolyzer. For both the PEM and AE unit, balance of plant (BOP) system voltage adapting or rectifying the electricity from the PV or wind plant into proper DC electricity required by the electrolyzer stack (where the water electrolysis reaction takes place). The stack electrolyzes the French average water grid's deionized water into H_2 and O_2.

Bareiß et al.[99] give a detailed technical description of the PEM unit under France's present market situation, from where this research cited the LCI of PEM electrolysis unit (both the BOP and stack) per kW. The LCI to AE unit electrolyzer stack per kW is diverted from the parameterized model developed by Romain Besseau[81].

As pointed out by Bareiß et al.[99], it can be challenging to give a sound estimation for LCI of electrolysis unit BOP. Considering the similarity between BOP for both alkaline and PEM stack[100-102], and due to data limitation, LCI of BOP in both PEM and AE unit in this research is cited from Bareiß et al., which is identical to each other[99]. Equation (6) clarifies the electrolysis unit capacity calculation installed in each RE-to-H_2 pathway.

$$P_{electrolyzer} = \frac{\dfrac{Elec_{H2}}{\eta_{unit}}}{\dfrac{Elec_{H2}}{\eta_{unit}} + Elec_{compress}} \times P_{REplant} \quad (6)$$

Where,

$P_{electrolyzer}$: Capacity of electrolysis unit installed in the solar-to-H_2 pathway or wind-to-H_2 pathway, in kW of stack and BOP.

$Elec_{H2}$: Net water splitting electricity demand with a 100 % efficiency in LHV kWh/ kg H_2, which is 33.3 kWh/kg H_2 (literally equals LHV of 1 kg H_2)[103].

η_{unit} : Water electrolysis unit global efficiency covering all facilities in the system, e.g., stack, electronics, pumps, safety equipment, infrastructure. Both for PEM, AE unit (average, 60 %) are summarized based on data from Bareiß et al.[99] and ecoinvent v3.6 database.

$Elec_{compress}$: Compressor electricity demand for compressing H_2 from stack outlet pressure (30 bar) to 200 bar, which is 3.91 kWh/kg H_2 as maximum[104].

$P_{REplant}$: Total installed capacity of one unit 2019 French national benchmark wind farm or PV plant, in kW, which equals to $P_{PVplant}$ or $P_{Windplant}$ in Equation (2) and (5), respectively.

The electrolyzer stack efficiency and electrolysis unit global efficiency determine H_2 productivity. One can calculate the total H_2 production amount in the RE-to-H_2 pathway, e.g., total H_2 produced during the entire lifespan of one unit French national benchmark wind farm or PV plant utilizing Equation (7).

$$H_{2yield_{LHV}} = H_{2yield_{kg}} \times LHV_{H2} = \frac{Elec_{gen}}{\frac{Elec_{H2}}{\eta_{unit}} + Elec_{compress}} \times LHV_{H2} \quad (7)$$

Where,

$H_{2yield_{LHV}}$: Accumulative H_2 production amount within one unit of French national benchmark wind farm or PV plant whole lifespan, in MJ, based on low heating value (LHV).

LHV_{H2} : LHV of H_2 in MJ/ kg H_2, which is 120 MJ/kg and equals 33.3 kWh/kg[12].

$Elec_{gen}$: Accumulative electricity generation within one unit of French national benchmark wind farm or PV plant whole lifespan, in kWh, e.g., $Elec_{PV}$ or $Elec_{wind}$ in Equation (2) and (4).

$H_{2yield_{kg}}$: Accumulative H_2 production amount within one unit of French national benchmark wind farm or PV plant whole lifespan, in kg.

2.3.3.2 H_2 compressor and compression process

This research uses the most common H_2 compressor type, the diaphragm compressor, to model the H_2 compression process in the RE-to-H_2 pathway.

LCI per unit diaphragm compressor can be found in the premise database[105]. Since no exact data for diaphragm compressor lifespan inconsistent with the diaphragm compressor LCI is found during the research. To better represent the actual situation in France 2019, we estimate the diaphragm compressor unit needed in the RE-to-H_2 pathway with Equation (8).

$$Compressor_{unit} = \frac{Life_{sys}}{Life_{Compressor}} = \frac{Life_{sys}}{\frac{H_{2unit}}{H_{2yield_{kg}}} \times Life_{sys}} = \frac{H_{2yield_{kg}}}{H_{2unit}} \quad (8)$$

Where,

$Life_{sys}$: Lifespan of one unit French national benchmark wind farm or PV plant, which equals to $Life_{PV}$ and $Life_{wind}$ in Equation (2) and Equation (4), respectively.

$Life_{Compressor}$: Lifespan of the diaphragm H_2 compressor.

H_{2unit} : Total amount of H_2 can be compressed during the entire lifespan of one compressor unit, which is 5906475.3 kg/unit on average, as shown in the premise database[105, 106].

Furthermore, this research applies the two-stage transportation process defined above to each component involved in the AE and PEM electrolysis unit (e.g., the electrolysis stack, BOP, and H_2 compressor) to cover the transportation process from the component production site to the installation site. The mass of each component can be found from their corresponding LCI as shown in the supplementary materials.

2.3.4 Parametric LCA model & scenarios analyzed

Compared with other LCA software, due to their open-sourced features, the Python framework for LCA: Brightway2 (BW2) and the algebraic parametrization BW2 layer: *lca_algebraic* library[107, 108] enable a more flexible environment for LCA research innovation, e.g., dynamic LCA, life cycle model simplifying[108], as well as the quick check for impact categories correlation[107]. To perform a complete and systematic life cycle interpretation, thus meeting the goal and scope of the research, this research builds the French national benchmark parametric LCA model of all the pathways studied together with the corresponding LCI and data source in BW2 based *lca_algebraic* library. The LCA results and the related interpretation are then generated by simulating the parametric LCA model. To summarize, as shown in Fig. 4, based on the renewable electricity type (wind or solar electricity) and electrolysis unit type (AE or PEM unit), four RE-to-H_2 pathways were analyzed in this research.

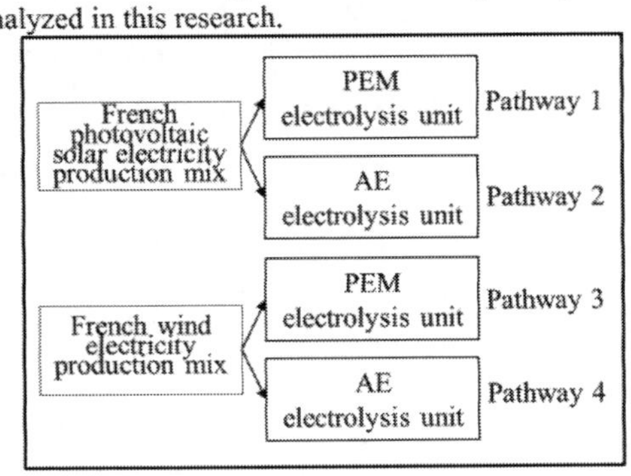

Fig. 4. RE-to-H_2 pathways analyzed in this research.

2.4 Life Cycle Impact Assessment

2.4.1 Life cycle impact categories

The choice of LCIA categories varies among different H_2 production LCA research, as pointed out by some H_2 production LCA overviews[49]. Based on the highest suggestion from the EU Joint Research Centre (climate change, ozone depletion, respiratory inorganics)[109], previous research[49], and the specific patterns of H_2 production pathway this research studied, the following 'ILCD 2.0 2018 midpoint' impact categories in Table 3 were analyzed by this research. Normalization, grouping, and weighting of the LCIA score are not included in the goal and scope of this research.

Table 3. LCIA categories analyzed in this research.

ILCD midpoint analyzed	Corresponding LCIA endpoint
Climate change total (GWP)	Climate change
Freshwater and terrestrial acidification (FTA)	Ecosystem quality
Carcinogenic effects (CE)	Human health
Ionizing radiation (IR)	Human health
Ozone layer depletion (OLD)	Human health
Photochemical ozone creation (POC)	Human health
Respiratory effects, inorganics (REI)	Human health
Fossils (FD)	Resources
Land use (LU)	Resources
Minerals and metals (MD)	Resources
Dissipated water (DW)	Resources

2.5 Life cycle result interpretation

2.5.1 Sensitivity and Uncertainty analysis

SA and UA can reveal the robustness of LCIA results in an LCA study (e.g., their sensitivity to the uncertainty of life cycle parameters). This research paid great attention to data uncertainty collection when building the French national benchmark parametric life cycle model for each pathway studied following the goal and scope.

Information and data collected enable us to define the probability distributions (PD) for every upstream parameter (e.g., parameters determining H_2 production feedstock market) and downstream parameter (e.g., electricity demand during H_2 compression) involved in each RE-to-H_2 pathway studied best representing French resource availability and technology market share.

Then, the uncertainty, e.g., PD, for each life cycle parameter is propagated holistically into the corresponding parametric LCA model to each pathway studied. Thus, becoming the first detailed and holistic SA and UA for renewable-electricity-based water electrolysis H_2 production LCA. The PD type, PD definition value, and PD representativeness for several most critical life cycle parameters in all RE-to-H_2 pathways identified by this research are illustrated in Table 4 below.

Table 4. PD model for critical life cycle parameters in RE-to-H_2 pathway.

Parameter	PD model	PD representativeness
Annual solar irradiance among France mainland (Irr, kWh/m^2/yr)	PD type: BETA Min: 1.365 E+06 Max: 2.401 E+06 α =1.735 β =6.822	Min, Max, α, β fitted by oracle crystal ball software based on data from AGATE dataset as stated above.
Performance ratio of the PV panels (PR, %)	PD type: Triangle Min: 50 Default: 80 Max: 90	Estimated based on IEA PV guideline[82, 112]. 50 %- 75 % for PV model installed in 1980s; 70 %- 80 % for 1990s; 80 %- 90 % for newly installed PV.
PV power plant lifespan ($Life_{PV}$, years)	PD type: Triangle Min: 20 Default: 30 Max: 40	Thirty years for model installed during the 1990s until now[82]. Fourth years for the very new model[82]. Twenty years[82] to cover the possibility of accident or overuse.
Onshore wind turbine capacity factor ($Cf_{onshore}$, %)	PD type: Normal Default: 25 Std: 0.01	Default and Std fitted by oracle crystal ball software based on capacity factor from Statista dataset[96] and corresponding land area[113] of 13 administrative regions in France.
Wind farm lifespan ($Life_{wind}$, year)	PD type: Triangle Min: 15 Default: 20 Max: 25	Default stands for the average[81]. Min[81] stands for frequently overload turbine with long working time. Max[81] stands for turbine under good maintenance.

Water electrolysis unit global efficiency (η_{unit} , %)	PD type: Triangle Min: 50 Default: 60 Max: 70	The stack efficiency for both the PEM and AE units was approved to be identical[99, 114]. Default stand for the average[99, 114]. The mix and Max were estimated based on the reference literature[99, 114].
BOP lifespan ($Life_{BOP}$, year)	PD type: Triangle Min: 18 Default: 20 Max: 22	Default stands for average[99]. Min stands for often overload facilities with long working time[99]. Max stands for the situation under good maintenance[99].

And the PD type and PD definition value for all the remaining background and foreground life cycle parameters is summarized in Table 5 and Table 6 below. Besides, the parameterized LCA model built in the *lca_algebraic* library applies the Sobol indices analyzing theory[110], namely the variance-based sensitivity analysis[111] as the SA method for LCIA results obtained.

The Sobol indices analyzing is a form of global sensitivity analysis that can decompose and attribute the variation and uncertainty of model output to each input's uncertainty. For one life cycle parameter, the closer its Sobol indices to 1 (or 100 %), the more uncertainty of one specific LCIA score is induced by this life cycle parameter's uncertainty[111].

The UA and SA for impact category LCIA score are performed through running Monte Carlo simulation (10,000 iterations) for each parametric LCA model. The Monte Carlo simulation simultaneously obtains the standard deviation (SD), and the arithmetic mean value (μ) for each LCIA score. This research chooses the coefficient of variation (CV, SD divided by μ) as indices for consistent uncertainty comparison of LCIA scores with different units.

Table 5. PD model for life cycle parameters studied in RE-to-H_2 pathway feedstock market mix.

Parameter and PD model
Nominal maximum power of PV panels (P_{peak} , Wp/m^2)   PD type: Triangle, Min: 86.3, Default: 115.1, Max: 131.5
PV panels yearly loss ratio (YLR , %)   PD type: Triangle, Min: 0.5, Default: 0.7, Max: 0.9
Electricity consumption during Mono Si ingot production, Czochralski process ($Elec_{s\text{-}ignot}$, kWh /Kg Si ingot)   PD type: Triangle, Min: 10, Default: 32, Max: 150
Electricity consumption during Multi Si ingot production, casted ($Elec_{m\text{-}ignot}$, kWh /Kg Si ingot)   PD type: Triangle, Min: 7, Default: 7, Max: 15
PV plant ground coverage ratio (GCR , %)   PD type: Triangle, Min: 36, Default: 36, Max: 45
Market share of monocrystalline silicon panel in total French PV panel installed 2019 ($R_{monoSiPV}$, %)   PD type: Triangle, Min: 30.1, Default: 33.04, Max: 49
Market share of the thin-film panel in total French PV panel installed 2019 (R_{filmPV} , %)   PD type: Triangle, Min: 4.83, Default: 4.83, Max: 10
Off-shore wind turbine installed capacity in France 2019 ($P_{offshore,total}$, GW)   PD type: Linear, Min: 0.01, Default: 0.01, Max: 0.48
Share of <1MW capacity level turbine assembly in total French onshore wind turbine assembly installed 2019 ($R_{turbine,<1MW}$, %)   PD type: Linear, Min: 6, Default: 11.27, Max: 11.27
Share of >3 MW capacity level turbine assembly in total French onshore wind turbine assembly installed 2019 ($R_{turbine,>3MW}$, %)   PD type: Linear, Min: 0.4, Default: 6.57, Max: 6.57
Off-shore wind turbine capacity factor in France 2019 ($Cf_{offshore}$, %)   PD type: Triangle Min: 33 Default: 38 Max: 55

3 Results & Discussion

3.1 Multi-environmental influence

3.1.1 Unit process contribution of LCIA result

The representative patterns of unit process contribution to LCIA result for each RE-to-H_2 pathway are shown in Fig. 5. Histogram height stands for the arithmetic mean value obtained through Monte Carlo simulation, while the LCIA score uncertainty is shown in the corresponding box plot (e.g., P10, P25, P50, P75, P90). The corresponding graph of unit process contributions of the LCIA categories remaining is shown in the supplementary materials.

• *Global warming potential, Photochemical ozone creation*

The production, transportation, and construction of electrolysis unit plant infrastructure (e.g., the electrolyzer BOP, H_2 compressor) are the most critical contributor to GWP and POC for all the RE-to-H_2 pathways studied, which contribute 58.9 %- 70.5 %, 53.5 %- 62.9 % to GWP and POC respectively. This high contribution can be furtherly attributed to the massive concrete consumption during electrolysis unit foundation construction, where industrial concrete with 40 MPa compressive strength is applied. And the production of clinker (primary raw material for electrolysis unit concrete foundation) is very CO_2 and NO_x emission-intensive under the current French national context[115].

The remaining GWP and POC score for the solar-to-H_2 pathway and wind-to-H_2 pathway mainly contributed from the French national average solar and wind electricity supply chain. Under the current French context, most multi-Si and mono-Si PV panels are produced and imported from China[78], which are still CO_2 and NO_x emission-intensive due to the current fossil-based Chinese energy structure 2019. Besides, PV panel mounting system production contributes equally to Si PV panels production and transportation due to the production of aluminum wrought alloy massively consumed in PV panel mounting system. Under the current French and average European context, aluminum wrought alloy production is very electricity-intensive, e.g., CO_2 and NO_x emission[115].

Inside the wind electricity supply chain, GWP and POC mainly come from the construction process of 1-3 MW wind turbine assemblies, which produce most onshore wind electricity under the current French context. This high contribution comes from the massive low-alloyed steel consumed during wind turbine assembly construction. Since the dominant steel production technology in France, e.g., converter steel production, is still CO_2 and pollutant emission-intensive under massive production. Besides, most wind farms are located in remote areas with low population density under the current French context. To access the wind farm site, specific roads should be built. The road construction process is another crucial contributor to GWP and POC scores. Since a large amount of diesel and petroleum is consumed for road construction and extensive road construction material transportation. Among the four pathways studied, pathway 1 shows the highest GWP, and pathway 2 shows the highest POC, namely 5.46 E-02 kg $CO_{2\text{-eq}}$ /MJ H_2, 1.53 E-08 kg $NMVOC^-$ /MJ H_2, respectively. Pathway 3 indicates the lowest result for GWP and POC, namely 2.62 E-02 kg $CO_{2\text{-eq}}$ /MJ H_2, 1.03 E-04 kg $NMVOC^-$ /MJ H_2, respectively.

Table 6. PD model for other life cycle parameters studied in RE-to-H2 pathway.

Parameter and PD model
The lifespan of network connection & auxiliaries for PV plant, which equals to plant DC/AC inverter lifespan ($Life_{\text{solar,aux}}$, yr)
PD type: Triangle Min: 10 Default: 15 Max: 30
The specific weight of PV plant DC/AC inverter (M_{inverter} , kg/KVA)
PD type: Triangle Min: 1 Default: 2 Max: 4
The lifespan of network connection & auxiliaries for wind plant ($Life_{\text{wind,aux}}$, yr)
PD type: Triangle Min: 15 Default: 20 Max: 25
PEM stack lifespan ($Life_{\text{PEM}}$, hour)
PD type: Triangle Min: 40000 Default: 50000 Max: 60000
AE stack lifespan ($Life_{\text{AE}}$, hour)
PD type: Triangle Min: 60000 Default: 75000 Max: 90000
The total amount of H_2 can be compressed during the lifespan of one H_2 compressor ($H_{2\text{unit}}$, kg/unit)
PD type: Triangle Min: 5.021 E+06 Default: 5.906 E+06 Max: 6.792 E+06
H_2 compressor electricity consumption for 1 kg H_2 compression ($Elec_{\text{compress}}$, kWh/Kg)
PD type: Triangle Min: 2.7 Default: 3.91 Max: 3.91
Lorry transport distance for each component in the two-stage transportation model ($Dist_{\text{lorry}}$, km)
PD type: Triangle Min: 40 Default: 700 Max: 1400
Train transport distance in the two-stage transportation model ($Dist_{\text{train}}$, km)
PD type: Triangle Min: 0 Default: 300 Max: 600
Oversea transportation distance for off-shore wind turbine component ($Dist'_{\text{oversea}}$, km)
PD type: Triangle Min: 0 Default: 10 Max: 100

- *Freshwater and terrestrial acidification, Ionizing radiation, Fossil resources depletion, Respiratory effects*

For FTA, IR, FD, and REI, French national average solar electricity supply chain, electrolysis unit plant infrastructure is the major contributor in the solar-to-H_2 pathway and wind-to-H_2 pathway, respectively. These two significant contributors take 44.95 %- 56.88 %, and 36.81 %- 56.95 % in all impact category scores among solar-to-H_2 and wind-to-H_2 pathways, respectively. The remaining half of FTA, IR, FD, and REI in solar-to-H_2 is mainly contributed by electrolysis unit plant infrastructure and French national average wind electricity supply chain in the wind-to-H_2 pathway. For FTA, the AE electrolysis stack contributes equally to the wind electricity supply chain, which is one-third of the contribution from the French national average solar electricity supply chain.

Like GWP, the production of aluminum wrought alloy used in PV panel mounting system; Chinese multi-Si and mono-Si PV panels production and transportation process is the main contributor to these four LCIA scores of French solar electricity supply chain., These LCIA scores for plant infrastructure mainly come from the clinker production process like GWP. And the contribution from hard coal & diesel burning during clinker and cement production; the cement transportation process is almost identical.

Besides, the electricity consumed by cement production under the current French context endows huge IR, contributing most of IR from electrolysis unit plant infrastructure. Since nuclear electricity contributes around 70 % of French electricity to the national electricity grid. And pressurized water reactors (PWR) generate almost all the nuclear electricity in France (as ecoinvent v3.6 database[115] indicated), which will produce a large pile of radioactive waste during PWR unit operation. Inside the wind electricity supply chain, the production process of copper (the primary component for 1-3 MW wind turbine assembly) is the main contributor to the FTA score. Under the current French context, major copper production technology is flash smelting furnaces reducing and refining, which is sulfur dioxide emission-intensive. The main contributor for IR can be further attributed to the production of electricity consumed during wind turbine assembling, mounting, and turbine end-of-life (EoL) removal. FD and REI mainly contributed from massive low-alloyed steel consumed during the construction of 1-3 MW wind turbine assembly. This low-alloyed steel consumption also ranked as the second significant contributor of IR and FTA. The production process of polyamide injection reinforced plastic glass fiber (main component of > 1 MW wind turbine blade[115]) is another significant contributor for FD. The primary raw material for glass fiber reinforced plastic production, Nylon6-6, is very fossil resources intensive during its production process[115]. The road-building process to access the wind farm is the third most significant contributor to FTA, IR, FD, and REI.

Pathway 2 shows the highest FTA, IR, FD, and REI results among each pathway studied, e.g., 3.61 E-04 mole H^+_{eq} /MJ H_2, 2.98 E-03 kg $U_{235\text{-}eq}$ /MJ H_2, 5.89E-01 MJ /MJ H_2, 3.16 E-09 disease i. /MJ H_2, respectively. Pathway 3 shows the lowest result of these LCIA scores, namely 1.37 E-04 mole H^+_{eq} /MJ H_2, 1.38 E-03 kg $U_{235\text{-}eq}$ /MJ H_2, 2.68 E-01 MJ /MJ H_2, 1.42 E-09 disease i. /MJ H_2, respectively.

• *Land use, Minerals and metals depletion, Dissipated water*

Regarding LU, MD, DW, the French national average solar electricity supply chain contributes around 62.20 %-89.66 % for the solar-to-H_2 pathway, ranking the most important contributor. Most of the LCIA score remained comes from the electrolysis unit plant infrastructure. However, for the wind-to-H_2 pathway, 42.21 %- 61.87 % of these 3 LCIA scores comes from electrolysis unit plant infrastructure, which is closely followed by the contribution from the French national average wind electricity supply chain.

Construction of PV panel mounting system will occupy large land areas[115], which contributes most of LU inside average French solar electricity supply chain. Besides, a lot of zinc is consumed during PV panel mounting system construction, which explains the MD score. The corresponding DW score is due to the massive multi-Si PV panel produced and imported from China[78]. Since current multi-Si wafer production in China is very water-consuming[78]. The road construction process occupies a large land area, which explains the LU score for the French national average solar electricity supply chain.

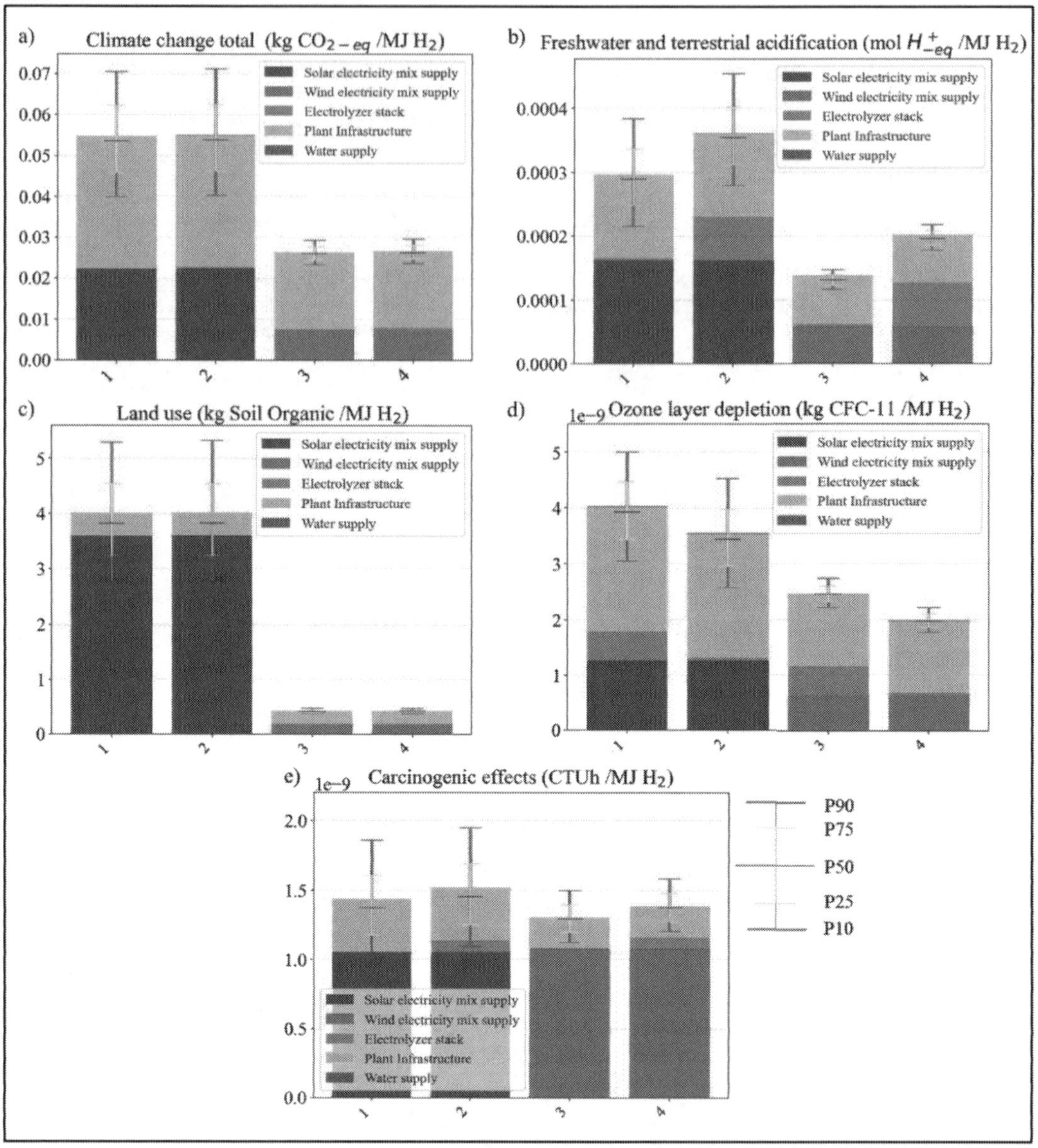

Fig. 5. Representative patterns of unit process contribution to LCIA result marked with uncertainty box plot. Pathway 1 to 4 corresponds to the rank in Fig. 4.

Under the current French context, Zinc coating is massively applied to wind turbine towers corrosion against, which contributes to most of the MD score. Regarding DW, the production of polyamide injection reinforced plastic glass fiber used in wind turbine blades and the low-alloyed steel used in the wind turbine tower contribute equally. Nylon6-6 for glass fiber reinforced plastic production is very water consumption intensive during its production process[115]. So is the same for the production process of low-alloyed steel production under the current French context[115].

Regarding the electrolysis unit plant infrastructure, most DW score is contributed from the concrete production process, which can be furtherly attributed to the water-intensive raw material extraction for cement production[78]: Clinker (Limestone rotary kiln calcination), gravel, and sand (extracted from river bed) production under current French context. Besides, the limestone rotary kiln calcination is limestone consumption intensive.

And this process will consume lots of zinc[78], which explains the corresponding MD score. The large amount of sand consumed in the electrolysis unit concrete foundation presents the most LU score, followed by the limestone consumption during clinker production and cement transportation.

Among each pathway studied, pathway 2 shows the highest result of LU, MD, DW, namely 4.02 kg Soil Organic /MJ H_2, 6.10 E-06 kg Sb_{-eq} /MJ H_2, 3.11E-02 m^3 /MJ H_2, respectively. Pathway 3 shows the lowest result of these LCIA scores: 4.16 E-01 kg Soil Organic /MJ H_2, 2.11 E-06 kg Sb_{eq} /MJ H_2, 5.79E-3 m^3 /MJ H_2, respectively.

•*Ozone layer depletion*

Regarding OLD, the electrolysis unit plant infrastructure contributes around 51.9 %- 64.3 % for all pathways studied. The remaining OLD in the solar-to-H_2 pathway and the wind-to-H_2 pathway is mainly contributed by the French national average solar and wind electricity supply chain, respectively. And in pathway 1 and pathway 3, where PEM electrolyzer is applied, 13.2 % and 21.6 % of OLD come from the production, transportation, and EoL of the PEM electrolysis stack.

Like GWP, the production of aluminum wrought alloy used in PV panel mounting system, Chinese multi-Si and mono-Si PV panels production and importation, contribute equally to OLD in French national average solar electricity supply chain. The electrolysis unit plant infrastructure's OLD score mainly comes from clinker production (primary raw material for electrolysis unit concrete foundation). And cement and clinker transportation process; the heavy fuel oil, petroleum coke, diesel burning during clinker and cement production rank as the main contributors.

Inside the wind electricity supply chain, the road construction process, with similar reason discussed in GWP, explain the OLD of the average French wind electricity supply chain. And low-alloyed steel consumed in wind turbine towers under the current French context is the second contributor to OLD of French national average wind electricity supply chain. Pathway 1 shows the highest OLD result among each pathway studied, 4.03 E-09 kg CFC-11 /MJ H_2. Pathway 4 indicates the lowest OLD result, namely 1.99 E-09 kg CFC-11 /MJ H_2.

•*Carcinogenic effect*

Regarding CE, the French national average solar and wind electricity supply chain contributes 69.4 % to 82.7 % for all pathways studied. The remaining CE is mainly contributed by electrolysis unit plant infrastructure.

The primary aluminum wrought alloy massively consumed by PV panel mounting system is mainly refined by aluminum oxide (mainly from red mud bauxite digestion) through the Hall-Héroult process[115]. The Hall-Héroult process is very electricity-intensive, and the bauxite digestion process is very pollutant emission-intensive, which explains most CE of the French average solar electricity supply chain. Besides, the PV panel mounting system consumed a lot of reinforcing steel mostly made from low-alloyed steel under the current French context, which explains most of the CE score remained.

Similarly, the production of low-alloyed steel and chromium steel used in the wind turbine tower is the main contributor to CE from the average French wind electricity supply chain. And clinker from limestone rotary kiln calcination is the main contributor of CE from the water electrolysis unit plant infrastructure. Among each pathway studied, pathway 2 shows the highest result of CE: 1.54 E-09 kg CTUh /MJ H_2. And pathway 3 indicates the lowest CE result, 1.30 E-09 kg CTUh /MJ H_2.

•*Particular focus: contribution from electrolysis stack*

Variation of environmental impact induced by different electrolysis stacks is illustrated in Fig. 6. Applying PEM electrolyzer stack has a much lower environmental impact than AE electrolysis stack for all impact categories studied, except OLD.

The production of manganese (the critical component of the cathode material), mainly through electrothermal and electrolysis treatment of manganese ore, is pollutant emission-intensive in terms of NH_3, HCl, VOC, as well as heavy metals[115]. This results in a higher CE score of the AE electrolysis stack than the PEM.

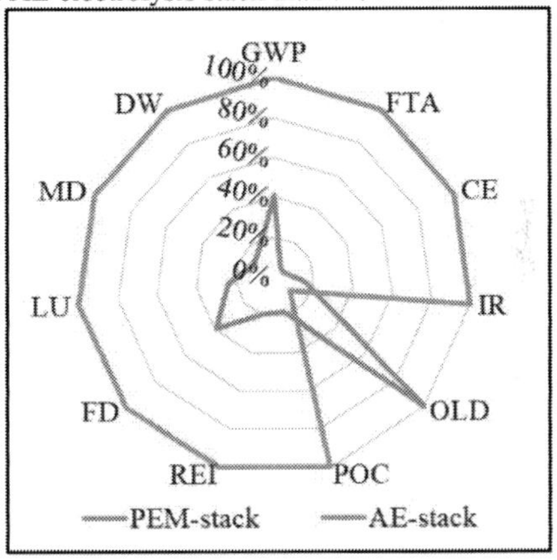

Fig. 6. Environmental influence difference between PEM stack and AE stack.

Besides, the production process of 99.5 % Nickel (another critical electrode material mainly comes from hydrometallurgical and pyrometallurgical nickel ore refining[115]) will emit large amounts of dust containing PM10 and VOC[115]. This will lead to a higher GWP, FTA, POC, and REI of AE electrolysis stack production.

The underground nickel ore mining infrastructure and heat consumption during the ore mining process explain the higher FD, LU, and MD for the AE electrolysis stack. Furthermore, a large amount of electricity is consumed during AE electrolysis stack assembling, explaining a higher IR and DW score. Regarding OLD, the solid polymer electrolyte production from trichloromethane is very CFC emission-intensive, explaining the higher OLD of the PEM electrolysis stack compared with the AE electrolysis stack.

3.1.2 Scenario variation and further comparison

Fig. 7 illustrated the LCIA result (arithmetic mean value obtained through Monte Carlo simulation) comparison among each RE-to-H_2 pathway studied. It's evident that whether from the view of comprehensive trade-off or only considering GWP, water electrolysis H_2 production with average French wind electricity through PEM electrolysis unit is the optional choice among all RE-to-H_2 pathways studied.

The total mass of the electrolysis unit (e.g., stack, electronics, pumps, safety equipment) is enormous, which can be up to 9.6 tonnes for a PEM electrolysis unit[99]. Thus, a foundation with high compressive strength is required for the stable installation of the electrolysis unit on the ground. And this research applies 40 MPa concrete when building the electrolysis unit foundation.

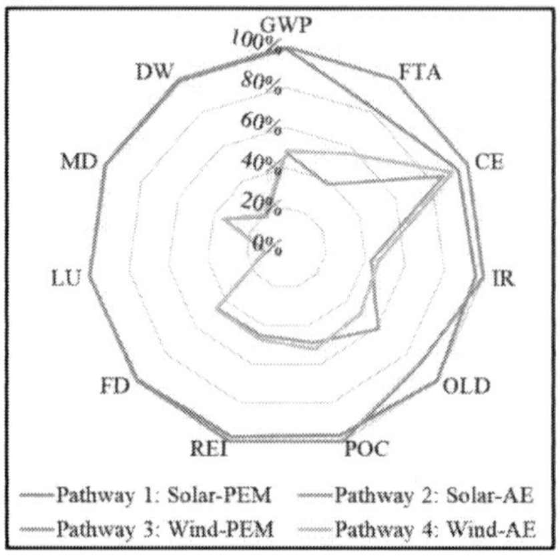

Fig. 7. LCIA result comparison among RE-to-H_2 pathways.

Fig. 8 a) illustrated RE-to-H_2 LCIA result variation among different electrolysis unit concrete foundations, various solar radiation, and wind energy availability conditions. Pathway 3 is compared with wind-to-H_2 applying PEM electrolysis unit built on 20MPa; wind-to-H_2 using PEM electrolysis unit under the maximum and minimum French wind turbine capacity factor. Pathway 2 is compared with solar-to-H_2 under maximum annual solar irradiance in France, applying PEM and AE electrolysis units, respectively.

The average capacity factor of the French wind turbine, both for onshore and off-shore applied in pathway 3 and pathway 4, is already close to the maximum capacity factor. Thus, wind-to-H_2 under the maximum capacity factor only reduces less than 7 % to each impact category compared to pathway 3.

But transferring the 40MPa concrete foundation to 20MPa concrete foundation will lead to a higher environmental impact reduction: 7.03 %-13.96 % reduction to GWP, FTA, IR, POC, FD, MD, and DW, as well as 1.47 %- 2.52 % reduction to CE, OLD, and REI. However, LU will increase around 10.68 % due to a higher sand consumption during 20MPa cement production[115].

An interesting result is revealed that while under maximum annual solar irradiance, all LCIA scores to solar-to-H_2 will be reduced around 20 % compared with pathway 2, but the environmental impact of solar-to-H_2 pathways is still more significant than the wind-to-H_2 pathways studied, as shown in Fig. 8 a).

Fig. 8 b) compares the environmental influence of RE-to-H_2 utilizing different electricity sources during the electrolysis process. RE-to-H_2 with PEM water electrolysis unit powered by electricity from Norwegian and French grid (94.1 % Norwegian electricity generated by hydropower and 70 % French electricity comes from PWR as stated above[115]), as well as European-average electricity from ENTSO-E is compared with pathway 2 and 3.

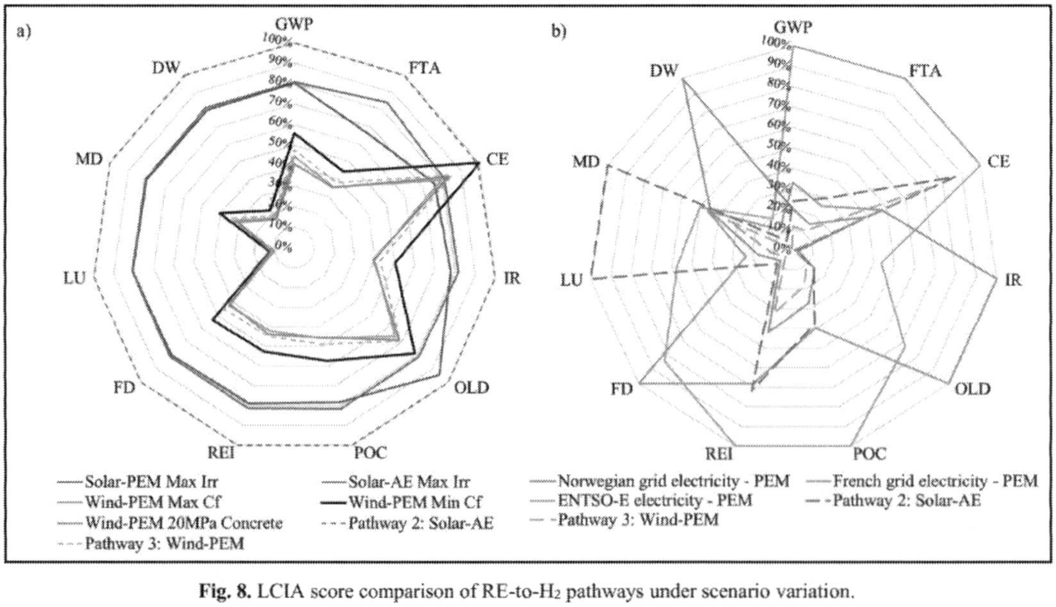

Fig. 8. LCIA score comparison of RE-to-H_2 pathways under scenario variation.

Compared with Norwegian, French, and ENTSO-E electricity, utilizing French national average wind electricity for water electrolysis H_2 production is still a wiser choice for most impact categories studied. Among them, PEM RE-to-H_2 through Norwegian electricity is most comparable to pathway 3 regarding most of the impact categories investigated. Despite the high water footprint of Norwegian electricity production, it will significantly reduce the CE compared with pathway 3. Applying the average French electricity will also reduce the CE (36.59 %) compared with pathway 3. However, it will induce a considerable increase in IR, OLD, REI, and

DW (981.52 %- 19188.59 %) and moderate to all the other categories (32.36 %- 190.22 %).

Applying the ENTSO-E electricity will increase the score of all the impact categories (34.32 %- 1688.00 %) compared with pathway 3. Another tricky point is that, for RE-to-H_2 with AE water electrolysis unit, applying the Norwegian, French, and ENTSO-E electricity will reduce the MD and LU (42.64 %- 76.78 %) significantly compared to pathway 2.

3.2 Sensitivity and uncertainty analysis

3.2.1 LCIA result sensitivity to life cycle parameter

Fig. 9 illustrated the SA result, i.e., First-order indices (S1 indices) of LCIA score to each life cycle parameter in pathway 3. For each impact category investigated, 33 %- 65 % LCIA score variation (or in another word, uncertainty) comes from the uncertainty of electrolysis unit global efficiency. Other critical life cycle parameters determining the LCIA score are wind turbine lifetime (contribute to 53 % CE variation, and 8 %- 29 % variation of all the other 11 impact categories), onshore wind turbine capacity factor (contribute 12 %- 19 % to score variation of all impact categories), electrolysis unit BOP lifetime (contribute 4 %- 11 % to score variation of all impact categories), and share of >3MW wind turbine assembly in French national average onshore wind turbine market (contribute 1 %- 9 % to score variation of all impact categories).

However, under the current French context, as shown in Fig. 9 b), uncertainty to each impact category score induced by all life cycle parameters' variation is relatively small. LCIA score uncertainty generated from electrolysis unit global efficiency variation is around 6-7 % regarding all impact categories.

Uncertainty of CE induced from wind turbine lifetime variation is about 8 %. Besides, uncertainty to each impact category score caused by the remaining life cycle parameters is below 5 %. Corresponding SA results for pathway 4 (in supplementary materials) are very close to pathway 3 as shown in Fig. 9.

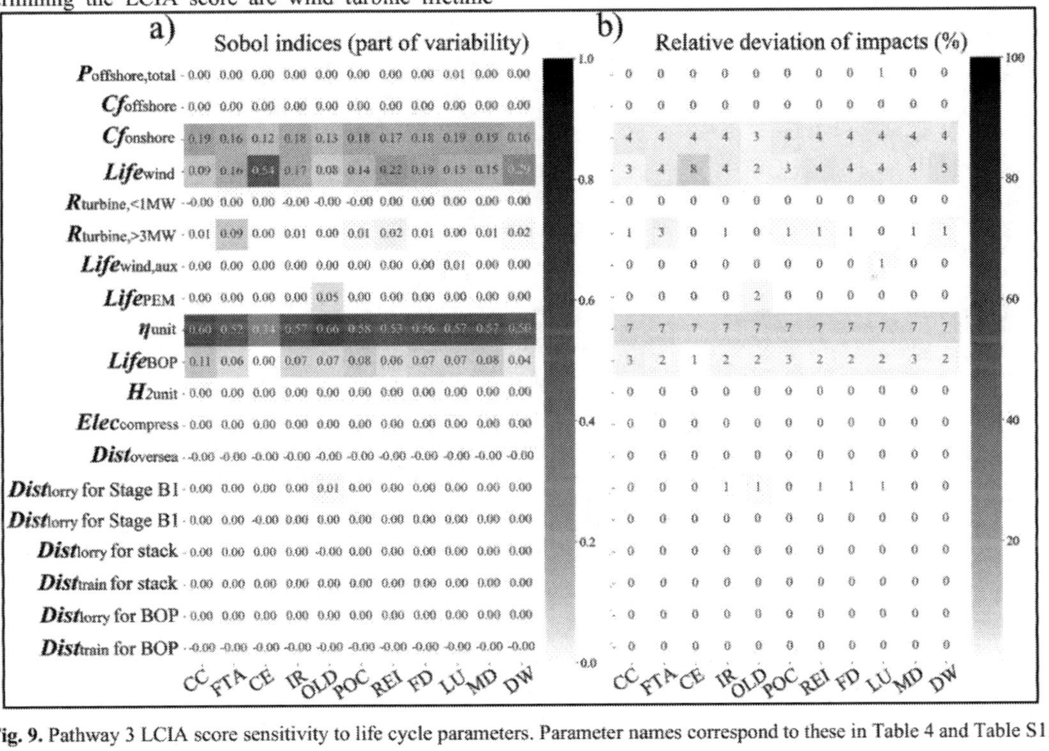

Fig. 9. Pathway 3 LCIA score sensitivity to life cycle parameters. Parameter names correspond to these in Table 4 and Table S1.

Fig. 10 a) shows the LCIA score S1 indices to each life cycle parameter in pathway 2. For each impact category studied, under French national average resource availability and technology mix 2019, around 38 %-50 % LCIA score uncertainty comes from the variation of French annual solar irradiation available on PV panels. 23 %-30 % LCIA score variation comes from the uncertainty of the PV panel performance ratio.

The remaining LCIA score variation mainly comes from PV panel lifetime uncertainty (contribute 19 %-28 % to the variation of CE, DW, and LU; 6 % to 16 % to variation of all the other eight impact categories), electrolysis unit global efficiency (6 %-12 % to variation of all impact categories), nominal maximum power of PV panels (1 %-6 % to the variation of all impact categories).

Inside the French national average solar electricity supply chain, critical life cycle parameters are identified with more considerable uncertainties than those in the French national average wind electricity supply chain. Thus, the LCIA score's uncertainty in the solar-to-H_2 pathway is more significant than the wind-to-H_2 pathway. For all impact categories, as illustrated in Fig. 10 b), LCIA scores uncertainty induced from French annual solar irradiation variation is around 13-16 %; Uncertainty caused from the PV panels performance ratio is 10-13 %; Uncertainty generated from PV panel lifetime is 6-13 %. Besides, uncertainty to each impact category induced by life cycle parameter remaining is below 7 %. Corresponding SA results for pathway 1 listed in the supplementary material are very close to these in Fig. 10.

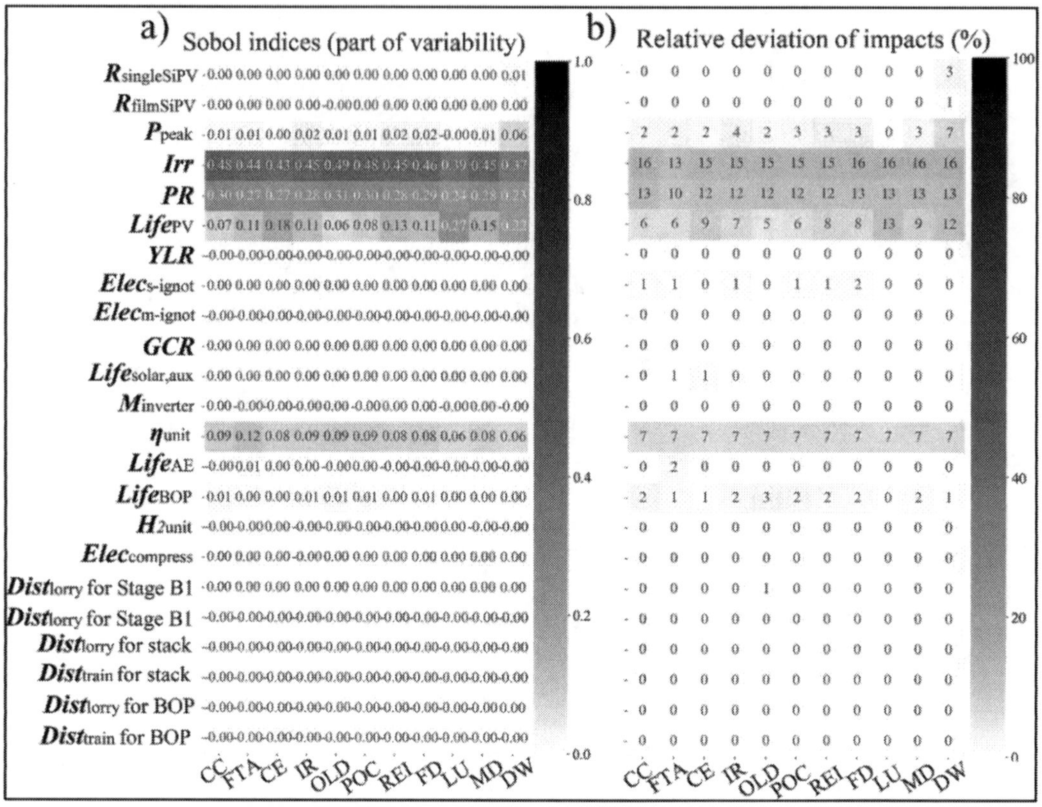

Fig. 10. Pathway 2 LCIA score sensitivity to life cycle parameters. Parameter names correspond to these in Table 4 and Table S1.

3.2.2 LCIA result uncertainty

Due to more life cycle parameters identified with relatively more significant uncertainties in French national average solar electricity supply chain stated above, as illustrated in Fig. 11, CV among all impact categories score of the solar-to-H_2 pathway is around 9.57 %-16.44 % higher, compared with the corresponding CV of the wind-to-H_2 pathway.

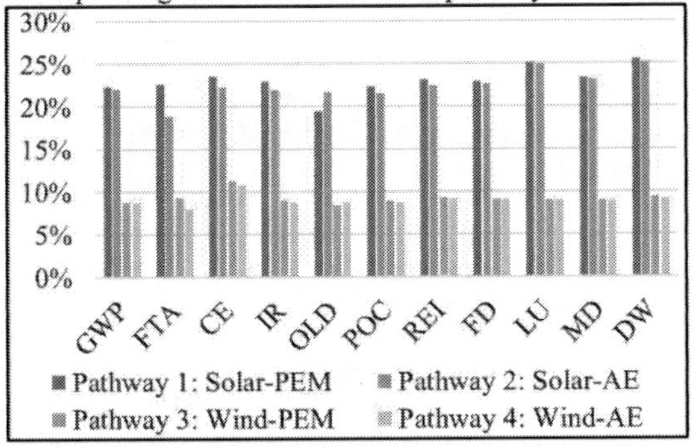

Fig. 11. LCIA results' uncertainty for each pathway.

Under the current French context, applying PEM electrolysis unit in the solar-to-H_2 pathway is embodied with more significant life cycle parameters uncertainty (especially the variation of stack lifespan[116]) than using the AE electrolysis unit. Thus, all impact categories CV for pathway 1 is 0.16 %- 3.76 % larger than those for pathway 2, except CV for OLD. OLD CV for pathway 1 is 2.21 % smaller than CV for pathway 2. LCIA results' CV for pathway 3 and pathway 4 is also comparable, where the difference among each impact category is smaller than 1.3 %.

3.3 Guidance for improving LCA sustainability

3.3.1 French solar electricity supply chain

As discussed above, in the current French context, many PV panels, especially the multi-Si and mono-Si panels, are manufactured and assembled in China, then imported to France and Europe. Manufacturing the Si PV panels related materials, e.g., solar gride silicon and silicon ingot, in China is very fossil energy, water, and emission (CO_2, NO_x, and other pollutants) intensive due to the current fossil energy-based Chinese energy structure. However, the renewable and sustainable energy transition is flourishing in China[117-119]. And remarkable technology advances have achieved in solar gride silicon production process depollution, attributed to the vigorous promotion and policy implementation of the Chinese government during last decade[120]. Besides, the gradual nationwide promotion of advanced technology, e.g., panel lead-free soldering, higher level automation production, has significantly improved the PV panel component manufacturing efficiency[121]. Thus, the environmental impact embodied in Chinese PV panels is visibly progressively decreasing[115]. This will continuously guarantee China as a primary PV panels provider for the French solar electricity market.

Regarding the PV mounting system, renewable energy-based zinc and aluminum refining should receive more research focus to reduce GWP, OLD, FTA, CE, and MD scores. And special attention should be paid to sustaining these materials' recycling process during PV mounting system EoL. Besides, bauxite digesters in Europe are still embodied with relatively large desulfurization and denitrification potential to reduce CE embodied in aluminium production.

Employees should receive qualified training regarding the PV plant operation phase to achieve a high enough PV panel performance ratio and lifetime under site-specific strategies. And innovation should primarily focus on irradiance tracing systems and PV panels site distribution optimization, which can achieve a lower LU.

3.3.2 French wind electricity supply chain

As mentioned above, chromium steel, low-alloyed steel, copper, and zinc coating are massively consumed in the wind turbine tower under the current French context. And the large-scale application of suitable & environmentally friendly metal alternatives (e.g., high-strength glass fibers, basalt, and aramid fibers[122]) is still far from strong enough economic competitiveness. Thus, improving the renewable energy permeation and material recycling ratio during the production of these metal materials is critical to bringing down the environmental influence of the French wind electricity supply chain in the foreseeable future. Besides, there is still enough improving space for desulfurization optimization to copper production in Europe to furtherly reduce the FTA embodied in copper production. Furthermore, more sustainable, e.g., biomass-based[123, 124], nylon 6-6 alternatives indicate significant DW and FD reduction potential during wind turbine blade production.

Innovation on less electricity-intensive wind turbine assembling, mounting, and EoL removal technology will achieve a lower DW and FTA. To fulfill more environmentally-friendly access to wind power plants in terms of OLD, LU, GWP, FTA, IR, FD, and REI, relative road building should highlight biomass-based alternatives[125, 126] to fossil asphalt.

And the particular focus should be paid on road construction raw material transportation distance management and diesel consumption efficiency during road construction. Operators should receive qualified training during the wind plant operation phase to achieve a high enough wind turbine lifetime, which is critical regarding all the impact categories studied.

3.3.3 Water electrolysis unit and operation

Similarly, to pursue a more environmental-friendly AE stack, improving the renewable energy permeation and material recycling ratio when producing the electrode material (manganese and 99.5 % nickel) is essential. And relatively high NH_3, HCl, VOC emission from the electrothermal manganese ore reforming (major manganese production technology) indicate the potential for further depolluting optimization. Meeting the huge heat requirement during manganese ore reforming from renewable biomass heat rather than fossil-based heat serves as another critical strategy. Similar regulations are also recommended for pyrometallurgical nickel ore refining, the primary technology for 99.5 % nickel production, to reduce PM10 and VOC emissions. Furthermore, attention should be paid to the innovation of more renewable solid polymer electrolytes, e.g., algal-based polysaccharides[127], to reduce the OLD score embodied in the PEM stack.

Furthermore, innovation on less electricity-intensive assembling, mounting, and EoL removal of AE electrolysis unit will achieve a lower DW and FTA sore. And the particular focus should be paid to improving stack efficiency, i.g., innovation on membrane temperature optimization technology[128], as well as the addition of more efficient electrolyte and electrocatalysts[129]. And during electrolysis unit operation, employees should receive qualified training to achieve higher electrolysis efficiency and BOP lifetime. Regarding BOP and H_2 compressor, compressive strength for qualified concrete that meets the electrolysis unit ground stabilization requirement should be as low as possible. To reduce GWP, POC, CE, OLD, and REI score, a high recycling rate of raw material (gravel and sand) and renewable electricity penetration during concrete production is indispensable.

Besides, innovation on improving the efficiency of significant clinker production technology in Europe, limestone rotary kiln calcination, should be highlighted. And renewable biomass sourced heat is a better choice rather than fossil-based heat during limestone rotary kiln calcination.

4 Conclusions

This study performs a comprehensive environmental impact investigation and comparison among renewable-electricity-based water electrolysis H_2 production in France, from the angle of average French resource availability and technology market share in 2019. With the corresponding French national benchmark parametric life cycle model built in BW2 based *lca_algebraic* library, this study becomes the first detailed and holistic SA and UA of RE-to-H_2 LCA both on the foreground and background life cycle parameters.

Four RE-to-H_2 pathways were analyzed based on different technical details and life cycle hypothesis. Based on the result and discussion discussed above, the following main conclusions are drawn:

1) The water electrolysis H_2 production with average French wind electricity via PEM electrolyzers is always the most environmentally-beneficial choice in all RE-to-H_2 pathways studied, whether in terms of impact categories comprehensive trade-offs or considering only GWP.

2) Compared with Norwegian, French, and ENTSO-E electricity, utilizing French national average wind electricity for water electrolysis H_2 production remains a more sensible option in terms of most impact categories studied.

3) Under the current French context, the uncertainty for solar-to-H_2 pathways is moderate, with CVs between 19% and 26% for all impact categories. LCIA score uncertainty primarily derived from variations of French annual solar irradiation available on PV panels; PV panel performance ratio; PV panel lifetime; electrolysis unit global efficiency; the nominal maximum power of PV panel.

4) In the current French context, the wind-to-H_2 pathway is embodied with relatively small uncertainties, with CVs between 8%-11% for all impact categories. Critical life cycle parameters determining the LCIA score are electrolysis unit global efficiency; wind turbine lifetime; onshore wind turbine capacity factor; electrolysis unit BOP lifetime; and share of >3MW wind turbine assembly in French national average onshore wind turbine market.

5) Several essential strategies should be highlighted to achieve a more renewable and sustainable RE-to-H_2 pathway. It's critical to improve the renewable energy permeation and material recycling ratio during the production of basic materials for infruscture construction (e.g., zinc, aluminum, copper, 99.5 % nickel, cement). Innovation in more sustainable substitution, such as biomass-based alternatives for infruscture materials and further depolluting optimization for infruscture materials production serve as other critical strategies. In addition, special emphasis should be placed on developing less electricity-intensive assembling, mounting, and EoL removal of both background and foreground infruscture (e.g., PV panels and mounting system, wind turbine assembly, and water electrolysis unit).

Acknowledgments

This research is supported by the National Natural Science Foundation of China (No. 51576087), the Foundation of State Key Laboratory of Coal Combustion (No. FSKLCCA1902), and the Double first-class research funding of China-EU Institute for Clean and Renewable Energy (No. 3011120016).We also would like to express our sincere thanks to professor Philippe BLANC, doctor Romain BESSEAU, and all the colleagues in Centre Observation, Impacts, Energie for their selfless help and valuable suggestions during the research and paper revision.

Supplementary Materials

Other supplementary materials, e.g., the '.ipynb' jupyter notebook files contenting the French national benchmark parametric LCA model of all the pathways studied together with the corresponding LCI and data source; accumulative annual solar irradiance 2019 for French mainland's 13 administrative regions taken from the CAMS radiation service dataset "AGATE"; Unit process contribution to LCIA results with uncertainty box plot for each impact categories studied; the LCIA score S1 indices to each life cycle parameter and each impact category score's uncertainty for all pathway studied can be found at the authors' GitHub branch at:

github.com/ZHANG-Zongyue/FR_H2_LCA/tree/EREGCE-proceeding-2022.

References

1. International Energy Agency. (2020) Key World Energy Statistics 2020. www.iea.org/reports/key-world-energy-statistics-2020.
2. Tilman, D., Fargione, J., Wolff, B., D'antonio, C., Dobson, A., Howarth, R., Schindler, D., Schlesinger, W.H., Simberloff, D., Swackhamer, D. (2001) Forecasting agriculturally driven global environmental change. Sci, 292: 281-284.
3. Vitousek, P.M., Mooney, H.A., Lubchenco, J., Melillo, J.M. (1997) Human domination of Earth's ecosystems. Sci, 277: 494-499.
4. Ravishankara, A., Daniel, J.S., Portmann, R.W. (2009) Nitrous oxide (N2O): the dominant ozone-depleting substance emitted in the 21st century. Sci, 326: 123-125.
5. Rockström, J., Steffen, W., Noone, K., Persson, Å., Chapin, F.S., Lambin, E.F., Lenton, T.M., Scheffer, M., Folke, C., Schellnhuber, H.J. (2009) A safe operating space for humanity. Nature, 461: 472-475.
6. Rissman, J., Bataille, C., Masanet, E., Aden, N., Morrow, W.R., Zhou, N., Elliott, N., Dell, R., Heeren, N., Huckestein, B., Cresko, J., Miller, S.A., Roy, J., Fennell, P., Cremmins, B., Koch Blank, T., Hone, D., Williams, E.D., Sisson, B., Williams, M., Katzenberger, J., Burtraw, D., Sethi, G., Ping, H., Danielson, D., Lu, H., Lorber, T., Dinkel, J., Helseth, J. (2020) Technologies and policies to decarbonize global industry: Review and assessment of mitigation drivers through 2070. ApEn, 266: 114848-114882.
7. Kakoulaki, G., Kougias, I., Taylor, N., Dolci, F., Moya, J., Jäger-Waldau, A. (2021) Green hydrogen in Europe – A regional assessment: Substituting existing production with electrolysis powered by renewables. Energy Convers. Manage., 228-247: 113649.
8. International Energy Agency. (2019) The Future of Hydrogen. https://www.iea.org/reports/the-future-of-hydrogen.
9. International Renewable Energy Agency. (2019) Hydrogen: A renewable energy perspective. https://www.irena.org/publications/2019/Sep/Hydrogen-A-renewable-energy-perspective.
10. Fuel Cells and Hydrogen 2 Joint Undertaking. (2019) Hydrogen roadmap europe: a sustainable pathway for the european energy transition. https://www.fch.europa.eu/news/hydrogen-roadmap-europe-sustainable-pathway-european-energy-transition.
11. Imprimerie de Montligeon. (2005) From Hydrogen to Energy Production. https://www.cea.fr/english/Documents/thematic-publications/hydrogen.pdf.
12. Michael, P.R. (2004) The Hydrogen Economy: Opportunities, Costs, Barriers, and R&D Needs. The National Academies Press, Washington D.C.
13. European Commission. (2020) Hydrogen generation in Europe: Overview of costs and key benefits. https://op.europa.eu/en/publication-detail/-/publication/7e4afa7d-d077-11ea-adf7-01aa75ed71a1/language-en.
14. Fasihi, M., Breyer, C. (2020) Baseload electricity and hydrogen supply based on hybrid PV-wind power plants. J. Clean. Prod., 243: 118466-118497.
15. Le Duigou, A., Miguet, M., Amalric, Y. (2011) French hydrogen markets in 2008 – Overview and future prospects. IJHE, 36: 8822-8830.
16. Fuel Cell and Hydrogen Observatory. (2021) Hydrogen Supply Capacity. https://fchobservatory.eu/observatory/technology-and-market/hydrogen-supply-capacity.
17. Fuel Cell & Hydrogen Energy Association. (2020) Clean Hydrogen Monitor 2020. https://hydrogeneurope.eu/reports/.
18. France Hydrogen Association. (2018) Rapport d'activités de France Hydrogène, édition 2021. www.france-hydrogene.org/publication/rapport-dactivites-de-france-hydrogene-edition-2021/.
19. United Nations Framework Convention on Climate Change. (2015) The Paris Agreement. https://unfccc.int/process-and-meetings/the-paris-agreement/the-paris-agreement.
20. Minstry for the Ecological and Solidary Transition. (2021) Long-term low-emission development strategies for France: The ecological and inclusive transition towards carbon neutrality. https://unfccc.int/sites/default/files/resource/en_SNBC-2_complete.pdf.
21. Yang, Q., Zhou, H., Bartocci, P., Fantozzi, F., Mašek, O., Agblevor, F.A., Wei, Z., Yang, H., Chen, H., Lu, X., Chen, G., Zheng, C., Nielsen, C.P., Mcelroy, M.B. (2021) Prospective contributions of biomass pyrolysis to China's 2050 carbon reduction and renewable energy goals. Nature Communications, 12: 1698-1710.
22. Antonini, C., Treyer, K., Streb, A., Van Der Spek, M., Bauer, C., Mazzotti, M. (2020) Hydrogen production from natural gas and biomethane with carbon capture and storage – A techno-environmental analysis. Sustainable Energy Fuels, 4: 2967-2986.
23. Conseil général de l'environnement et du développement durable. (2018) Plan de deploiement de l'hydrogene pour la transition energetique. https://www.ecologie.gouv.fr/sites/default/files/2018.06.01_IP_presentation_plan_hydrogene.pdf.
24. Ministère de l'Écologie, du Développement Durable et de l'Énergie. (2020) National energy and climate plans for France. ec.europa.eu/energy/topics/energy-strategy/national-energy-climate-plans_en#public-consultation-on-necps.
25. Ministère de la Transition écologique. (2020) French strategy for energy and climate-multi annual energy plan. https://www.ecologie.gouv.fr/sites/default/files/PPE-Executive%20summary.pdf.
26. Ministère de la Transition écologique. (2021) Assemblée nationale. 17 février 2021 relative à

l'hydrogène. www.legifrance.gouv.fr/eli/ordonnance/2021/2/17/TRER2018536R/jo/texte.

27. EU Commission. (2019) Hydrogen for climate action: Zero Emission Urban Delivery @ Rainbow Unhycorn. https://static1.squarespace.com.
28. EU Commission. (2020) Hydrogen: for climate Action project group. https://www.hydrogen4climateaction.eu/projects.
29. EU Commission. (2017) Hydrogen Law and removal of legal barriers to the deployment of fuel cells and hydrogen applications. www.hylaw.eu/.
30. Mission Innovation Initiative. (2015) IC8-renewable and clean hydrogen. mission-innovation.net/our-work/innovation-challenges/renewable-and-clean-hydrogen/.
31. Hydrogen Fuel News Inc. (2018) France to demonstrate the production of hydrogen from biomass. www.hydrogenfuelnews.com/france-to-demonstrate-the-production-of-hydrogen-from-biomass/8536346/.
32. Fuel Cell Works Inc. (2021) In France, the VHyGO Project (GrandOuest Hydrogen Valley) will Develop Green Hydrogen in Three Different Regions. https://fuelcellsworks.com/news/in-france-the-vhygo-project-grand-ouest-hydrogen-valley-will-deploy-green-hydrogen-in-three-different-regions/.
33. Meerman, J.C., Hamborg, E.S., Van Keulen, T., Ramírez, A., Turkenburg, W.C., Faaij, A.P.C. (2012) Techno-economic assessment of CO2 capture at steam methane reforming facilities using commercially available technology. Int. J. Greenh. Gas Control., 9: 160-171.
34. Lhyfe Inc. (2020) Lhyfe: Producteur et fournisseur d'hydrogène renouvelable. https://lhyfe.com/.
35. OffshoreWIND biz Inc. (2021) World's First Offshore Green Hydrogen Plant to Go Online in France. www.offshorewind.biz/2021/06/04/worlds-first-offshore-green-hydrogen-plant-to-go-online-in-france/.
36. AirLiquide Inc. (2021) Air Liquide makes a strategic investment to support large scale renewable hydrogen production in France. https://energies.airliquide.com/air-liquide-makes-strategic-investment-support-large-scale-renewable-hydrogen-production-france.
37. Fuel Cell and Hydrogen Joint Undertaking. (2021) Fuel Cell and Hydrogen Joint Undertaking projects. https://www.fch.europa.eu/page/energy.
38. International Energy Agency. (1999) Case studies of integrated hydrogen systems Task 11 - Integrated Systems. https://digital.library.unt.edu/ark:/67531/metadc715026/.
39. EU Commission. (2021) Water electrolysis: a promising remedy for the off grid solar energy torage problem. https://cordis.europa.eu/article/id/418019-water-electrolysis-a-promising-remedy-for-the-off-grid-solar-energy-storage-problem.
40. Valente, A., Iribarren, D., Dufour, J. (2017) Harmonised life-cycle global warming impact of renewable hydrogen. J. Clean. Prod., 149: 762-772.
41. Valente, A., Iribarren, D., Dufour, J. (2018) Harmonising the cumulative energy demand of renewable hydrogen for robust comparative life-cycle studies. J. Clean. Prod., 175: 384-393.
42. Koroneos, C., Dompros, A., Roumbas, G., Moussiopoulos, N. (2004) Life cycle assessment of hydrogen fuel production process. IJHE, 29: 1443-1450.
43. Valente, A., Iribarren, D., Candelaresi, D., Spazzafumo, G., Dufour, J. (2020) Using harmonised life-cycle indicators to explore the role of hydrogen in the environmental performance of fuel cell electric vehicles. IJHE, 45: 25758-25765.
44. Valente, A., Iribarren, D., Dufour, J. (2021) Harmonised life-cycle indicators of nuclear-based hydrogen. IJHE, 46: 29724-29731.
45. Valente, A., Iribarren, D., Dufour, J. (2019) How do methodological choices affect the carbon footprint of microalgal biodiesel- A harmonised life cycle assessment. J. Clean. Prod., 207: 560-568.
46. TotalEnergies Inc. (2021) Total and Engie partner to develop France's largest site for the production of green hydrogen from 100% renewable electricity. https://totalenergies.com/media/news/press-releases/total-and-engie-to-develop-france-s-largest-site-of-green-hydrogen.
47. Solar Media Inc. (2021) Engie, Neoen building subsidy-free 1GW solar project with storage, electrolyser in France. https://www.energy-storage.news/engie-neoen-building-subsidy-free-1gw-solar-project-with-storage-electrolyser-in-france/.
48. Cetinkaya, E., Dincer, I., Naterer, G.F. (2012) Life cycle assessment of various hydrogen production methods. IJHE, 37: 2071-2080.
49. Bhandari, R., Trudewind, C.A., Zapp, P. (2014) Life cycle assessment of hydrogen production via electrolysis–a review. J. Clean. Prod., 85: 151-163.
50. Dincer, I., Acar, C. (2015) Review and evaluation of hydrogen production methods for better sustainability. IJHE, 40: 11094-11111.
51. Parra, D., Zhang, X., Bauer, C., Patel, M.K. (2017) An integrated techno-economic and life cycle environmental assessment of power-to-gas systems. ApEn, 193: 440-454.
52. Al-Qahtani, A., Parkinson, B., Hellgardt, K., Shah, N., Guillen-Gosalbez, G. (2021) Uncovering the true cost of hydrogen production routes using life cycle monetisation. ApEn, 281: 115958-115970.
53. Zhang, X., Bauer, C., Mutel, C.L., Volkart, K. (2017) Life Cycle Assessment of Power-to-Gas: Approaches, system variations and their environmental implications. ApEn, 190: 326-338.
54. Salkuyeh, Y.K., Saville, B.A., Maclean, H.L. (2018) Techno-economic analysis and life cycle assessment of hydrogen production from different biomass gasification processes. IJHE, 43: 9514-9528.
55. Khojasteh Salkuyeh, Y., Saville, B.A., Maclean, H.L. (2017) Techno-economic analysis and life

cycle assessment of hydrogen production from natural gas using current and emerging technologies. IJHE, 42: 18894-18909.
56. Eggemann, L., Escobar, N., Peters, R., Burauel, P., Stolten, D. (2020) Life cycle assessment of a small-scale methanol production system: A Power-to-Fuel strategy for biogas plants. J. Clean. Prod., 271: 122476-122488.
57. Lee, B., Heo, J., Choi, N.H., Moon, C., Moon, S., Lim, H. (2017) Economic evaluation with uncertainty analysis using a Monte-Carlo simulation method for hydrogen production from high pressure PEM water electrolysis in Korea. IJHE, 42: 24612-24619.
58. Lee, B., Lee, H., Heo, J., Moon, C., Moon, S., Lim, H. (2019) Stochastic techno-economic analysis of H2 production from power-to-gas using a high-pressure PEM water electrolyzer for a small-scale H2 fueling station. Sustainable Energy Fuels, 3: 2521-2529.
59. Heng, L., Xiao, R., Zhang, H. (2018) Life cycle assessment of hydrogen production via iron-based chemical-looping process using non-aqueous phase bio-oil as fuel. Int. J. Greenh. Gas Control., 76: 78-84.
60. Khzouz, M., Gkanas, E.I., Shao, J., Sher, F., Beherskyi, D., El-Kharouf, A., Al Qubeissi, M. (2020) Life Cycle Costing Analysis: Tools and Applications for Determining Hydrogen Production Cost for Fuel Cell Vehicle Technology. Energies, 13: 3783-3802.
61. Tőke, P.M., Hortay, O. (2021) Simulation-based sensitivity analysis of an on-site hydrogen production unit in Hungary. IJHE, 46: 4881-4889.
62. Zhang, C., Xu, Y. (2020) Economic analysis of large-scale farm biogas power generation system considering environmental benefits based on LCA: A case study in China. J. Clean. Prod., 258: 120985-120995.
63. Zhang, Y., Wang, L., Wang, N., Duan, L., Zong, Y., You, S., Maréchal, F., Van Herle, J., Yang, Y. (2019) Balancing wind-power fluctuation via onsite storage under uncertainty: Power-to-hydrogen-to-power versus lithium battery. Renew. Sust. Energ. Rev., 116: 109465-109479.
64. Lee, B., Heo, J., Kim, S., Kim, C.H., Ryi, S.K., Lim, H. (2019) Integrated techno-economic analysis under uncertainty of glycerol steam reforming for H2 production at distributed H2 refueling stations. Energy Convers. Manage., 180: 250-257.
65. Zhao, G., Kraglund, M.R., Frandsen, H.L., Wulff, A.C., Jensen, S.H., Chen, M., Graves, C.R. (2020) Life cycle assessment of H2O electrolysis technologies. IJHE, 45: 23765-23781.
66. Reiter, G., Lindorfer, J. (2015) Global warming potential of hydrogen and methane production from renewable electricity via power-to-gas technology. Int J LCA, 20: 477-489.
67. Chen, J., Xu, W., Zuo, H., Wu, X., E, J., Wang, T., Zhang, F., Lu, N. (2019) System development and environmental performance analysis of a solar-driven supercritical water gasification pilot plant for hydrogen production using life cycle assessment approach. Energy Convers. Manage., 184-198: 60-73.
68. Rajabi Hamedani, S., Villarini, M., Colantoni, A., Moretti, M., Bocci, E. (2018) Life Cycle Performance of Hydrogen Production via Agro-Industrial Residue Gasification—A Small Scale Power Plant Study. Energies, 11: 675-694.
69. Sarkar, O., Katakojwala, R., Mohan, S.V. (2021) Low carbon hydrogen production from a waste-based biorefinery system and environmental sustainability assessment. Green Chem., 23: 561-574.
70. Tugnoli, A., Landucci, G., Cozzani, V. (2008) Sustainability assessment of hydrogen production by steam reforming. IJHE, 33: 4345-4357.
71. Xu, D., Li, W., Ren, X., Shen, W., Dong, L. (2020) Technology selection for sustainable hydrogen production: A multi-criteria assessment framework under uncertainties based on the combined weights and interval best-worst projection method. IJHE, 45: 34396-34411.
72. Boyano, A., Blanco-Marigorta, A.M., Morosuk, T., Tsatsaronis, G. (2011) Exergoenvironmental analysis of a steam methane reforming process for hydrogen production. Energy, 36: 2202-2214.
73. Yadav, D., Banerjee, R. (2020) Net energy and carbon footprint analysis of solar hydrogen production from the high-temperature electrolysis process. ApEn, 262: 114503-114518.
74. International Organization for Standardization. (2006) Environmental management—Life cycle assessment—Principles and framework. https://www.iso.org/standard/37456.html.
75. International Organization for Standardization. (2006) Environmental management — Life cycle assessment — Requirements and guidelines. https://www.iso.org/standard/38498.html.
76. Pérez-López, P., Gschwind, B., Blanc, P., Frischknecht, R., Stolz, P., Durand, Y., Heath, G.A., Ménard, L., Blanc, I. (2017) ENVi PV: an interactive Web Client for multi criteria life cycle assessment of photovoltaic systems worldwide. Prog Photovolt, 25: 484-498.
77. Deloitte Touche Tohmatsu Inc. (2021) Fueling the future of mobility: hydrogen electrolyzers. https://www2.deloitte.fr/formulaire/pdf/fueling-the-future-of-mobility-hydrogen-electrolyzers.pdf.
78. International Energy Agency. (2020) Life Cycle Inventories and Life Cycle Assessments of Photovoltaic Systems. https://iea-pvps.org/key-topics/life-cycle-inventories-and-life-cycle-assessments-of-photovoltaic-systems/.
79. Agence de la Transition Ecologique. (2017) Solar photovoltaic: STRATEGIC ROADMAP. https://fixpower.pk/wp-content/uploads/2019/05/fdr_solar_32p_engl_web.pdf.
80. Global Market Insights Inc. (2020) Solar PV Module Market Size By Technology: Industry Analysis Report, Regional Outlook, Price Trends, Competitive Market Share & Forecast, 2020 – 2026.

https://www.gminsights.com/industry-analysis/solar-pv-module-market.
81. Besseau, R. (2019) Analyse de cycle de vie de scénarios énergétiques intégrant la contrainte d'adéquation temporelle production consommation. https://pastel.archives-ouvertes.fr/tel-02732972.
82. International Energy Agency. (2019) Methodology Guidelines on Life Cycle Assessment of Photovoltaic 2020. https://iea-pvps.org/key-topics/methodology-guidelines-on-life-cycle-assessment-of-photovoltaic-2020/.
83. Doubleday, K., Choi, B., Maksimovic, D., Deline, C., Olalla, C. (2016) Recovery of inter-row shading losses using differential power-processing submodule DC–DC converters. SoEn, 135: 512-517.
84. Paul, B. (2016) Photovoltaics in positive energy buildings. http://www.diva-portal.org/smash/record.jsf?pid=diva2%3A902245&dswid=6411.
85. Celik, I., Lunardi, M., Frederickson, A., Corkish, R. (2020) Sustainable End of Life Management of Crystalline Silicon and Thin Film Solar Photovoltaic Waste: The Impact of Transportation. J Appl Sci (Faisalabad), 10: 5465-5478.
86. Le Centre Observation, Impacts, Energie. (2021) CAMS radiation service dataset "AGATE" over Europe computed with McClear version 3 and CAMS radiation bias correction. http://www.soda-pro.com/help/cams-services/cams-radiation-service/download-europe-volume.
87. Energypedia consult. (2017) Standard test conditions. https://wiki.openmod-initiative.org/wiki/Standard_test_conditions#:~:text=STC%20is%20an%20industry%2Dwide,5)%20spectrum.
88. SMA Solar Technology Inc. (2021) Performance ratio - SMA Solar Technology AG. files.sma.de/downloads/Perfratio-TI-en-11.pdf.
89. Perez-Lopez, P., Gschwind, B., Frischknecht, R., Stolz, P., Mehl, C., Payeur, M., Heath, G., Blanc, I. (2019) Combining region-specific supply chains with geo-located PV electricity production for Life Cycle Assessment of worldwide crystalline silicon photovoltaic systems in ENVI-PV 2.0. In: 36th European Photovoltaic Solar Energy Conference and Exhibition. Marseille. pp. 1-7.
90. Jordan, D.C., Kurtz, S.R., Vansant, K., Newmiller, J. (2016) Compendium of photovoltaic degradation rates. Prog. Photovoltaics Res. Appl., 24: 978-989.
91. France Energie Eolienne Inc. (2019) WIND OBSERVATORY: Analysis of the French wind power industry: market, jobs and challenges. fee.asso.fr/wp-content/uploads/2019/10/2019-wind-observatory-final.pdf.
92. France Energie Eolienne Inc. (2020) Wind Observatory 2020 Analysis of the French wind power industry: market, jobs and challenges. https://fee.asso.fr/wp-content/uploads/2020/10/Observatoire-2020-VF-finale-ENG.pdf.
93. WindEurope Inc. (2021) Offshore Wind in Europe: Key trends and statistics 2020. https://windeurope.org/intelligence-platform/product/offshore-wind-in-europe-key-trends-and-statistics-2020/.
94. Renewables Now Inc. (2020) Renewables share in power consumption hits 23% in France in 2019. https://renewablesnow.com/news/renewables-share-in-power-consumption-hits-23-in-france-in-2019-686911/.
95. The Wind Power Inc. (2021) France wind farms database. https://www.thewindpower.net/store_country_en.php?id_zone=1.
96. STATISTA Inc. (2020) Share of the average plant load factor (PLF) of wind electricity in France in 2019, by region. https://www.statista.com/statistics/761018/wind-energy-average-load-factor-france-region/.
97. WindEurope Inc. (2020) Wind energy in Europe in 2019. https://windeurope.org/wp-content/uploads/files/about-wind/statistics/WindEurope-Annual-Statistics-2019.pdf.
98. International Energy Agency. (2019) Offshore Wind Outlook 2019. https://iea.blob.core.windows.net/assets/495ab264-4ddf-4b68-b9c0-514295ff40a7/Offshore_Wind_Outlook_2019.pdf.
99. Bareiß, K., De La Rua, C., Möckl, M., Hamacher, T. (2019) Life cycle assessment of hydrogen from proton exchange membrane water electrolysis in future energy systems. ApEn, 237: 862-872.
100. H2 Energy S.r.l. (2021) Alkaline Water Electrolysis PEM - H2 Energy. https://www.h2e.it/H2E-Presentation.pdf.
101. National Renewable Energy Laboratory. (2010) Current (2009) State-of-the-Art Hydrogen Production Cost Estimate Using Water Electrolysis: Independent Review. https://www.hydrogen.energy.gov/pdfs/46676.pdf.
102. Guo, Y., Li, G., Zhou, J., Liu, Y. (2019) Comparison between hydrogen production by alkaline water electrolysis and hydrogen production by PEM electrolysis. IOP Conf, 371-376: 042022.
103. Agence de la Transition Ecologique. (2020) Rendement de la chaîne hydrogène. https://librairie.ademe.fr/mobilite-et-transport/1685-rendement-de-la-chaine-hydrogene.html.
104. Makridis, S. (2017) Hydrogen storage and compression. ET Digital Library, Kozani.
105. Paul Scherrer Institute. (2021) Coupling Integrated Assessment Models Output with Life Cycle Assessment. github.com/romainsacchi/premise.
106. PDC Machines Inc. (2013) Introduction to Diaphragm Compressors. https://www.pdcmachines.com/introduction-diaphragm-compressors/.
107. Paul Scherrer Institute. (2021) Brightway2 LCA framework. https://brightway.dev/.
108. Le Centre Observation, Impacts, Energie. (2021) Layer over brightway2 for algebraic definition of parametric models and super fast computation of LCA. github.com/oie-mines-paristech/lca_algebraic.

109. Publications Office of the European Union. (2018) Supporting information to the characterisation factors of recommended EF Life Cycle Impact Assessment methods: New methods and differences with ILCD. https://publications.jrc.ec.europa.eu/repository/handle/JRC109369.
110. Sobol, I.M. (2001) Global sensitivity indices for nonlinear mathematical models and their Monte Carlo estimates. Math Comput Simul, 55: 271-280.
111. Saltelli, A., Ratto, M., Andres, T., Campolongo, F., Cariboni, J., Gatelli, D., Saisana, M., Tarantola, S. (2008) Global sensitivity analysis: the primer. John Wiley & Sons, England.
112. Wilfried, V.S., Nils, R., Björn, M., Alfons, A., Klaus, K., Christian, R. (2012) Review of PV performance ratio development. In: World Renewable Energy Forum. Colorado. pp. 1-6.
113. EU Commission. (2021) Glossary: Nomenclature of territorial units for statistics (NUTS). https://ec.europa.eu/eurostat/statistics-explained/index.php?title=Glossary:Nomenclature_of_territorial_units_for_statistics_(NUTS).
114. Ecoinvent Inc. (2019) Ecoinvent v3.6 cut-off: Fuel cell production, stack solid oxide, 125kW, electrical, future. https://v36.ecoquery.ecoinvent.org/Account/SessionExpired.
115. Wernet, G., Bauer, C., Steubing, B., Reinhard, J., Moreno-Ruiz, E., Weidema, B. (2016) The ecoinvent database version 3 (part I): overview and methodology. Int J LCA, 21: 1218–1230.
116. Schmidt, O., Gambhir, A., Staffell, I., Hawkes, A., Nelson, J., Few, S. (2017) Future cost and performance of water electrolysis: An expert elicitation study. IJHE, 42: 30470-30492.
117. Ma, T., Chen, Y., Pavlenko, A.N., Wang, Q. (2021) Heat and mass transfer advances for energy conservation and pollution control in a renewable and sustainable energy transition. Renew. Sust. Energ. Rev., 145-147: 111087.
118. Ren, J. (2021) China's Energy Security: analysis, assessment and improvement. World Scientific Publishing, London.
119. Liu, J., Yin, M., Xia-Hou, Q., Wang, K., Zou, J. (2021) Comparison of sectoral low-carbon transition pathways in China under the nationally determined contribution and 2°C targets. Renew. Sust. Energ. Rev., 149: 111336-111349.
120. Xu, L., Zhang, S., Yang, M., Li, W., Xu, J. (2017) Environmental effects of China's solar photovoltaic industry during 2011–2016: A life cycle assessment approach. J. Clean. Prod., 170: 310-329.
121. National Development and Reform Commission of China. (2016) Cleaner Production Evaluation Index System in Photovoltaic Cell Industry. https://www.ndrc.gov.cn/xxgk/zcfb/gg/201610/t20161014_961167.html?code=&state=123.
122. Mishnaevsky, L., Branner, K., Petersen, H.N., Beauson, J., Mcgugan, M., Sørensen, B.F. (2017) Materials for Wind Turbine Blades: An Overview. Materials (Basel), 10: 1285-1309.
123. Kind, S., Neubauer, S., Becker, J., Yamamoto, M., Völkert, M., Abendroth, G.V., Zelder, O., Wittmann, C. (2014) From zero to hero – Production of bio-based nylon from renewable resources using engineered Corynebacterium glutamicum. Metab. Eng., 25: 113-123.
124. Radzik, P., Leszczyńska, A., Pielichowski, K. (2020) Modern biopolyamide-based materials: synthesis and modification. Polym. Bull., 77: 501-528.
125. Mahssin, Z.Y., Abdul Hassan, N., Yaacob, H., Puteh, M.H., Ismail, C.R., Putra Jaya, R., Mohammad Zainol, M., Mahmud, M.Z.H. (2021) Converting Biomass into Bio-Asphalt – A Review. IOP Conf, 682: 012066.
126. Ragab, A.A. (2018) Asphalt Modified with Biomaterials as Eco-Friendly and Sustainable Modifiers. IntechOpen, London.
127. Torres, F.G., De-La-Torre, G.E. (2021) Algal-based polysaccharides as polymer electrolytes in modern electrochemical energy conversion and storage systems: A review. Carbohydr. Polym. Technol. Appl., 2: 100023-100040.
128. Scheepers, F., Stähler, M., Stähler, A., Rauls, E., Müller, M., Carmo, M., Lehnert, W. (2021) Temperature optimization for improving polymer electrolyte membrane-water electrolysis system efficiency. ApEn, 283: 116270-116281.
129. Mahmood, N., Yao, Y., Zhang, J., Pan, L., Zhang, X., Zou, J. (2018) Electrocatalysts for Hydrogen Evolution in Alkaline Electrolytes: Mechanisms, Challenges, and Prospective Solutions. Adv. Sci., 5: 1700464-1700487.

Practical Research on the Fact that Clear Waters and Lush Mountains are Invaluable Assets in the Ecological Civilization Experimental area of Guizhou Province

Baishan Li[1,2,*], *Yaoshun* Song[3], *Rui* Yang[4], *Song* Hong[1], *Jingquan* Peng[5] and *Ying* Su[6]

[1]School of Resources and Environmental Sciences, Wuhan University, Hubei Provincial Key Laboratory of Biomass Resources Chemistry and Environmental Biotechnology, Wuhan, Hubei 430079, 573104852@qq.com, China
[2]The Party School of the CPC Guiyang Municipal Committee, Guiyang, Guizhou 550005, lbs1113@126.com, China
[3]Guizhou Environmental Monitoring Central Station, Guiyang, Guizhou, 550002, 573104852@qq.com, China
[4]4.School of Geography and Environmental Sciences, Guizhou Normal University, Guiyang, Guizhou, 55000, 3539678282@qq.com, China
[5]Guizhou Environmental Science Research and Design Institute, Guiyang, Guizhou, 550002, 419773451@qq.com, China
[6]General Medical 300 Hospital (Guihang Guiyang Hospital), Guiyang, Guizhou, 550009, 382008439@qq.com, China

Abstract. Guizhou is one of the first national ecological civilization pilot zones selected, which needs to play a leading role in the road of ecological civilization construction.The good ecological environment status quo and the continuous high growth rate of economic development provide a research basis for the project. In-depth analysis of the relationship between ecological environment and economic development, to provide scientific suggestions for the solid progress.The main achievements of the research are as follows: (1) Construct the practice evaluation index system of "Two mountains" in ecological civilization pilot area in Guizhou Province, and the overall evaluation is "high"; (2) the evaluation result of "Two mountains" in Bijie pilot area shows that the evaluation result of "Two mountains" practice index in Bijie pilot area increased from 0.056 in 2015 to 0.107 in 2019, the comprehensive index of ecological civilization in Bijie pilot area increased steadily, and the total value of green water and green mountains in Bijie pilot area was 4201235.478 billion yuan in 2019(3) The fitting curve of the relationship between environmental and economic benefits in Chishui River Basin in Guizhou Province shows that the ecological environment protection work in Guizhou Province effectively promotes the rapid improvement of environmental benefits and economic benefits; (4) the accounting of ecosystem service value in Chishui City shows that the total value of Chishui city in 2019 was 4,2170.788 billion yuan; (5) Summarize the "two mountains theory" in Guizhou innovation practice experience and model, put forward "speed up the construction of green ecological system, improve environmental carrying capacity" "speed up the construction of characteristic industry system, improve urban economic strength" "speed up the construction of modern urban system, improve the urban agglomeration ability" "vigorously develop circular economy, efforts to promote production mode change" "actively guide form green low carbon consumption pattern" and so on five countermeasures.

1 introduction

Due to the overexploitation of the natural world by human beings in the period of industrial civilization, which produced worldwide problems such as environmental pollution, ecological destruction, population explosion, and energy crisis, people began to gradually realize that they should respect nature, conform to nature, and protect nature. Foreign scholars paid early attention to ecological issues. Between 1962 and 1997, American scientist Rachel Carson touched the bud of ecological civilization in Silent Spring (Ruilan et al., 2008), Limits of Growth (Li Tao et al., 2013), and The Reasons of Nature (Zang Peihong, 2003)[1-2]. Research on the governance countermeasures of freshwater lakes (Yang Ya, 2008)[3].Li Bo (2010) [4] put forward countermeasures such as formulating plans to increase investment, building a characteristic economic and industrial system, and accelerating urbanization construction. From 1983 to 2009, the adoption of Our Common Future, Agenda 21, World Environment Day in 2007 with the theme of "melting glaciers and worrying consequences" , and the Copenhagen Climate Change Conference in Denmark in 2009, indicating that the construction of ecological civilization has become an inevitable result of global human development.Foreign scholars analyze ecological problems from a relatively unique perspective, which provides useful viewpoints and methods for our research.

Domestic scholars on the green mountains and silver mountains began in the 20th century. In terms of practice, Guizhou Province construction of ecological civilization pilot demonstration area, Guiyang City issued the "construction of ecological civilization city index system and monitoring methods", "Guiyang City to promote

* Corresponding author: 573104852@qq.com;

ecological civilization Construction Regulations" and so on[5-8]. The index system of Guiyang city is more operable. Ability to conduct both horizontal and longitudinal comparisons. It shows that Guizhou Province's ecological civilization construction is now at the forefront of the country. In the new era, it is of great scientific significance and application prospect to deeply study the practice that clear waters and green mountains in Guizhou Ecological civilization experimental zone are gold and silver mountains.

2 Practice evaluation of "Two Mountains" in Guizhou Provincial Ecological Civilization Experimental Zone

Reference domestic related national ecological civilization experimental area ecological civilization construction practice, according to the ecological environment about "two mountains" index, national ecological civilization city and county evaluation requirements, around ecological civilization in Guizhou green water castle peak is the jinshan yinshan practice, build ecological civilization experimental area in Guizhou "two mountains" practice index system.Combined with Fujian province, Jiangxi province national ecological civilization experimental area construction results and Practice results of two mountains in Guizhou Province, build Guizhou ecological civilization test area in Guizhou "two mountains" practice index system,Take Bijie pilot area and Chishui City as an example for evaluation and analysis, and strive to open up the channel for the transformation of clear water and green mountains into gold and silver mountains. Which provided technical support for the construction of ecological civilization pilot zone in Guizhou Province.

2.1 Indicator system construction

According to the meaning of ecological civilization, the Ministry of Ecology and Environment's construction index system of demonstration cities and counties and the index evaluation system of "two mountains". According to the principles of integrity, operability, relative independence, guidance, forward-looking, practicality and importance of the index system construction. In combination with the requirements of ecological environment, development and reform, natural resources, forestry and other departments. According to the main goals of economic and social development of the province and all cities and states during the 14th Five-year Plan period. Depending on the accessibility of each index value, Select the allocation and efficient utilization of natural resources, green development mechanism, clean production process, terminal control, ecosystem protection and waste comprehensive utilization, green development performance assessment and other 8 tasks as the standard layer. The evaluation index system of per capita GDP, urban sewage treatment rate and public satisfaction were selected as the index layer. Practice evaluation method of "Two Mountains" in the Ecological Civilization Pilot Zone of Guizhou Province: First, calculated according to the determined index to obtain the target value of each index, and then the corresponding target degree is weighted, and then calculate the criterion degree according to the weight of the three criterion layers, and calculate the comprehensive evaluation index of "Two Mountains" practice in Guizhou Ecological Civilization Pilot Zone according to the corresponding criterion degree.

Classify the factors involved in the comprehensive evaluation of "two mountains" in Guizhou Province,Constructs a hierarchical hierarchy of interconnected elements. Construct a judgment matrix to judge the relative importance of each indicator. Constructed the comprehensive evaluation index system of "two mountains" in Guizhou Province. The hierarchical analysis method is used to assign the weight of each index of "two mountains" evaluation in Guizhou Province.Obtain the original indicator data right, because each indicator data has different dimensions and magnitudes, direct use can not truly reflect the "two mountains" practice of the situation.Therefore, the data needs to be normalized before its practical application to eliminate the influence of the dimension and magnitude between the data.

The index value has a positive correlation with the comprehensive evaluation index of "two mountains" practice, and shows a single increase distribution, that is, the higher the index value, the higher the practice level of "two mountains". For example, the proportion of days with good ambient air quality and the standard quality rate of centralized drinking water sources, etc.

The index value has a negative correlation with the comprehensive evaluation index of "two mountains" practice, and shows a single reduction distribution, that is, the higher the index value, the lower the practice level of "two mountains",Such as water consumption per unit of GDP, sulfur dioxide emissions, the urban registered unemployment rate at the end of the year, etc.

The data preprocessing formula is as follows:

forward pointer:

$$x_{ij}^{*} = 1 - \frac{S_i - x_{ij}}{S_i - \min(x_{ij})} \quad (1)$$

$$x_{ij}^{*} = 1 - \frac{x_{ij} - S_i}{\max(x_{ij}) - S_i} \quad (2)$$

Negative indicators:

x_{ij}^{*} :The value after the preprocessing of the j th index in the i th index layer;

In the formula:

x_{ij} :The original value of the j th index of the i th index layer;

S_i: The target value of the i th index layer;

$\max(x_{ij})$: The maximum value of the j th index in all i th index layers is multiplied by 1.05;

$\min(x_{ij})$: The minimum of the j th indicator in all i indicator layers is divided by 1.05.

In the actual calculation process, the practice evaluation index and grade standard value of "Two Mountains" are jointly standardized by formula (1) and (2). The larger the index value of the conversion result, the higher the practice level of "Two Mountains".

2.2 Calculation model of the practice comprehensive evaluation index of "Two Mountains" in the Ecological Civilization Experimental Zone of Guizhou Province

The comprehensive evaluation index of "Two Mountains" practice in the Ecological Civilization Pilot Zone of Guizhou Province comprehensively reflects the ability of "two mountains" practice to promote society, economy, population and ecology and its coordination degree with environmental consequences.Therefore, the calculation model of the practice comprehensive evaluation index of "Two Mountains" in Guizhou Provincial Ecological Civilization Pilot Zone is as follows:

$$E = \sum_{i=1}^{n} \omega_i E_i \tag{3}$$

The criterion layer is calculated using the model of:

$$A_i = \sum_{j=1}^{m} \omega_{ij} B_{ij} \tag{4}$$

Formula E is the size of the comprehensive evaluation index of "Two mountains" in Guizhou Ecological civilization pilot area; Ai is i criterion index; Bij is the value of i criterion j index, the weight of i criterion j index; n is the number of criterion indicators; m is the number of index layer indicators.

2.3 Determination of index weight and evaluation methods

According to some internationally recognized indicators and the Basic Standards for the National People's Well-off Life and the National Ecological City Construction Standards already issued by China. Referring to the social and economic development of Guizhou Province in recent years, it is determined that the highest score of the standardized index in the accounting table is 1 point, and the lowest theoretical score is 0 point. Each index between 0~1 is divided into the worst value, pass value and the optimal value. The optimal value can be considered as the best value for maintaining the good resilience of the ecosystem in Guizhou Province. The passing value can be considered as the internationally or domestically recognized risk-free safety value of an indicator, and the worst value is considered as the critical value that cannot maintain ecological elasticity. This is easy to analyze the dynamic change characteristics of urban ecological environment, and can meet the needs of comparative research to the greatest extent.

The weight of each index in the "two mountains" practice evaluation index system in the Ecological Civilization pilot Zone of Guizhou Province is determined by hierarchical analysis method. Hierarchical analysis method is a combination of qualitative and quantitative. Hierarchical analysis method is a kind of qualitative and quantitative combination, systematic and hierarchical analysis method. The relationship between each indicator can be well described. First of all, the index system is hierarchical. According to the mutual relationship and subordination between the factors, the factors are combined according to different levels to form a multi-level analysis structure model. It finally boils down to the relative importance of the bottom level and the highest level. Ultimately, it comes down to the question of the weights of the lowest level relative to the relative importance of the top layer or the relative order of superiority and inferiority. Construct the judgment matrix of the criterion layer to the target layer, construct the judgment matrix of the indicator, standardize the matrix and solve the feature vectors, and determine the consistency of the matrix, so as to determine the weight value of each criterion layer. Guizhou Province is one of the regions in the world where the karst region is seriously affected by rocky desertification. The development and utilization of resources in the province is complex, and some indicators are far behind compared with international indicators. In order to reflect the actual situation within the region, on the basis of relevant domestic data and existing research results, the target value, status value and weight calculation of the "Two Mountains" practice evaluation index system.

Classification of the evaluation index:

At present, there is no clear unified standard for the practice evaluation of "two mountains".The traditional evaluation criteria generally include questionnaire survey method, standard method and reference frame method.According to the rating criteria of the above indicators, this study divides the influence of the above factors on the practice of "Two Mountains" in the Guizhou Ecological Civilization Pilot Zone into 5 grades.

2.4 Practice evaluation and analysis of "Two Mountains" in Guizhou Provincial Ecological Civilization Experimental Zone

According to Equations 1, 2, 3, and 4,Evaluation and analyze the practice of "two mountains" in ecological civilization pilot zone of Guizhou Province,As shown in Figures 1, 2, and 3.

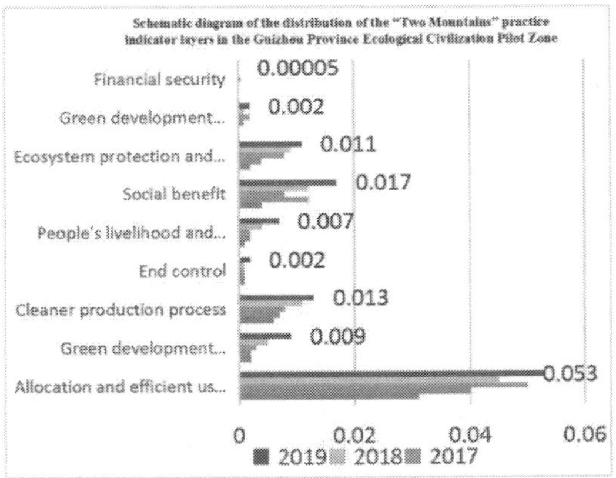

Figure 1 Schematic diagram of "Two Mountains" in the Ecological Civilization Experimental Zone of Guizhou Province

From Figure 1, Figure 2 and Figure 3, the comprehensive index of building green waters and green mountains, environmental protection, social progress norms and ecological civilization construction has shown a wave upward trend. It has grown rapidly since 2005. It shows that the economic, social and environmental conditions of Guizhou province have been greatly improved, which is inseparable from the ecological civilization construction in Guizhou Province. In terms of economy, the per capita GDP of Guizhou province in the past 10 years has increased from 4,957 yuan to 26,393, more than five times, with gratifying results,However, under the national economy facing the new normal, the economic development index of Guizhou Province in 2014 is still in the "medium" level; In terms of environmental protection, in the decade since 2005, Guizhou province has made chemical oxygen demand, ammonia nitrogen, soot, industrial dust and industrial solid waste emissions controlled within the national plan targets. Maintaining the overall water quality of the eight major river systems in the two major river basins of Quansheng is maintained well, controlling soil erosion, greening the city, etc., which has brought the province's environmental protection level to a new level. In many hardware and software facilities are not lagging behind other provinces, so its environmental protection level is "higher"; In terms of social and people's livelihood, Guizhou province has continuously increased its investment in science and technology, culture, education and health, the per capita disposable income of urban residents and the net income of farmers have been steadily increased, and the improvement degree of social and people's livelihood has reached a "higher" level. Since 2015, with the gradual implementation of the specific measures, the economic development index and environmental protection index have shown a steady upward trend, and the social livelihood index has shown a leaping upward trend since 2007. It shows that after 2007, Guizhou province has gradually increased the contribution of social society and people's livelihood by vigorously carrying out the construction of ecological civilization city. The comprehensive evaluation index of "Two Mountains" practice in Guizhou Provincial Ecological Civilization Pilot Zone increased from 0.047 in 2015 to 0.114 in 2019. The overall evaluation was "high". It shows that Guizhou Province's national ecological civilization construction shows a steady rising trend.

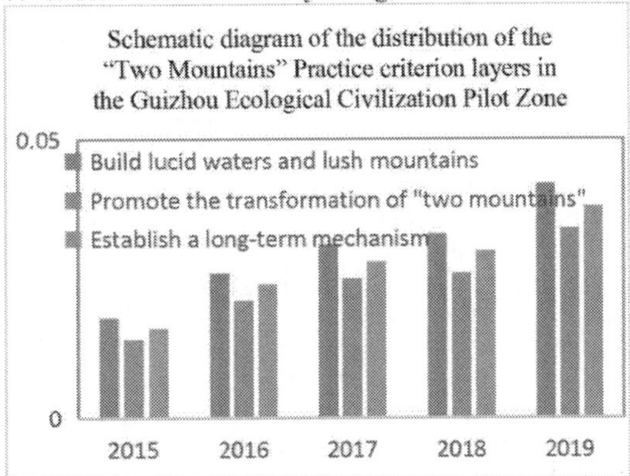

Figure 2 Schematic diagram of the layer distribution of the "Two Mountains" experimental criteria in the Ecological Civilization Experimental Zone of Guizhou Province

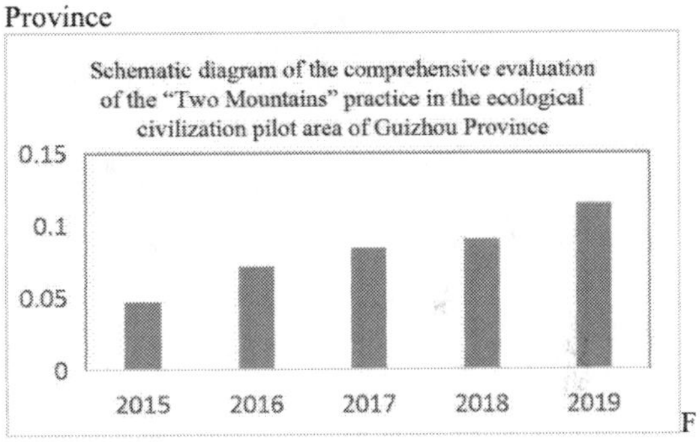

Figure 3 Schematic diagram of the comprehensive evaluation of the practice of "Two Mountains" in the Ecological Civilization Experimental Zone of Guizhou Province

3 Countermeasures and Suggestions for the Practice of "Two Mountains" in the Ecological Civilization Experimental Zone of Guizhou Province

(1)Accelerate the construction of a green ecological system and comprehensively improve the carrying capacity of the environment

Establish and improve the ecological space control mechanism. Strengthen ecological protection to ensure the safety of ecosystems. Strengthen environmental pollution control. Improving the systems and mechanisms for promoting ecological progress.

(2)Accelerate the construction of a characteristic industrial system, and comprehensively enhance the city's economic strength and development. Deepen a new type of industrialization.

Deepen a new type of industrialization. Accelerate the development of health care tourism. Accelerate the development of characteristic and efficient agriculture.

(3)Accelerate the construction of a modern urban system, and comprehensively enhance the capacity of urban agglomeration

Speed up the improvement of urban planning. Speed up the construction of municipal facilities. Innovate urban management services. Accelerate efforts to improve urban quality.

(4)Vigorously develop the circular economy and strive to promote the reform of the production mode

Under the guidance of Xi Jinping Thought on Socialism with Chinese Characteristics for a new era, we will firmly establish the five new development concepts, promoting rural economic construction, political construction, cultural construction, social construction, ecological civilization construction and the party's construction, comprehensively promote the new era "four beautiful rural countryside" construction. Which accelerating the modernization of rural governance system and management ability, and the modernization of agriculture and rural areas, accelerate the realization of "strong agriculture, rural beauty, farmers rich". We will scientifically formulate a strategic plan for rural revitalization, and optimize and integrate the spatial layout of villages in light of local conditions. Promote the development of the whole agricultural industrial chain, vigorously develop rural tourism, and vigorously implement the plan to improve the living environment, water, electricity, telecommunications, housing, roads, rural education, rural health, and basic rural security. The quality of rural living environment has been greatly improved, and the ecology is more livable. The whole industrial chain of "ten million" leading industries and the agricultural industry development pattern of "one township one special, one village one product" have basically taken shape. Domestic sewage and garbage treatment facilities in rural areas have basically covered all rural areas, and remarkable results have been made in the development of demonstration sites for integrated rural areas with distinctive features. The spiritual outlook of rural residents has been comprehensively improved, the civilized village style has become a trend, and the neighborhood relations are more harmonious. We will comprehensively eliminate absolute poverty in rural areas, increase the per capita disposable income of rural residents above 15,500 yuan, and enrich rural residents. A number of "Chishui experiences" that can be replicated and extended, and establish and improve long-term normal measures are further promoted.

We will fully implement the basic state policy of resource conservation and environmental protection, vigorously develop a circular economy, and build a circular industrial chain in response to local conditions, and reduce energy and resource consumption and waste production in all links of the production process at the source and actively promote cleaner production and promote the use of bulk industrial solid waste and agricultural and forestry waste resources. We will vigorously develop circular agriculture and promote highly efficient water-saving irrigation technologies.Give full play to the core supporting role of your various economic development zones, further integrate resources, promote the integrated development of the service industry and other industries, promote clean services and reduce resource consumption, improve the comprehensive utilization rate of energy and energy, and encourage the recycling of water resources.

The study is financially supported by the Environmental protection special Foundation of Guizhou Provincial Environmental Protection Bureau in 2018(ZFCG-LBLH2018Z016-23) .

References

1. L.D. Yang, Preliminary Discussion on Forestry Development, —— and Discussion on Ecological Force and Ecological Relations in Forestry Production, Guizhou Social Science, no. 3,1990.
2. S.G. Shen, Ecological Civilization and its Theoretical and Practical Basis. Journal of Peking University (Philosophy and Social Sciences edition). **(3)** 31~37 (1994)
3. Y. Yang, Research on Problems and Countermeasures in China's Ecological Civilization Construction, Journal of Nanjing Forestry University (Humanities and Social Sciences edition), **3** (2008)
4. B. Li, Theoretical and Practical Exploration of Ecological Civilization Construction in Ethnic Areas of Guizhou Province, Journal of Guizhou University (Social Science edition), **2** (2010)
5. G.N. Sun, J.P. Sun, Y. Lv, et al. Evaluation of recreational value in Forest Park on the North Slope of Qinling Mountains. Journal of Shaanxi Normal University (Natural Science edition), **32(1)** 116-120 (2010_
6. Y.Z. Hou, S.R.Wu, Review of the value evaluation and compensation of forest ecological services World Forestry Research, **18(3)** 1-5 (2005)
7. E.K. Guo, Z.H. Wang, Research on the influence of environmental management on environmental regulation and sports consumption upgrading under the perspective of ecological civilization construction [J]. Journal of Natural Science, Hunan Normal University, **44 (6)** 62-69 (2021)
8. B.H. Li, X. Li, S. Wang, et al. Research on the Transformation and Development of Traditional Village Living Environment from the View of Rural Revitalization [J]. Journal of Natural Science of Hunan Normal University, **45 (1)** 1-10 (2022)

A Study on the Performance of IrOx Electrodes Prepared by Double-titanium Cyclic Voltametric Electrodeposition Mode

Dandan Wu[1], *Xi* Wang[1,2], and *Xu* Wu[1,*]

[1]School of Environmental Science and Engineering, Huazhong University of Science and Technology, Wuhan, 430074, China
[2]Hubei HuaDeLai (HDL) Co. Ltd, Wuhan, 430079, China

Abstract. The electrodes were fabricated in double-titanium cyclic voltametric electrodeposition system, corresponding to the anode and cathode of the system. For further study the practicality of this mode, the performance of the resulting electrodes was investigated. Through a series of physical and electrochemical characterization, it is concluded that the IrOx electrode corresponding to the anode of the system has a flat and uniform deposit layer, and shows better electrolytic durability (its actual working life is more than 3072 h), while the cathode electrode has better performance in electrocatalytic activity.

1 Introduction

Iridium-coated titanium anodes have the advantages of low oxygen evolution potential, high current density carrying capacity (>100A dm^{-2}), excellent catalytic activity, and electrical durability [1-3]. Since their inventation, they have been widely used in water electrolysis, cathodic protection, electroplating, non-ferrous metal extraction, water treatment, etching solution recovery and other fields. In 1973, Bianchi obtained the patent right of IrO_2/Ta_2O_5 coated electrode[4]. Since then, the core position of thermal decomposition method for preparing iridium-tantalum-coated titanium electrode has been established. Recently, many creative fabrication methods of IrO_2 electrodes, such as chemical bath deposition [5], sputtering deposition [6, 7], pulsed laser deposition [8], chemical bath deposition [9] and electrodeposition [10-12], etc have been investigated.

With the continuous expansion of the market demand for DSA (Dimensionally Stable Anode) and the country's emphasis on the upgrading of the industrial system, the traditional thermal decomposition method can no longer meet the requirements of industrial production and conform to the industrial development of the times. Although the new method such as sol-gel, chemical bath, and sputtering deposition, etc. can provide Nano-scale coatings and thin iridium oxide film, but due to its complex equipment and strict process requirements, it is incompatible with the preparation of iridium-coated titanium electrodes, cannot be used for the preparation of large-scale iridium-based coated titanium electrodes in the actual industrial system, seems no operability. Due to its mild preparation conditions, high degree of mechanization, and strong operability, electrodeposition method is very suitable for the preparation of iridium-coated titanium electrodes in the actual industrial system. Relevant studies [13-15] have shown that the iridium oxide titanium electrode prepared by the electrodeposition method has better electrocatalytic activity than that prepared by the traditional thermal decomposition method, it shows relatively simple preparation process and controllable parameters, having research value and development potential.

In recent years, iridium oxide electrodeposition technology is mainly used to prepare pH transmission micro-sensing electrodes, polymer membrane electrodes and DSA anodes. In most cases, these electrodes are mainly prepared by potentiostatic, galvanostatic and cyclic voltammetry electrodeposition techniques. The author's research group mainly has done research on the electrodeposition preparation technology of DSA anode. When using cyclic voltametric electrodeposition technology to prepare the electrode, it was found that when iridium oxide was electrodeposited on the anode, a black layer of iridium compound material was also generated on the cathode. Based on this, a double-titanium cyclic voltametric electrodeposition mode was proposed, and the physical characterization and electrochemical properties of the obtained anode, and cathode electrodes were studied. It is expected to reduce unnecessary waste of iridium elements while ensuring the performance of the electrodes, thereby improving the electrodeposition efficiency of iridium.

2 Materials and Methods

To prepare the electrodeposition solution, 0.1mM iridium chloride ($IrCl_3$) was dissolved in 250mL deionized water (DI water) and mixed for 10min under magnetic stirring. Then, 0.5mM oxalic acid ($(COOH)_2 \cdot 2H_2O$) were added, and the resulting solution was mixed for 10min (the molo ratio of $IrCl_3$ and $H_2C_2O_4$ is 1:5). After the solution pH was adjusted to 10.0~10.5 by the addition of anhydrous potassium

* Corresponding author: Profxuwu@hust.edu.cn

carbonate (K_2CO_3), a light green solution was obtained. Finally, the electrodeposition solution was kept at 40℃ in thermostatic water bath for 4 days to allow stabilization, achieving a dark blue stable solution.

The anodic electrodeposition process was completed in a three-electrode system in which a saturated calomel electrode (SCE) was used as the reference electrode. Industrially pure titanium (grade TA2) with oxalic acid etching treatment (effective deposition area: 1.0 cm^{-2}) was used as the counter and working electrodes, respectively. Cyclic voltametric electrodeposition technology was adopted to prepare the IrOx film. Fig 1 shows a typical production process executed. After the electrodeposited anode and cathode were prepared, the IrOx coatings were completely converted into metal oxides through a thermal decomposition process: heating at 450℃ for 1h in a muffle furnace.

The morphologies of deposits on anode and cathode were observed by SCE (scanning electron microscopy) with the magnification of 10k, and was investigated by EDS (Energy Dispersive Spectthe component element type and contentrometer). Furthermore, X-ray diffraction (XRD) was utilized to analyze the microstructure of IrOx films. Electrochemical characterization was performed in a three-electrode system by an electrochemical workstation (CS310H, CorrTest Ltd.). Cyclic voltammetry (CV) and linear sweep voltammetry (LSV) were performed in a 0.5 M H_2SO_4 solution with a potential scan rate of 20 mV s^{-1} and 1 mVs^{-1}, respectively. Electrochemical impedance spectroscopy (EIS) was determined at 1.35V (vs. SCE) with an amplitude of 5mV, and the frequency of that from 0.01Hz to 10kHz. An accelerated life test (ALT) on these electrodeposited IrOx/Ti electrodes was carried out in an electrochemical cell with Pt cathode, referring to previous studies, the potential-time curve was monitored at an applied current density of 0.25 A cm^{-2}. These tests were run until the potential increased up to 10V.

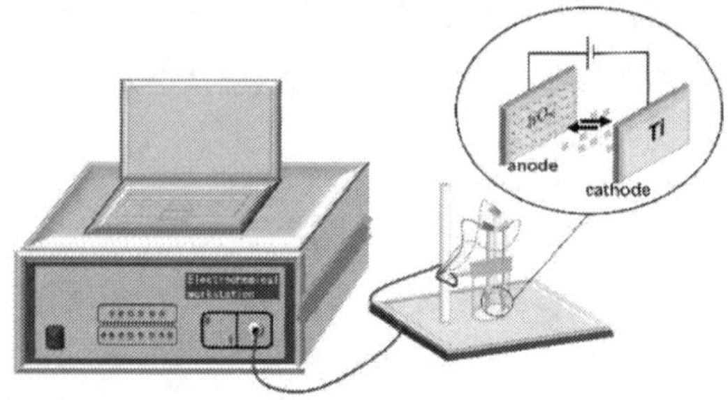

Fig. 1. Schematic diagram of preparing IrOx electrode by double-titanium cyclic voltametric electrodeposition mode.

3 Results and Discussion

Cyclic voltametric electrodeposition technology is mainly used to prepare iridium oxide electrodes based on the substate of Pt, CP (carbon paper). In this study, the iridium oxide electrode was fabricated by this method where both the anode and the cathode were based on titanium. The deposition bath is a complicated iridium complex based on oxalate. Scheme 1,2 shows the typical reaction on anode[16], but the related reaction on cathode is uncleared.

$$[Ir(C_2O_4)(OH)_4]^{3-} \rightarrow [Ir(C_2O_4)(OH)_4]^{2-} + e- \quad (1)$$

$$[Ir(C_2O_4)(OH)_4]^{2-} \rightarrow IrO_2 + 2H_2O + 2CO_2 + e^- \quad (2)$$

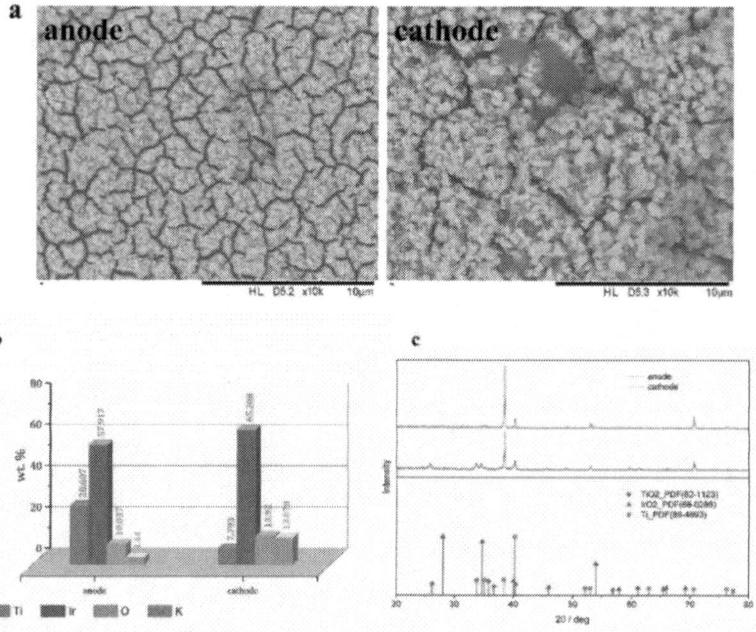

Fig. 2. (a) SEM images, (b) EDS results, and (c) XRD patterns of IrOx electrodes corresponding to the anode and cathode of the cyclic voltametric electrodeposition system.

3.1 Physical characterization

The resulting physical characterization of the IrO_X electrodes corresponding to the anode and cathode prepared by the cyclic voltametric electrodeposition technology are shown in Fig. 2. The morphologies of the deposits based on the anode is varied from that of the cathode, the former is film-like, smooth and flat, while the latter is pile-like, rough and uneven, and both cracks generating. The component element type and content of the deposits was summarized in Fig. 2b. In addition to the main Ir and O elements, there are Ti elements and a small amount of K elements. The presence of K indeed suggest that the redox deposition reaction occurs with the salt adsorption. Compared to the anode, the Ir content is lower than that of the cathode, indicating that the electrodeposition of Iridium on the cathode is much more efficient, resulting in higher coverage of the titanium substrate. obviously, two typical crystalline microstructures can be observed in the anode and cathode samples (Fig. 2c), no rutile IrO_2 is found, probably IrO_X deposits is too thin to be characterized. The deposits of cathode are more complicated, with pronounced TiO_2 crystallization, which suggests that the IrO_X deposits are not tightly bonded to the substrate, resulting in the generation of TiO_2 during the thermal decomposition.

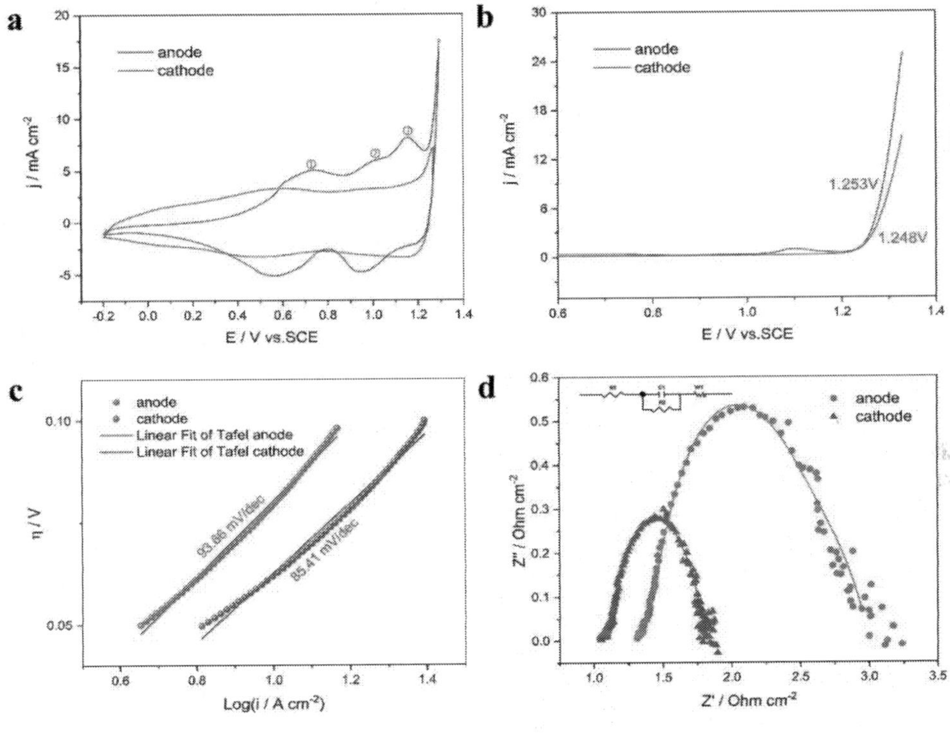

Fig. 3. (a) Cyclic voltammograms, (b) polarization curves, (c) Tafel curves, and (d) Bode diagrams of IrO_X electrodes corresponding to the anode and cathode of the cyclic voltametric electrodeposition system.

Table 1. the fitting result of EIS plots with electrical equivalent circuit.

	anode	cathode
R1	1.299	1.051
C1	0.131	0.245
R2	0.725	0.272
W1-R	0.975	0.475
W1-T	2.257	0.331
W1-P	0.315	0.398

https://doi.org/10.1051/e3sconf/202235001023

3.2 Electrochemical characterization

The voltammetry behaviour was summarized in Fig. 3. Obviously, the cyclic voltammograms (shown in Fig. 3a) of these IrOx/Ti electrodes (derived from anode and cathode electrodeposited by the cyclic voltametric technique) are very different from each other. These curves are obtained by cycling the potential in the range between -0.2 and 1.3V. the CV curve of the anode sample is similar with that of the TDIROF (an IrO2 electrode prepared by thermal decomposition of a precursor), showing an axial symmetry (around the potential axis). However, the CV curve of the cathode sample shows more complexity. Three anodic peaks appear at 0.70V(region ① in Fig. 3a), 1.0V(region ② in Fig. 3a), 1.15V(region ③ in Fig. 3a). the reversible peak ① (reported as pre-peak) is the first to appear followed by the reversible peaks ② and ③ before oxygen evolution. According to previous reports[17], the surface reaction related to redox couples② and ③ is correspond to the Ir(Ⅲ)/Ir(Ⅳ) and Ir(Ⅳ)/Ir(Ⅴ)redox, which are believed to be the following:

At 1.0V, peak ② on Fig. 3a:

$$Ir(OH)_3 \leftrightharpoons IrO(OH)_2 + H^+ + e- \quad (3)$$

At 1.15V, peak ③ on Fig. 3a:

$$IrO(OH)_2 \leftrightharpoons IrO_2(OH) + H^+ + e- \quad (4)$$

The polarization curves of the IrOx deposits based on anode and cathode samples were measured to compare its electrocatalytic characters. In Fig. 3b, before the onset of oxygen evolution, there is a clear Ir(Ⅳ)/Ir(Ⅴ)redox for the cathode sample, corresponding to ③ peak (Fig. 3a). The oxygen evolution overpotential (obtained by linear fitting of the potential over the oxygen evolution section) of the cathode sample is larger than that of the anode sample, indicating that the anode sample is more practical because of the lower energy consumption. As shown in Fig. 3c, the Tafel plots for the polarization curves provided a straight line for each sample and the Tafel slops (lines shown in Fig. 3c) are calculated. The value of the cathode sample is smaller (85.41mV/dec <93.66 mV/dec), which means the cathode sample shows better electrocatalytic performance. This might be result of the rough deposits on cathode, providing more active area.

Intuitively, the electrochemical impedance spectroscope of the anode and cathode sample was roughly similar in Bode diagrams (Fig. 3d), which means that they have the same circuit model[R1(C1-R2) Ws]. The circuit was utilized to acquire the data obtained from the analysis of EIS spectra, presented in Table 1. The value of R1 reflects the difficulty of charge transfer between electrode and electrolyte. Therefore, the large the R1 value, the more difficult it is for the redox reaction occur on the electrode surface, at the same time, means better corrosion resistance. The R2 value represent the charge transfer situation between IrOx deposits and titanium substrates, larger R2 means larger charge transfer resistance, indicating a better adhesion of deposits for the titanium substrate. Where, C is capacitance, has a positive relationship with the active surface area. In summary, the anode sample has larger R1, R2, and smaller C, so its shows worse electrocatalytic activity but better electrolytic durability compared to the cathode sample. This is consistent with previous analysis of voltammetry behaviour such as LSV. Ws consists of three parts: Ws-R, T, P. Thereinto, Ws-R is relevant to the corrosion resistance of IrOx deposits. the larger the value, the more complete and uniform the deposits is, and the less likely it is to be damaged.

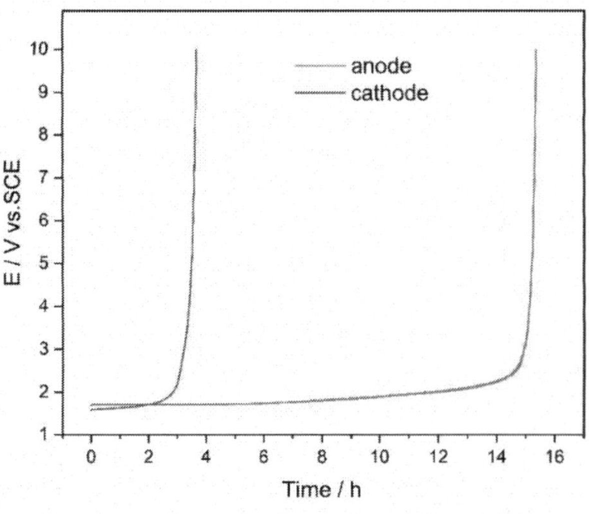

Fig. 4. Accelerate life of IrOx electrodes corresponding to the anode and cathode of the cyclic voltametric electrodeposition system.

3.3 Accelerate life test

For DSA, the actual working life is longer due to the working condition at low current density and weak acid-base strength. So, its lifetime is expressed in terms of accelerate life time, which is measured in strong acid solution at high current density and temperature. The accelerate life time of the samples are shown in Fig. 4. The result shows that the initial potential of the anode sample is higher than that of the cathode, but its accelerate life time is longer (15.82h, more than the value estimated in the literature for 4.16h [18]). Furthermore, the SEM of the anode and cathode sample measured by the accelerate life test for 2h were investigated. The IrOx deposits on the anode sample appears unchanged, while the IrOx film on the cathode sample is visibly damaged, showing spilt blocks-like, implying a worse working life for the cathode sample. This confirms above physical characterization and electrochemical performance analysis for these two electrodes. According to the Hideo Tamura experience formula, the accelerate lifetime is 200-300 times of the actual service lifetime, concluding that the actual working life of the anode and cathode sample is at least 732h and 3072h, respectively, indicating that the IrOx electrode corresponding to the anode of the cyclic voltametric electrodeposition system has value in practical fields such as wastewater treatment, electrodeposition industry.

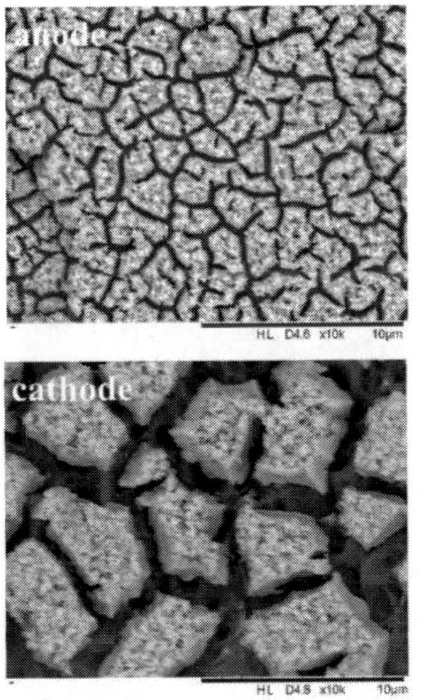

Fig. 5. SEM of the IrOx films on anode and cathode after electrodeposited in 0.5 M H_2SO_4 at 0.25 A cm^{-2} for 2h.

4 Conclusion

The physical characterization and electrocatalytic activity of the IrO_X electrodes corresponding to the anode and cathode of the cyclic voltametric electrodeposition system are compared. The cathode has better electrocatalytic activity, but worse lifetime, no practical value. This may be ameliorated by pretreating the cathode titanium substrate, such as anodizing, hydrogen heat treatment to enhance the adhesion of the deposited IrO_x coating to the substrate. The anode IrO_X electrode exhibits superiority in electrolytic durability, and its actual working lifetime (at least 3072h) indicates that it can be used in practical fields such as water electrolysis and wastewater treatment, etc. Besides, it provides a good idea for pre-coating a layer of iridium oxide film on the titanium substrate because of its perfect wrapping. Therefore, the successful preparation of these two electrodes shows that the double-titanium cyclic voltametric electrodeposition mode has certain research value and development potential.

References

1. S. Choe, et al., Electrodeposited IrO2/Ti electrodes as durable and cost-effective anodes in high-temperature polymer-membrane-electrolyte water electrolyzers. Applied Catalysis B: Environmental, **226**, 89-294 (2018)
2. C.A. Huang, S.W. Yang and P.L. Lai, Effect of precursor baking on the electrochemical properties of IrO2-Ta2O5/Ti anodes. Surface and Coatings Technology, **350**, 896-903 (2018)
3. E. Prats-Alfonso, et al., Iridium oxide pH sensor for biomedical applications. Case urea-urease in real urine samples. Biosens Bioelectron, **39(1)**, 163-169 (2013)
4. Yamanaka and Kazusuke, Anodically Electrodeposited Iridium Oxide Films (AEIROF) from Alkaline Solutions for Electrochromic Display Devices. **28**, 632-637 (1989)
5. L.L. Tran, J. Kim, and J. Yoon, Facile chemical bath deposition to fabricate RuO2 electrodes for electrochemical chlorine evolution. Desalination and Water Treatment, **99**, 204-210 (2017)
6. Thanawala, S., et al., Characterization of iridium oxide thin films deposited by pulsed-direct-current reactive sputtering. Thin Solid Films, 2007. 515(18): p. 7059-7065.
7. D. Labou, et al., Performance of laboratory polymer electrolyte membrane hydrogen generator with sputtered iridium oxide anode. Journal of Power Sources, **185(2)**, 1073-1078 (2008)
8. Y. Gong, et al., Microstructure and properties of annealed IrO2 thin films prepared by pulsed laser deposition. Materials Chemistry and Physics, **116(2-3)**, 573-577 (2019)
9. J.Y. Chen, et al., Chemical bath deposition of IrO2 films on ITO substrate. Ceramics International, **40(9)**, 14983-14990 (2014)
10. I.G. Casella, M. Contursi and R. Toniolo, Anodic electrodeposition of iridium oxide particles on glassy carbon surfaces and their

electrochemical/SEM/XPS characterization. Journal of Electroanalytical Chemistry, **736**, p. 147-152 (2015)

11. M. Mehdipour, S.H. Tabaian and S. Firoozi, Anodic electrodeposition of ligand-free iridium oxide on titanium with high mass loading and study of electrochemical treatments. Journal of Electroanalytical Chemistry, **858**, (2020)
12. Y.J. Park, et al., Electrodeposition of High-Surface-Area IrO2 Films on Ti Felt as an Efficient Catalyst for the Oxygen Evolution Reaction. Front Chem, **8**, 593272 (2020)
13. B. Liu, et al., Energy efficient electrodeposition of metallic manganese in an anion-exchange membrane electrolysis reactor using Ti/IrO2–RuO2–SiO2 anode. Journal of Cleaner Production, 258 (2020)
14. B. van Dijk, et al., The Influence of the Ligand in the Iridium Mediated Electrocatalyic Water Oxidation. ACS Catal, **10(7)**, 398-4410 (2020)
15. R.A. Herrada, et al., IrO2-Ta2O5|Ti electrodes prepared by electrodeposition from different Ir:Ta ratios for the degradation of polycyclic aromatic hydrocarbons. Electrochimica Acta, **263**, 353-361 (2018)
16. Y. Zhang, et al., Electrodeposited nanometer-size IrO2/Ti electrodes with 0.3 mg IrO2 cm−2 for sludge dewatering electrolysers. Electrochimica Acta, **265**, 507-513 (2018)
17. L. Ouattara, et al., Electrochemical comparison of IrO2 prepared by anodic oxidation of pure iridium and IrO2 prepared by thermal decomposition of H2IrCl6 precursor solution. Journal of Applied Electrochemistry, **39(8)**. , 1361-1367 (2009)
18. M. Yousefpour and A. Shokuhy, Electrodeposition of TiO2-RuO2-IrO2 coating on titanium substrate. Superlattices Microstruct, **51(6)**, 842-853 (2012)

A Critical Review on Regional Ecological Environment Assessment

Xinghan Wang[1], *Pengfei* Jia[2] and *Mingxiang* Yang[1,*]

[1]China Institute of Water Resources and Hydropower Research, Beijing; 10038, China
[2]CITIC CONSTRUCTION, CO., LTD

Abstract. With the continuous advancement of industrialization and urbanization, the relationship between mankind and the ecological environment has become increasingly tense, and the ecological environment assessment has become a research hotspot in recent years. The article summarizes the research content and development process of ecological environment assessment, lists various mainstream assessment methods and introduces their application characteristics, and then divides the weight determination methods into subjective weighting, objective weighting and subjective and objective combination, and analyses their advantages and disadvantages; Meanwhile, the application of remote sensing technology in ecological environment assessment research is analyzed. Finally, the main problems of ecological environment assessment work are summarized and its future development direction is pointed out.

1 Introduction

In order to meet the development needs of industrialization and urbanization, the scope and intensity of human activities continue to expand, ecosystem areas such as urban and rural natural environments and nature reserves have been shrinking, resulting in a substantial shrinkage of ecological resources. At the same time, environmental problems of different scales such as land and water pollution, heat island effect, and global warming caused by human activities have become more and more serious, the degradation of the ecological environment has become a bottleneck restricting the development of human economy. The ecological environment assessment(EEA) could contribute to provide a scientific basis for formulating ecological environmental protection policies and measures, as well as ecological coordination and sustainable development[1].Therefore, it is of great significance to carry out regional ecological environment monitoring and evaluation timely. The article summarizes the content and development of EEA, discusses the main methods of current EEA and the application of remote sensing in EEA, and finally summarizes the shortcomings of current EEA.

2 The content and development of EEA

2.1 Contents of EEA

Ecological environment refers to the entirety of various ecosystems composed of biological communities and non-biological natural factors. The purpose of EEA is to analyze the ecological environmental impact of a region due to human activities and the impact on humans, mainly involving ecological risk assessment, ecological environmental vulnerability assessment, ecological health assessment, ecological footprint, etc, this attributes to carry out ecological management and ecological protection policies, etc[2-3].There are many types of EEA objects based on different perspectives, from the assessment area, they can be divided into administrative EEA and thematic EEA. Administrative ecological environmental assessment studies the impact of various human social activities on a certain administrative area[4-5]. Thematic EEA refers to the assessment of the ecological environment of a certain type of geographic area, such as plateaus[6], lakes[7], highways[8], cities[9],mine [10], etc. According to the element attributes of the assessment object, it can be divided into comprehensive element assessment and partial single element assessment. Comprehensive element assessment takes the study area as an ecosystem and uses landscape ecology[11], ecosystem services[12] and other methods to evaluate the ecological environment from a macro perspective. Partial single element assessment is the evaluation of key elements in the ecosystem, such as diatoms [13], heavy metal toxic elements[14-15], etc.

2.2 The development of EEA

In 1964, the International Conference on

* Corresponding author: yangmx@iwhr.com

Environmental Quality first proposed the concept of environmental impact assessment. In 1969, the United States promulgated the National Policy Act (NEPA) and established the world's first environmental assessment system, it is also the beginning of the world's EEA work. Subsequently, the U.S. Environmental Commission further improved it and added relevant content of environmental impact assessment (EIA). EIA is a process that proposes the potential environmental impacts of the development project and proposes appropriate measures to avoid, reduce or compensate for these impacts (called mitigation measures) [16-17].

In the mid-1970s, some developed countries in Europe and the United States improved the shortcomings of EIA and extended its application to the planning level and policy level, this is the strategic environmental assessment (SEA). SEA is used to solve the environmental impact of strategic decisions and designed to incorporate environmental sustainability into strategic decision-making[18]; In 1989, the World Bank stipulated that all major projects under its supervision required ecological environmental assessment[19]. So far, SEA has been widely used worldwide, mainly in the countries of the Organization for Economic Co-operation and Development (OECD) and the client countries of the World Bank, these countries have established the SEA system in the form of legislation, and have formulated the SEA framework to a certain extent[20]. In 2001, the United Nations Foundation and other organizations launched the Millennium Ecosystem Assessment (MA) project, which was implemented from 2001 to 2005.MA uses a new conceptual framework to record, analyze and understand the impact of environmental changes on ecosystems and human activities, it is a summary of the comprehensive evaluation of global ecosystems[21-22]; For the first time, MA has systematically and comprehensively revealed the status and change trends of various ecosystems on a global scale. Since 2005, MA has conducted hundreds of ecosystem assessments of different scales and different themes[23], which has also brought EEA research into a new stage of development.

3 EEA method and weight determination

3.1 Assessment method

After years of eco-environmental assessment and research by experts all over the world, there are currently some more mature eco-environmental assessment methods, such as fuzzy discrimination, biodiversity evaluation, ecological footprint, Index evaluation,etc. Fuzzy discrimination is an assessment method that transforms qualitative evaluation into quantitative evaluation based on the principles of fuzzy mathematics, finally, a certain value between 0 and 1 is used to indicate the relationship between each assessment index and the system, this method is applicable to both large-scale and small-scale areas, Sami comprehensive environmental assessment of the farm system with the help of fuzzy discriminant method[24].The Biodiversity assessment method is based on the investigation of the on-site ecosystem, use Shannon-Wiener index to represent biodiversity, this method can better reflect the relationship between the biological community and the ecological environment; Wang estimated the economic value of China's forest species diversity based on the biodiversity evaluation method[25].The ecological footprint method evaluates the sustainable development of the ecosystem by calculating the profit and loss difference between the ecological carrying capacity and the size of the ecological footprint; this method is suitable for urban EEA with good basic data; Świąder use the ecological footprint method to evaluate the ecological environment carrying capacity in Wroclaw, Poland[26].Index evaluation method is to evaluate each index participating in the evaluation separately, and finally use the weighted sum method to achieve the effect of comprehensive evaluation of the ecological environment. Index evaluation method has the characteristics of difficulty in weighting and quantitative evaluation, because it needs to establish an evaluation system, but it can be more comprehensively evaluated,so it is widely used in ecological environment evaluation, for example, the Ecological Environment Index (EI) proposed in the "Technical Specifications for Evaluation of Ecological Environment Condition (Trial)" promulgated and implemented by the State Environmental Protection Administration of China,the Environmental Quality Index (NWF) proposed by the United States and Canada's Total Environmental Quality Index (EQI)[3].There are many eco-environmental assessment methods and each has its own focus; therefore, when dealing with different eco-environmental assessment issues, it should be determined according to the specific assessment objectives.

3.2 Index weight determination method

The indicator weighting methods in EEA are mainly divided into subjective weighting method (SW) and objective weighting method (OW). SW determines the index weight based on the researcher's prior knowledge and subjective judgment. It is highly subjective. Commonly used methods are as follows: Analytic Hierarchy Process (AHP)[27], Delphi Method[28], Fuzzy Mathematics (FM)[29], etc. OW determines the weight of each indicator based on the correlation between various indicator data. This type of weighting method is less artificially affected and can objectively reflect the relationship between indicators, common OW methods are: Entropy Method (EWM)[29], Principal Component Analysis (PCA)[30], Random Forest (RF)[31],Convolutional Neural Network (CNN) [32], etc. At present, many studies combine the advantages of subjective and objective weighting methods to determine the index weights, which not only avoids the excessive subjectiveness of objective weighting methods, but also combines prior knowledge to judge the importance of the evaluation indicators in the study area, effectively reducing the evaluation Error of result, for example, Li used the PCA-AHP-TOPSIS method to estimate the ecological environment index of the area along the

Beijing-Hangzhou Grand Canal [33], avoiding the subjectivity and extensiveness of conventional multi-factor decision analysis.

4 Application of Remote Sensing in EEA

EEA focuses more on the integration of data and information[34], so the accuracy of data acquisition is very important, in the early days, the evaluation index data was easily restricted by time and space scales, With the development of space science and computer technology, 3S technology has been widely used in ecological environment evaluation. The spatial resolution, time resolution, and spectral resolution of remote sensing data continue to increase, which improves the real-time and operability of remote sensing technology; combines remote sensing image data with long-term statistics and observation data to strengthen the research on dynamic evaluation of the ecological environment. The spatial resolution, time resolution, and spectral resolution of remote sensing data continue to increase, which improves the real-time and operability of remote sensing technology; combines remote sensing image data with long-term statistics and observation data to strengthen the research on the dynamic evaluation of ecological environment. Yao based on Landsat8 data, selected vegetation coverage, bare soil index and slope as evaluation indicators, using Index evaluation method to evaluate the highway ecological environment quality[35].Ying uses ASTER GDEM and Landsat data to extract the three indicators of vegetation coverage, soil index and soil moisture, and determines the weight of each indicator through the Delphi method, and then uses the index evaluation method to analyze the ecological environment quality of the Wujiang River Basin in Guizhou Province, China[36].

In 2013, Xu[30] improved China's EI index and constructed a new remote sensing ecological index RSEI based on PCA using greenness, humidity, heat, and dryness indicators,and verified its effective application in EEA. Based on high-resolution remote sensing images, Hao[37] selected the fractional vegetation cover (FVC), water density (WD), impervious surface coverage (ISC), net primary production (NPP) and land surface temperature (LST) evaluated and analysis of the ecological environmental impact caused by urban expansion in Beijing, China. Based on Landsat data, Zhang[38] used the RSEI index to evaluate the ecological environmental quality of Nanjing, China from 1990 to 2013, and predicted the development trend of ecological environmental quality. Wu[39] selected fractional vegetation cover (FVC), leaf area index (LAI), total primary productivity (GPP), land surface temperature (LST), and wet (Wet) retrieved from MODIS data as evaluation indicators. assessment the ecological environment of China's Tibet from 2006 to 2016.

With the development of network big data, Google has launched a global-scale cloud computing platform for processing earth science data — Google Earth Engine (GEE), compared with traditional geospatial data processing methods, GEE has unprecedentedly improved computational efficiency. It also has advantages such as free and parallel, it is now widely used in remote sensing image classification, land use change monitoring, etc.Chen[40] used Landsat data from the GEE platform to extract four indicators including vegetation index, humidity component, heat and dryness to evaluate and monitor the quality of the ecological environment in China's Three-River Source Region. Based on Landsat images from the GEE platform, Mahdianpari[41]used random forests to assess the spatial dynamics of wetlands in Newfoundland, Canada from 1985 to 2015.In the future, remote sensing technology based on cloud computing and big data analysis will be widely used in EEA, on this basis, with the help of artificial intelligence, knowledge base, etc. to mine the ecological environment big data, enhance the use value of big data, and provide more accurate and efficient services for the decision-making management of the ecological environment.

5 Conclusion

In the era of global climate change and information exchange, monitoring regional ecological environmental changes is of great significance for humans to solve ecological problems. The maturity of 3S technology, the development of big data, and the optimization of machine learning algorithms have further improved the methods and theories of EEA research. Eco-environmental assessment objects are also more targeted, from early large-scale macro-evaluation to partial thematic regional eco-environmental assessment, which provides a theoretical basis for solving typical regional eco-environmental problems. However, due to the regional differences in different research areas, although many EEA work combines local geography, ecology, and social conditions to construct an assessment system, there is still a lack of complete EEA systems suitable for various objects and scales. In addition, the weights of evaluation indicators and the classification of evaluation results tend to be subjective or objective, the use of a single subjective or objective weighting method will lead to deviations in the evaluation results, and inevitable errors in the grading process will cause the evaluation results to be subjective. The future ecological environment evaluation work should fully consider the actual situation of the study area, construct an appropriate evaluation system and determine the index weights according to local conditions, and use big data, emerging technology platforms, etc. to conduct ecological environment evaluation from a multidisciplinary perspective.

References

1. Sun D, Zhang J, Zhu C, et al. An assessment of China's ecological environment quality change and its spatial variation, Acta Geographica Sinica, **67(12)**, 1599-1610 (2012)

2. J. Sarukhan, A. Whyte, R. Hassan, et al. Millenium Ecosystem Assessment: Ecosystems and human well-being, (2005)
3. B. Gao and M. Chen, University H A . A review of eco-environmental quality assessment[J]. Forestry and Ecological Sciences, 2018.
4. Wei, Li, Yan-ju, et al. Preliminary Strategic Environmental Assessment of the Great Western Development Strategy: Safeguarding Ecological Security for a New Western China[J]. Environmental Management, 2012.
5. Liu W, Zhang B . Regional Ecological Security Assessment Based on Long Periods of Ecological Footprint Analysis[C]// Remote Sensing, Environment and Transportation Engineering (RSETE), 2012 2nd International Conference on. 2012.
6. Yue D X, Zeng J J, Yang C, et al. Ecological risk assessment of the Gannan Plateau, northeastern Tibetan Plateau[J]. Journal of Mountain Science, 2018, 15(006):1254-1267.
7. Bain M B, Singkran N, Mills K E . Integrated Ecosystem Assessment: Lake Ontario Water Management[J]. PLoS ONE, 2008, 3(11):e3806.
8. Box J D, Forbes J E . Ecological considerations in the environmental assessment of road proposals[J]. Highways and Transportation, 1992, 39(4):16-22.
9. Gontier M . Integrating landscape ecology in environmental impact assessment using GIS and ecological modelling[J]. Annan Naturresursteknik, 2006, 12:345-354.
10. Xueming Du, Hongyuan Fang, Kang Liu, Binghan Xue, Xin Cai. Environmental Evaluation of Coal Mines Based on Generalized Linear Model and Nonlinear Fuzzy Analytic Hierarchy[J]. Geofluids, 2020, 2020(3).
11. Moertberg U M, Balfors B, Knol W C . Landscape ecological assessment: a tool for integrating biodiversity issues in strategic environmental assessment and planning.[J]. Journal of Environmental Management, 2007, 82(4):457-470.
12. Carpenter S R, Mooney H A, Agard J, et al. Science for managing ecosystem services: Beyond the Millennium Ecosystem Assessment[J]. Proc Natl Acad Sci U S A, 2009, 106(5):1305-1312.
13. Bennett J R, Sisson D R, Smol J P, et al. Optimizing taxonomic resolution and sampling effort to design cost - effective ecological models for environmental assessment[J]. Journal of Applied Ecology, 2015, 51.
14. Maanan M, Saddik M, Maanan M, et al. Environmental and ecological risk assessment of heavy metals in sediments of Nador lagoon, Morocco[J]. Ecological Indicators, 2015, 48:616-626.
15. Xu D, Gao B, Peng W, et al. Boron toxicity coefficient calculation and application for ecological risk assessment in reservoir sediments[J]. Science of The Total Environment, 2020:139703.
16. Burris R K, Canter L W . Cumulative impacts are not properly addressed in environmental assessments[J]. Environmental Impact Assessment Review, 1997, 17(1):5-18.
17. Munn E, Munn R E . Environmental impact assessment:principles and procedures[M]. SCOPE, 1979.
18. Awange J L, Kiema J K . [Environmental Science and Engineering] Environmental Geoinformatics || Protection and Conservation of Animals and Vegetation[J]. 2013, 10.1007/978-3-642-34085-7(Chapter 25):415-435.
19. Alshuwaikhat H M . Strategic environmental assessment can help solve environmental impact assessment failures in developing countries[J]. Environmental Impact Assessment Review, 2005, 25(4):307-317.
20. Jia H, Bao C K, Shu T F, et al. Framework for integration of urban planning, strategic environmental assessment and ecological planning for urban sustainability within the context of China[J]. Environmental Impact Assessment Review, 2011, 31(6):549-560.
21. Jie F B, Liu S L, Ming M K . The contents and methods of integrated ecosystem assessment (IEA)[J]. Acta Ecologica Sinica, 2001.
22. Carpenter S R, Mooney H A, Agard J, et al. Science for managing ecosystem services: Beyond the Millennium Ecosystem Assessment[J]. Proc Natl Acad Sci U S A, 2009, 106(5):1305-1312
23. Allison H, Brown C . A review of recent developments in ecosystem assessment and its role in policy evolution[J]. Current Opinion in Environmental Sustainability, 2017, 29(DEC.):57-62.
24. Sami, M., et al., Environmental comprehensive assessment of agricultural systems at the farm level using fuzzy logic: A case study in cane farms in Iran. Environmental Modelling & Software, 2014. 58: p. 95-108.
25. Wang B, Zheng Q H, Guo H . Economic Value Assessment of Forest Species Diversity Conservation in China Based on the Shannon-Wiener Index[J]. Forest Research, 2008, 21(2):268-274.
26. Świąder, M., et al., The application of ecological footprint and biocapacity for environmental carrying capacity assessment: A new approach for European cities. Environmental Science & Policy, 2020. 105: p. 56-74.
27. Ekumah B, Armah F A, Afrifa E, et al. Geospatial assessment of ecosystem health of coastal urban wetlands in Ghana[J]. Ocean & Coastal Management, 2020, 193(1):105226.
28. Dwa B, Dck B, Ar C, et al. Validation of a rapid wetland ecosystem services assessment technique

using the Delphi method - ScienceDirect[J]. Ecological Indicators, 125.

29. Cheng W, Xi H, Sindikubwabo C, et al. Ecosystem health assessment of desert nature reserve with entropy weight and fuzzy mathematics methods: A case study of Badain Jaran Desert[J]. Ecological Indicators, 2020, 119:106843.
30. Rsei, Xu H Q . A remote sensing index for assessment of regional ecological changes[J]. China Environmental Science, 2013, 33(5):889-897.
31. Gou R, Zhao J . Eco-Environmental Quality Monitoring in Beijing, China, Using an RSEI-Based Approach Combined With Random Forest Algorithms, IEEE Access, **8**, 196657-196666 (2020)
32. L. Wang and X. Bi, Risk assessment of knowledge fusion in an innovation ecosystem based on a GA-BP neural network, Cognitive Systems Research, **66(46)** (2020)
33. Y. Li, et al., Estimation of remote sensing based ecological index along the Grand Canal based on PCA-AHP-TOPSIS methodology. Ecological Indicators, **122**, 107214 (2021)
34. European Ecosystem Assessment - concept, data, and implementation. 2015.
35. A.D. Yao, W.K. Guan and Y.M. Feng, Research on Environmental Quality Assessment of Shaowu-Guangze Speedway Based on Landsat 8 RS Image, The Administration and Technique of Environmental Monitoring, (2017)
36. Y. Lu, X.D. Li and Z. Yang, Evaluation on Eco-environmental Quality Change of Wujiang River Basin in Guizhou Province from 1990 to 2015, Bulletin of Soil and Water Conservation, (2018)
37. W. Hao, X. Ning, W. Zhu, et al. Comprehensive Evaluation of Urban Sprawl on Ecological Environment Using Multi-source Data: A Case Study of Beijing, Isprs International Archives of the Photogrammetry Remote Sensing & Spatial Information Sciences, (2016)
38. H. Zhang, Ecological Change Analysis of Nanjing City Based on Remote Sensing Ecological Index, Geospatial Information, (2017)
39. Y.J. Wu, X. Zhao, X.I. Yue, et al. Comprehensive evaluation and spatial-temporal changes of eco-environmental quality based on MODIS in Tibet during 2006-2016, Acta Geographica Sinica, (2019)
40. W. Chen, H. Huang, Y. Tian, et al. Monitoring and Assessment of the Eco-Environment Quality in the Sanjiangyuan Region based on Google Earth Engine, Journal of Geo-information Science, (2019)
41. M. Mahdianpari, H. Jafarzadeh, J.E. Granger, et al. A large-scale change monitoring of wetlands using time series Landsat imagery on Google Earth Engine: a case study in Newfoundland, GIScience&Remote Sensing, (2020)

Analysis on Temporal and Spatial Pattern Change and Driving Force of Land Use in Agro Pastoral Ecotone in Northern Shanxi

Wenchao Liu[1], *Mei* Liu[2,*], *Wenhui* Jia[1], *Xiaoyi* Miao[1] and *Meihong* Ma[1]

[1]Tianjin Normal University, School of geography and environmental sciences, 300387 TianJin, China
[2]Tianjin Chengjian University, School of economics and management, 300384 TianJin, China

Abstract. Land use change is a considerable representation of mutual effect between human and natural activities. Northern Shanxi is located in the northern agro pastoral ecotone, which the land use pattern changes violently. This paper focuses on the dynamic change characteristics of cultivated land and grassland in the agro pastoral ecotone of northern Shanxi from 2000 to 2015, and explores the main driving factors of their change. It can provide support for land planning and management and relevant policy-making in northern Shanxi and promote regional sustainable development. The results showed that: ① There was a close interrelation between the change of cultivated land and grassland I n northern Shanxi, and the whole showed the opposite dynamic change trend. ②The main land transfer type of cultivated land and grassland is construction land. And grassland is also the main transfer type of cultivated land. And cultivated land contributes greatly to the growth of grassland area. ③Economic factors are the main driving force which affecting the area change of cultivated land and grassland, followed by the benefit factors of agricultural development, and finally the labor factors. ④Economic factors can explain the shift of cultivated land and grassland in different districts and counties in northern Shanxi. The benefit factor of cultivated land development plays a negative driving role on cultivated land, the benefit of grassland development plays a promoting role on the growth of grassland area, and the driving force of labor factor is relatively small.

1 Introduction

Land use change can directly reflect the results of the interaction between human and natural activities. It is an important research topic related to environmental change and sustainable development. Northern Shanxi is located in the northern agro pastoral ecotone. The natural environment is fragile and transitional, and it is the key construction area of ecological engineering. The land use situation is special and complex, and the land use pattern changes violently.

Among the global environmental change issues, land use change is the most closely issue of the intersection of natural and humanistic processes. Since 1990, the research field of global environmental change has gradually strengthened the research on Land Use and Land Cover Change (LUCC), mainly focusing on the process, driving force and eco-environmental effect of land use change[1]. The international community mainly focuses on the relationship between land use change and global environmental change and sustainable development, focusing on the research on the driving mechanism of global and intercontinental macro-scale land use change and eco-environmental effects, especially the impact on climate change and gas emissions[2]. Relevant research scales in China are relatively wide, it is involved from the macro national scale to provinces, cities and counties[3]. The Loess Plateau, the Yellow River Basin [4] and the ecotone between agriculture and animal husbandry[4] are the research hotspots. Research methods are mostly based on 3S technology to analyze the magnitude, speed and regional differences of quantitative change of land use types in different research areas, and then explore the main types, distribution characteristics and regional direction of land use spatial change[6]. The driving force research mostly adopts mathematical statistical methods such as principal component analysis and regression analysis[7]. The driving force factors, including socio-economic driving force, natural driving force and land use policy driving force.

The ecological environment in northern Shanxi is fragile and the contradiction between man and land is prominent. Based on the typical characteristics of farming pastoral ecotone in northern Shanxi and RS and GIS technology, this paper focuses on the change characteristics and driving forces of two representative land use types of cultivated land and grassland in the study area from 2000 to 2015. This study helps to ensure the sustainable development of the region, so does to help land planning and management and the formulation of relevant policies.

* Corresponding author: hebeiliumei@126.com

2 Overview of the study area and data sources

2.1 Overview of the study area

Northern Shanxi is located in the north of Shanxi Province, including Datong City, Shuozhou City and Xinzhou City. Northern Shanxi has complex and changeable landform, widely distributed loess, poor ecological environment and infrastructure. And the development of social economy lags behind. Although coal mine related industries are developed, the intensity of air pollution and environmental damage is great. Since 2000, the study area has taken a series of major measures to actively control the ecological environment, and the quality of the ecological environment has been improved. However, due to the great natural conditions and man-made pressure, there are still common environmental problems such as land desertification and sparse vegetation.

2.2 Data source and processing

The basic data in this paper are mainly the land use data and socio-economic data in 2000, 2005, 2010 and 2015. The land use data comes from the "national land use 1:100000 database" provided by the Chinese Academy of Sciences. The Landsat remote sensing image and Google Earth image are used for data processing, and finally the land use data with an accuracy of more than 95% is obtained. The process and spatial pattern of land use change in each county in the study area are obtained through ArcGIS spatial processing. The economic and social statistical data comes from China county statistical yearbook, and used SPSS to standardize the data and PCA principal component analysis to obtain the main socio-economic driving forces of land use pattern change in northern Shanxi, and then explore the driving forces of different principal components.

3 Analysis of temporal and spatial pattern change characteristics of land use

3.1 Temporal and spatial trajectory analysis of land use change

The farming and pastoral areas are distributed in a zonal crisscross manner, roughly from southwest to northeast. Moreover, most of the cultivated land is concentrated in the abdomen of the study area, while the grassland is distributed in the marginal area (Fig.1), which is related to the mountainous terrain in northern Shanxi and also reflects its obvious characteristics of farming and pastoral crisscross zone.

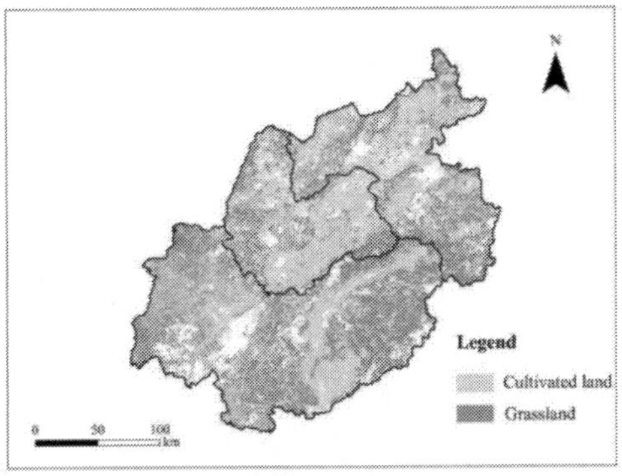

Fig. 1. Distribution of land use status in Northern Shanxi in 2015

It is found that the cultivated land area in the study area generally showed a decreasing trend from 2000 to 2015 (Table 1), and the change was the most drastic from 2005 to 2010, with a decrease of 1.76% compared with the original area. By analyzing the trajectory of land use change through the transfer matrix, it can be seen that grassland and construction land contributed the most to the reduction of cultivated land area from 2000 to 2005, forest land and construction land from 2005 to 2010 and construction land from 2010 to 2015. It is found that the cultivated land occupied by construction land has been serious, and the implementation effect of ecological conversion of farmland and conversion of farmland to forest from 2000 to 2010 is remarkable. From 2000 to 2015, the grassland area in the study area increased first and then decreased, and increased significantly from 2000 to 2005, an increase of 0.84% compared with the initial area. From 2000 to 2010, cultivated land contributed the most to the growth of grassland area, far exceeding the area converted from grassland to cultivated land. After 2010, the area of grassland converted to cultivated land has increased significantly, the area of cultivated land and grassland transferred to each other is equivalent, and the area of grassland converted to construction land has more than doubled compared with before.

Overall, the change degree of cultivated land is more intense than that of grassland, and the change trend is roughly opposite. From 2000 to 2010, the policy of returning farmland to forest had an obvious effect on the area change of cultivated land, grassland and forest land. The cultivated land showed a decreasing trend. A large number of cultivated land was converted to grassland and forest land, while the grassland was in a fluctuating and increasing trend. After 2010, the decreasing trend of cultivated land area slowed down significantly, and the trend of cultivated land turning into grassland and forest land decreased significantly, while the grassland area showed a decreasing trend after 2010, but the change range was small, only 0.18% less than the initial area. Construction land has always been the main transfer type of cultivated land and grassland, which is related to the massive mining of coal, the continuous development and construction of cities and towns and the improvement of urbanization level in northern Shanxi.

Table 1. Statistics of Land Use Type Change

Time period (year)	2000-2005		2005-2010		2010-2015	
Land use type	Cultivated land	Grassland	Cultivated land	Grassland	Cultivated land	Grassland
Area change (%)	-1.33	0.84	-1.76	0.04	-0.36	-0.18

3.2 Spatial dynamic analysis of land use change

According to the dynamic degree calculation, the cultivated land area of most counties and districts in northern Shanxi showed a decreasing trend from 2000 to 2015, among which the absolute value of the dynamic degree of cultivated land change in the mining area and urban area of Datong City was the largest. The urban construction in these two districts developed rapidly, resulting in the fastest reduction of cultivated land. The dynamic degree of cultivated land change in the Southern suburb and Pinglu district is relatively low, which are -1.095% and -0.827% respectively. Both districts are important coal producing counties. Many cultivated land has been converted to industrial and mining land and workers' living areas, while the agricultural development is poor, resulting in the relatively rapid reduction of cultivated land. Ying county has the highest dynamic degree of cultivated land change, which is 0.069%. Its agricultural development conditions are good, which is conducive to cultivated land reclamation. Among the 31 counties, there are 15 positive and 16 negative grassland change dynamics. From the number of counties, the grassland area increases and decreases, and the number of counties increases and decreases is the same. In terms of change degree, Kelan county has the largest grassland change dynamic degree, which is 1.00%, far higher than other counties and regions. The grassland change dynamic degree of Guangling, Hequ, Pinglu, Shanyin, Wuzhai and Youyu counties is relatively high, all around 0.20%. The county dynamic degree of grassland area with negative trend change is generally low, and the absolute value of change dynamic degree is within 0.17%.

By comparing the spatial distribution of the area change dynamics of cultivated land and grassland, it can be seen that the utilization changes of cultivated land and grassland in northern Shanxi are closely related, showing the opposite dynamic change trend as a whole. The counties with high Grassland Change Dynamics generally have good animal husbandry development, while the spatial dynamic changes of cultivated land change dynamics are greatly affected by mineral resources (Fig.2). The dynamic degree of cultivated land change in Pinglu district and Kelan county is low, and the grassland is relatively high; the area of cultivated land and grassland in Guangling County, Pianguan County, Baode County and Wutai County also showed the opposite dynamic change trend; several counties in the urban area, mining area and southern suburb of Datong are special, and the dynamic degree of cultivated land and grassland is low.

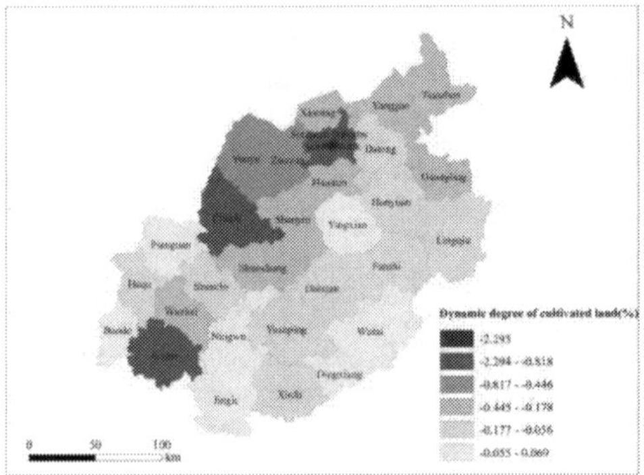

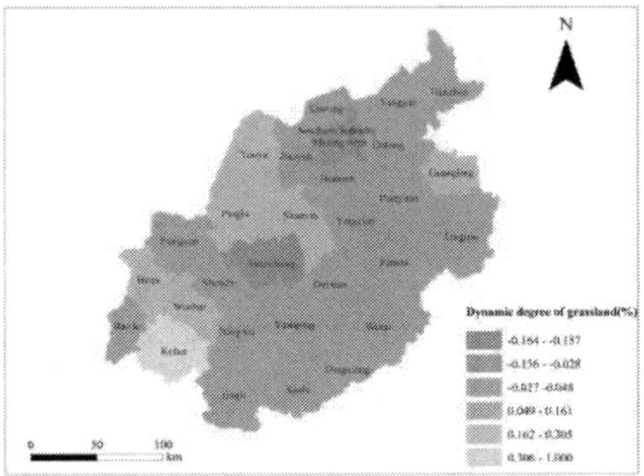

Fig. 2. Spatial distribution of dynamic degree of cultivated land and grassland Change

4 Analysis on driving forces of land use change

According to the existing analysis, natural factors and relevant policies are the main factors leading to the change of cultivated land and grassland in the agro pastoral ecotone in northern Shanxi. This paper uses the principal component analysis method to comprehensively analyze the socio-economic factors and explore the main driving forces affecting the change of cultivated land and grassland in the region.

4.1 Determine the driving force index and driving force principal component

The influencing factors of land use and cover change include natural factors and socio-economic factors, among which the socio-economic driving force plays a strong role. This paper makes a principal component analysis on 103 samples from 29 counties and 4 years in northern Shanxi. The total population, agricultural population, regional GDP, primary industry output value, secondary industry output value, tertiary industry output value, agricultural output value, animal husbandry output value, forestry output value, main grain output, total meat output, total power of agricultural mechanization, industrial sales output value, total retail sales of social consumer goods The general budget expenditure and the number of people engaged in agriculture are 16 socio-economic indicators, and the kmo test value of the correlation matrix is 0.794, which is suitable for principal component analysis. Three principal components are extracted from the 16 indicators. The first principal component (F1) economic factor, the second principal component (F2) agricultural development benefit factor and the third principal component (F3) labor force factor are the main driving factors of cultivated land use change. The contribution rates are 55.962%, 17.374% and 9.469% respectively, and the cumulative contribution rate reaches 82.805%, which can represent the social and economic development of northern Shanxi.

4.2 Analysis on driving forces of land use pattern

4.2.1 Quantitative analysis of driving force

In order to clearly judge the driving effect of each principal component on the change of cultivated land and grassland utilization in northern Shanxi, the extracted principal components were quantitatively calculated. According to the factor score coefficient, combined with the standardized value of each influencing factor, the principal component scores of driving force of each district and county in northern Shanxi in different years are calculated. The analysis shows that in 2000, the third principal component of the driving force, the labor force factor, was the main driving force for the change of cultivated land use at that time, but since 2010, the first principal component and the second principal component have gradually developed into the main socio-economic driving force (Fig.3).

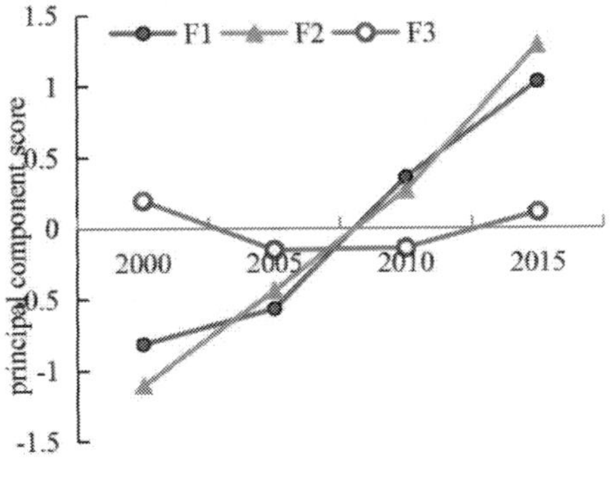

Fig. 3. Statistical chart of principal component scores from 2000 to 2015

4.2.2 Principal component analysis of driving force

(1) Economic factors

From 2000 to 2015, the first principal component scores of districts and counties in northern Shanxi showed an upward trend year by year. In 2000, the overall scores of districts and counties were low, but with the increase of years, the score gap of economic factors between regions was significantly widened, among which Shanyin County, Huairen County and Hequ County were several counties with rapid increase in economic factor scores. The abundance of coal resources in northern Shanxi determines the driving force direction of economic factors to a great extent. For districts and counties that rely on coal to develop economy, represented by Yuanping City, mainly rely on the development of industry to drive economic growth, and economic factors play a negative driving role in the increase of cultivated land and grassland. for Youyu County, Wuzhai County and Kelan County, represented by Shanyin County, which lack coal resources and focus on the development of animal husbandry economy, economic factors play a role in promoting the grassland area and a negative driving force on the cultivated land area.

(2) Benefit factors of agricultural development

From 2000 to 2015, the overall agricultural output value and animal husbandry output value in northern Shanxi showed an upward trend, and the agricultural output value was higher than the animal husbandry output value. While the score of agricultural development benefit factors is rising, the cultivated land area in northern Shanxi is greatly decreasing, while the grassland area is increasing. It can be seen that although the area of cultivated land has decreased, its output value is increasing. It shows that with the improvement of agricultural science and technology level, the output benefit per unit of cultivated land is also increasing. Therefore, the benefit factors of cultivated land development play a negative driving role in the increase of cultivated land. At the same time, for the districts and counties with developed animal husbandry, the benefit of developing animal husbandry is higher than that of developing agriculture, so as to promote the transfer of cultivated land to grassland and accelerate the reduction of cultivated land and the increase of grassland. Therefore, the benefit factors of grassland development play a promoting role in grassland area and a negative driving role in cultivated land area.

(3) Labor factor

From 2000 to 2005, the urbanization rate in northern Shanxi developed rapidly, a large number of rural population transferred to urban life and employment, and the number of agricultural labor force decreased to a certain extent. From 2005 to 2015, the number of agricultural labor force increased from 10246 million to 11624 million, and the decline of cultivated land area decreased significantly from 2010 to 2015. The increase of rural labor force engaged in agriculture has played a certain role in promoting the increase of cultivated land. The labor force factor plays a positive driving role in the area of cultivated land, while it plays a negative driving role in the increase of grassland area. The increase of population will drive the growth of food demand, and then attract labor force to engage in agriculture, plough a pile of wasteland or reclaim new cultivated land, promote the increase of cultivated land, but also promote the transfer of grassland to cultivated land.

5 Conclusion

Firstly, this paper studies the land use change pattern and its socio-economic driving forces in the agro pastoral ecotone in northern Shanxi from 2000 to 2015, and draws the following main conclusions:

(1) From the time dimension, the cultivated land and grassland showed the opposite trend around 2010. From 2000 to 2010, the cultivated land in northern Shanxi decreased rapidly, which was mainly transformed into construction land and grassland, and the grassland area showed an increasing trend. The main source of increase was cultivated land. From 2010 to 2015, the reduction rate of cultivated land area decreased significantly, and the grassland area began to decrease. Labor factors and cultivated land protection policies are the main driving forces for the change of cultivated grassland.

(2) From the perspective of county spatial dimension, the distribution of coal resources and animal husbandry development conditions in the county of northern Shanxi Province largely determine the driving direction of economic factors, and then affect the spatial change dynamics of cultivated land and grassland. For districts and counties that rely on coal to develop economy, economic factors play a negative driving role in the growth of cultivated grassland area. For districts and counties that develop animal husbandry, economic factors promote the growth of grassland area and play a negative driving role in the growth of cultivated land.

(3) On the whole, firstly, the policy of returning farmland to forest and grassland directly induced the reduction of cultivated land and the increase of grassland in northern Shanxi. The cultivated land protection implemented later slowed down the reduction rate of cultivated land and indirectly led to the reduction of grassland. Secondly, the driving effect of socio-economic driving force on the change of cultivated grassland is more intense, in which economic factor is the main driving force, followed by the benefit factor of agricultural development, and finally the labor factor.

Funding

This work was supported by Deep learning-based research on the inducing factors and early warning of flash floods (42101086).

References

1. H Zhao, He H, Bai C, et a l. Characteristics of land use change and its environmental effects on the Loess Plateau. China land science. **(07)**: 51-59 (2018).
2. Watson R T, Noble I R, Bolin B, et al. Land Use, Land Use Change, and Forestry: a Special Report of the IPCC (2000).
3. J Wang, L Chen, Y Wang . Review on the evolution of landscape pattern in the Loess Plateau. Progress in geographical science. **29 (05)**: 535-542 (2010).
4. Y Luo, S Yang, X Liu, et al. Characteristics of land use change in Hekou town Tongguan section of the Yellow River from 1998 to 2010. Journal of geography. **(01)**: 44-55 (2014).
5. S Yang, H Yan, L Guo . Land use change and its eco-environmental effects in the northern agro pastoral ecotone -- a case study of Yulin City, Northern Shaanxi. Progress of geographical science. **(06)**: 49-55 (2004).
6. H Zhu, X Li, S He, et al. Analysis of temporal and spatial changes of land use around the Bohai Sea. Journal of geography. **(03)**: 253-260 (2001).
7. J Liu, W Kuang, et al. Basic characteristics and spatial pattern of land use change in China since the late 1980s. Journal of geography. **69 (01)**: 3-14 (2014).

E3S Web of Conferences **350**, 01026 (2022) https://doi.org/10.1051/e3sconf/202235001026
EREGCE 2022

Applicability Analysis of Different Runoff Schemes of the Gulang River Basin in Arid Region of Gansu Province, China

Xiaojing Wang[1], *Lianguo* Zhao[1,*], *Jingnan* Zhang[1], *Tianyuan* Zou[1] and *Nan* Zhang[2]

[1]PIESAT International Information Technology Limited, Beijing, 100195, China
[2]Tianjin Normal University, Tianjin, 300387, China

Abstract. [**Background**] Mountain torrent disasters are one of the deadliest weather-related natural disasters in the world, often causing extremely serious economic and property losses and casualties; in recent years, due to heavy rainfall and complex geological and geomorphological conditions, mountain torrent disasters have occurred frequently in Gansu Province. An effective rainfall-runoff model is the key to prevent them. There are still many problems about how to establish suitable hydrological models in the arid basin of China. [**Methods**] Therefore, the Gulang River Basin in Gansu Province was selected as the research area to construct the HEC-HMS rainfall-runoff model; through the screening of different runoff methods, different runoff schemes were established to explore the optimal runoff algorithm for the arid basin; there are mainly SCS-CN method, initial and constant method, Green and Ampt method and exponential loss method in runoff generation; SCS unit hydrograph and Clark unit hydrograph methods are selected for slope transform, kinematic wave and lag algorithms for channel routing, totally 16 different runoff schemes. Three floods in the Gulang River Basin are selected to analyze the simulation effect of different runoff schemes in Gulang River Basin. [**Results**]1) If the Nash-Sutcliffe efficiency coefficient is selected as the evaluation index, the best simulation results are schemes 1 and 16 in the 20180826 flood, schemes 2 and 1 in the 20190626 flood and schemes 1 and 3 in the 20190911 flood. If the percentage of flood peak error is selected as the evaluation index, the best simulation results are schemes 16 and 1 in the 20180826 flood, schemes 9 and 6 in the 20190626 flood and schemes 13 and 4 in the 20190911 flood.If the percentage of runoff depth error is selected as indicator, the best simulation results are schemes 1 and 8 in the 20180826 flood, schemes 7 and 8 in the 20190626 flood and schemes 6 and 13 in the 20190911 flood. 2)The mean value of Nash-Sutcliffe efficiency coefficient obtained by the SCS-CN method for runoff generation, the SCS unit hydrograph for slope transform and the lag algorithm for channel routing are 0.8,0.65 and 0.65,respectively; the mean absolute percentage errors of flood peak are 9.29%,9.71% and 8.47%, respectively; the mean absolute percentage errors of runoff depth are 6.07%,7.17% and 7.74%, respectively; the mean time difference of flood peak of SCS unit hydrograph for slope transform and lag algorithm for channel routing are 1.21 hour and 1.5 hour, respectively.[**Conclusion**]The most suitable scheme is the combination of the SCS-CN method for runoff generation, SCS unit hydrograph for slope transform, and lag algorithm for channel routing. The results can provide a certain reference for the prevention of flood disasters in the arid region of Gansu Province.

1 Introduction

In recent years, due to heavy rainfall, and complex geological and geomorphic conditions, Gansu Province has experienced frequent flash flood disasters; among them, a flash flood disaster, occurred in Zhouqu County on August 7, 2010, causing 1,841 deaths and disappearances. On July 9-11, 2018, Gansu Province encountered the heaviest precipitation, among which the largest accumulated rainfall in Southeast Gansu reached 250 mm, inducing severe mountain torrent disasters and causing huge economy and property losses. In August 2020, Wenxian County suffered continuous heavy rains and severe waterlogging occurred in many county towns; Jingyuan County, suffered local torrential rains and severe mountain flood disasters occurred in some townships. Obviously, effective flash flood disaster prevention technology is especially important for Gansu Province.

HEC-HMS is a semi-distributed hydrological model developed by the U.S. Army Engineers, which has been widely used to realize the simulation of rainfall-runoff processes through modular operation. For example, Wang Ruimin et al. constructed a hydrological model based on HEC-HMS for the Xingying River basin in southwest China, and performed parameter rate determination and model validation for multiple floods, and the results showed that the model simulated well. Tian Jing et al. developed a HEC-HMS model for the Guanshan River basin and analyzed the sensitivity parameters, and the results showed that the model simulated well. Ruting Liao et al. applied the HEC-HMS model to simulate multiple floods in Wenyu River basin,

* Corresponding author: ezhao_cn@163.com

https://doi.org/10.1051/e3sconf/202235001026

and the results showed that the model simulation was well applicable. In China, the HEC-HMS model is commonly used for flood forecasting in watersheds, in which there are more studies on the application to wet areas and less studies on the application to arid areas. Therefore, in this paper, based on the HEC-HMS model, the Gulang River basin in Gansu Province is selected as the study area to explore the applicability of different production and confluence schemes in arid regions of China.

2 Study area and data

The Gulang River basin is located in the east end of the Hexi Corridor, in the middle of Gansu Province, between 102°42'~103°11'E and 37°9'~37°28'N. The terrain is high in the south and low in the north, with a temperate continental arid climate. Figure 1 shows the distribution of the water system and 25 rainfall observation stations in the Gulang River basin, with a basin area of 1,010 km2 and an elevation range between 2,080 and 4,298 m.

The DEM data of the study area were obtained from the geospatial data cloud platform, and analyzed by ArcGIS, to extract watershed characteristics, and the results are shown in Fig. 1. The land use types are mainly grassland, cropland and forested land (Fig. 2), and the soil types are mainly clay loam, sandy loam and loam (Fig. 3).

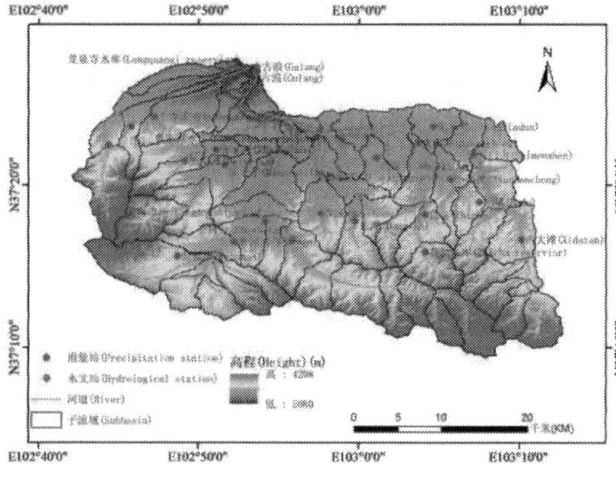

Fig. 1. Water system and station distribution map of Gulang River Basin

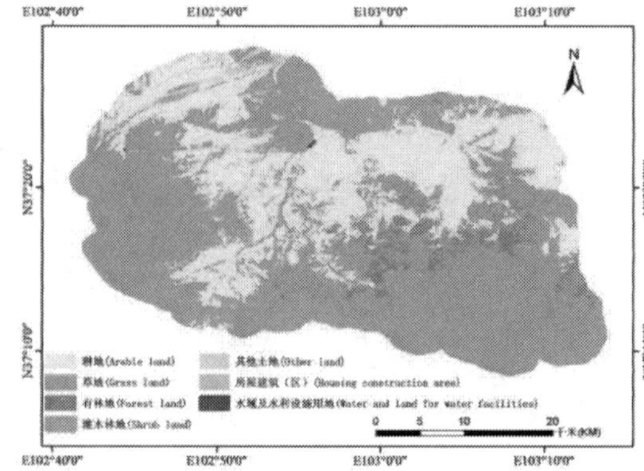

Fig. 2. Land use distribution map

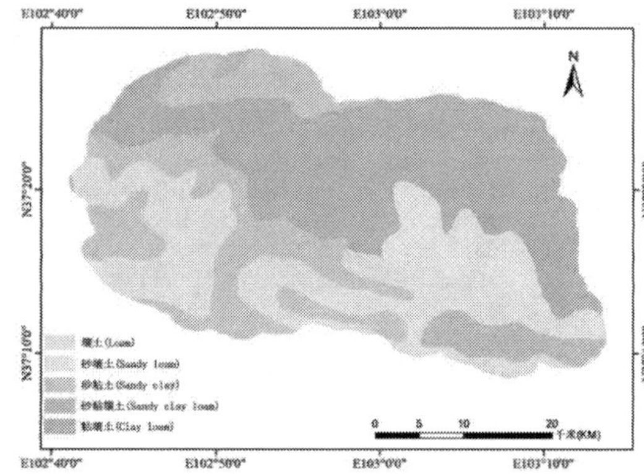

Fig. 3. Soil texture distribution map

3 Research Methodology

3.1 Production convergence program

The HEC-HMS model provides a variety of flow production and sink algorithms, so it can be combined into a variety of flow production and sink schemes by selecting different algorithms. In this paper, four flow-producing methods are selected, namely, exponential loss model, Green Amputation square model, initial loss followed by loss model and SCS-CN model; slope sink mainly selects Clark unit line and SCS unit line; river sink selects motion wave and hysteresis algorithm; base flow mainly adopts recession algorithm. According to the different methods selected for flow production, slope confluence and river confluence, a total of 16 sets of flow production and confluence schemes are established to analyze the applicability of different schemes in the Gulang watershed of Wuwei City, Gansu Province. For details, please see Table 1.

Table 1. Scheme table of the combination of different runoff methods

Schemes	Runoff generation	Slope transform	Channel routing	Base flow
Scheme 1	SCS-CN	SCS unit hydrograph	Lag algorithm	Recession
Scheme 2	SCS-CN	SCS unit hydrograph	Kinematic wave	Recession
Scheme 3	SCS-CN	Clark unit hydrograph	Kinematic wave	Recession
Scheme 4	SCS-CN	Clark unit hydrograph	Lag algorithm	Recession
Scheme 5	Initial and Constant	Clark unit hydrograph	Lag algorithm	Recession
Scheme 6	Initial and Constant	Clark unit hydrograph	Kinematic wave	Recession
Scheme 7	Initial and Constant	SCS unit hydrograph	Kinematic wave	Recession
Scheme 8	Initial and Constant	SCS unit hydrograph	Lag algorithm	Recession
Scheme 9	Green and Ampt	Clark unit hydrograph	Lag algorithm	Recession
Scheme 10	Green and Ampt	Clark unit hydrograph	Kinematic wave	Recession
Scheme 11	Green and Ampt	SCS unit hydrograph	Kinematic wave	Recession
Scheme 12	Green and Ampt	SCS unit hydrograph	Lag algorithm	Recession
Scheme 13	Exponential loss	Clark unit hydrograph	Lag algorithm	Recession
Scheme 14	Exponential loss	Clark unit hydrograph	Kinematic wave	Recession
Scheme 15	Exponential loss	SCS unit hydrograph	Kinematic wave	Recession
Scheme 16	Exponential loss	SCS unit hydrograph	Lag algorithm	Recession

In this paper, three floods occurred in the Gulang River basin, recorded as 20180823, 20190626, 20190911, with precipitation of 218 mm, 110 mm and 88 mm, and the basin outlet flood flows of 20.3 m3/s, 23.3 m3/s and 9.3 m3/s, with flood peaks of 3, 2 and 1, respectively; where rainfall data The Tyson polygon and weighting method were mainly used to calculate the average rainfall of sub-basins.

3.2 Error and accuracy evaluation

The simulation results are evaluated by three indexes, mainly the relative error of peak flow RQ, the Nash efficiency coefficient (N) and the peak-present time difference Δt; where the Nash efficiency coefficient reflects the degree of fit between the simulated runoff process and the actual measurement.

The evaluation indicators are calculated by the following formula.

$$R_Q = \frac{Q_s - Q_o}{Q_o} \times 100 \quad (1)$$

$$N = 1 - \frac{\sum\left[Q_S(i) - Q_o(i)\right]^2}{\sum\left[Q_S(i) - Q_{o,mean}\right]^2} \quad (2)$$

$$\Delta t = T - T' \quad (3)$$

Q_S , Q_o are the simulated and measured flow rate, m^3/s; $Q_{o,mean}$ is the average of measured flow rate, m^3/s; T, T' are the measured flood occurrence time and simulated flood occurrence time, h. Where the closer R_Q approaches to 0 means the better of fitting effect, the closer N approaches to 1 means the better of fitting effect, the closer Δt approaches to 0 means the simulated peak occurrence time domain is closer to the measured peak occurrence time.

4 Results and Analysis

Firstly, the initial values of the parameters of each scenario model are determined using the HEC-HMS model; then the measured flood data are used to rate and optimize the parameters of each scenario. According to Tables 2 and 3, it can be seen that if the larger Nash efficiency coefficient is selected as the evaluation index, the best results are scenario 1 and scenario 16 in the simulated flood of field 20180826; the best results are scenario 2 and scenario 1 in the simulated flood of field 20190626; the best results are scenario 1 and scenario 3 in the simulated flood of field 20190911. If the percent error of flood peak is selected as the evaluation index, the best results are scenario 16 and scenario 1 in the simulated flood of field 20180826; scenario 9 and scenario 6 in the simulated flood of field 20190626; and scenario 13 and scenario 4 in the simulated flood of field 20190911. If the percent error of runoff depth is selected as the evaluation index, the best results are scenario 13 and scenario 4 in the simulated 20180826 field floods, the best are scenario 1 and scenario 8; in the simulated 20190626 field floods, the best are scenario 7 and scenario 8; in the simulated 20190911 field floods, the best are scenario 6 and scenario 13.

According to the flood simulation results in Table 2 and Table 3, it can be seen that the average Nash efficiency coefficients for the three floods of SCS-CN, initial-loss-post-loss, Greenampot and exponential simulation for the flow production method selection are

0.8, 0.5, 0.56 and 0.68, respectively; the average absolute flood error percentages are 9.29%, 11.65%, 7.97% and 8.69%, respectively; the average absolute runoff depth error percentages 6.07%, 7.09%, 13.24% and 7.64%, respectively; therefore, the SCS-CN flow production method has the highest average Nash efficiency coefficient, while the average absolute flood error percentage flow production is the smallest using the Greene Amputation method and the average absolute runoff depth error percentage flow production is the smallest using the SCS-CN method.

The average Nash efficiency coefficients for the three floods were obtained for the slope confluence method by choosing SCS unit line and Clark unit line as 0.65 and 0.61; the average absolute flood peak error percentages were 9.71% and 9.09%; and the average absolute runoff depth error percentages were 7.17% and 9.85%, respectively. Therefore, the average Nash efficiency coefficients of the slope confluence method using SCS unit line and Clark unit line are not significantly different, while the average absolute flood peak error percentage is the smallest for Clark unit line and the average absolute runoff depth error percentage is the smallest for SCS unit line.

The average Nash efficiency coefficients of the river confluence methods selected are 0.62 and 0.65 for the kinematic wave and hysteresis algorithms, respectively; the average absolute flood peak error percentages are 10.33% and 8.47%, respectively; and the average absolute runoff depth error percentages are 9.28% and 7.74%, respectively. Therefore, the average Nash efficiency coefficients of the two methods of river confluence method kinematic wave and lagging algorithm are not very different, while the average absolute flood peak error percentage of lagging algorithm is the smallest and the average absolute runoff depth error percentage of lagging algorithm is the smallest.

By comparing the peak present time differences of the three floods simulated by different schemes, it can be seen that the average peak present time difference for the slope confluence using the SCS unit line method is 1.21 hours; the average peak present time difference obtained using the Clark unit line method is 1.18 hours; the average peak present time difference obtained using the river confluence using the motion wave method is 1.54 hours, and the average peak present time difference obtained using the lagging algorithm is 1.5 hours.

Table 2. Statistical table of flood simulation results for different floods

	Floods	Observation	Scheme 1	Scheme 2	Scheme 3	Scheme 4	Scheme 5	Scheme 6	Scheme 7	Scheme 8
Peak flow /($m^3 \cdot s^{-1}$)	20180826	20.30	20.60	21.50	18.80	18.10	19.50	21.80	23.10	22.60
	20190626	23.30	20.80	21.50	20.60	19.30	27.60	22.60	26.30	25.10
	20190911	9.30	10.20	10.60	8.50	8.70	10.70	11.70	10.50	10.00
Runoff depth /mm	20180826	53.35	53.74	53.98	46.61	46.43	47.99	50.68	53.77	53.65
	20190626	15.18	16.24	16.29	14.62	14.61	13.51	11.71	15.68	15.66
	20190911	13.98	14.67	14.73	13.06	13.04	13.77	13.79	12.20	12.19
Percentage of flood peak error/%	20180826		1.48	5.91	-7.39	-10.84	-3.94	7.39	13.79	11.33
	20190626		-10.73	-7.73	-11.59	-17.17	18.45	-3.00	12.88	7.73
	20190911		9.68	13.98	-8.60	-6.45	15.05	25.81	12.90	7.53
Percentage of runoff depth error/%	20180826		0.74	1.19	-12.62	-12.96	-10.04	-5.00	0.80	0.58
	20190626		6.93	7.27	-3.72	-3.77	-11.04	-22.87	3.27	3.16
	20190911		4.96	5.39	-6.55	-6.73	-1.47	-1.35	-12.73	-12.79
Time difference of flood peak /h	20180826		1	1	1	2	1	3	2	2
	20190626		1	0	2	2	1	2	2	2

	20190911	0	0	1	1	4	4	3	3
N	20180826	0.91	0.80	0.86	0.87	0.75	0.80	0.72	0.77
	20190626	0.87	0.93	0.64	0.54	0.72	0.58	0.35	0.32
	20190911	0.85	0.74	0.83	0.83	0.31	0.22	0.21	0.23

Table 3. Statistical table of flood simulation results for different floods

	Floods	Scheme 9	Scheme 10	Scheme 11	Scheme 12	Scheme 13	Scheme 14	Scheme 15	Scheme 16
Peak flow /($m^3 \cdot s^{-1}$)	20180826	21.5	19.3	19.1	19.9	21.1	22	20.8	20.5
	20190626	23.8	24.6	24.2	26.5	25.3	25.3	29.3	25.9
	20190911	10.2	10.3	10.9	10.6	9.3	10.6	10.2	10.3
Runoff depth /mm	20180826	55.66	50.47	50.19	52.19	54.25	49.62	49.69	49.49
	20190626	11.15	11.22	11.64	13.89	13.11	13.17	13.75	13.69
	20190911	11.77	11.84	12.16	12.24	13.74	12.93	14.94	14.91
Percentage of flood peak error/%	20180826	5.91	-4.93	-5.91	-1.97	3.94	8.37	2.46	0.99
	20190626	2.15	5.58	3.86	13.73	8.58	8.58	25.75	11.16
	20190911	9.68	10.75	17.20	13.98	0.00	13.98	9.68	10.75
Percentage of runoff depth error/%	20180826	4.33	-5.39	-5.92	-2.16	1.70	-6.97	-6.85	-7.23
	20190626	-26.59	-26.08	-23.32	-8.51	-13.63	-13.29	-9.46	-9.86
	20190911	-15.79	-15.30	-13.04	-12.42	-1.71	-7.47	6.85	6.67
Time difference of flood peak /h	20180826	1	2	2	2	3	2	1	1
	20190626	1	1	1	1	2	2	1	1
	20190911	1	1	0	1	1	3	0	1
N	20180826	0.58	0.68	0.67	0.71	0.80	0.82	0.83	0.88
	20190626	0.73	0.73	0.76	0.72	0.52	0.51	0.75	0.73
	20190911	0.27	0.25	0.17	0.41	0.63	0.28	0.69	0.70

the flood simulation of field 20180826, for the first flood peak, the simulation results of scenarios 1-4 are better than those of scenarios 5-16, and also the overall simulation results of this flood are the best for scenarios 1-4. For the second flood peak of the 20190626 field flood, scenarios 5, 6 and 9-16 failed to simulate the second flood peak of this flood; while for the 20190911 field flood, the simulation results of scenarios 1-4 are better than those of scenarios 5-16. In conclusion, in the Gulang River basin of Gansu Province, the SCS-CN method is the best choice for the flow production method, followed by the exponential loss method, the SCS unit line is better than the Clark unit line for the slope confluence, and the river confluence lag algorithm is better than the motion wave method.

5 Conclusion

In this paper, using the HEC-HMS model, 16 production and confluence flow schemes were established using different production and confluence methods to simulate three rainfall flood processes in the Gulang River basin. The results show that among the production flow methods, the SCS-CN method produces the best simulation results; among the slope confluence methods, the SCS unit line method produces better simulation results than the Clark unit line method; among the river confluence methods, the hysteresis algorithm produces better simulation results than the motion wave method. Therefore, for the simulation of rainfall flood process in the Gulang River basin, it is suggested that the SCS-CN method is chosen for the flow production method, the SCS unit line is chosen for the slope confluence, and the lag algorithm is chosen for the river confluence. Since the object of this study is flooding in the hilly area in the arid region of northwest China, which usually has low vegetation cover and floods are usually sudden and short-lived, and the three hydrological processes of interception, depression filling and evaporation are neglected in this study when building the hydrological model, the subsequent study will further consider the three hydrological processes of interception, depression filling and evaporation.

Acknowledgements

This work was supported by Deep learning-based research on the inducing factors and early warning of flash floods (42101086), and Research on Key Technologies of Flash Flood Prevention Based on Multi-machine Learning Model (2019KJ086).

References

1. B. Yong, W.C. Zhang, D.Z. Zhao et al. Application of hydrological modeling system HEC-HMS to Baohe catchment of Hanjiang Basin, Bulletin of Soil and Water Conservation, **2006(3)**, 86 (2006)
2. Z.Y. Liang, Y.W. Jia, K.J. Li, et al. Summary of application and study on distributional hydrologic model to flood forecasting, Yellow River, **29(2)**, 29 (2007)
3. R.M. Wang, H.X. Li, Q. Huang, et al. Flood forecasting of small-medium river in Southwest China using HEC-HMS model, Water Resources and Power, **39(1)**, 79 (2021)
4. J. Tian, J. Xia, Y.J. Zhang, et al. Application of HEC-HMS model in the Guanshan River Basin, Engineering Journal of Wuhan University, **54(1)**, 8 (2021)
5. R.T. Liao, S.S. Hu, L.G. Du, et al. Hydrological simulation of Wenyu River Basin based on HEC-HMS model[J].South-to-North Water Transfer and Water Science & Technology, **16(6)**,15 (2018)
6. Y. Li, X.T. Chen and C.X. Zhu, Application of HEC-HMS in flood forecasting, Yellow River, **30(4)**, 23 (2008)
7. Y.F. Kang, J.Z. Li and Q.S. Ma, Using HEC-HMS model to simulate flooding in Zijingguan watershed, Journal of Irrigation and Drainage, **38(4)**, 108 (2019)
8. L. Zhang, J. Li, X.R. Huang, et al. Application of HEC-HMS model for mountain flood forecasting in Qingxi River Basin, China Rural Water and Hydropower, **2020(1)**, 130 (2020)
9. Y. Zou, G.H. Hu, Z.X. Yu, et al. Application of HEC-HMS model for mountain flood forecasting in the Wushui Basin, Science of Soil and Water Conservation, **16(2)**, 95 (2018)
10. J.F. Li, W.C. Wang, P.P. Che, et al. Application research on HEC-HMS model and TOPMODEL model in mountain flood forecasting of Dongzhuang watershed, Water Resources and Power, **37(3)**, 50 (2019)
11. G.A. Tang, The experiment course of ArcGIS geographic information system spatial analysis, Beijing: The Science Publishing Company, (2006)
12. Z.G. Ma and L.L. Li, Hydrology information extracting of Yang River watershed based on GIS and DEM[J]. Journal of Hebei North University (Natural Science Edition), **24(1)**, 69 (2008)
13. G.M. Qin and R. Dai, Acquisition of basin characteristics based on GIS and DEM-A case study of Jitaiqu watershed, Northwest Hydropower, 2010 (13):4 (2010)
14. R. Dai and Z.F. Wang Application of GIS Technology in Calculation of Design Runoff of Basins, Northwest Hydropower, **2012(3)**, 1 (2012)
15. L.M. Zhang, W.H. Liao, Z.K. Yin, et al. Application of the HEC-HMS model for calculating the local inflow of giant reservoir group in Xijiang River Basin, China Rural Water and Hydropower, **2017(6)**, 65, (2017)
16. M. Saidi, A. Shalamu, J.L. Ding, et al. Application of HEC-HMS in data-insufficient mountainous watersheds-A case study from Urumqi River Basin, Bulletin of Soil and Water Conservation, **35(6)**, 140 (2015)
17. J.T. Wang, Simulation and research of mountain flood early warning in typical small watershed in North China based on HEC-HMS model, University of Jinan, (2018)
18. A.D. Feldman, Hydrologic Modeling System HEC-HMS, Technical Reference Manual. Technical Reference Manual, (2000)

The Ways to Improve the Ecological Benefits of Carbon Sequestration of Garden Green Space

Zhixuan Chu[2], *Yanhuan* Guan[2], *Jinyi* Ma[2] and *Yunjue* Yuan[1,2,*]

[1]Dongguan Polytechnic, Department of Architecture, 523808, NO.3, Daxue Road, Dongguan, China
[2]Nanjing Agricultural University, Department of Landscape Architecture, 210095, No. 1, Weigang, Nanjing, China

Abstract. Carbon dioxide is one of the greenhouse gases responsible for global warming, and how to reduce the amount of carbon dioxide in the air is a major issue. Scholars have found that herbaceous flowers can fix 12.16 $g{\bullet}m^{-2}{\bullet}\ D^{-1}$ carbon dioxide per unit leaf area in green space, which has a good ecological benefit of carbon sequestration. Therefore, how to further improve the ecological benefits of carbon sequestration has become a concern of landscape architecture industry. This paper summarizes seven kinds of garden green land carbon ecological concrete approaches to improve, including selecting high carbon sequestration ability of plant species, increase high carbon sequestration ability of plant population, rich garden plant hierarchy, increase use native plants, increasing the vertical greening, choose low maintenance plants landscape, choose the appropriate maintenance management way, In order to deal with global warming phenomenon on the full play of green landscape fixed carbon dioxide capacity.

1 Introduction

Under the background of global warming, climate change has become a current hot topic[1]. The rapid urbanization process has brought some pressure to the reduction of greenhouse gas emissions, coupled with China's increasingly important position in the global climate governance, and the gradual deepening of the understanding of global climate change, it is urgent to actively explore various ways to deal with climate change[2-3].

Plants can absorb carbon dioxide and release oxygen through photosynthesis, which has the function of carbon fixation[4-5]. By improving the ecological benefits of carbon sequestration of garden green space plants and relying on vegetation to increase carbon sink, the greenhouse gas (carbon dioxide) content in the air can be effectively reduced. This article summarizes how to through the selection of plant species, plant population, plant community hierarchy, plant landscape design form and concept as well as the maintenance management method to improve the botanical garden green land vegetation to absorb carbon dioxide in the air (Fig. 1), in order to reduce the greenhouse gas content, and response to the global ecological crisis of climate warming.

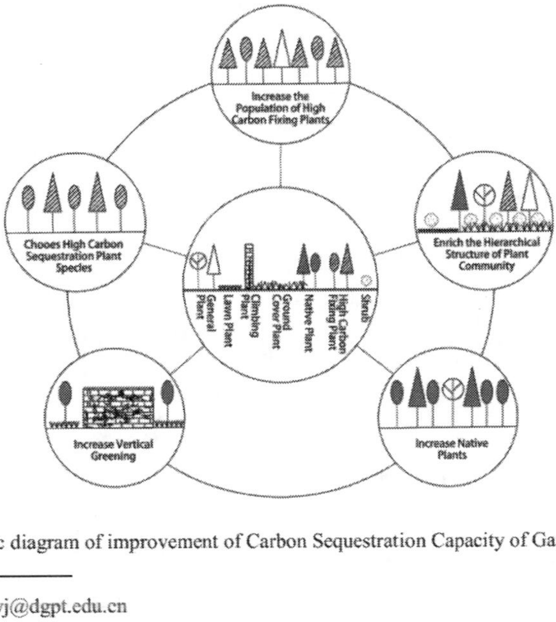

Fig. 1. Schematic diagram of improvement of Carbon Sequestration Capacity of Garden Green Space.

* Corresponding author: yuanyj@dgpt.edu.cn

2 Choose high carbon sequestration plant species

The ecological plant landscape of garden green space emphasizes ecology and diversity, and species, age and density of plant play a decisive role in their annual carbon sequestration capacity[6-7]. During the planning of urban green space system, the species of key trees and backbone trees are the main vegetation of garden green space, which not only reflect the vegetation characteristics of the city, but also can significantly improve the overall carbon sequestration benefit of urban garden green space if select the tree species with high carbon sequestration ability as the key tree and backbone tree species. Plants with different life forms have different carbon sequestration capacities. The research found that the order of daily carbon sequestration per unit leaf area from large to small was herbaceous flowers > deciduous shrubs > deciduous trees > evergreen shrubs > evergreen trees > lianas, and the order of daily carbon sequestration per unit cover area was herbaceous flowers > deciduous trees > evergreen shrubs > deciduous shrubs > evergreen trees > lianas (Table 1) [8]. Therefore, in addition to selecting trees with high carbon sequestration capacity for key trees and backbone trees, appropriately increasing the planting area of herbaceous flowers with low cost and strong adaptability can not only improve the value of the landscape, but also improve the ecological benefits of carbon sequestration of green space. For example, *Liriope spicata*, *Trifolium repens*, *Hemerocallis middendorffii* and *Iris tectorum* are herbacean flowers with high economizing index, which can maximize the ecological benefits of carbon sequestration under the premise of cost saving[9].

Table 1. Different daily carbon sequestration of different life forms [8].

Life form	Number of species	Average daily carbon sequestration per unit leaf area ($g \cdot m^{-2} \cdot d^{-1}$)	Average daily carbon sequestration per unit cover area ($g \cdot m^{-2} \cdot d^{-1}$)
Herbaceous flowers	81	12.16	83.12
Deciduous shrubs	56	10.05	36.24
Deciduous trees	118	9.75	53.84
Evergreen shrubs	80	7.99	37.10
Evergreen trees	77	7.81	31.72
Lianas	33	3.70	19.48

3 Increase the number of plants with high carbon sequestration capacity

Increasing the number of plant populations with high carbon sequestration ability and expanding the coverage area of plant populations with high carbon sequestration ability are helpful to increase the ecological benefits of carbon sequestration of garden green space. Key trees and backbone trees are the main vegetation landscape in urban garden green space, especially key trees, which have few species, large population and wide distribution range. Selecting tree species with high carbon sequestration ability will undoubtedly greatly improve the carbon sequestration ability of garden green space [8]. Forest landscape is a popular landscape at present, as well as an important part of "urban forest". It has a large population of dominant tree species, most of which are tall trees, with relatively strong carbon sequestration capacity and a long carbon fixation period[10-11]. If deliberately select trees with higher carbon sequestration ability, such as *Metasequoia glyptostroboides*, *Taxodium distichum* and *Glyptostrobus pensilis*, the per unit area of biomass and carbon sink are large, so they are used in many places [8-12].

4 Enrich the hierarchy of garden plant community

Create abundant hierarchy of garden plant community and improve the carbon sequestration ability of garden green space by increasing the ecological benefits of carbon sequestration of garden green scape. Related studies have shown that the ecological benefit of green space with multi-layer planting structure of tree, shrub and herb is the highest, which can significantly improve the carbon sequestration capacity of green space, followed by shrub and herb, and the single herb is the smallest [13-15]. For example, near natural plant community configuration, by protecting and induced natural more planted, to tree, shrub and herb cladding mode to enhance the complexity of the plant community, increase leaf area per unit area flora composite index, promote the plant photosynthesis, gradually to close to the natural state of community growth and development, so as to further enhance the comprehensive ability of carbon sequestration garden greenbelt [16]. It is the carbon density which is the carbon storage per unit area of the stand that the higher value, the stronger the carbon sequestration capacity per unit area of plant communities. Some scholars measured the carbon density of artificial plant communities in Gaojia manor in Shanghai and found that carbon density varies greatly between plant communities with different hierarchies [17]. The carbon density of evergreen deciduous broad-leaved mixed forest was 54.4 t/hm2, *Metasequoia glyptostroboides* forest was 31.4 t/hm2, bamboo forest was 13.2 t/hm2 and fruit-bearing forest was 9.2 t/hm2 [17] (Fig. 2). Carbon density is proportional to carbon sequestration capacity, so on the basis of rational use of various plants with high carbon sequestration ability, more tree-dominated composite vegetation cover structures are conducive to improving the ecological benefits of carbon sequestration of urban green space [18]. In addition, in different hierarchical structures, the collocation of fast-growing and slow-growing trees and evergreen and deciduous trees can also improve the ecological benefits of carbon sequestration of plant communities [19].

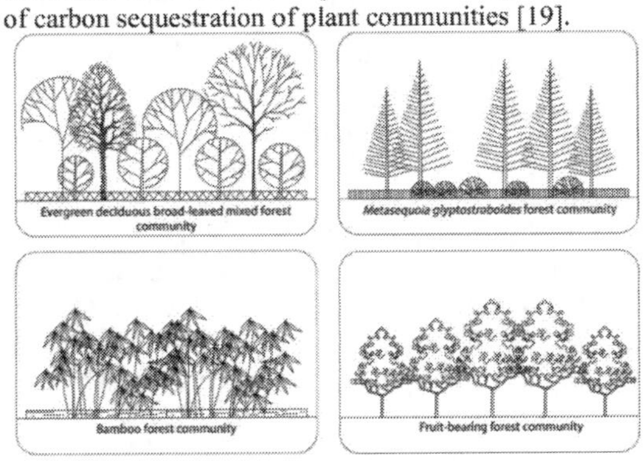

Fig. 2. Schematic diagram of plant community structures with different carbon densities.

5 Increase the use of native plants

Native plants have strong adaptability, are easy to survive, and are convenient for planting and maintenance, which is not only conducive to improving the environment, but also can reflect local cultural characteristics [20-21]. The construction of garden green space with native plants as the main body has gradually become one of the development directions in the industry [22]. In the study of the ecological benefits of carbon sequestration in green scape, relevant scholars concluded that native plants generally have high carbon sequestration ability through observation and analysis of portable photosynthetic apparatus [23-24]. Appropriately increasing the proportion and number of local dominant tree species in the planting planning and design of garden green space, and building a stable community structure with local plants as dominant species, not only has a strong ecological benefits of carbon sequestration, but also can improve the ecological service ability of green space and reduce carbon emissions [25].

6 Increase vertical greening

Vertical greening, as a supplementary form of green space, can effectively alleviate the problem of insufficient urban green space and increase the covered area of urban green space [26-27] (Fig. 3). Although the area of vertical greening is not large, the leaf area index of vertical greening plants is high and the carbon sequestration per unit area is considerable. Based on the improvement of carbon sequestration ability, different scholars have conducted in-depth studies on the plant species available for vertical greening, and found that some lianas used in vertical greening have higher leaf area index and higher ecological benefits of carbon sequestration, such as *Thunbergia laurifolia* (the leaf area index is 5.80 and the carbon sequestration per unit leaf area is 15.53 g•m-2•d-1), *Ficus pumila* (the leaf area index is 4.33 and the carbon sequestration per unit leaf area is 10.33 g•m-2•d-1), *Parthenocissus tricuspidat* (the leaf area index is 5.33 and the carbon sequestration per unit leaf area is 6.91 g•m-2•d-1)*a*, etc., which proves that vertical greening has good carbon sequestration potential [28-29]. There are also studies to establish a vertical greenery ecosystem carbon dioxide sequestration calculation model. Using *Zoysia Matrella*, *Salvia Nemorosa*, *Geranium Sanguineum* and other plant materials, the average annual carbon dioxide uptake of plant biomass was estimated to be 0.44-3.18 kg CO_2eq m^{-2}, demonstrating the advantages of vertical wall carbon fixation [30]. Therefore, increasing vertical greening becomes one of the ways to improve the ecological benefits of carbon sequestration of green space.

Fig. 3. Different forms of vertical greening.

7 Choose low-maintenance plant landscape

Urban green space is mostly constructed and managed manually, which requires a certain amount of maintenance input. Therefore, choosing low-maintenance or even self-maintenance plants can effectively reduce maintenance costs and reduce carbon dioxide emissions during maintenance, thereby indirectly improving the ecological benefits of carbon sequestration of green space [31]. Through fuzzy comprehensive evaluation and analysis, some scholars found that the herbaceous plants are arranged from high to low according to the conservation index: *Liriope spicata* > *Trifolium repens* > *Hemerocallis middendorffii* > *Iris tectorum* > *Poa pratensis*, indicating that the maintenance cost of *Poa pratensis* is too high and should not be used in a large area of green space from the perspective of reducing carbon dioxide [9]. In addition, some scholars said that the mosaic flower bed and other landscapes that need a lot of manual maintenance and management, especially the artificially cultivated annual and biennial herb, need to be replaced frequently. This kind of too exquisite plant landscape, even if the herb varieties have high ecological benefits of carbon sequestration, the high maintenance cost and high carbon emission will greatly offset the ecological benefits of carbon sequestration of green space. Therefore, it is not reasonable to consider from the perspective of increasing the ecological benefits of carbon sequestration [10].

8 Choose the appropriate cultivation, maintenance and management approach

In the maintenance and management stage in the later stage of garden green space planting, water and fertilizer management is the key point. In the soil environment with high water content, root respiration, litter and microbial decomposition will be inhibited, thus reducing the carbon emission of green space soil, which has an indirect effect on the improvement of the ecological benefits of carbon sequestration of garden green space [32]. Therefore, water-saving irrigation should be adopted to appropriately keep the soil moist, actively prevent diseases and insect pests in plants and soil, reduce the use of chemical fertilizers, recycle biological waste such as fallen leaves [33-34], and achieve low-carbon management to promote the building of high ecological benefits of carbon sequestration of garden green space. Pay regular attention to the growth dynamics of plants. According to different planting methods, reasonable pruning management is conducive to increasing the carbon sink efficiency. The artificial maintenance amount of natural plant landscape is generally lower than that of regular planting, so the carbon emission is relatively low and the ecological benefits of carbon sequestration is more lasting [35].

Due to the botanical garden green space carbon sequestration ability to corresponding contribution to slow global warming, so summarize the ways and methods to improve the ability of carbon sequestration, can design low-carbon landscape for the future provide a more scientific guidance, also let the landscape architecture profession in response to the global ecological crisis brought greenhouse gases play a proper role. In addition, although various approaches and methods to improve the ecological benefits of carbon sequestration have their own characteristics and functions, they are not isolated from each other in practice. Therefore, they should be rationally combined and comprehensively applied according to the specific situation of specific green space in order to improve the ecological benefits of carbon sequestration in green space more effectively. In the future, with the in-depth development of carbon sequestration research in the field of landscape architecture, new ways and methods will be explored to improve the ecological benefits of carbon sequestration in green space, so as to further develop and complete the research system of carbon sequestration benefit in green space, and contribute to improving the global ecological environment.

References

1. G.Chen, L. Chen, D. Pang, J. Ma, H. Wan, J. Wang, J. Li and X. Li, Study on carbon fixation and oxygen release capacity of 10 typical plants in Helan Mountain. Journal of Soil and Water Conservation, **35(03)**, 206-213+220 (2021) (in Chinese)
2. A. Lee, H. Jordan and J. Horsley, Value of urban green spaces in promoting healthy living and wellbeing: prospects for planning. Risk Management and Healthcare Policy, **8**, 131-137 (2015)
3. G. Zhuang, F. Bo and J. Zhang, China's role and strategic choice in global climate governance. World Economics and Politics, **2018(04)**, 4-27+155-156 (2018) (in Chinese)
4. Y. Tang, T. Shi, Y. Bu and Y. Shi, Carbon storage estimation and spatial distribution characteristics of urban green space. Chinese Journal of Ecology, **39(04)**, 1387-1398 (2020) (in Chinese)
5. W. Chen, The role of urban green infrastructure in offsetting carbon emissions in 35 major Chinese cities: A nationwide estimate. Cities, **44**, 112-120 (2015)
6. M. Wang and Q. Shi, Study on influencing factors and optimization of urban green carbon sink efficiency. Journal of Chinese Urban Forestry, **13(04)**, 1-5 (2015) (in Chinese)
7. J. Kil and H Park, Changes in Growth Rate and Carbon Sequestration by Age of Landscape Trees. Journal of the Korean Institute of Landscape Architecture, **45(5)**, 97-104 (2017)
8. Q. Gao, J. Ma, H. Shao and H. Chen, Statistical analysis of carbon sequestration ability of garden plants with different life forms. Journal of Jiangsu Forestry Science & Technology, **47(02)**, 44-47 (2020) (in Chinese)
9. F. Xia, K. Ji and J. Wang, Establishment of optimal selection system for low maintenance garden plants. Chinese Landscape Architecture, **28(06)**, 85-89 (2012) (in Chinese)
10. D. McKinley, M. Ryan, R. Birdsey, C. Giardina, R. Houghton, R. Jackson, J. Morrison, B. Murray, D. Pataki and K. Skog, A synthesis of current knowledge on forests and carbon storage in the United States. Ecological Applications, **21(6)**, 1902-1924 (2011)
11. Q. Zhang, The Carbon Sink Estimation and Carbon Sink Increase Potential of Jiangmen, China. 4th International Conference on Manufacturing Science and Engineering (ICMSE 2013), **712-715**, 3049-3058 (2013)
12. X. Liu, M. Liang, L. Chen, S. Wang, W. Zheng, X. Yu, R. Li, G. Zhang, F. Wang and H. Yang, Carbon storage, carbon density and spatial distribution of forest ecosystem in Hunan Province. Chinese Journal of Ecology, **36(09)**, 2385-2393 (2017) (in Chinese)
13. G. Yi, L. Zhuo, C. Wang and Y. Zhang, Analysis of community structure and ecological function of garden plants. Journal of Shanghai Jiaotong University (Agricultural Science), **27(03)**, 248-252 (2009) (in Chinese)
14. F. Zhao, S. Chen, X. Han, G. Yang, Y. Feng and G. Ren, Policy-Guided Nationwide Ecological Recovery: Soil Carbon Sequestration Changes Associated with the Grain-to-Green Program in China. Soil Science, **178(10)**, 550-555 (2013)

15. J. Zhang and Y. Sui, Selection and Design of Ornamental Plants for Low-carbon Urban Landscape: A Review. 2nd International Conference on Civil Engineering, Architecture and Building Materials (CEABM2012), **174-177**, 2314-2317 (2012)
16. M. Wang and Q. Shi, Evaluation index system and empirical study of green carbon sink efficiency in urban high-density area: a case study of Huangpu District, Shanghai. Chinese Landscape Architecture, **32(08)**, 18-24 (2016) (in Chinese)
17. J. Gao, G. Zhang, W. Sun and Q. Gong, Present situation analysis and Optimization Countermeasures of artificial plant community in Gaojia manor in Shanghai. Journal of Shanghai Jiaotong University (Agricultural Science), **32(05)**, 20-24+32 (2014) (in Chinese)
18. T. Shi, X. Pan, C. Gao and J. Sun, Study on oxygen release capacity of urban green space. Journal of Shenyang Jianzhu University (Natural Science), **29(02)**, 349-354 (2013) (in Chinese)
19. H. Xing and K. Qin, Analysis of low-carbon plant landscape construction. Northern Horticulture, **2013(05)**, 81-84 (2013) (in Chinese)
20. D. She and X. Hu, Landscape design based on low carbon concept -- Taking Changsha Jiwang garden as an example. Guangdong Agricultural Sciences, **38(06)**, 60-63 (2011) (in Chinese)
21. M. Irmak, Use of native woody plants in urban landscapes. Journal of Food Agriculture & Environment, **11(2)**, 1305-1309 (2015)
22. S. Li, GARDEN PLANTING DESIGN: Theory. China Agriculture Press, Beijing (2010) (in Chinese)
23. H. Guo, Analysis of carbon fixation and oxygen release capacity of common plants in park green space of Zhengzhou city, Henan Province. Jiangsu Agricultural Sciences, **45(15)**, 103-106 (2017) (in Chinese)
24. H. Xue, H. Tang, Y. Li, X. Li, J. Guo and J. Ma, Study on ecological regulation services of common green plants in Beijing. Journal of Beijing Normal University (Natural Science), **54(04)**, 517-524 (2018) (in Chinese)
25. Z. Ge, D. Zhou, Y. Hao, P. Chen and Q. Guan, Analysis of carbon storage status of tree layer in Xuzhou urban green space. Journal of Forestry Engineering, **27(06)**, 30-34 (2013) (in Chinese)
26. A. Medl, R. Stangl and F. Florineth, Vertical greening systems-A review on recent technologies and research advancement. Building and Environment, **125**, 227-239 (2017)
27. S. Zaid, E. Perisamy, H. Hussein, N. Myeda and N. Zainon, Vertical Greenery System in urban tropical climate and its carbon sequestration potential: A review. Ecological Indicators, **91**, 57-70 (2018)
28. G. Li, S. Ding and X. Zhou, Ecological effects of twelve vertical greening plants in South China. Journal of South China Agricultural University, **2008(02)**, 11-15 (2008) (in Chinese)
29. G. Liu and Q. Chen, Ecological effects of four vertical greening plants in Chengdu. Journal of China West Normal University (Natural Sciences), **2004(03)**, 259-262 (2004) (in Chinese)
30. M. Marchi, R. Pulselli, N. Marchettini, F. Pulselli and S. Bastianoni, Carbon dioxide sequestration model of a vertical greenery system. Ecological Modelling, **306**, 46-56 (2015)
31. L. Dong and X. Wang, Low intervention, low consumption, low maintenance, low emission -- Research on design strategy of low cost landscape Architecture. Chinese Landscape Architecture, **29(05)**, 61-65 (2013) (in Chinese)
32. B. Wang, C. Mou and B. Wang, Carbon storage of coniferous bog wetland ecosystem in Changbai Mountain. Acta Ecologica Sinica, **39(9)**, 3344-3354 (2019) (in Chinese)
33. Y. Ji, J. Luo, T. Wang and X. Liang, Carbon source and carbon sink quantification of landscape life cycle based on low-carbon concept: A case study of Shilinyuan Residential area in Tianjin. Chinese Landscape Architecture, **36(08)**, 68-72 (2020) (in Chinese)
34. N. Li, M. Chen, X. Gao, X. Long, H. Shao, Z. Liu and R. Zed, Carbon sequestration and Jerusalem artichoke biomass under nitrogen applications in coastal saline zone in the northern region of Jiangsu, China. Science of the Total Environment, **568**, 885-890 (2016)
35. H. Zhou, Z. Sun, F. Gao and S. Li Green landscape design of expressway based on carbon sink theory: A case study of Yuqing - Kali Expressway in Guizhou province. Journal of China & Foreign Highway, **35(03)**, 20-23 (2015) (in Chinese)

Decomposition of Maize Straw between Two Phosphate Solubilizing Fungi: *Aspergillus Niger* and *Penicillium Chrysogenum*

Liyan Wang[1,2], *Jun* Hu[1,2], *Hao* Guan[1,2], *Da* Tian[1,2,*] and *Hongjian* Gao[1,2]

[1]Anhui Province Key Laboratory of Farmland Ecological Conservation and Pollution Prevention, Key Laboratory of JiangHuai Arable Land Resources Protection and Eco-restoration, College of Resources and Environment, Anhui Agricultural University, Hefei 230036, China

[2]Research Centre of Phosphorus Efficient Utilization and Water Environment Protection along the Yangtze River Economic Belt, Anhui Agricultural University, Hefei 230036, China

Abstract. *Aspergillus niger* (*A. niger*) and *Penicillium chrysogenum* (*P. chrysogenum*) can significantly promote the degradation of maize straw and phosphorus release. Compared with *P. chrysogenum*, *A. niger* is more efficient in maize straw degradation and phosphorus releasing. After seven days of incubation, the highest degradation ratio and phosphorus content in *A. niger*+maize straw treatment is 25.8% and 2.3 mg/L, respectively. The mechanisms for maize straw decomposition between these two fungi are different. Oxalic acid is the primary organic acid secreted by *A. niger*, which is more function in the decomposition of maize straw compared with propionic acid secreted by *P. chrysogenum*. In addition, *A. niger* has higher acidic xylanase and lignin peroxidase enzymes activities, which is conducive to the degradation of more stable substances in maize straw, i.e., lignin. This study indicated that *A. niger* is the primary candidate for the reuse of crop straw in the way of return to the field.

1 Introduction

Crop straw (CS) is a common residual material produced by the agricultural industry. The production of CS is rapidly increased during the development of agriculture, especially in great agricultural countries. Due to CS contains more than half of the photosynthesis products by crops, it stores large amounts of cellulose and lignin. In addition, CS is also rich in nitrogen (N), phosphorus (P), potassium (K), and other nutrients, which has been recognized as a renewable biological resource with multiple uses [1].

Straw return to fields (SRF) has become an important mode of agricultural resource reuse. In China, more than 5% of CS was treated by the way of SRF every year [2]. This mode not only increases soil fertility (organic matter and NPK) but also improve soil structure and microbial activity [3]. However, the SRF mode is usually limited in the agricultural production process due to the low degradation of CS. The cellulose, lignin and hemicellulose in CS are stable and recalcitrant to degradation, especially for lignin. The degradation ratio of lignin is approximately 30% after one year in nature, which lead to soil pathogens and aggravation of crop diseases [4]. Therefore, the accelerating degradation of straw is an important factor for improving straw utilization and sustainable agricultural development.

Physical, chemical, and biological treatments have been conducted to improve straw digestibility in various researches. Compared with physical and chemical methods, the biological method is environmentally friendly and potentially economically viable alternatives, especially via microbial degradation (e.g., fungi) [5]. On the one hand, the fungal mycelium can destroy the CS cell wall structure, grown inside the parenchyma cells and fibroblasts. On the other hand, the mycelium utilized intracellular substances to secrete the cellulose-degrading enzymes and lignin-degrading enzyme, decomposed cell wall and fibrous tissue, then the cell wall is separated from other substances [6]. In addition, straw incorporation with the fungal is beneficial to the release of available nutrients of N, P and K, hence increasing plant height and yield [7].

The phosphate solubilizing fungi (PSF) *Penicillium* and *Aspergillus* are the common fungi in the soil, which function in the solubilization of insoluble phosphates via secretion of large amounts of organic acids [8]. More importantly, these two fungi can also promote straw degradation [6]. However, the CS decomposition mechanism and P release ratio between these two fungi is scant. Therefore, both *Aspergillus niger* (*A. niger*) and *Penicillium chrysogenum* (*P. chrysogenum*) were selected for the maize straw degradation and its P release ratio in this research. The content of P was determined by an inductively coupled plasma emission spectrometer (ICP-OES). The organic acids secreted by *A. niger* and *P. chrysogenum* were analyzed by high-performance liquid chromatography (HPLC). The enzyme activity of cellulase (CL), acidic xylanase (ACX) and lignin peroxidase (Lip) were determined. The morphology of maize straw and fungi was analyzed by X-ray diffraction (XRD) and scanning electron microscopy (SEM).

[*] Corresponding author: tianda@ahau.edu.cn

2 Materials and Methods

2.1 Preparation of maize straw and fungal strains

The stalk of fresh maize plants was collected and heated for 30 min at 105 °C and then dried at 65 °C to a constant weight. Subsequently, dried and cleaned maize straw was cut into 2 cm sections for use.

A. niger and *P. chrysogenum* were isolated from maize rhizosphere soil in Suzhou city, Anhui province, China, respectively. After incubated for five days at 28 °C, the spores of *A. niger* and *P. chrysogenum* were collected by sterile water from potato dextrose agar. The suspension was then filtered through three layers of sterile cheesecloth to eliminate mycelial fragments. Spore concentration was calculated and adjusted to 10^7 CFU mL^{-1} by using a hemocytometer [9].

2.2 Fungal incubation with maize straw

The experiment was performed with five treatments, i.e., maize straw (Mst), *A. niger* (ANG), *P. chrysogenum* (PCH), *A. niger*+straw (ANG+Mst) and *P. chrysogenum* +straw (PCH+Mst). Before the incubation, the PVK medium was sterilized at 121 °C for 20 min. Then, 0.2 g straw powder was added to 150 mL conical flasks with 50 mL PVK culture medium, respectively[10]. Subsequently, 1 mL spores of *A. niger* and *P. chrysogenum* were added in the respective treatments. All operations were performed under sterile conditions. The flasks were incubated for seven days at 28 °C, 180 rpm. After seven days of incubation, the PVK medium was filtered through a 0.22 μm polyethersulfone membrane. The straw and microbial hyphae were collected separately. The filtrate was collected for the test of organic acids, soluble P and degrading enzyme activities. The hyphae were frozen dried to determine biomass. The straw was repeatedly rinsed with sterile water, then was dried to degradation rate and to analyse XRD and SEM.

In parallel, a experiment was pursued to detect the P release and degradation of different organic acids on the straw. The experiment was performed with three treatments i.e., maize straw (Mst), oxalic acid+straw (OA+Mst) and propionic acid+straw (PA+Mst). The concentration of OA and PA in each treatment was 2000 mg/L. The culture conditions and subsequent processing were analyzed as described above.

2.3 Degradation rate and degrading enzymes

The straw degradation rate was determined by oven-dried at 65 °C to constant weight, the formula of the degradation rate is as follows:

$$DR(\%) = \frac{Wo-Wt}{Wo} \times 100\% \quad (1)$$

Wo was the original dry weight of the maize straw and Wt was the dry weight of the straw after incubated for seven days.

The filtrate was collected to analyze CL, ACX and Lip activities via CL, ACX and Lip activity assay kits, respectively (Comin Biotechnology Co., Ltd, Suzhou, China). All the samples were obtained and determined under relevant guidelines and regulations.

2.4 Instrumentation

The pH values were determined using a pH meter (FE28, Mettler Toledo, Shanghai, China). The concentration of the soluble P was analyzed by using ICP-OES (PerkinElmer Avio 200). A calibration curve (0.1, 0.5, 1, 2 and 5 mg/L) was prepared by using the P standard. The R square value of the external standard curve is 0.9999. The contents of organic acids in each treatment were analyzed by HPLC (Agilent 1260). Mineralogical characterization of the precipitates was examined by XRD by using D/Max-2500 X-ray diffraction. The surface morphology of the samples was studied under SEM (S4800 Hitachi) with an acceleration voltage of 10 kV. Detailed methods refer to the previous studies [10].

3 Results & Discussion

3.1 Fungal biomass and P concentration

The fungal dry biomass in Mst, PCH and ANG treatments were 20.3, 11.5 and 10.8 mg on 7th day of incubation, respectively. However, the dry biomass in PCH+Mst and ANG+Mst treatments significantly increased to 299.9 and 418.2 mg after seven days of incubation (Fig. 1). The addition of Mst can significantly promote the growth of fungi. Compared with *P. chrysogenum*, *A. niger* was more efficient in the utilization of Mst.

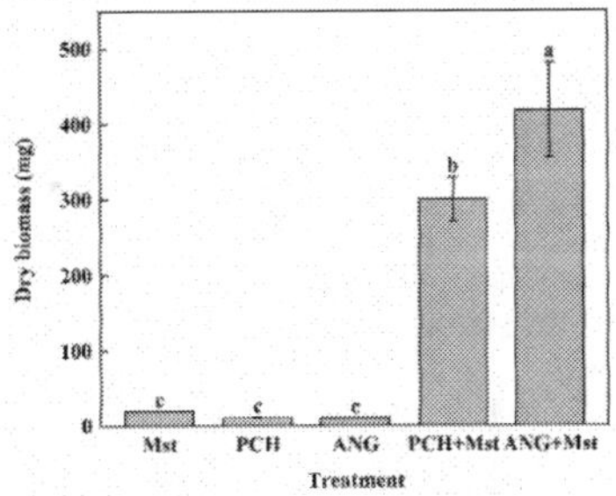

Fig. 1. The concentrations of dry biomass in each treatment. The error bars represent the standard deviations of three replicates.

The soluble P content in Mst, PCH and ANG treatments were 0.63, 0.5 and 0.6 mg/L after seven days of incubation, respectively (Fig. 2). In PCH+Mst and ANG+Mst treatments, the soluble P content was 2.0 and 2.3 mg/L, respectively (Fig. 2). Although both *P. chrysogenum* and *A. niger* promoted the release of P from Mst, *A. niger* showed a higher P solubilizing capacity.

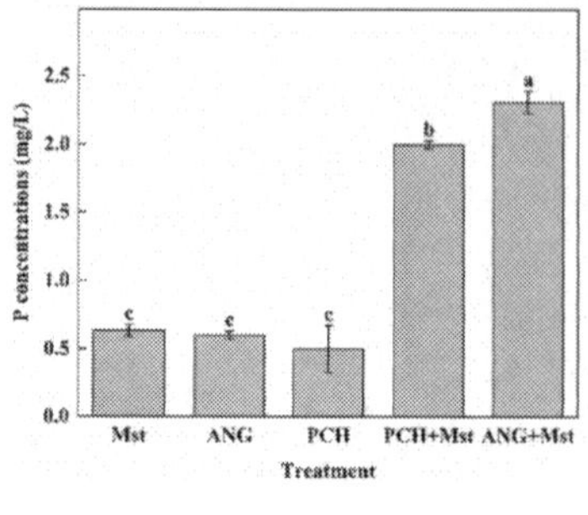

Fig. 2. The concentrations of soluble P in each treatment after seven days of incubation. The error bars represent the standard deviations of three replicates.

3.2 Degradation rate and degrading enzymes

The straw degradation rate in Mst and PCH+Mst treatments was 14.66% and 17.2%, respectively (Fig. 3). In ANG+Mst treatment, the straw degradation rate was 25.86%, significantly higher than other treatments (Fig. 3). Due to P was involved in the formation of the plant cell wall, hence the release of P was also accompanied by the destruction of crop straw [11]. The existence of *A. niger* and *P. chrysogenum* improved the degradation of straw and promoted the release of P.

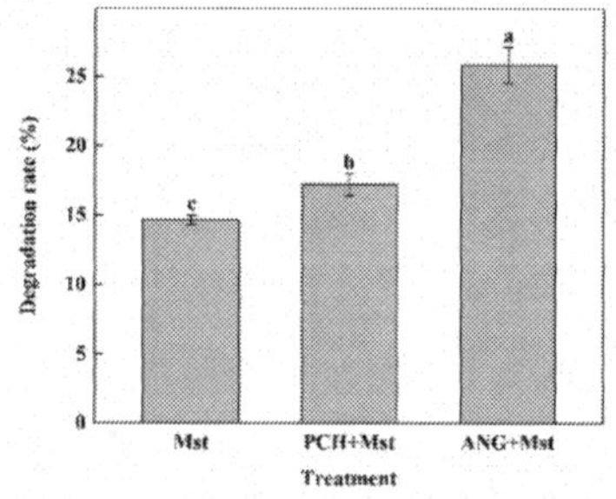

Fig. 3. The degradation rate in each treatment after seven days of incubation. The error bars represent the standard deviations of three replicates.

The activities of degrading enzymes in the treated group (ANG+Mst and PCH+Mst treatments) were significantly higher than others (Mst, ANG and PCH treatments). The CL activity in Mst, PCH and ANG treatments were 348.8, 447.2 and 382.5 nmol/min/mL, respectively (Fig. 4). While the CL activity in PCH+Mst and ANG+Mst treatments was 740.87 and 919.67 nmol/min/mL (Fig. 4). The ACX activity in Mst, ANG and PCH treatments were ranged from 23.3 to 38.6 nmol/min/mL (Fig. 4). In PCH+Mst and ANG+Mst treatments, the ACX activity increased to 45.9 and 169.3 nmol/min/mL treatment, respectively (Fig. 4). The Lip activities in Mst, ANG and PCH treatments were 375.3, 472 and 773 nmol/min/mL, while significantly increased to 1258.3 and 1666.7 nmol/min/mL in PCH+Mst and ANG+Mst treatments (Fig. 4). That is to say, *P. chrysogenum* showed a high CL activity for cellulose degradation, while *A. niger* had high ACX and Lip activities for lignin degradation. Due to lignin was more stable than cellulose, hence *A. niger* was more efficient in maize straw degradation and P dissolution compared with *P. chrysogenum*.

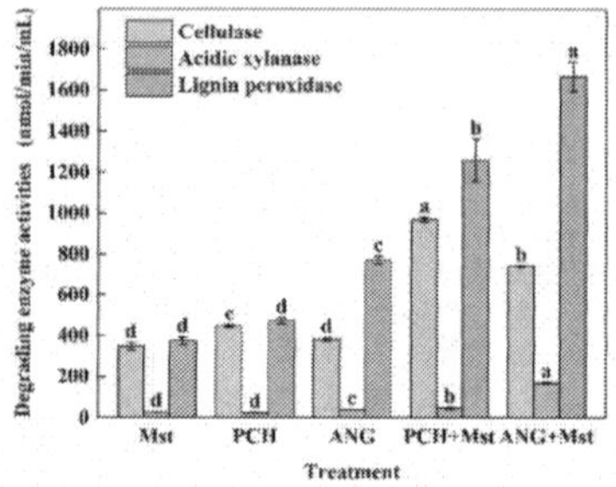

Fig. 4. The degrading enzyme activities in each treatment after seven days of incubation. The error bars represent the standard deviations of three replicates.

3.3 The organic acid secretion by *A. niger* and *P. chrysogenum*

The primary organic acids secreted by *A. niger* and *P. chrysogenum* were oxalic and propionic acid (Fig. 5). In Mst, PCH and ANG treatment, the content of OA and PA had a low value (lower than 200 mg/L) (Fig. 5). In PCH+Mst treatment, the contents of OA and PA secreted by *P. chrysogenum* were 12 and 882 mg/L (Fig. 5). In ANG+Mst treatment, the contents of OA and PA secreted by *A. niger* was 1183 and 200 mg/L (Fig. 5). These two fungi exhibited different organic acid secretion strategies for the same crop straw. The existence of Mst stimulated *P. chrysogenum* to secrete more PA, while *A. niger* secreted more OA. Although both OA and PA had a high capacity of medium acidification and strong metal-complexation activity ability, OA was more suitable for the lignocellulose degradation as the substrate like wheat straw [12]. In addition, oxalic acid was more efficient than propionic acid wood degradation via the formed calcium oxalate crystals in cell walls [13] .

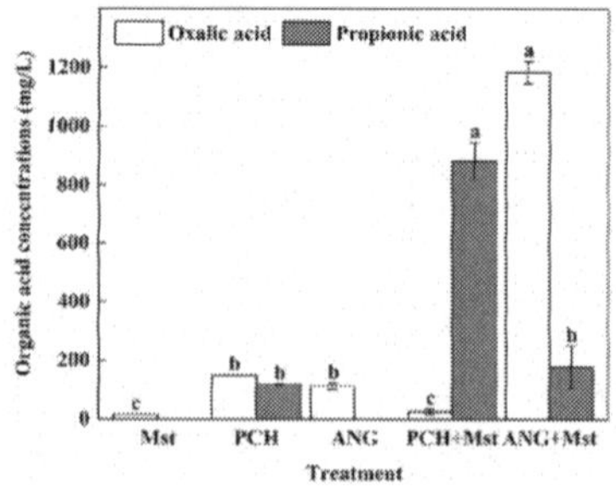

Fig. 5. The secretion of organic acids by *A. niger* and *P. chrysogenum* after seven days of incubation. The error bars represent the standard deviations of three replicates.

3.4 The maize straw degradation and P release by PA and OA

The P release ratio in PA+Mst and OA+Mst treatments was 59.34% and 68.18%, higher than Mst treatment (20 and 28.9 %), respectively (Fig. 6). The degradation ratio of maize straw in Mst treatment was 30.3% (Fig. 6). In PA+Mst and OA+Mst treatments, the degradation ratio was 44.4% and 51.0% (Fig. 6). Organic acids not only have an essential function in the release of P but also facilitate straw degradation in multiple ways [14, 15]. The initial reaction between organic acid and crop straw involves mild acid-catalyzed hydrolysis of the glycosidic bonds of hemicellulose and the α-ether linkage in lignin. The formed by cleavage of the labile ester groups, catalyze the hemicellulose hydrolysis [16]. PA can loosen the structure of lignocelluloses and cellulose, resulting in an improved overall rate due to the increased accessibility to enzymes [17]. In addition, OA has several roles in chelating unstable Mn^{3+} ions, providing H_2O_2, and lowering pH outside of the fungal hyphae, all of which are important factors for the performance of lignin-degrading peroxidases [12]. Our results confirmed that OA increased the Lip activity of *A. niger*, which was conducive to straw degradation and P release.

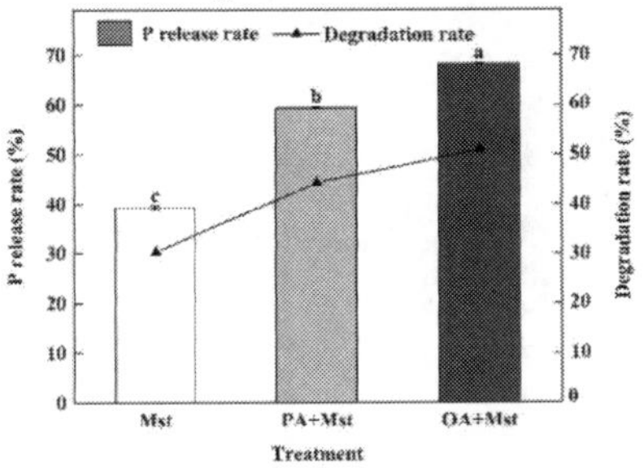

Fig. 6. P release rate and degradation rate of straw in propionic acid (PA) and oxalic acid (OA) treatments.

3.5 XRD and SEM analysis

The spectra of X-ray diffraction showed that the cellulose peaks located at 37.3 and 43.7° were detected in Mst, PA+Mst and OA+Mst treatments (Fig. 7). However, the cellulose peaks were weakly occurred in PCH+Mst treatment and disappeared in ANG+Mst treatment. The new peaks (16.9 and 19.7°) stand for cellobiose and cellotriose (31.03°) were identified in PCH+Mst and ANG+Mst treatments (Fig. 7). This result indicated that cellulose fractions could be transformed to cellotriose and cellotriose by *P. chrysogenum* and *A. niger*.

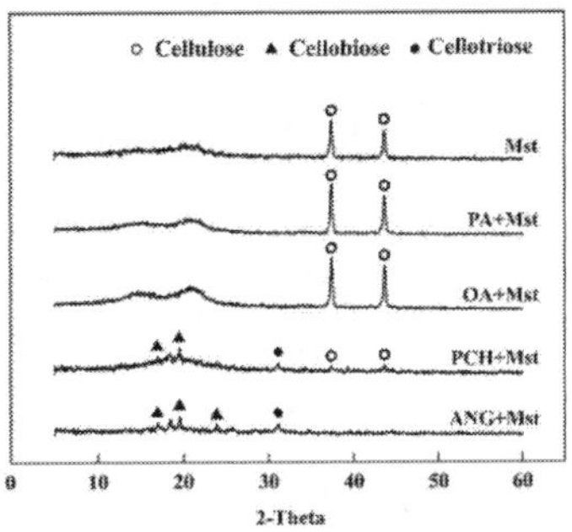

Fig. 7. XRD patterns of the straw in each treatment after seven days of incubation.

SEM imaging of maize straw in different treatments after seven days of incubation was demonstrated in Fig 8. The surface of the original maize straw was dense and smooth (Fig. 8A). After seven days of incubation, the straw also maintained its skeleton despite the presence of holes area in Mst treatment (Fig. 8B). In OA+Mst and PA+Mst treatments, the cellulose skeleton of straw began to collapse, and the broken and split fibres could be observed (Fig. 8C&D). In ANG+Mst and PCH+Mst treatment, the structure of the maize straw matrix was disrupted, both the fungal hyphae and fracture fragments were clearly visible within the straw (Fig. 8 E&F). The results indicated that *A. niger* and *P. chrysogenum* could degrade straw, which was also confirmed by XRD results.

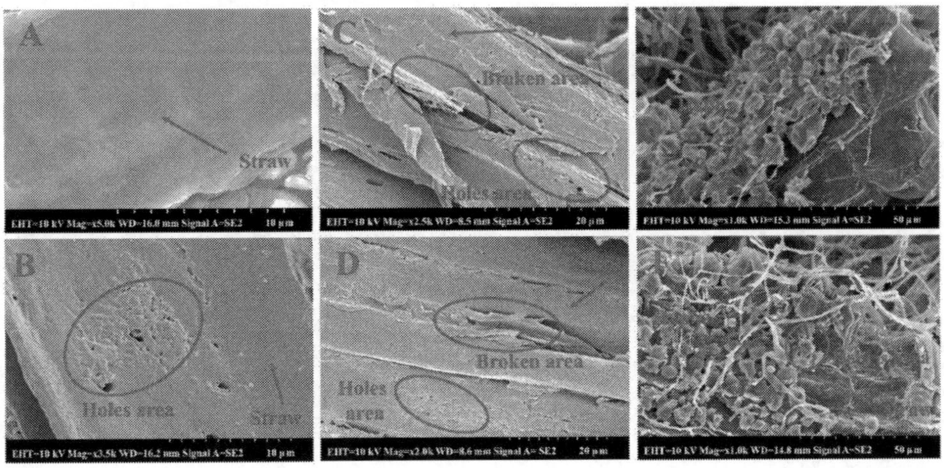

Fig. 8. SEM imaging of the precipitates from maize straw in different treatments after seven days of incubation. A: maize straw before incubation; B: maize straw after 7 days of incubation; C: OA+Mst; D: PA+Mst; E: ANG+Mst; F: PCH+Mst.

4 Conclusion

This study indicated that both *A. niger* and *P. chrysogenum* can promote maize straw degradation and P release. However, *A. niger* is more effective compared with *P. chrysogenum* due to the different enzymes and organic acids production. Although *P. chrysogenum* has a high CL activity, the ACX and Lip activity in *A. niger* is higher, which is more efficient in the degradation of stable lignin in maize straw. In addition, the dominant oxalic acid secreted by *A. niger* is also more efficient than propionic acid secreted by *P. chrysogenum* in maize straw degradation and P release. These results indicate that the mechanisms of maize straw decomposition and P release are different between *A. niger* and *P. chrysogenum*. *A. niger* should be considered as the primary candidate for the crop straw return to the field.

This work was supported by the program at Department of Natural Resources of Anhui Province (NO. 2021-K-11), the National Natural Science Foundation of China (NO. 42007030 and NO. 41877099) and the program at Anhui Agricultural University (NO. yj2019-20).

References

1. G. Kumar, P. Bakonyi, S. Periyasamy, S.H. Kim, N. Nemestóthy and K. Bélafi-Bakó, Lignocellulose biohydrogen: Practical challenges and recent progress. Renewable and Sustainable Energy Reviews. 44, 728-737 (2015)
2. J. Ren, P. Yu and X. Xu, Straw utilization in China—Status and Recommendations. Sustainability, **11**, 1-17 (2019)
3. M.S. Turmel, A. Speratti, F. Baudron, N. Verhulst and B. Govaerts, Crop residue management and soil health: A systems analysis. Agr. Syst., **134**, 6-16 (2015)
4. J. Prasad, C.S. Rao, K. Srinivas, C.N. Jyothi, B. Venkateswarlu, B.K. Ramachandrappa, G.N. Dhanapal, K. Ravichandra and P.K. Mishra, Effect of ten years of reduced tillage and recycling of organic matter on crop yields, soil organic carbon and its fractions in Alfisols of semi arid tropics of southern India. Soil. Till. Res., **156**, 131-139 (2016)
5. A. Machado and A. Ferraz, Biological pretreatment of sugarcane bagasse with basidiomycetes producing varied patterns of biodegradation. Bioresource. Technol., **225**, 17-22 (2017)
6. X. Chu, M.K. Awasthi, Y. Liu, Q. Cheng, J. Qu and Y. Sun, Studies on the degradation of corn straw by combined bacterial cultures. Bioresource. Technol., **320**, 124174 (2020)
7. C. Engels, E. Kirkby and P. White, Chapter 5-Mineral nutrition, yield and source-sink relationships. In: Marschner P. (Eds) Marschner's Mineral Nutrition of Higher Plants (Third Edition). London, Academic Press, 85-133 (2012)
8. Z. Li, T. Bai, L. Dai, F. Wang, J. Tao, S. Meng, Y. Hu, S. Wang and S. Hu, A study of organic acid production in contrasts between two phosphate solubilizing fungi: Penicillium oxalicum and Aspergillus niger. Sci. Rep., **6**, 25313 (2016)
9. D. Tian , Z. Jiang, L. Jiang, M. Su, Z. Feng, L. Zhang, S. Wang, Z. Li and S. Hu, A new insight into lead (II) tolerance of environmental fungi based on a study of Aspergillus niger and Penicillium oxalicum. Environ. Microbiol., **21**, 471-479 (2019)
10. D. Tian, L.Y. Wang, J. Hu, L.L. Zhang, N.N. Zhou, J.J. Xia, M.Y. Xu, K.K. Yusef, S.M. Wang, Z. Li, et al. A study of P release from Fe-P and Ca-P via the organic acids secreted by Aspergillus niger. J. Microbiol., **59**, 819-826 (2021)
11. Z. Dai, J. Lu, X. Li, M. Lu, W. Yang and X. Gao, Nutrient release characteristic of different crop straws manure. Transactions of the CSAE., **26**, 272-276 (2010)
12. U. Urzúa, P.J. Kersten and R. Vicuña, Manganese peroxidase-dependent oxidation of glyoxylic and oxalic acids synthesized by ceriporiopsis subvermispora produces extracellular hydrogen peroxide. Appl. Environ. Microbiol., **64**, 68-73 (1998)
13. F. Palmieri, A. Estoppey, G.L. House, A. Lohberger, S. Bindschedler, P.S.G.Chain and Junier, Chapter Two-Oxalic acid, a molecule at the crossroads of bacterial-fungal interactions. In: Gadd G.M, Sariaslani S. (Eds.) Advances in Applied Microbiology. Academic Press, Elsevier publishing Inc, **106**, 49-77 (2019)
14. Y. Yu, B. Park and S. Hwang, Co-digestion of lignocellulosics with glucose using thermophilic acidogens. Biochem. Eng. J., **18**, 225-229 (2004)
15. B. Zhang, P. He, F. Lu and L. Shao, Enhancement of anaerobic biodegradability of flower stem wastes with vegetable wastes by co-hydrolysis. J. Environ. Sci., **20**, 297-303 (2008)
16. S.S. Anette and B.T. Anne, Optimization of wet oxidation pretreatment of wheat straw. Bioresource. Technol., **64**, 139-151 (1998)
17. Q. Zhang, J. He, M. Tian, Z. Mao, L. Tang, J. Zhang and H. Zhang, Enhancement of methane production from cassava residues by biological pretreatment using a constructed microbial consortium. Bioresource. Technol., **102**, 8899-8906 (2011)

 https://doi.org/10.1051/e3sconf/202235001029

Flood Forecasting with Merged Satellite Precipitation and Hydrologic Model

Ningpeng Dong[1], *Mingxiang* Yang[1,*], *Jianqiu* Li[2] and *Shaokui* Hao[3]

[1] Department of Water Resources, China Institute of Water Resources and Hydropower Research, Beijing 100038, China
[2] CSG Power Generation Co., Ltd., Guangzhou 510950, China
[3] Water Authority of Tongzhou District, Beijing 101100, China

Abstract. Flood forecasting has been an effective way to reduce the potential flood hazards for a sustainable socio-economy. However, the lack of in-situ precipitation records has limited the applicability of flood forecasting with hydrologic models in poorly gauged basins. To address this problem, we aim to develop a flood forecasting framework based on the merged satellite precipitation and a hydrologic model. The framework was then applied to a small basin in the upper Lequan River Basin, Hainan, China for flood forecasting experiments. Results indicate that the combination of merged satellite precipitation and hydrologic model can generally well reproduce the past major flood events occurred in the basin. Our approaches are expected to provide new insights into the flood forecasting in small and poorly gauged basins and can be used to support the sustainable development of the socio-economy.

1 Introduction

The massive advantages of hydrologic simulation in floods, droughts and other water hazards forecasting have been well acknowledged for a rather long time [1-2]. However, the model has not been widely used until recent decades due to the computational burden. Current development in the computational efficacy has allowed the hydrologic models to be used in real-time flood forecasting, which has the potential to reduce the flood hazard before it arrives [3-4]. Nowadays, state-of-art hydrological models are considered able to depict the runoff generation, runoff routing and many other hydrologic components in a relatively accurate way [5-7]. With the use of hydrologic models, multiple studies have reported increasing benefits in protecting the downstream area from inundation and guiding the dam operation for flood control [6-7].

Despite the hydrologic model has brought many advantages to the flood forecasting, the lack of in-situ observed precipitation data have limited the use of hydrologic model in flood forecasting. Normally, flood forecasting by hydrologic models require meteorologic forcing data such as precipitation and temperature for calibration and validation [8]. However, in areas with sparse population or inhabitable environment, gauged meteorological stations were rarely built, and historic meteorologic data are often unavailable. This is particularly the case for the upper Lequan River Basin in the Hainan Island, China. The basin is surrounded by mountains, and meteorologic stations were sparsely constructed. On the other hand, the need for flood forecasting has been increasing for the past years because the population and economy are growing in the downstream areas, which are prone to flood disasters. That being said, the lack of observed precipitation has posed a challenge to flood forecasting with hydrologic models due to the difficulty in calibrating and validating the model [9].

To address this problem, in this study we develop a flood forecasting framework based on the merged satellite precipitation and a hydrologic model. The merged satellite precipitation was originally developed by the National Meteorological Center and was employed in this study to drive the Xin'anjiang hydrologic model to reconstruct the flood events in the recent years. The simulated flood events are then compared with observed streamflow records for an extensive validation. Results indicate that the combination of the merged satellite precipitation and the Xin'anjiang hydrologic model can well reproduce the flood events in the past and can be used in real-time flood forecasting. Our approaches can support the government to prevent flood hazards, and to achieve the sustainable development of the socio-economy.

2 Materials and methods

2.1 Study area

The upper Lequan River Basin has a drainage area of 508 km^2 and its outlet is the Fucai hydrologic station (see Fig. 1). The basin is climatologically affected by the East Asia monsoon and sees a large temporal variability of precipitation. Historically, the basin experienced a

* Corresponding author: yangmx@iwhr.com

large amount of floods that brought significant economic losses to the downstream areas.

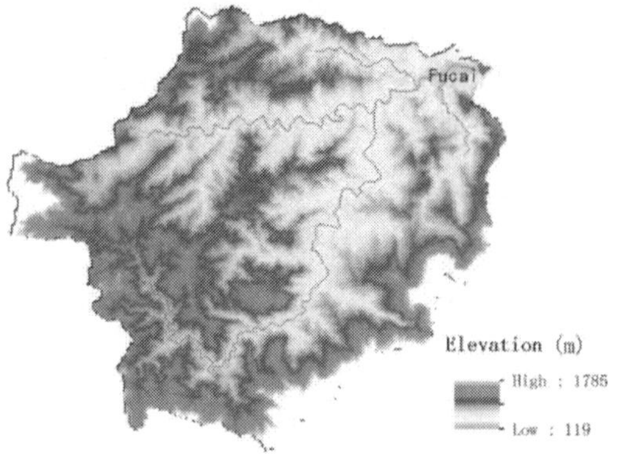

Fig. 1. The map of the upper Lequan River Basin.

2.2 Merged CMORPH satellite precipitation product

The merged CMORPH satellite precipitation product was developed from China's automatic stations and CMORPH precipitation products with a two-step data merging algorithm of probability density matching + optimal interpolation (PDF+OI), and the date product was developed at a resolution of 1 hour and 0.1°, covering the land area of the country since 2008. Its temporal and spatial resolution is high, which can meet the calibration and validation requirements of the flood forecasting model in the upper Lequan River Basin.

The merging steps of this dataset are as follows:

Step 1: Obtain hourly precipitation data from ground observations: extract the hourly precipitation data from the real-time database every hour after quality control at the national automatic station, and interpolate them into an hourly grid with a spatial resolution of 0.1°×0.1°.

Step 2: Obtain precipitation products retrieved by CMORPH satellite: Through FTP transmission, the CMORPH precipitation products developed by the Climate Prediction Center of the US Environmental Prediction Center are obtained (the original CMORPH data has a spatial resolution of 8km and a time resolution of 30 minutes) and resampled to derive a 1-hour, 0.1°×0.1° satellite precipitation product.

Step 3: Use the probability density function matching (PDF) error correction method to correct the systematic errors of the precipitation products retrieved by the CMORPH satellite in China based on the hourly ground observation precipitation data in China.

Step 4: Correct the initial estimation field from the satellite product obtained in Step 3 with the Optimal Interpolation (OI) method based on the observation-based product obtained in Step 1, thereby achieving the effective combination of the two datasets and obtaining the final merging results of hourly precipitation data products.

Step 5: Quantitative evaluation and analysis of product spatial distribution, error characteristics, etc.

2.3 Evaporation and streamflow data

The Xin'anjiang hydrologic model requires evaporation as an input in addition to precipitation. Due to no evaporation gauging stations within the basin, here we select two meteorological stations of Danzhou and Qiongzhong near the basin as the data source of potential evapotranspiration within the basin. The data is at a daily scale, and are processed by the weighted Tyson polygon method to calculate the daily evaporation, before downscaling to every hour.

The daily streamflow data of the Fucai station was collected from the Hydrologic Yearbook to calibrate and validate the hydrologic model.

2.4 Xin'anjiang hydrologic model

In this study, the Xin'anjiang hydrologic model is used as the base model for flood forecasting in the upper Lequan River Basin. Its concept is briefly introduced as follows:

In the 1970s, Professor Zhao Renjun of Hohai University designed the first complete watershed hydrological model in China—Xin'anjiang Two Water Source Model [10]. In the mid-1980s, drawing on the concept of hillside hydrology and research results at home and abroad, he proposed the Xin'anjiang Three Water Source model. The Xin'anjiang model is reliable for streamflow simulation in both humid and semi-humid areas, and has good simulation performance and a high forecast accuracy in most cases. Its application is not only limited to rainfall runoff simulation and flood forecasting, but also water-related fields such as water resources planning and management, disaster prevention, and agricultural development. It has also made a major contribution to the development of the national socio-economy.

The structure of this model is simple and clear and easy to realize through computer programming. Therefore, it is widely favoured in scientific research and engineering applications. However, due to the looped structure of the model, it is not possible for the model to comprehensively describe the flow generation and routing of the hydrological system, and it also cannot simulate the land surface process in the changing environment (such as land use, soil erosion, non-point source pollution, climate change impact assessment, etc.).

The model divides the watershed into many unit watersheds, and calculates the sink term of each unit watershed to obtain the outgoing flow process of the unit watershed. The model then performs the runoff and flood routing along the unit watersheds towards the basin outlet. The main purpose of dividing the unit basin is to deal with the spatial variability of rainfall, so the unit basin should be of appropriate size so that the rainfall distribution on each area is relatively uniform. If there are large and medium-sized reservoirs in the basin, the catchment area above the reservoir should be regarded as a unit basin.

The Xin'anjiang model can be divided into two parts: the runoff generation module and the routing module. The runoff module employs the Three Water Source runoff generation module. The module itself consists of three parts, namely the evapotranspiration calculation, the runoff calculation, and the water source division. Similarly, the Xin'anjiang model divides the flow routing processes into three parts: the land surface routing, the river network routing, and the groundwater routing. The land surface routing is calculated through unit hydrograph method, the river network routing is calculated through the Muskingum method simplified from the St. Venant equations, and the groundwater routing is calculated through the simple, effective linear reservoir method.

2.5 Particle Swarm Optimization Algorithm for Model Parameter Calibration

In this study, the parameters of the Xin'anjiang model are calibrated automatically by the Particle Swarm Optimization (PSO) algorithm [11]. The particle swarm optimization is an algorithm based on intelligence optimization, similar to the genetic algorithm derived from the perspective of biological evolution. This method was first produced when people observe the predation behaviour of birds/fish schools. This method is mainly based on the process of collaboration and competition with individuals to find the optimal solution in the space. The difference between the PSO and another parameter optimization algorithm of genetic algorithm is that the PSO mainly relies on the cooperation between individuals to find the optimal solution, while genetic algorithm seeks the optimal solution based on the survival of the best mutation noted in Darwin's evolution theory.

The PSO algorithm assumes the following scenario: a group of birds are randomly distributed in a certain natural space. They don't know where the food is, but they can feel how far the food is from them. All the birds are searching randomly for food, and the optimal strategy depends on two aspects: search for the area closest to the food, and search based on your own flight experience.

A major concept of the PSO algorithm is that when people make decisions, the main effect is people's own experience and group experience. In other words, people always make their own decisions based on their own experience and group experience. The particle swarm algorithm is simple to apply based on the above principles, and the algorithm is easy to implement.

3 Results and discussion

3.1 Parameter sensitivity analysis

There are in all 16 parameters that can be calibrated (see Table. 1). To reduce the equifinality of parameters, a model parameter sensitivity analysis is first performed to identify the sensitivity analysis to reduce the number of calibrated parameters.

Table 1. Parameters of the Xin'anjiang model

Parameter	Abbreviation	Unit	Values
Soil water cappacity, upper	WM0(1)	mm	5~60
Soil water cappacity, lower	WM0(2)	mm	10~140
Soil water cappacity, deeper	WM0(3)	mm	10~120
Evaporation factor	K		0.3~1.8
Deep evaporation factor	Cd		0.1~0.5
Impermeable area ratio	IMP		0.01~0.5
Storage capacity curve parameter	B		0.2~0.8
Mean free water storage capacity	SM	mm	3~65
Free water storage capacity curve	EX		1~1.5
Free water groundwater discharge parameter	KG		0.1~0.85
Free water soil water discharge parameter	KSS		0.1~0.85
Groundwater parameter	KD		0.3~0.8
Soil flow subsidence parameter	KKSS		0.1~0.999
Fast groundwater subsidence parameter	KKGF		0.7~0.999
Low groundwater subsidence parameter	KKGS		0.7~0.999
Unit hydrograph parameters 1~5	UH1~5		sum = 1

We randomly sample the above 20 parameters, and obtain 10,000 parameter groups. These parameters were input into the model to carry out flood simulation of typical flood events at Fucai Station. The results show that the upper, lower and deep soil water storage capacity WM0, the mean free water storage capacity of the basin SM, the free water groundwater discharge parameter KG, the free water soil discharge parameter KSS, the soil flow subsidence parameter KKSS, the fast groundwater subsidence parameter KKGF, slow groundwater subsidence parameter KKGS, and unit hydrograph parameters UH (1~5) are more sensitive, and the other parameters are not sensitive. Therefore, we select the above 14 parameters for calibration.

3.2 Flood forecasting with the calibrated model

This study selects 10 typical flood events that occurred in the upper Lequan River basin, and divided them into the calibration group and the validation group to calibrate and validate the parameters of the Xin'anjiang flood forecasting model. The calibration and validation results of the flood simulation are presented in the form of the percentage bias (PB) and the determination of coefficient (DC) Table 2.

Table 2. Calibration and validation of flood simulations

Flood event	RB	DC	Note
20081011	-10.5%	0.88	calibration
20091008	-9.6%	0.95	validation
20101001	0.0%	0.83	calibration
20101014	-21.8%	0.76	calibration
20130801	18.3%	0.83	calibration
20131109	-7.4%	0.85	validation
20140715	11.1%	0.66	calibration
20140914	20.6%	0.92	validation
20160816	33.0%	0.82	validation
20161012	-15.7%	0.91	validation

The results indicate that the model can well reproduce the selected flood events, as the RB are fewer than 20% and the DC are higher than 0.8 for most of the flood events, indicating the model and the merged precipitation have a good capability of simulating and forecasting floods. Here, we further present a few representative flood events that were caused by severe storms and typhoons for illustration.

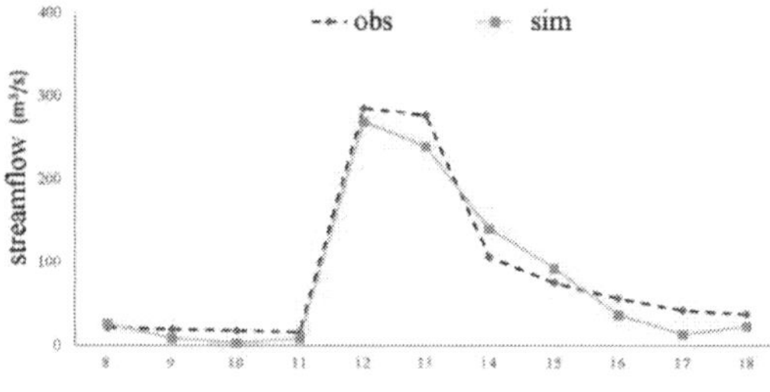

Fig.2. The observed and simulated flood event during Oct.8 and Oct. 18, 2009, where the Typhoon Bama passed by. Obs means observed streamflow, and Sim means simulated streamflow, same for all figures.

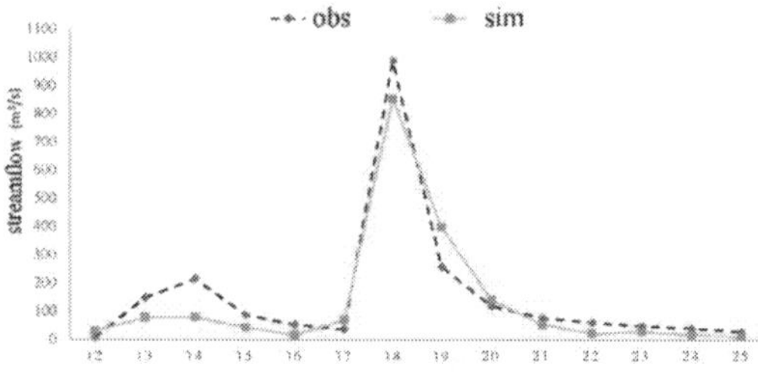

Fig. 3. The observed and simulated flood event during Oct.12 and Oct. 25, 2016, where the Typhoon Sarika passed by.

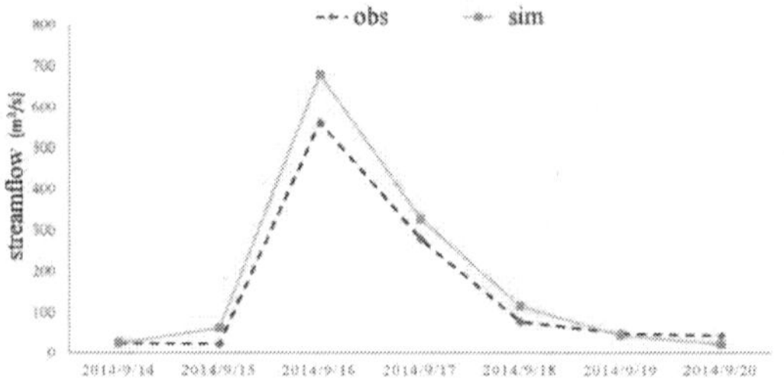

Fig. 4. The observed and simulated flood event during Sep.14 and Sep. 20, 2014, where the Typhoon Kalmaegi passed by.

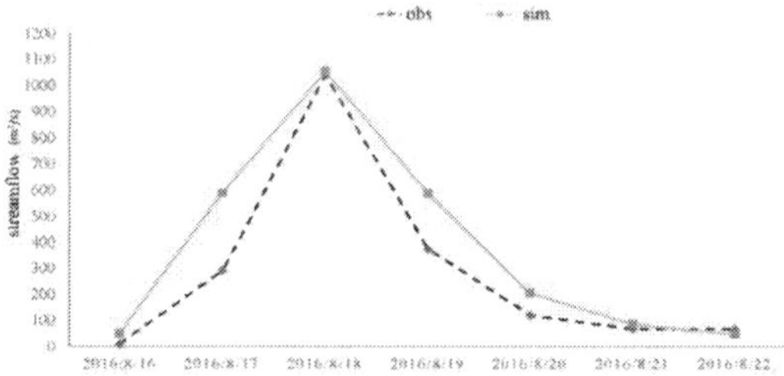

Fig. 5. The observed and simulated flood event during Aug.16 and Aug. 22, 2016, where the Tropical Storm Dianmu passed by.

As is shown in Fig.2, 3, 4, and 5, the flood forecasting model well captures the flood peak and the flood processes for the flood events. This again confirms that the merged satellite precipitation is a reliable data source and can serve as a substitute of in-situ precipitation in case that the gauged precipitation is unavailable. Our results also confirm the applicability of Xin'anjiang model in the flood forecasting in the upper Lequan River Basin and similar basins that are affected by strong storms and typhoons.

4 Conclusions

In this study, we develop a flood forecasting framework based on the merged satellite precipitation and a hydrologic model. The framework was then applied to a small basin, namely the upper Lequan River Basin, Hainan, China for validation of flood forecasting. Results indicate that the combination of merged satellite precipitation and hydrologic model can generally well reproduce the past major flood events occurred in the basin in terms of both the flood peak and flood volume. This confirms the applicability of merged precipitation and Xin'anjiang model in flood forecasting for basins that are of small size and are often affected by storms and typhoons. Our approaches and findings are expected to provide new insights into the flood forecasting in these basins, and can be used to support the sustainable development of the socio-economy.

This study is supported by CSG Power Generation Co., Ltd. (0200002019030304SG00003) and the National Natural Science Foundation of China (U1865102).

References

1. S Grimaldi, Y. Li, J. P. Walker, V. R. N. Pauwels. Effective representation of river geometry in hydraulic flood forecast models. *Water Resour Res.* **54**, 2: 1031-1057.(2018)
2. S Han, P Coulibaly, Bayesian flood forecasting methods: A review. *J. Hydrol.* **551**: 340-351. (2017)
3. J Karsten, J Gurtz, H Lang. Advanced flood forecasting in Alpine watersheds by coupling meteorological observations and forecasts with a distributed hydrological model. *J. Hydrol.* **267**: 40-52. (2002)
4. H.O. Sharif, L Sparks, A.A. Hassan., J Zeitler, H Xie, Application of a distributed hydrologic model to the November 17, 2004, flood of Bull Creek watershed, Austin, Texas. *J. Hydrol. Eng.* **15**, 8: 651-657. (2010)
5. L.C. Chang, M.Z.M. Amin, S.N. Yang, F.J. Chang, Building ANN-based regional multi-step-ahead flood inundation forecast models. *Water.* **10**, 9: 1283. (2018)
6. H Shi, T Li, R Liu, J Chen, J Li, A Zhang, G Wang. A service-oriented architecture for ensemble flood forecast from numerical weather prediction. *J. Hydrol.* **527**: 933-942. (2015)
7. G Jean-Pierre, M Gleizes, P Glize, C Régis. Real-time simulation for flood forecast: an adaptive multi-agent system staff. In *Proceedings of the AISB.* Aberystwyth. 109-114. (2003)
8. K Roman. The case for probabilistic forecasting in hydrology. *J. Hydrol.* **249**, 2-9.. (2001)
9. H.L. Cloke, F Pappenberger. Ensemble flood forecasting: A review. *J. Hydrol.* **375**: 613-626. (2009)
10. R Zhao. The Xinanjiang model applied in China. *J. Hydrol.* **135**: 371-381. (1992)
11. K James, R Eberhart. Particle swarm optimization. In *Proceedings of ICNN'95-international conference on neural network.* Perth. 1942-1948. (1995)

Long-Distance Transport of Green Power via High Voltage Direct Current Submarine Cable

Leiqiong Cai[1], *Li* Chen[3], *Fabrice* Devaux[3], and *Yan* Long[1,2,*]

[1]Huazhong University of Science and Technology, China-EU Institute for Clean and Renewable Energy, 430074, Wuhan, PR. China
[2]Huazhong University of Science and Technology, School of Energy and Power engineering, 430074, Wuhan, PR. China
[3]TotalEnergies SE, 2 place Jean-Millier, 92078 Paris la Défense cedex, France

Abstract. This study aims to perform a technical, environmental, and economic assessment of long-distance transport (around 10000 km) of green power from Australia to Japan through High Voltage Direct Current (HVDC) submarine power cables by literature studies. A PV power plant generates green power; 8000 GWh annual production is chosen as capacity. For the HVDC value chain, according to the assessments, energy efficiency is 74%. Power loss during cable transport is the key contributor. GHG footprint of power delivered is 112 kg CO2e/MWh compared to 50 kg at the outlet of the PV plant. Capital expenditure (CAPEX) is 29058 M€. At the end of the 10th and 20th years, another 4500 M€ investment is required for battery replacement. Operating expense (OPEX) is 166 M€/y. The technical cost of power produced is 428 €/MWh. Results of sensitivity analysis show that submarine cables length, power loss and lifetime, battery storage system sizing, and power plant availability have a significant impact on the economic and environmental performance of the whole HVDC value chain.

1 Introduction

As a consequence of the Paris Agreement, the global energy system must reduce its greenhouse gas (GHG) emissions. Energy generated from renewable sources will play an increasingly vital role as renewable energy can be supplied with low GHG emissions[1].

From a global point of view, the spatial offer of renewable energy such as solar power differs significantly due to regions of renewable energy surplus on the one hand and regions with high energy demands on the other. There are various ways of green power transport, either by cables or by chemical molecules, etc. High Voltage Direct Current (HVDC) transmission lines are one option for power transport[2].

The first submarine power HVDC cable used for electricity transmission was commissioned in 1954, connecting the electric grid of Gotland Island to Sweden's mainland grid. The cable was rated at 20 MW, traversing a submarine route length of 98 km[3]. Among existing submarine power transmission projects, NorNed Link has a 580 km length of submarine cables, the longest up-to-date power submarine cable. Among the planned submarine power cables in the world, two cables are longer than or equal to 1000 km: Ice Link (1170 km) and Euro-Asia Interconnector (1000 km)[4].

This study evaluates the efficiency and performance of transporting 1 GW of green power generated by PV plant via HVDC submarine cable with the reference case of Australia - Japan. It includes a general description of the HVDC value chain (from power production to grid in the end-user country), technical assessments of each process involved in the value chain, and environmental and economic performance, as well as energy efficiency in order to provide a state-of-art of HVDC application in long-distance power transport and support further R&D development/business case reflection.

2 Methodology

2.1 Technical assessment

When using submarine HVDC cable to transport electricity from Australia to Japan, the whole value chain can be divided into five sections, which is showed in Fig. 1, including renewable power generation, equipped with energy storage system (ESS) to ensure 1 GW stable power (Australia), converter station (Australia), subsea transmission (between Australia and Japan), converter station (Japan) and AC grid (Japan).

Fig. 1. Simplified Block Flow Diagram of HVDC chain between Australia to Japan.

2.1.1 Green power production

In 2020, 27.7% of electricity was generated from renewable energy in Australia. Solar energy accounts for 34.4%, which is the main contributor to renewable

* Corresponding author: ly_hust@163.com

generation[5]. Hence, a PV plant in New South Wales (NSW), the eastern part of Australia, combined with an energy storage system, is assumed to provide 8000 GWh power per year (with an average power production of 1 GW and 8000 hours as max operational hours).

Availability of PV is one key parameter that reflects its intermittency, which can be defined below. Considering the location of the PV plant, 27% availability is chosen[6], which can be used to obtain the installed capacity of the PV plant, 3.7 GW.

2.1.2 Energy storage system

Lithium-ion battery technology is relatively mature and has a high energy density, and is considered one of the most promising battery storage approaches, which is chosen as the energy storage system in our base case, assuming an efficiency of 90%[7]. Several batteries are connected in parallel to meet the required storage capacity. The energy storage system should allow storing the difference between the power produced by the power plant and average power output.

In our study, by simplification, we assume an optimist case as base case: battery full charge phase and use phase can be finished in a single day. Averaging charging hours per day of renewable power plants are simplified as average full load hours of power plants per day. Hence, the average full load hours and averaging charging hours for the PV plant are 2365 hours per year and 6.5 hours/day, respectively. After considering the efficiency of batteries, the installed capacity of battery installed capacity can be calculated by equation (1).

$$C_{EES} = (P_{EES} * T_{charing})/\eta_{EES} \qquad (1)$$

Where $C_{EES}$is the installed capacity of energy storage system (GWh), P_{EES} is the power of energy storage system (GW), is the charging hours of energy storage system (hours/day), is the efficiency of energy storage system (%). Key assumption of the installed capacity of PV plant and energy storage system is shown in Table 1.

Table 1. Key assumptions of installed capacity of PV plant and energy storage system.

Key assumptions	Value	Unit
Power generation of PV plant	8000	GWh
Installed capacity of PV plant	3.7	GW
Power of energy storage system	2.7	GW
Installed capacity of energy storage system	19	GWh

2.1.3 Converter station

There are two converter stations in the HVDC value chain: one is between the PV plant and submarine cable (located in Australia), and the other is between submarine cable and AC grid (located in Japan).

For the converter station in Australia, the primary function is to convert low voltage into high voltage. It is worth mentioning that, for a PV plant, a DC/DC converter is required (as PV produces DC power). For the converter station in Japan, its function is to convert DC into AC and high voltage into low voltage. In our study, we choose Voltage Source Converter (VSC) as converter technology because it is newer and more flexible than the traditional technology of Line Commutate Converter (LCC)[8].

There are three key parameters for a converter station: power loss, rated power, and voltage. 1% of power losses is considered in the current study[8]. The rated power of the converter is set as 1000 MW, based on the design of the power production and battery storage blocks. 320 kV is chosen as its output voltage[9]. The annual operating hours of converter stations are assumed as 8760 hours. Key assumptions of converter stations are listed in Table 2.

Table 2. Key assumptions of converter stations.

Key assumptions	Value	Unit
Converter technology	VSC	
Converter rated power	1000	MW
Converter voltage	320	kV
Converter power loss	1	%
Annual operating hours	8760	hours

2.1.4 Subsea transmission

Commonly used topologies of HVDC transmission systems are DC monopoles and bipoles. Considering the high reliability and large market share of the HVDC configuration, a bipolar configuration is chosen in our base case[8].

As one of the extruded cables, cross-linked polyethylene cable (XLPE) is chosen in our base case, which developed rapidly in the past 20 years due to the benefit of VSC technology's increasing market share[8].

Cable arrangement is related to the installation of submarine cables. In our study, the cable's layout of two single-core cables bundled is chosen[4]. It requires only one trench to lay out 2 submarine cables, which is benefited to cost reduction.

320 kV of cable voltage and 500 MW power rating are assumed for each submarine cable[4]. The cross-sectional area of submarine cable with a rated power of 500 MW is about 630 mm2[10].

Most existing transmission power cables are laid in relatively shallow water, i.e., at less than 500 m depth. The deepest cables are installed around 1500 m[11]. The water depth from Australia to Japan is around 4000-6000 m. Since there is no submarine power cable installed in such a water depth, one limit of the current project is the maturity of deepwater installation.

There is no existing submarine power cable project with such a long distance between Australia and Japan, but many submarine optical fiber cable projects, such as AJC network (12700 km) and JGA Cable system (9700 km)[12, 13], globally have achieved lots of long-distance telecommunication transmission between Australia and Japan, which indirectly increase the feasibility of long-distance power transmission through HVDC submarine cables, especially in terms of water depth and transmission distance. In the current study, 9700 km is estimated as the length of submarine cables.

Cable joint is necessary, but its impact on cost and environmental performance is neglected in the current study[4].

Based on literature data, the subsea transmission power losses range is roughly 1.6-3.5%/1000 km[6, 14]. In our base case, 2.5%/1000 km is assumed as the power losses of submarine cables. Key assumptions of subsea transmission are presented in Table 3.

Table 3. Key assumptions of subsea transmission.

Key assumptions	Value	mm
Submarine cable configuration	Bipole	
Submarine cable type	XLPE	
Submarine cable power rating	500	MW
Submarine cable numbers	2	
Submarine cable voltage	320	kV
Cable conductor cross section	630	mm^2
Submarine cable arrangement	Two single-core cables bundled	
Submarine cable length	9700	km
Water depth	4000-6000	m
Cable joint	Required	
Subsea transmission power losses	2.5	%/1000 km
Annual operating hours	8760	hours
Submarine cable lifetime	30	years

2.1.5 AC grid

After HVDC submarine cable transmission, green power generated by renewable energy in Australia will be connected to the AC grid in Japan and then be supplied to users through a power distribution system. The impact of power import on the Japanese's grid is excluded in the current study.

2.1.6 Energy efficiency

Energy efficiency is defined as the ratio between the power input of the AC grid and the power output of the PV plant, which is expressed by equation (2).

$$\eta_{HVDC} = \frac{P_{in(AC\ grid)}}{P_{out(PV)}} \quad (2)$$

Where η_{HVDC} is the energy efficiency of the HVDC chain (%), is the power input of the AC grid (GW) and is the power input of the PV plant (GW).

2.2 Environmental assessment

For environmental assessment, we focus mainly on the greenhouse gas (GHG) emissions, with the terms of carbon dioxide equivalent (CO2e). A complete assessment or a focus on the marine ecosystem can be carried out as a separate study in the future.

GHG emissions can be divided into direct GHG emission (emitted during operation) and indirect GHG emission (related to construction). It is the sum of contributions from each block. For the PV plant, the GHG footprint has the range of 1-218 kg CO2/MWh with a mean of 49.91 kg CO2/MWh, and here it is expressed as 50 kg CO2e/MWh produced[15]. This value already takes into account plant availability and lifetime. The annual GHG emission of the PV plant can be calculated by equation (3).

$$GHG_{PV} = E_{output} * GHG_{PV} \quad (3)$$

Where is the GHG emission of PV plant (kg CO2e), is the annual energy output of PV plant (MWh), is the GHG footprint of PV plant (kg CO2e/MWh).

For the lithium-ion battery storage system, available data from the literature review is in the range of 40-110 kg CO2e/kWh installed[16]. We take the median value of 75 kg CO2e per kWh installed capacity as the base case. Hence, the annual GHG emission of the energy storage system can be calculated by equation (4).

$$GHG_{EES} = (C_{EES} * GHG_{unit})/T \quad (4)$$

Where is the GHG emission of energy storage system (kg CO2e/year), is the installed capacity of energy storage system (MWh), is the GHG footprint per installed capacity of energy storage system (kg CO2e/MWh), and T is the lifetime of energy storage system (year).

The GHG emissions of submarine cables come mainly from the construction of the submarine cables, which is related to the materials of the cables. We assume that the submarine cables in our base case have cable weights of 35 kg/m Copper[17]. The GHG footprint of cable conductor is obtained from the Ecoinvent 3.6 database, which is 7.99 kg CO2e/kg Copper. The GHG emissions of submarine cable is shown by equation (5).

$$GHG_{cable} = N * G * L * GHG_{Copper} \quad (5)$$

Where is the GHG emission of subsea transmission (kg CO2e), is the number of submarine cables, is the submarine cable weight (kg Copper/m), is the submarine cable length (m), and is the conductor GHG footprint (kg CO2e/kg Copper).

The GHG footprint of the HVDC chain shows the CO_2e emission per delivered power, which can be obtained by equation (6).

$$GHG_{HVDC\ chain} = \frac{GHG_{PV}+GHG_{EES}+GHG_{cable}}{P_{AC\ grid}} \quad (6)$$

Where is the GHG footprint of the HVDC chain (kg CO2e/MWh), is the GHG emission of PV plant (kg

CO2e), is the GHG emission of energy storage system (kg CO2e), is the GHG emission of submarine cables (kg CO2e), and is the power input of AC grid (MWh).

2.3 Economic assessment

For the PV plant, its CAPEX can be derived from the CAPEX of the base plant that can be found in the literature, according to equation (7)[1].

$$C_{design} = C_{base}(\frac{S_{design}}{S_{base}})^{SF} \quad (7)$$

Where represents the cost of the designed plant, is the cost of base plant in the literature, and are the capacities of designed plant and base plant respectively. specifics the scaling factor, and we take 1 for the PV plant.

The PMT function in Excel can obtain the annual payments of loans (CAPEX) based on the fixed discounted rate and equal instalments, which is described by equation (8).

$$Function = PMT(rate, nper, pv) \quad (8)$$

Where is the discounted rate (in our study, it is 7%), is the total amount paid on loan (in our study, it is 30 years for all parts except that 10 years for the energy storage system) the and is the total present value of a series of future payments (in our study, it is CAPEX of the whole chain).

An operational expenditure (OPEX) is the sum of variable OPEX and fixed OPEX. Variable OPEX varies with the quantity produced, as opposed to fixed OPEX.

$$OPEX = variable\ OPEX + fixed\ OPEX \quad (9)$$

In our study, variable OPEX is neglected, as there is no significant consumption of chemicals or energy. Fixed OPEX covers annual expenses such as labor cost, maintenance, or insurance. It is generally assumed as a certain percentage of CAPEX.

It is found in the literature that PV plants can last for about 25-30 years[18]. The lifetime of the battery storage part is assumed a replacement every 10 years. For cables, the main longest cables which were decommissioned had this operation done after 30-40 years of function[4]. Here, we assume that the lifespan of the HVDC chain is 30 years (except for 10 years of energy storage system).

3 Results

3.1 Energy efficiency

Power input and output of each block can be obtained, and then the energy efficiency of the HVDC chain can be calculated in Table 4.

Table 4. Calculation results of energy efficiency.

Key assumptions	Value	Unit
Installed capacity of PV plant	3.7	GW
Power output of plant	1	GW
Power output of converter station (Australia)	0.99	GW
Power output of subsea transmission	0.75	GW
Power output of converter station (Japan)	0.74	GW
Power input of AC grid	0.74	GW
Energy efficiency of HVDC chain	74	%

Therefore, the energy efficiency of the HVDC chain is 74%, delivering 5920 GWh/y to Japan. The key contributor to power loss is long-distance subsea transmission, accounting for 93%, shown in Fig. 2.

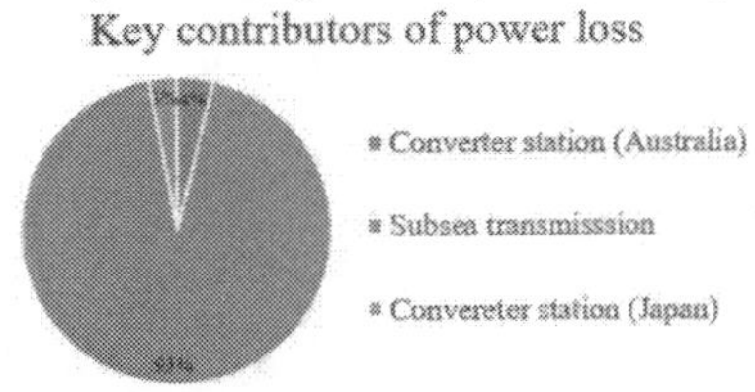

Fig. 2. Key contributors of power loss.

3.2 GHG emission

The GHG emissions of each block of the HVDC chain are listed in Table 5. According to the reference[19], the conversion losses during operation account for 95% of climate impact. In the current study, operation loss of converter represents 1%, which means GHG contribution from converter construction can be neglected in front of submarine cables.

Table 5. GHG emissions of each block of HVDC chain.

GHG emisssions	Value	Unit
PV plant	400	kta CO_2e
Energy storage system	145	kta CO_2e
Submarine cable	181	kta CO_2e

Fig. 3 shows the GHG emissions of the HVDC chain with the copper conductor of submarine cables. The PV plant has the most significant proportion of GHG emissions (around 55%), energy storage system has a share of 20%, and submarine cable shares 25%.

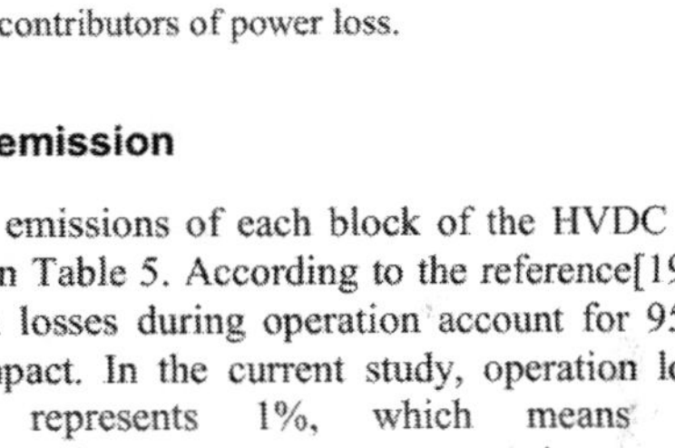

Fig. 3. GHG emissions of HVDC chain.

The GHG footprint of the delivered power is 112 kg CO_2e/MWh chain (compared to 50 kg CO_2e/MWh without battery storage and transport).

3.3 Cost estimation

To deliver 5920 GWh/y in Japan, CAPEX is 29058 M€, OPEX is 166 M€/y. At the end of the 10th year and 20th year, we need to invest another 4500 M€ for battery replacement.

As illustrated in Fig. 4, The construction and operation of submarine cables are the main components of the technical cost (around 65%). Renewable power plant and energy storage system are also essential parts (around 35%), and the technical cost of converter stations are too small so that it can be ignored.

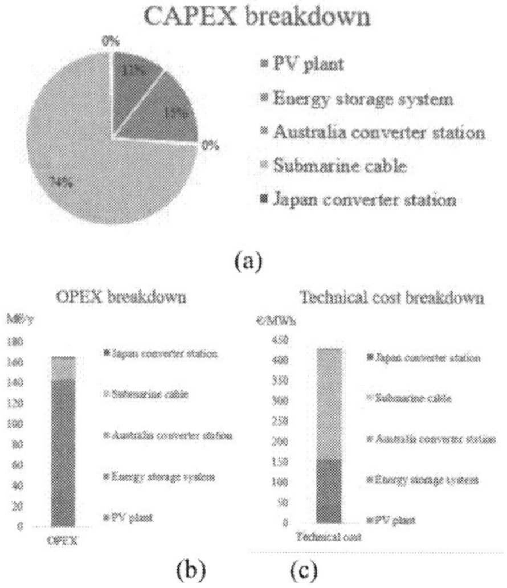

Fig. 4. (a) CAPEX breakdown, (b) OPEX breakdown, (c) Technical cost breakdown.

4 Sensitivity analysis

There are uncertainties in the assumptions used in this analysis. Hence, sensibility analysis is conducted to assess the influence degree of key parameters on the performance of our HVDC value chain. The selected key parameters and the range of their values for sensitivity analysis are listed in Table 6.

Table 6. Base case, better case and worse case for sensitivity analysis.

Key parameters	Better	Base	Worse	Unit
Submarine transport power loss	1.6	2.5	3.5	%/1000 km
PV plant availability	30	27	15	%
Battery storage installed capacity	19	19	38	GWh
GHG footprint of energy storage plant	40	75	110	kg CO_2e/kWh
CAPEX of submarine cables	0.73	1.7	3.1	M€/km
Submarine cable length	9700	9700	12700	km
Power plant lifetime	40	30	20	years
Submarine cable lifetime	40	30	20	years

The results of sensitivity analysis on the technical cost of the HVDC chain are shown in Fig. 5. CAPEX of submarine cables has an enormous impact on the technical cost of the HVDC chain, followed by submarine cable length, power plant availability, energy storage installed capacity of PV plant, and submarine transport power loss. By contrast, the impacts of PV plant lifetime and submarine cable lifetime are limited.

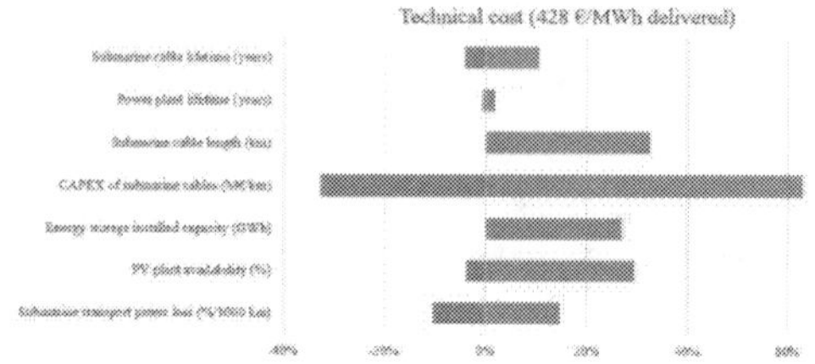

Fig. 5. Sensitivity analysis on technical cost of HVDC chain.

The results of sensitivity analysis on GHG emission of the HVDC chain are shown in Fig. 6. Submarine transport power loss and PV plant availability are the major factors influencing GHG footprint. Other factors, including submarine cable length, the GHG footprint of the energy storage plant, and energy storage installed capacity, have similar impacts on the GHG footprint of the whole chain.

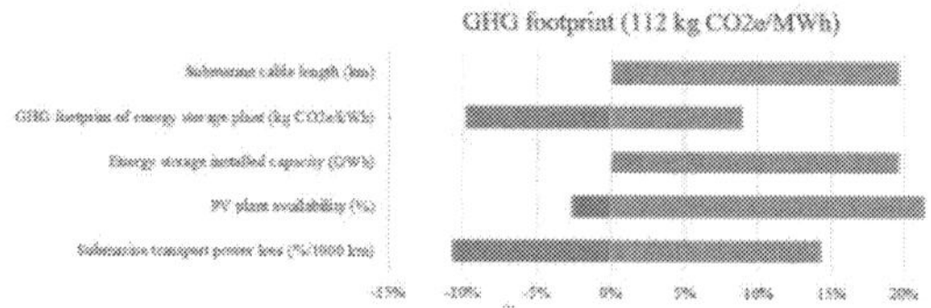

Fig. 6. Sensitivity analysis on GHG footprint.

To sum up, submarine cable length, power loss and lifetime, battery storage system sizing, and power plant availability have a significant impact on the whole chain's cost and GHG performance. These parameters should be carefully investigated if we would like to develop an HVDC chain for power transport.

5 Conclusion

This paper aims to evaluate the long-distance transport of green power through HVDC submarine cables based on technical, economic, and environmental assessments.

8000 GWh/y power is generated by a PV power plant, of which 5920 GWh is delivered in Japan. The energy efficiency of the HVDC chain is 74%, and 93% of power loss is due to long-distance submarine transmission.

Considering 5920 GWh delivered to Japan, the GHG footprint is around 112 kg CO_2e/MWh. PV plant has the most significant proportion of 55%, followed by 25% of submarine cable and 20% of the energy storage system.

The CAPEX is 29058 MW. The materials and installation of submarine cables are primary components of CAPEX, which account for 74%. Annual OPEX is 166 MW/y. Renewable power plants and battery storage systems are the main proportions. The technical costs are 428 €/MWh. Submarine cable cost is the primary source of the HVDC chain in technical cost, around 65% of submarine cable cost followed by 30% of the PV plant and energy storage cost.

Sensitivity analyses show that, for economic performance, CAPEX of submarine cables has a dominant impact, and for environmental performance, power plant availability has the most considerable impact.

References

1. M. Niermann, S. Timmerberg, S. Drünert and M. Kaltschmitt. "Liquid Organic Hydrogen Carriers and alternatives for international transport of renewable hydrogen." Renewable and Sustainable Energy Reviews, **135**: 110171 (2021).
2. T.S. Kishore and S.K. Singal. "Optimal economic planning of power transmission lines: A review." Renewable and Sustainable Energy Reviews, **39**: 949-974 (2014).
3. K. R. Schell, J. Claro and S.D. Guikema. "Probabilistic cost prediction for submarine power cable projects." International Journal of Electrical Power & Energy Systems, **90**: 1-9 (2017).
4. A. M. Ardelean and P. Minnebo. "HVDC Submarine Power Cables in the World." (2015)
5. Clean-Energy-Council. "Clean Energy Australia Report." https://assets.cleanenergycouncil.org.au/documents/resources/reports/clean-energy-australia/clean-energy-australia-report-2021.pdf. (2021)
6. G. Ashish, B. Dmitrii, F. Mahdi and B.J.S. Christian. "Can Australia Power the Energy-Hungry Asia with Renewable Energy?", **9**: 233 (2017).
7. A. N. Abdalla, M.S. Nazir, H. Tao, S. Cao, R. Ji, M. Jiang and L. Yao. "Integration of energy storage system and renewable energy sources based on artificial intelligence: An overview." Journal of Energy Storage, **40**: 102811 (2021).
8. A. Alassi, S. Banales, O. Ellabban, G. Adam and C. Maciver. "HVDC Transmission: Technology Review, Market Trends and Future Outlook." Renewable Sustainable Energy Reviews **112**: 530-554 (2019).
9. ABB. "HVDC Light (VSC) (2021)."
10. Prysmian. "Extruded Cables for HVDC Power Transmission." (2013)
11. R. d'Amore-Domenech, T.J. Leo and B.G. Pollet. "Bulk power transmission at sea: Life cycle cost comparison of electricity and hydrogen as energy vectors." Applied Energy, **288**: 116625 (2021).
12. SubmarineNetworks. "The Japan-Guam-Australia (JGA) Cable System." https://www.submarinenetworks.com/en/systems/asia-australia/jga. (2020)
13. Australia-Japan-Cable. "The AJC Network." https://ajcable.com/network-connectivity/ (2001).
14. B. Miao, L. Giordano and S.H. Chan. "Long-distance renewable hydrogen transmission via cables and pipelines." International Journal of Hydrogen Energy, **46**: 18699-18718 (2021).
15. D. Nugent and B.K. Sovacool. "Assessing the lifecycle greenhouse gas emissions from solar PV and wind energy: A critical meta-survey." Energy Policy, **65**: 229-244 (2014).
16. J. F. Peters, M. Baumann, B. Zimmermann, J. Braun and M. Weil. "The environmental impact of Li-Ion batteries and the role of key parameters – A review." Renewable and Sustainable Energy Reviews, **67**: 491-506 (2017).
17. ABB. "XLPE Submarine Cable Systems-Attachment to XLPE Land Cable Systems-User´s Guide." https://new.abb.com/docs/default-source/ewea-doc/xlpe-submarine-cable-systems-2gm5007.pdf (2014).
18. J. Deign. "Most Solar Panel Warranties Last 25 Years. Europe's Plant Owners Are Counting on Much More." https://www.greentechmedia.com/articles/read/europes-solar-market-grapples-with-35-year-plant-lifespans (2020).
19. ABB. "What is LCA, how it is done and why it is relevant?" https://www.hitachiabb-powergrids.com/no/no/offering/product-and-system/hvdc/hvdc-light (2020)

Prediction and Analysis of Household Energy Consumption by Machine Learning Algorithms in Energy Management

*M.*Rambabu[1], *N.S.S.*Ramakrishna[1,*], *P Kumar* Polamarasetty[1]

[1] Department of Electrical & Electronics Engineering, GMR Institute of Technology, Rajam, India

Abstract. Now the world is becoming more sophisticated and networked, and a massive amount of data is being generated daily. For energy management in residential and commercial properties, it is essential to know how much energy each appliance uses. The forecast would be more clear and practical if the task is based purely on energy usage data. But in the real world, it's not the case, energy consumption is strongly dependent on weather and surroundings also. In a home appliances network when measured/observed data is available then algorithms of supervised-based machine learning provide an immeasurable alternative to the annoyance associated with many engineering and data mining methodologies. The patterns of household energy consumption are changing based on temperature, humidity, hour of the day, etc. For predicting household energy consumption feature engineering is performed, and models are trained by using different machine learning algorithms such as Linear Regression, Lasso Regression, Random Forest, Extra Tree Regressor, XG Boost, etc.. To evaluate the models R square is used as the forecasting is based on time. R square tells how much percentage of variance in the dependent variable can be predicted. Finally, it is suggested that tree-based models are giving best results.

1 Introduction

In recent decades, studying energy consumption issues has been a popular academic area. Energy issues are critical to society's security and well-being [1]. Predicting energy consumption in industry and the energy sector is an essential part of macro-planning in the industry sector, according to economic theories. [2]. Energy supply-demand planning on a long-term basis must meet the needs of countries' long-term development. Making accurate estimates about future energy usage can help decision-makers plan and schedule supply system operations more effectivel. Forecasting the energy consumption load is a key aspect of power distribution systems' economic and safe operations planning. According to the forecasting horizon, there are primarily three types of energy consumption studies in the literature [3]. Resource management and development spending account for the majority of long-term forecasts (5–20 years). The distribution network's scheduling and analysis are frequently aided by short-term forecasting (for periods ranging from an hour to a week), where as mid-term forecasting (a month to 5 years) is mostly used for planning power production resources and rates [4][5]. The focus of the paper is on short-term forecasting. Because energy demand is affected by a variety of factors such as time, climate, socio economic, and demographic factors, precisely anticipating consumption is both crucial and difficult.

Regression models, as well as ANNs and time-series forecasting, are the most widely utilized methods. The focus of this study is primarily on regression models [6][7]. This study uses regression models such as linear regression, lasso regression, ridge regression, random forest, XGBoost, and Extra tree regressor [8].

The dataset contains measurements of house temperatures and humidity recorded at 10-minute intervals over 4.5 months using a ZigBee Wireless sensor network; the entire experiment is based on this dataset. Our experiment is anticipating the following instances energy usage based on the previous data and providing it to the power plant for generating reasons [3]. It would be simple to anticipate energy use if energy consumption was just based on the appliances in the home. However, the amount of energy used in the home is often determined by the temperature and humidity of the day [7]. As a result, to create good predictions with the data given, these experiments are carried out to find the optimal model with the best attributes. The basic goal is to anticipate the energy consumption of a future instance using current and historical data on temperatures, humidity, and other environmental variables.

The second section of the paper is dedicated to data collection and analysis of study area. The research provides answers to inquiries about how energy consumption varies hourly, day by day, and month by

* Corresponding author: nramkrishna231@gmail.com

month. The third section would concentrate on feature engineering and selection. The fourth section would involve modeling and comparing many models to choose the best one. The dataset is good enough and only requires scaling for data processing [9].

2 Study Area and Data Analysis

The average temperature outside is about 7.5 degrees centigrade. The temperature fluctuates from -6 to 28 degrees centigrade. Inside the building, the average temperature has been approximately 20 degrees centigrade in all rooms. It has a temperature range of 14 to 30 degrees centigrade. Warming appliances have been put in place to keep the insides of the building warm, according to this. As a result, there would be a link between indoor temperature and energy use. Outside the building, the average humidity has been higher than within. The average humidity in the children's and parent's rooms is significantly higher, indicating that the residents of this building spend most of their time in these areas. Because the humidity levels are so high, some appliances, such as dehumidifiers, may be necessary. As a result, there would be a link between humidity and energy use.

There are two instances of peak hour activity. The first is at 11 a.m., and the second is at 6 p.m. The peak at 11 a.m. is shallow and low, whereas the peak at 6 p.m. is taller and sharper. The energy usage of appliances is roughly 50Wh during the sleeping hours (10 PM - 6 AM). After 6 a.m., energy use gradually increases until 11 a.m. (probably due to morning chores). At roughly 3 p.m., it progressively drops to around 100 Wh. After that, until 6 p.m., energy use skyrockets (probably due to requirement lights in rooms). However, as night falls and everybody in the house retires to bed around 10 p.m., appliance energy consumption drops to 50 Wh. In Belgium, it appears that the month has no impact on energy use. However, due to the warm weather, may month's energy use may be lower. Energy trend over months for every hour are shown in figure 1.

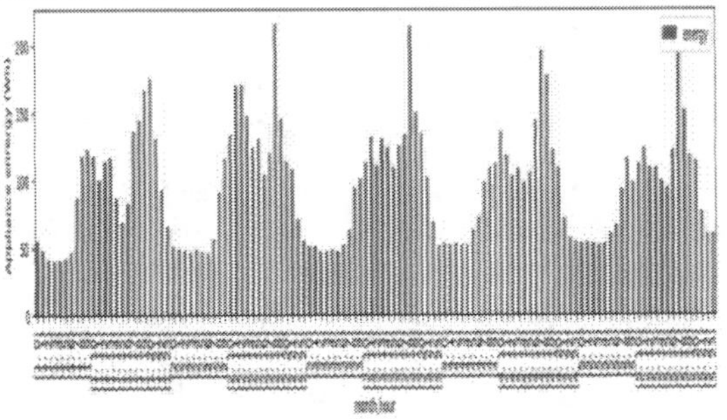

Fig. 1. Energy trend over months for every hour in Belgium

Weekend energy use is higher than weekday energy usage.

This could be because weekends are holidays, and most of the building residents stay at home and use appliances. Hourly energy trend on weekdays and weekends are depicted in figure 2.

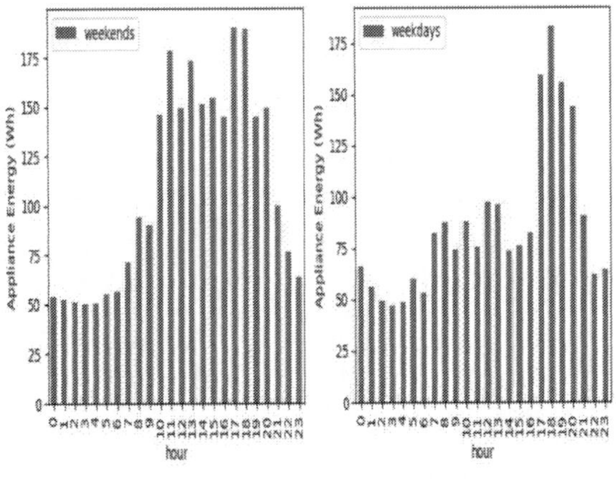

Fig. 2. Hourly energy trend on weekdays and weekends in Belgium

The data does not follow a normal distribution. Because the data is skewed, log10 is used. Because many models have the underlying assumption of normal distribution, this type of transformation has a positive impact on modeling.

The Pearson correlation is used to determine how interdependent the features are. It has been discovered that characteristics have a range of -0.09 to 0.02 with energy, which is lower than expected. This necessitates a concentration on feature engineering to obtain a greater number of linked features with the dependent feature for modelling purposes. Energy distribution before and after log transformation respectively are shown in figure 3.

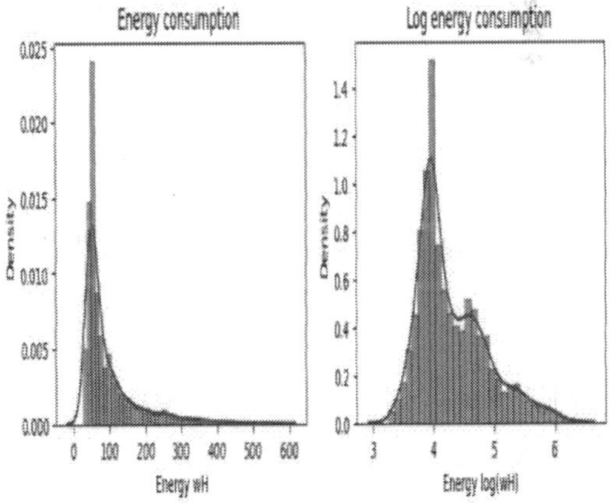

Fig. 3. Energy distribution before and after log transformation respectively in Belgium

3 Feature Engineering

According to the above analysis, we discuss about feature engineering. Different significant features are retrieved from the features provided in this section. Because the data is skewed, a new feature named log energy is added to the energy feature by performing the log10 transformation to the energy feature. Temperature and humidity features are highly reliant on their own features, thus by averaging them, a new feature is created from one of the temperature features and one from the humidity features. Temperature and humidity are also inversely connected, according to the data analysis. This reason can be used to multiply each of the temperature and humidity features of the same room, yielding an additional 9 features. Apart from features like temperature and humidity levels in rooms and outside, there is another crucial column date, which contains timestamps for each of the sensor data samples. The hour data for each sample is derived from the timestamps, and the goal is to see if there is a pattern of energy usage over a day. For 4.5 months, the figure 4 below depicts the average energy usage of appliances at a given hour of the day.

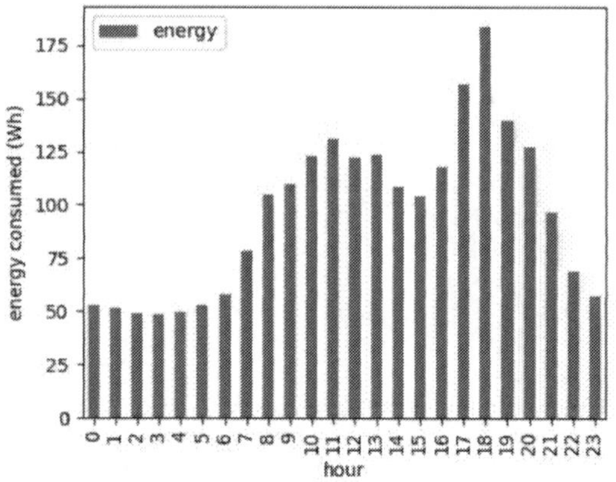

Fig. 4. Average hourly energy consumption trend

Hours and their relationship with energy consumption were explored in the analysis section. The 24 hour day can be divided into three classes based on energy consumption patterns: Class 1: 10 PM - 6 AM, Class 2: 6 AM - 3 PM, and Class 3: 3 PM - 10 PM so that the model can learn from multi-class classifier [10].

An additional 13 features are retrieved by feature engineering, and observed characteristics are now more dependent on energy use, with correlations ranging from -0.21 to 0.55.

4 Modelling

Now having good features, The next stage is to model using different regression models and compare them to find the optimum regression model for energy prediction. Because this is a regression problem, the metric utilized will be R2 (R squared), which gives a measure of the variance of the target to a variable that can be explained using the given characteristics. It is a measure of how well the model fits the data in a practical sense. It can be expressed numerically as equation (1):

The percentage of R^2 score is

$$R^2 = (1 - \frac{SS_{Regression}}{SS_{Total}})*100 \quad (1)$$

where,

$SS_{Regression}$ = Total Residual Sum of Squares computed from the best fit line

SS_{Total} = Total residual sum of squares measured from the mean

Using the 'r2_score()' function of the metrics module of the scikit-learn library.

4.1 Linear Models

Linear Regression is the most basic Regression algorithm. There is no need for further complexity if a Linear model can adequately describe the facts[11]. We can use Regularization approaches to penalize the coefficient values of the features as a modification to the original Least Squares Regression because higher values contribute to over fitting and loss of generalization. Linear models benefit tremendously from regularization approaches. In addition, there are only a few circumstances in which a Linear model may fit the data effectively without Regularization. Regularization transforms the problem of Linear Regression into Lasso or Ridge Regression, depending on whether we add the absolute values of coefficients or their squares to our loss function. The R2 score of linear regression without feature engineering, which is used as a baseline model, is 17%.

4.1.1 Linear Regression :

The bias term is added to the weighted sum of the input features in a linear model to produce a forecast in equation (2) (also called the intercept term)[12]

Model prediction

$$\hat{y} = \theta_0 + \theta_1 x_1 + \theta_2 x_2 + ...\theta_n x_n \quad (2)$$

$\hat{y}$ = predicted value.

n = number of features.

x_i = ith feature value.

θ_j = jth model parameter (including the bias term θ_0 and the feature weights $\theta_1, \theta_2, \cdots, \theta_n$)

Setting the parameters of a model so that it best fits the training set is referred to as training it. To do so, we'll need a metric to determine how well (or poorly) the model matches the training data.

Since it may be used to a wide range of problems, Gradient Descent is an extremely versatile optimization technique. Gradient Descent's general concept is to iteratively change parameters to minimize additional cost functions. The learning rate hyper parameter determines the size of the steps, which is an important parameter in Gradient Descent [12].

Regularization of a linear model is often accomplished by restricting the model's weights. Ridge Regression, Lasso Regression, and which implement alternative techniques to restrict the weights.

4.1.2. Ridge Regression

Ridge Regression (also known as Tikhonov regularization) is a regularized variant of Linear Regression in which the cost function is given a regularization term of equation (3)

$$\alpha \sum_{i=1}^{n} \theta_i^2 \quad (3)$$

Learning algorithms are compelled to adjust to the data and keep weights as low as feasible when this is enforced. Remember that the cost function should only be subjected to regularisation when it is being trained. To evaluate the model's performance using the unregularized performance metric once it's been trained from equation (4).

$$F(\theta) = MSE(\theta) + \alpha \frac{1}{2} \sum_{i=1}^{n} \theta_i^2 \quad (4)$$

The type of regularization term to utilize is determined by the penalty hyper parameter. When you choose "l2," SGD will add a regularization term to the cost function equal to half the square of the l2. Ridge Regression is simply the norm of the weight vector [12][13].

4.1.3. Lasso Regression

Least Absolute Shrinkage and Selection Operator Regression (also known as Lasso Regression) is another regularized variant of Linear Regression: it adds a regularization factor to the cost function, same as Ridge Regression, but it employs the l1 norm instead of half the square of the l2 norm, uses the weight vector. The fact that Lasso Regression tries to fully exclude the weights of the least significant variables is a key aspect from equation (5) (i.e., set them to zero) [13].

$$F(\theta) = MSE(\theta) + \alpha \sum_{i=1}^{n} \theta_i \quad (5)$$

4.2 Models based on trees

Tree-based Regression models are the next type of method [12]. When compared to Linear models, Tree-based models have a significant advantage in terms of robustness against outliers. Because no linear link has been found between any attribute and the target variable, regression trees are likely to outperform linear models.

Given a large number of characteristics, a Decision Tree will almost certainly overfit the data. As a result, it could be skipped and instead go straight to the ensemble methods listed below, which include building multiple regressors on copies of the same training data and combining their output either through mean, median, mode (Bagging), or growing trees sequentially (i.e. each tree is built from the data of the previous tree) and using a weighted average of these weak learners (a learner that performs just as well as the strongest learner) [14].

Random Forests is a common Bagging technique that works well with high-dimensional data like ours. Extra Trees Regressor takes it a step further by randomizing splits. A form of boosting approach is Gradient Boosting Machines. It creates an additive model in which performance is continually improving[13].

4.2.1 Random Forests

Using the bagging method (or occasionally pasting), Random Forests are an ensemble of Decision Trees with maximum samples set to the training set's size.[15][16]. Instead of searching for the best feature, the Random Forest approach introduces additional randomization by searching for the best feature among a random collection of features. As a result, there is a greater variety of trees, which reduces the variance while increasing the bias.

4.2.2 Extra-Trees

When building a tree in a Random Forest, only a random subset of the characteristics are examined for splitting at each node (as discussed earlier). Instead of searching for the best possible thresholds, it is feasible to make trees even more random by employing random thresholds for each feature (like regular Decision Trees do) . Extremely Randomized is the name given to a forest made up of such extremely random trees (or Extra-Trees for short) [17]. Once again, higher bias is exchanged for reduced variance. It also makes Extra-Trees significantly faster to train than standard Random Forests, because one of the most time-consuming jobs of tree growth is determining the best feasible threshold for each feature at each node. The ExtraTreesRedgressor class in Scikit-Learn can be used to generate an Extra-Trees classifier. It has the same API as the RandomForestRegressor class.

4.2.3 Gradient Boosting

Gradient Boosting is another prominent boosting strategy. Gradient Boosting, like Ada Boost, operates by successively adding predictors to an ensemble, with each one correcting the one before it. Unlike Ada Boost, this method seeks to adapt the new predictor to the residual errors created by the prior predictor rather than modifying the instance weights at each iteration. The paper [18] has clearly shown the mathematical procedure behind gradient boosting. A subsample hyper parameter is also supported by the Gradient Boosting Regressor class, which sets the fraction of training instances to be used for each tree's training. If subsample=0.25, for example, each tree is trained on 25% of the training cases, chosen at random. As you would have guessed, a bigger bias is exchanged for lower variance. It also makes training much more efficient [19][20].

4.2.4 XG Boost

XGBoost is a machine learning method that has recently won Kaggle contests for structured or tabular data. XGBoost is a high-speed and high-performance implementation of gradient boosted decision trees.

Along with these models neural networks can also be used. With the help of PCA dimensional reduction can be done. [22] address PCA to select the transformed features that can be more related to the target variable. The discipline of tweaking the hyper parameters of these algorithms to achieve optimal performance is known as hyper parameter optimization. One of the most obvious methods to perform is manually selecting hyper parameters but it is not feasible as we cannot select values that would help in the optimization. Because there are so many choices and options, this technique does not scalable [23]. Many strategies for hyper parameter optimization have been proposed. Grid search, Random search, and Bayesian optimization are three well-known strategies .

All of the models are modeled using the steps below:

The final set of features used are really important. These features are solely responsible for the better performance of the model [24[25]]. With the help of a correlation plot, these features are chosen. These are the final set of features taken into consideration for modeling.

final_features = ['th_kitchen', 'th_living', 'th_laundry','th_office', 'session', 'th_teen', 'th_parents', 'press_mm_hg', 'windspeed', 'avg_house_temp', 'hour', 'avg_house_hum', 'temp_outside', 'humid_outside', 'weekday', 'month','log_energy']. Here for the train-test split 80-20 is used.Steps for modeling of ML algorithm are shown in figure 5.

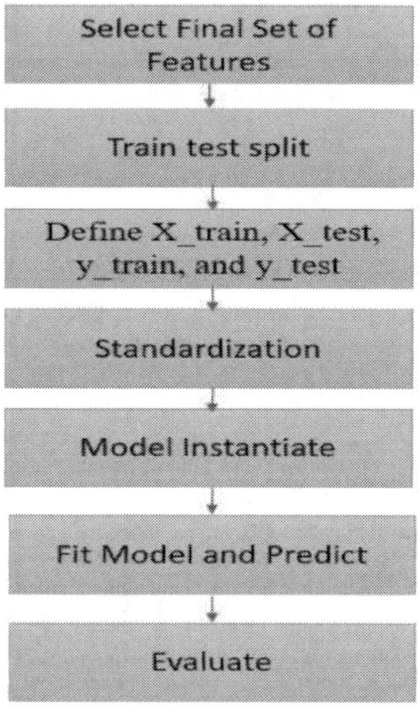

Fig. 5. Flow chart for modeling of ML algorithm

5 Result Analysis

5.1 Case1: Results for Baseline model

Table-1 shows the baseline model results. For the baseline model, the selected models are linear models and the modelling is done before feature engineering with the raw data. When the modelling is done with raw data only 17% of variance is only predicted.

Table 1. R2 score of Linear Models before feature engineering

Name	Train_Time	Train_R2_Score	Test_R2_Score
Lasso	0.00996	0.0000	-0.0777
Linear	0.021046	16.9169	16.9655
Ridge	0.008213	16.9168	16.9659

Fig 6. Shows the prediction graph for 100 samples. Blue colour line indicates the target value which is actual value and the prediction done with linear regression algorithm is shown by yellow line. It is observed that the model was not able to perform well.

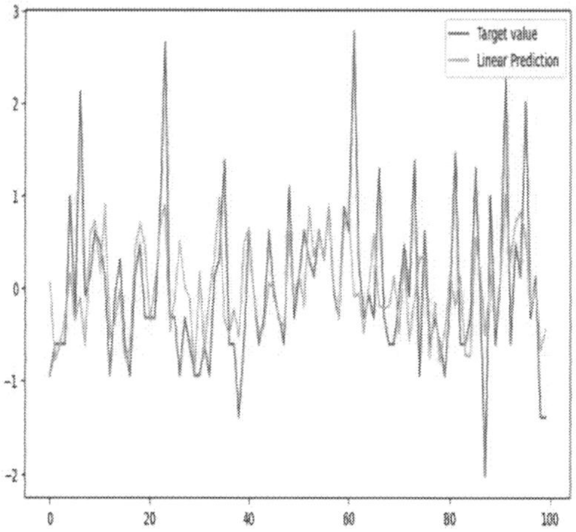

Fig. 6. Energy Prediction of Baseline model(Linear Regression) before feature engineering

5.2 Case2: Results after feature engineering without hyper parameter tuning

Table 2. shows the results for different algorithms after feature engineering. Linear and Ridge has 28% of R2 score on test data. Random forest has 67% and Extra-tree regressor has 71%. of line.

Table 2. R2 score of different algorithms after feature engineering

Name	Train_Time	Train_R2_Score	Test_R2_Score
Lasso	0.010408	0.0000	-0.0110
Linear	0.017416	30.0561	28.7402
Ridge	0.007331	30.0561	28.7407
K-Neighbors	0.250363	75.7320	61.0030
Random Forest	22.978521	95.6592	67.4624
Extra Trees	6.360532	100.0000	71.000213

Regressor			
Gradient Boosting	7.515340	47.1071	42.2821
XGBoost	1.823460	46.9366	42.1626

Fig. 7. Demonstrates the bar plot visualizing the results.

It can be seen that extra-trees is performing high than all other algorithms. It can be seen that linear and ridgeare having very close test and train scores. Fig 8. Illustrates the prediction graph for 100 samples. Blue colour line indicates the target value which is actual value. The prediction done with extra tree algorithm is shown by green and random forest with red color. It can observed that extra tree algorithm is able to mimic the original values and was able generalize more than all the other models.

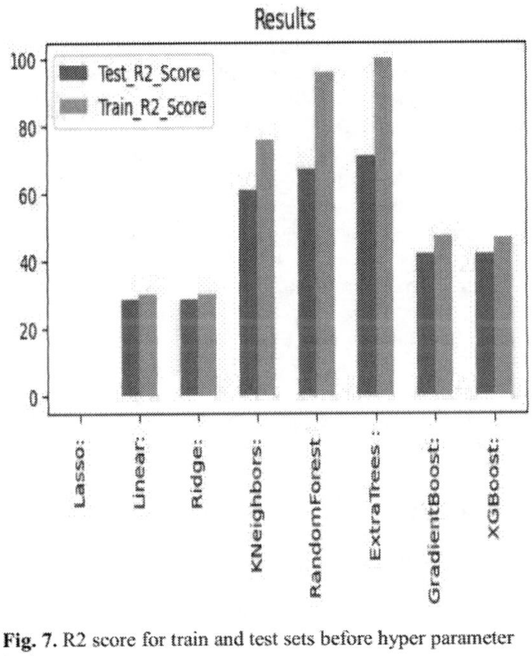

Fig. 7. R2 score for train and test sets before hyper parameter tuning

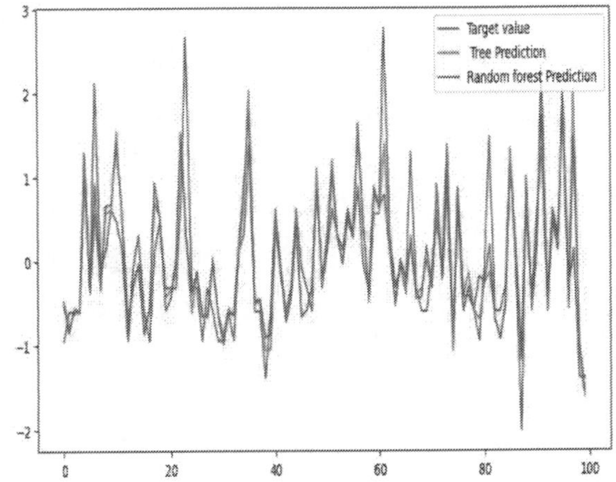

Fig. 8. Energy Prediction for Extra Trees and Random forest algorithms.

5.3 Case 3: Results with Hyper parameter tuning

The results for different algorithms after feature engineering and hyper parameter tuning are demonstrated as shownin the table 5. Linear and Ridge has 27% R2 score on test data. Random forest has 70% and Extra-tree regressor has 74% of scores.

Table 3. R2 score of different algorithms after hyper parameter tuning

Name	Train_Time	Train_R2_Score	Test_R2_Score
Lasso	0.013325	0.0000	-0.0110
Linear	0.01014	26.9782	27.1519
Ridge	0.007331	26.9782	27.1520
K-Neighbors	0.035377	77.3739	64.0166
Random Forest	15.88206	95.7882	70.1538
Extra Trees Regressor	4.9309	100.0000	74.4962
Gradient Boosting	5.104979	43.06144	40.2229
XGBoost	1.23693	43.13197	39.9348

The bar plot visualizing the results are illustrated inFig. 9. It can be seen that extra-trees is performing high than all other algorithms. It can be seen that linear and ridges are having very close test and train scores. After hyper parameter tuning it is observed that the test and train scores have come slightly closer. It seems the models more promising now than before.

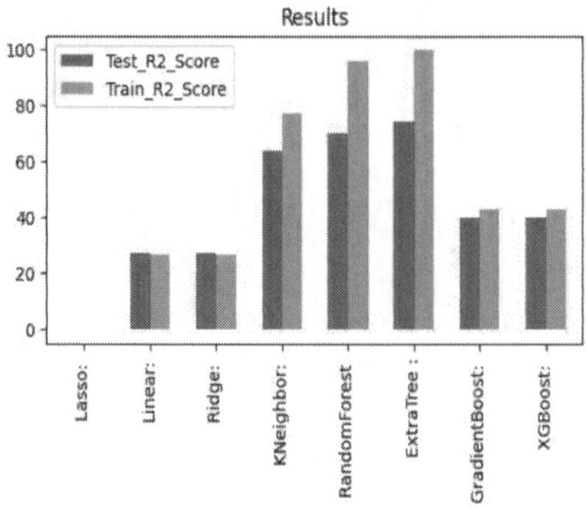

Fig. 9. R2 score for train and test sets after hyper parameter tuning

Fig 10 shows the prediction graph for 100 samples. Blue color line indicates the target value which is actual value. The prediction done with extra tree algorithm is shown by green and random forest with red color. It can observed that extra tree algorithm is able to mimic the original values and was able generalize more than all the other models. The winner model is extra-tree regressor. The prediction rate has increased by 4% after hyper parameter tuning. However, the sudden peaks are difficult to predict as they are not happening in general

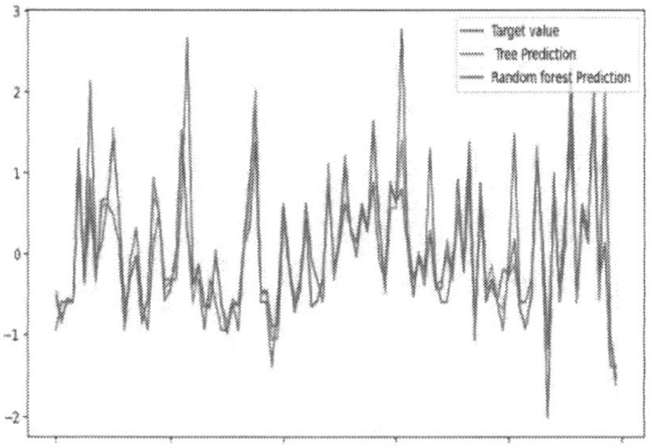

Fig. 10. Energy Prediction for Extra tree and Random forest algorithms after tuning.

Features important here are average house humidity and multiplicative features engineered to have more importance as shown in Fig. 11.Extra tree regressor is performing incredibly well, and Random forest regressor is also working well, as shown in the table above. Additional hyper parameter tuning was done on the Extra tree regressor, which resulted in an R^2 score of 74.5%of which utilising the optimal parameters as {'max_depth': 80, 'max_features': 'sqrt', 'n_estimators': 250}. As a result, predictions can be made using these models.

Lasso is at worst in this case this might be because features that are highly correlated are discarded by making the coefficients zero except only the one which is selected. As a result, the model is not at all able to capture the pattern here.

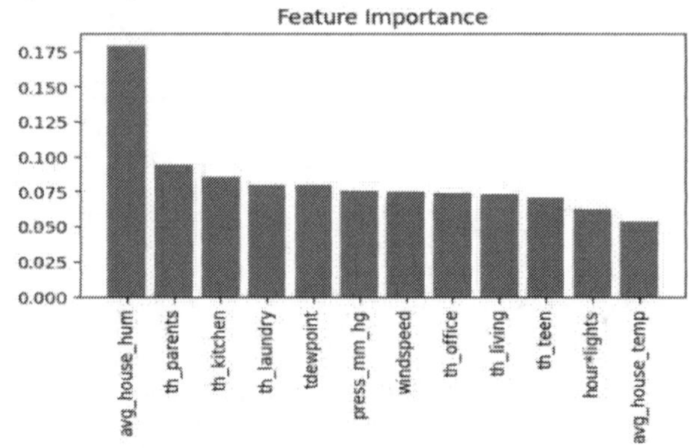

Fig. 11. Feature importance for Extra Trees regressor model

6 Conclusion

Despite being substantially linearly associated within themselves, the temperature/humidity features exhibited little to no linear (Pearson) connection with the objective variable. Although fascinating characteristics were engineered by modifying features, when it comes to energy use, the time of day is crucial. Extra Trees Regressor is the best algorithm to deploy for this dataset (tree-based algorithm). The untuned model could explain 71% of the variance (R^2 score = 0.710)) on the test set, while the tuned model could explain 74.5% of the variance (R^2 score = 0.7449). Average house humidity and other features which are engineered are showing more feature importance. When dealing with a data set with most of its features having no linear correlation with the objective variable, tree-based models are by far the best model.

References

1. Md. Sumon Shahriar and M. Sabbir Rahman. Urban Sensing and Smart Home Energy Optimisations: A Machine Learning Approach. In Proceedings of the 2015 International Workshop on Internet of Things towards Applications (IoT-App '15). Association for Computing Machinery, New York, NY, USA, 19–22. (2015).
2. B Dutta, V Adhinarayanan, and W C Feng. GPU power prediction via ensemble machine learning for DVFS space exploration. In Proceedings of the 15th ACM International Conference on Computing Frontiers (CF '18). Association for Computing Machinery, New York, NY, USA, 240–243. (2018).
3. Feinberg E.A., Genethliou D. Load Forecasting. In: Chow J.H., Wu F.F., Momoh J. (eds) Applied Mathematics for Restructured Electric Power Systems. Power Electronics and Power Systems. Springer, Boston, MA. (2005)

4. Ahmad, T., Chen, H., Huang, R., Yabin, G., Wang, J., Shair, J., Akram, H.M.A., Mohsan, S.A.H. and Kazim, M., Supervised based machine learning models for short, medium and long-term energy prediction in distinct building environment. Energy, **158**, 17-32. (2018).
5. Ghalehkhondabi, I., Ardjmand, E., Weckman, G.R. and Young, W.A., An overview of energy demand forecasting methods published in 2005–2015. Energy Systems, **8(2)**, 411-447. (2017).
6. Sarker, I.H., Machine learning: Algorithms, real-world applications and research directions. SN Computer Science, **2(3)**, 1-21. (2021)
7. Ng, A.,. Advice for applying machine learning. In Machine learning (2011)
8. Sarker, I.H., Alqahtani, H., Alsolami, F., Khan, A.I., Abushark, Y.B. and Siddiqui, M.K., Context pre-modeling: an empirical analysis for classification based user-centric context-aware predictive modeling. Journal of Big Data, **7(1)**, 1-23. (2020).
9. Azuaje, Francisco & Witten, Ian & E, Frank.. Witten IH, Frank E: Data Mining: Practical Machine Learning Tools and Techniques. Biomedical Engineering Online - BIOMED ENG ONLINE. **5**. 1-2. (2006)
10. Basu, K., Debusschere, V. and Bacha, S., Appliance usage prediction using a time series based classification approach. In IECON 2012-38th Annual Conference on IEEE Industrial Electronics Society (1217-1222). IEEE. October (2012)
11. D Bishwajit & A Vignesh & F Wu.. GPU power prediction via ensemble machine learning for DVFS space exploration. 240-243. (2018) 10.1145/3203217.3203273.
12. G Hackeling.. Mastering Machine Learning With scikit-learn. Packt Publishing. (2014)
13. Géron, A. 'Hands-on machine learning with Scikit-Learn and TensorFlow concepts, tools, and techniques to build intelligentsystems' , O'Reilly Media , Paperback . (2017)
14. A. Zeng, H. Ho, Y. Yu, Prediction of building electricity usage using Gaussian Process Regression, Journal of Building Engineering (2019) doi: https://doi.org/10.1016/ j.jobe.2019.101054
15. Chen, Y. and Tan, H., Short-term prediction of electric demand in building sector via hybrid support vector regression. Applied energy, **204**, 1363-1374. (2017)
16. Ahmad T, Chen H, Nonlinear autoregressive and random forest approaches to forecasting electricity load for utility energy management systems, Sustainable Cities and Society (2018)
17. Geurts, Pierre & Ernst, Damien & Wehenkel, Louis. Extremely Randomized Trees. Machine Learning. 63. 3-42. (2006) 10.1007/s10994-006-6226-1.
18. Chen, T. and Guestrin, C., August. Xgboost: A scalable tree boosting system. In Proceedings of the 22nd acm sigkdd international conference on knowledge discovery and data mining (785-794). (2016)
19. Wang, R., Lu, S. and Feng, W., A novel improved model for building energy consumption prediction based on model integration. Applied Energy, **262**, 114561. (2020)
20. Touzani, S., Granderson, J. and Fernandes, S. Gradient boosting machine for modeling the energy consumption of commercial buildings. Energy and Buildings, **158**, 1533-1543. (2018)
21. Fabian Pedregosa, Gaël Varoquaux, Alexandre Gramfort, Vincent Michel, Bertrand Thirion, Olivier Grisel, Mathieu Blondel, Peter Prettenhofer, Ron Weiss, Vincent Dubourg, Jake Vanderplas, Alexandre Passos, David Cournapeau, Matthieu Brucher, Matthieu Perrot, and Édouard Duchesnay. 2011. Scikit-learn: Machine Learning in Python. J. Mach. Learn. Res. 12, null (2/1/2011), 2825–2830.
22. Kangji Li, Chenglei Hu, Guohai Liu, Wenping Xue, Building's electricity consumption prediction using optimized artificial neural networks and principal component analysis, Energy & Buildings (2015)
23. Putatunda, S. and Rama, K., 2018, November. A comparative analysis of hyperopt as against other approaches for hyper-parameter optimization of XGBoost. In Proceedings of the International Conference on Signal Processing and Machine Learning (6-10). (2018)
24. Dodier, R. H., and Henze, G. P. (February 12, 2004). "Statistical Analysis of Neural Networks as Applied to Building Energy Prediction ." ASME. J. Sol. Energy Eng. **126(1)**: 592–600, February (2004)
25. Alawadi, S., Mera, D., Fernández-Delgado, M., Alkhabbas, F., Olsson, C.M. and Davidsson, P. A comparison of machine learning algorithms for forecasting indoor temperature in smart buildings. Energy Systems, 1-17. (2020)

Development, Application and Prospect of Comprehensive Logging Technology in Geothermal Drilling

Delong Zhang[1,2,3,4], *Qiang* Guo[3,4], *Peng* Yang[3,4], *Tong* Lu[3,4], *Yunchao* He[3,4], *Wei* Weng[3,4], and *Baolin* Liu[1,2,*]

[1]China University of Geosciences (Beijing), School of Engineering and Technology, 100083 Beijing, China
[2]Ministry of Natural Resources, Key Laboratory on Deep Geo-Drilling Technology, 100083 Beijing, China
[3]Beijing Institute of Exploration Engineering, Research Department, 100083 Beijing, China
[4]Ministry of Natural Resources, Technology Innovation Center for Geothermal & Hot Dry Rock Exploration and Development, 050060 Shijiazhuang Hebei, China

Abstract. With the continuous acceleration of the exploration and development progress of geothermal (hot dry rock) resources, the workload of geothermal resource investigation and drilling has increased sharply, but the research and application of comprehensive logging technology is relatively lagging behind in China. This article analyses the response relationship of various logging data to geothermal geological conditions item by item, introduces the development and application status of logging technology at home and abroad, and takes HR-1 well as an example to give a brief introduction and analysis of the comprehensive logging technology. Theory and practice have proved that the comprehensive logging can provide an important reference and scientific basis for the evaluation of the potential of geothermal reserves. The development and application status of the geothermal logging technology in China can no longer meet the needs of geothermal resource exploration and development. It is imperative to carry out in-depth research on geothermal logging technology and further increase its application.

1 Introduction

Geothermal resources are a kind of renewable, carbon-free clean energy. With the rapid development of the economy, facing the urgent need of dual-carbon goals, the development needs of China on geothermal resources are increasing rapidly. The exploration process of geothermal resources mainly includes the stages of geophysical exploration, geochemical exploration, drilling investigation and evaluation. The geophysics, geochemistry and other exploration methods are important for delineating the key target areas, but drilling is the only method to accurately obtain the geological parameters of deep geothermal formations. In recent years, a large amount of funds has been invested in the exploration of high-temperature geothermal resources in China, a number of key target areas for high-temperature geothermal and hot dry rocks are delineated, and a number of geothermal resource survey and drilling projects are organized and implemented. The core samples and logging data obtained through geothermal drilling are important basis for comprehensive evaluation of thermal reservoir characteristics and reservoir modelling analysis [1,2]. However, because geothermal logging has not been paid enough attention, the analysis of logging data in the drilling process is not timely enough, resulting in the lack of a large number of important data, which has a very negative impact on the accuracy of geothermal resource evaluation.

2 The response of logging data to geothermal geological conditions

Logging technology is widely used in oil and gas drilling projects, which is very important for discovering reservoirs and arranging drilling plans. And it is of great significance to reservoir research and evaluation. In geothermal drilling, logging technology also plays an important role in the investigation and potential evaluation of geothermal resources [3].

The logging data can be used to effectively judge the lithology, stratification, fracture development, geothermal gradient, water and gas characteristics and other important information of the formation [4]. Timely and accurate analysis of the logging data can be used to effectively discover and identify reservoirs, provide guidance for the follow-up drilling plan, improve the accuracy and efficiency of the geothermal reservoir evaluation.

In addition, the interpretation of completion logging data often requires a large amount of regional drilling data as a reference, while the previous geothermal drilling data is usually very scarce, which makes the accuracy and reliability of the interpretation of completion logging is difficult to guarantee. Analysing

* Corresponding author: tgszdl@126.com

completion logging data in combination with comprehensive logging data can further improve the accuracy and reliability of the research and evaluation result of geothermal reservoir.

In addition to the significance for geothermal reservoir evaluation, comprehensive logging data also has a strong guiding significance for drilling project. By analysing comprehensive logging data, the lithology and temperature changes of formation can be discovered in real time, which can provide a basis for the subsequent drilling, well completion, testing and other plans. By recording the flowrate of drilling fluid inlets and outlets, it is possible to accurately locate water inrush or mud loss formations, which is also an important reference for complex formation treatment and cementing operation. through logging data, formation pressure can be analysed, so as to prevent blowout accidents in advance. With mud logging data, formation pressure can be analysed in real time, so as to prevent blowout accidents in advance.

2.1 Cutting logging

Cuttings logging mainly includes lithology naming, layering depth, and lithology description. Its main work is to collect rock samples, calculate the depth and describe the lithology [5]. Through the cuttings, the stratigraphic sequence, buried depth, thickness, and lithological characteristics of the formations can be accurately identified, the structural deformation of the formation can be found, and the geophysical survey results can be verified and revised to guide subsequent drilling operations, which can provide greater probability for reservoir identification and discovery.

2.2 Mud logging

The main data of drilling mud includes drilling mud properties, blowout, kick, lost circulation, and inlet and outlet temperature. The mud parameter includes density, viscosity, loss, sand content, and static shear force. The changes and anomalies of drilling fluid parameters have important guiding significance for deep formation hydrogeological research.

(1) Affected by the rock and fluid in the formation, the properties of the drilling mud will change; therefore, the lithology of the formation and the characteristics of the fluid contained in the formation can be judged by the changes in the properties of the drilling mud. The lithology, pressure, fluid content of formation are important data for the evaluation and research of geothermal reservoir. It is essential for the accuracy of the research of geothermal reservoir to detect and analyse the changes in drilling mud properties in real time.

(2) The leakage and water outflow of the formation can be reflected by the inlet and outlet flowrate of drilling mud effectively. By measuring the inlet and outlet flowrate, the leakage and water outflow of the formation can be calculated accurately, so as to calculate the water content, pressure and fracture development of the formation. Leakage is the most common downhole complications in geothermal wells. The geothermal reservoir can be accurately judged and located through the temperature change and leakage of the drilling mud circulation. After the reservoir is discovered through logging, the reservoir test and the next operation plan can be discussed and decided in time. Affected by the heat transfer between the upper and lower layers with the drilling mud in the well, it is difficult to accurately locate the top and bottom depths of the reservoir only by completion logging. The method of identifying reservoirs by mud logging is more time-efficient and accurate than completion logging.

(3) Formation thermal anomalies can be detected in time through temperature changes at the inlet and outlet of the drilling mud. The heat transfer between the drilling mud and formation occurs during the mud circulation process, the temperature changes of the drilling mud is an important indicator of the temperature change of the downhole formation. The change of the geothermal gradient can be accurately calculated by the dynamic change of the drilling mud [6-8], which is an important basis for locating the reservoir.

(4) The change in the resistivity of the drilling mud is manifested as an abnormal response of formation. Studies have shown that the hydrochemical characteristics of the fluids in geothermal reservoir have a good correlation with the characteristics of geological structural units and geothermal reservoirs [9]. The content of each ion at a certain point in the geothermal well is a comprehensive reflection of the conditions of the replenishment, streamflow, and seepage of the fluid in the reservoir, and is related to its structural location, fracture nature, lithology, burial depth, temperature and other factors [10].

2.3 ROP logging

The ROP Logging data mainly includes well depth, footage, rate of penetration (ROP), weight on bit (WOB), suspending weight, drill pipe rotation speed, torque, drilling tool assembly, drilling bit, mud flowrate, standpipe pressure, etc. Drilling parameters logging is also called engineering logging [11-13]. Drilling parameters such as WOB and torque characterize the drill ability and integrity of the formation. By comparing the drilling efficiency under the same drilling parameters, the difference in the drill ability of the formation can be distinguished, so as to determine the changes in lithology, structure and other aspects along the depth of the well [3]. At the same time, changes in drilling parameters such as WOB and torque can also effectively reflect the development of the cracks and fractures in formation, which play an important role in the analysis of reserves and productivity of geothermal reservoirs.

The Dc-exponent is usually used to predict the formation pressure in oil and gas drilling. The Dc-exponent of the formation can be calculated by parameters such as WOB, ROP, drill pipe rotation speed, bit diameter, drilling fluid performance, etc., and the change of formation pressure can be derived from the Dc-exponent [14,15]. There are usually very few existing drilling data in the geothermal resource exploration block, and the formation pressure is usually unknown. There is no doubt that formation pressure is very important to ensure drilling safety, and it is one of the important parameters in the evaluation and development of geothermal reservoirs.

2.4 Core logging

Core logging is a process of analysing, testing, and comprehensively researching drilling cores. Cores can provide more accurate geological information than cuttings, including the thermal conductivity, heat generation rate, and geochemical characteristics of rocks in different formations, provide a scientific basis for the geothermal structure and heat source mechanism research of the reservoir. In some important formations, it is necessary to carry out sidewall supplementary coring, to verify the analysis results of the mud logging and completion logging more accurately, and to conduct more accurate and scientific evaluation of the reservoir.

2.5 Gas logging

As one of the important works in the oil and gas drilling process, the gas logging is to collect the gas in the drilling fluid through a degasser and perform component analysis. Its main purpose is to identify and judge oil and gas formations by analysing the gas composition and content in the drilling mud, and to detect the harmful gas composition and content at the same time to ensure drilling safety. In the process of research and evaluation of geothermal reservoirs, the analysis of the composition and content of gas in the formation is equally important for the study of the reservoir characteristics [16].

3 Application of comprehensive mud logging technology

As a necessary work, comprehensive mud logging technology is widely used in oil and gas drilling, and plays an important role in the process of oil and gas resource exploration and evaluation. The importance of comprehensive mud logging for the exploration and evaluation of geothermal reservoirs has been realized by European and American countries a long time ago. In these countries, comprehensive mud logging runs through the entire process of geothermal drilling (or dry hot rock drilling), reservoir reconstruction, enhanced geothermal systems construction, and reservoir development. At present, the domestic research and application of comprehensive mud logging technology in geothermal drilling is still in its infancy, and a complete research, application, and evaluation system based on comprehensive mud logging has not yet been formed.

3.1 The application of comprehensive mud logging in geothermal exploration abroad

Comprehensive mud logging has been used in geothermal drilling for a long time in European and American countries. In the early days, Italy proposed the method of geochemical logging, which is essentially to study the changes in the chemical composition of the drilling mud during the drilling process to determine the possible thermal reservoir. Practical application results show that this method is very effective [17]. The combined analysis of mud logging and completion logging data was applied in the Soultz and Rittershoffen geothermal fields in France, which provided scientific basis and guidance for the later construction of artificial reservoir reconstruction [18,19]. Iceland is currently the country with the highest degree of geothermal development and utilization. Comprehensive mud logging technology has played an important role in the exploration and development of Iceland's geothermal resources [20, 21].

The USA has applied comprehensive mud logging technology very early in hot dry rock development, and the selection of logging technology and equipment is one of the important works of geothermal drilling engineering design [22]. In addition to ground logging equipment, downhole geological logging while drilling tools are also used in the geothermal drilling process [1, 20]. The parameters recorded by the logging equipment mainly include drilling depth, drilling time, drilling pressure, drill pipe rotation speed, torque, inlet and outlet mud flowrate, inlet and outlet mud temperature, H2S gas content, standpipe pressure, annulus pressure, mud loss, drilling fluid parameters, etc. The temperature resistance ability of downhole logging while drilling tools can reach above 150℃, and the acquired parameters mainly include the temperature, pressure, flowrate, fracture distribution of the reservoir, and the radioactivity and electrical properties of the rock [1, 22, 23]. Taking the Geysers geothermal field in California as an example, the gamma data collected by the downhole logging tool played an important role in the research and evaluation of the volcanic fracture [24].

Similarly in the Fenton Hill geothermal field, comprehensive mud logging technology was applied during the processes of drilling, fracturing test, connection test, water injection test, and production test, which provided a good support for the development of geothermal resources. During the drilling process of Well GT-2, the loss analysis based on logging data provided a sufficient basis for judging and locating the lost section, and provided effective guidance for cement plugging operations. Before the hydraulic fracturing test of the GT-2 well, the selection of the 6 fractured sections was entirely based on the analysis of the comprehensive mud logging and completion logging data. During the connection test between Well GT-2 and Well EE-1, logging technology was used to monitor and analyse the injection flowrate of Well GT-2 and the return flowrate of Well EE-1. During the connection test between Well EE-3A and Well EE-2A, the chemical logging of drilling mud was also used to monitor and analyse the whole process. During the water injection test of Well EE-3A, the injection time, pressure, and flowrate were recorded by the logging equipment, and the working status and time of the downhole packer were judged through related data. Continuous steady-state observation of production data was carried out during 1992-1995, the data such as pressure, flowrate, temperature of injection and production well were completely recorded, which has provided important contributions for scientists to carry out research on geothermal energy mining technology [25].

3.2 Application of logging technology in Chinese geothermal development

With the continuous acceleration of the exploration and development of geothermal resources, the invest in geothermal drilling is increasing year by year, and the recognition of the comprehensive mud logging in Chinese geothermal resources exploration has gradually increased.

In 2014, the former Ministry of Land and Resources promulgated the "Technical Regulations for Geothermal Drilling" and Tianjin City issued the "Regulations for Mid-low Temperature Geothermal Drilling"[11, 12]. The technical regulations set clear requirements for logging work. On November 4, 2019, the National Energy Administration issued the "Geothermal Logging Technical Specifications", which will be implemented on May 1, 2020. The issuance and implementation of this specification is a milestone in the application and development of geothermal logging technology. In addition to the definition of the logging items and series, the specification also explains the work content and requirements of each logging item [13].

However, in most of the current geothermal drilling projects, the logging work still stays at the cuttings logging and simple hydrological observation level [3, 24]. Although the specification has been implemented in 2020, there has not yet been a fully equipped geothermal logging equipment and technical system, and there is no way to talk about geothermal reservoir evaluation methods based on the comprehensive mud logging technology. In the current geothermal drilling regulations, there are still the problems of inconsistent parameters and requirements for logging work, which also fully shows that the recognition of the comprehensive mud logging has not been unified, and the attention of it is far from enough.

Although the specification has clarified the geothermal logging items and technical requirements, due to the technology and cost factors, the application cases of comprehensive mud logging in actual geothermal drilling project are extremely rare; even if a drilling project is equipped with logging equipment, there is still a lack of systematic analysis methods to interpret the logging data in real time and efficiently, which severely restrict the development of geothermal logging and reservoir evaluation technology.

4 Application case of geothermal logging technology

Well HR-1 is a geothermal exploration well deployed by the China Geological Survey in Guangdong Province. A comprehensive logging system was applied Well HR-1, which mainly included cuttings logging, core logging, ROP logging, drilling mud logging, production test logging. It provided important data and scientific basis for the drilling and reservoir evaluation.

4.1 Cutting logging

During the drilling process of Well HR-1, a total of 588 packs of cuttings were collected. Based on the analysis of cuttings, the formations were accurately divided, and the regional thermal metamorphism sequence from shallow to deep was clearly recognized. At the same time, the well structure and coring plan were optimized according to the changes in the formations. Take the No. 1-6 cuttings obtained from the 2410m-2415m depth as an example (Figure 1). No. 1~3 are monzonitic granites with dense lithology and strong abrasiveness; No. 4-6 cuttings have darker colour and low hardness, which is determined to be alteration hornfels according to mineral analysis, indicating that the lithology of the formation is dominated by altered minerals after the depth of No. 4 cutting.

Fig. 1. 2410m-2415m cuttings samples from Well HR-1

4.2 Core logging

The coring plan of Well HR-1 was arranged based on the changes in the lithology of formation. There were 13 coring trips in total. The cores taken are extremely representative, which is of great significance to the study of regional geological structure and geothermal reservoir. For example, when drilling to the depth of 2913m, the ROP is unstable and the drilling tool bouncing is severe. According to the core taken from 2917m-2920m (Figure 2), it can be seen that the formation fractures are developed, and fluid crystalline minerals can be seen on the cores.

Fig. 2. 2917m-2920m cores of Well HR-1

4.3 Mud logging

The Well HR-1 locates the temperature abnormal section and water content formation through the mud logging. And the leakage of drilling mud was recorded in detail and accurately. When drilling to the depth of 2572m, a sudden loss occurred, the leakage rate was 8 cubic meters per hour.

The sealing while drilling was carried out and succeeded after a total of 55 cubic meters of mud was lost. The second severe mud loss occurred at the depth of 2777.12m, the drilling mud lost return. After all possible methods failed, the drilling mud was forced to be changed to water. From the depth of 2777.12m to 3006.9m, a total of 1399.8 cubic meters of drilling mud and 7922.2 cubic meters of water were pumped into the well. This loss accident showed that fractures of the formation from the depth of 2777.12m to 3006.9m are well developed.

4.4 ROP logging

The drilling time of the upper granite formation ss about 50-70min/m, then it suddenly decreases to 13-20min/m when drilled to the depth of 2413m, the WOB and standpipe pressure begins to change significantly, and the temperature of outlet drilling mud gradually increase. According to the completion logging data, the resistivity of the depth of 2413m-2420m is reduced from 1000Ω·m to 140-150Ω·m, and the natural gamma is reduced from 400API to 20-30API. The temperature curve of completion logging shows a positive anomaly. The acoustic travel time increases slightly, which indicates that the bottom layer of the section from the depth of 2413m-2420m is a typical high-temperature altered formation. Analysis of the cuttings logging, mud logging, ROP logging and completion logging data shows that the fractures of the formation in this section are developed, which proves that the formation must be a good water-producing layer.

4.5 Production test

A total of 6 production tests are carried out in Well HR-1, and the temperature, flowrate, and pressure data are recorded real-time (Figure 3), which has provided an important scientific basis for the evaluation, productivity analysis, and economic evaluation of the geothermal reservoir [26].

Fig. 3. Data collection system of thermal storage test of Well Huire-1

5 Conclusions and prospects

Application practices at home and abroad have proved that comprehensive logging technology can provide an important scientific basis for the evaluation and research of geothermal reservoirs.

At present, the research and application of comprehensive logging technology in geothermal exploration is still in its infancy in China, and practitioners still have insufficient awareness of the importance of it. It is urgently needed to further increase the investment in the research and application of geothermal comprehensive logging technology.

With the rapid development of information technology, the technical problems of integrating geophysical prospecting, drilling, comprehensive logging, and completion logging data no longer exist. The concept of integrated exploration and development system has been put forward in the field of oil and gas exploration. The development of an integrated exploration and development system is of great significance, for improving the of the exploration and promoting the development and utilization of geothermal resources.

The development of an integrated exploration and development system is of great significance, which will help to improving the scientificity and efficiency of exploration, and provide an important impetus for the development and utilization of geothermal resources.

This work was supported by the National Natural Science Foundation of China (No. 41802197), the project of China Geological Survey (No. DDDD20221681, DD20201102) and Research Foundation of Key Laboratory of Deep Geodrilling Technology, Ministry of Natural Resources(No.KF202103).

References

1. A. Jelacic, R. Fortuna, R. Lasala, et al. An evaluation of enhanced geothermal systems technology. (2008).
2. M. Aleardi, A. Mazzotti, A. Tognarelli, et al. Seismic and well log characterization of fractures for geothermal exploration in hard rocks. Geophysical Journal International, **203(1)**:270-283, (2015).
3. H. Tenzer. Development of Hot Dry Rock Technology. (2001).
4. R. R. Pendon. Borehole geology and hydrothermal mineralisation of well HE-22, Ikelduháls field, Hengill area, SW-Iceland.
5. S. Wang, J. Liu, P. Lin, et al. Some problems to be noted in geological logging during geothermal prospecting. Urban Geology, **3(1)**: 33-36 (2008).
6. F. Kong, Y. Liu. Study and application of the MWD pressure and temperature system. Petroleum Geology & Oilfield Development in Daqing, **24(1)**:76-79 (2005).
7. B. Zhong, D. Fang, T. Shi. A new model for predicting the temperature distribution in the well during drilling. Journal of Southwest Petroleum University (Science & Technology Edition), **21(4)**:53-56 (1999).
8. W. Zhu, D. Zhang, M. Li, et al. Numerical calculation analysis of cyclic temperature field during hot dry rock drilling process. The 18th National Annual Conference on Prospecting Engineering (Rock & Soil Drilling and Tunneling) Academic Exchange, 372-379 (2015).
9. M. Song, J. Liu, L. Qin, et al. Analysis on the hydrochemical characteristics and isotope of geothermal fluid in Tianjin. Geological Survey and Research, **v.41; No.162(02)**:60-66 (2018).
10. Y. Hu, B. Gao, B. Jin, et al. Chemical Distribution and Forming Mechanism of the Geothermal Water in Tianjin. Geological Survey and Research, **30(3)**:213-218 (2007).
11. DZ/T0260-2014. Technical Regulations for Geothermal Drilling. Beijing: China Standard Press, (2014).
12. DB12/T 541-2014. Medium and low temperature geothermal drilling regulations. Tianjin: Tianjin Municipal Market and Quality Supervision and Administration Commission, (2014).
13. NB/T 10268—2019. Technical specification for geothermal logging. Beijing: National Energy Administration of the People's Republic of China, (2019).
14. Z. Chen, J. Deng, B. Yu. Pore-Pressure Composite Forecast in Exploration Well. Special Oil & Gas Reservoirs, **23(01)**:25-27+145+152 (2016).
15. S. Fu, W. Jia. Computing method improvement for formation pressure monitoring while drilling. Mud Logging Engineering, **16(4)**:48-53 (2005).
16. P. Zhao, Duoji, T. Liang, et al. Gas geochemical characteristics of the Yangbajing geothermal field in Tibet. Chinese Science Bulletin, **20(7)**: 691-696 (1998).
17. Y. Zheng. Geochemical logging-a bit of experience in a geothermal investigation in Italy. Foreign Geoexploration Technology, **(10)**:23-24 (1983).
18. H. Tenzer, C. H. Park, O. Kolditz, et al. Application of the geomechanical facies approach and comparison of exploration and evaluation methods used at Soultz-sous-Forêts (France) and Spa Urach (Germany) geothermal sites. Environmental Earth Sciences, **61(4)**:853-880 (2010).
19. J. Vidal, A. Genter, J. Schmittbuhl. Pre- and post-stimulation characterization of geothermal well GRT-1, Rittershoffen, France: insights from acoustic image logs of hard fractured rock. Geophysical Journal International, **206(2)**:845-860, (2016).
20. X. Li, J. Wang, J. Sun. Iceland's geothermal development and the enlightenment to our country for geothermal resources management. Conservation and Utilization of Mineral Resources, **(01)**:6-9 (2014).
21. BM Sveinbjörnsson. Success of High Temperature Geothermal Wells in Iceland. Iceland: Iceland GeoSurvey, (2014).
22. J. Finger, D. Blankenship. Handbook of best practices for geothermal drilling. New mexico: Sandia National Laboratories, (2010).
23. I. Stober, K. Bucher. Geothermal Energy From Theoretical Models to Exploration and Development. Springer-Verlag Berlin Heidelberg, (2013).
24. S. Sanyal, M. Che, R. Dunlap, M. Twichell. Qualitative Response Patterns on Geophysical Well Logs from The Geysers, California, Geothermal Resources Council TRANSACTIONS, **(6)**:313-316 (1982).
25. D. W. Brown, D. V. Duchane, G. Heiken, et al. Mining the Earth's Heat: Hot Dry Rock Geothermal Energy. Springer Geography, (2012).

26. L. Hu, G. Zheng. A discussion on the sieve residue logging techniques and methods in geothermal well drilling. Urban Geology, **2(3)**:48-50 (2007).
27. T. Li, W. Lin, H. Gan, et al. Research on the genetic model and exploration progress of hot dry rock resources on the southeast coast of China. Journal of Geomechanics, **26(2)**:187-200

Effects of Mesoporous Size and Structure on Solidification Characteristics of Solar Salt

Rui Mao[1,*], *Qirong* Yang[1], and *Zhaoying* Li[1]

[1]Qingdao University, College of Mechanical and Electrical Engineering, 266071, Qingdao, China

Abstract. The effects of mesoporous size and structure on the solidification characteristics of mixed nitrate were simulated by molecular dynamics (MD). The solar salt models of different scales are established by using material Studio software, and the models are exported to the Lammps software package for simulation calculation. The changes of parameters such as radial distribution function, potential energy temperature curve, freezing point, and phase transformation latent heat are summarized, and the micro mechanism of solar salt solidification characteristics at the nanoscale is analyzed. The results show that the freezing point of solar salt increases first and then decreases with the increase of nanopore size, and finally tends to be stable with the increase of nanopore size. The enthalpy of solidification decreases with the increase of scale.

1 Introduction

Phase change materials(PCM) absorb a lot of heat, increase the internal molecular energy at an almost constant temperature, and absorb and release heat in the process of phase change. This characteristic makes it attractive in the medium-temperature thermal energy storage (TES) system of solar thermal utilization and waste heat recovery[1-4]. In recent years, great progress has been made in solar photothermal power generation using molten salt. Solar salt (60wt%NaNO3+40wt%KNO3) is one of the two most popular molten salt media in solar power plants. However, its low thermal conductivity, easy leakage, and undercooling affect the utilization efficiency of solar energy[5-8].

Goitandia et al.[9] analyzed the effects of a eutectic mixture of butyl and fatty acids as PCM confined cycle stability and an undercooling degree from micropores(greater than 0.6nm) to macropores(<200μm). It was found that the nano constraint of silicon increased the undercooling degree by 1-6℃. Kota et al.[10] studied the crystallization behavior of ET in and around two-dimensional mesoporous silica with three different pore sizes(7.5nm, 8.3nm, and 9.2nm). The results show that the phase transition of ET can be controlled by different pore sizes of micro constraints. The above research shows that the pore size of the substrate will affect the thermal properties of CPCM.

In this paper, the MD simulation method is used to explore the effects of mesoporous scale and cooling rate on the solidification characteristics of solar salt. Based on the simulation process of solidification of nano metal particles and microencapsulated phase change materials within the confines, the solar salt models with different ion numbers are established by using Material Studio software. Through the simulation calculation of lammps software package, and potential energy temperature curve, Gibbs free energy, freezing point The micro mechanism of solidification characteristics of solar salt at the nanoscale is analyzed.

2 Simulation

2.1 Model establishment

Build the nitrate molecular model in the Materials Studio software, as shown in Figure 1, and mix $NaNO_3$ and KNO_3 with a mass ratio of 6:4 to build the binary mixed nitrate model. Solar salt models at different scales are established by proportionally increasing or decreasing the number of molecules, as shown in Fig. 2(a)-(f), which are solar salt models with 460, 920, 1380, 1840, 2300, and 2760 molecules in turn. The number of atoms and model scale is shown in Table 1.

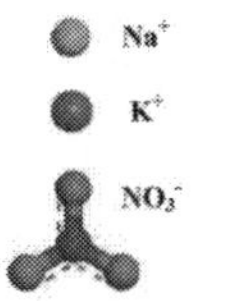

Fig. 1. $NaNO_3$ and KNO_3

Table 1. Molecular numbers of NaNO3 and KNO3 in solar salts(NaNO3:KNO3=6:4)[11]

* Corresponding author: 420698897@qq.com

https://doi.org/10.1051/e3sconf/202235002004

Molecular number Type	460	920	1380	1840	2300	2760
Na	60	120	180	240	300	360
K	32	64	96	128	160	192
N	92	184	276	368	460	552
O	276	552	828	1104	1380	1656
Model Size	5nm	6nm	7nm	8nm	9nm	10nm

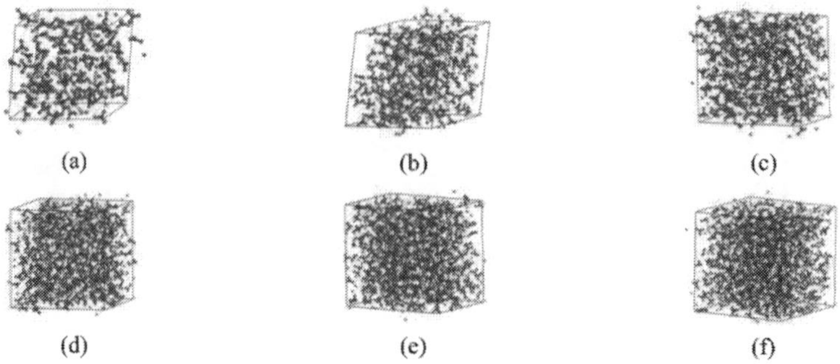

Fig. 2. Solar salt model (a)460; (b)920; (c)1380; (d)1840; (e)2300; (f)2760.

2.2 simulation method

The geometry optimization of the forcite module is used for geometric optimization. The method to evaluate the atomic charge of electrostatic interaction is used current. In the simulation process, atom-based[12] and Ewald[13] methods are used for van der Waals (VDW) and electrostatic interaction (Coulomb) respectively. The temperature is set at 453K-573K, the temperature interval is 10K, and the pressure is set at 0.1GPa. Under the condition of the canonical ensemble (NVT), the method to evaluate the atomic charge of electrostatic interaction is forcefield assigned. PPPM and Ewald are still selected for electrostatic interaction force and van der Waals force, and continue to operate for 70ps to reach equilibrium. Analyze the mean square displacement (MSD) using analysis under the forcite module. The preliminary work of the research group has confirmed the effectiveness of this method[11].

The solar salt model established in MS software is exported to Lammps software package, and the conjugate gradient algorithm is used to minimize the energy. After structural minimization[12], the temperature rises rapidly from 300K to 1000K within 300ps under the NVT ensemble. The electrostatic interaction force is calculated by the PPPM method, and the time step is set to 1fs. L-J and harmonic potential function parameters are shown in Table 2.Using NPT ensemble, the model is first heated and then cooled from 323K to 773K. The simulation parameters are summarized in Table 3. Potential energy, kinetic energy, enthalpy, and total energy are obtained.

Table 2 Potential parameters of Solor Salt composites[14-15]

Atom	Q[e]	E[eV]	σ[Å]
Na	1	6.6373×10^{-3}	2.407
K	1	4.336×10^{-3}	3.188
N	0.95	4.017509×10^{-3}	3.431
O	-0.65	3.469129×10^{-3}	3.285

Table 3 Simulation parameters

Number of Atoms	460、920、1380、1840、2300、2760
Timestep (fs)	1
Pressure (Pa)	1×10^5
Ensemble	NPT
Cooling Rates (K/ps)	0.1

In this simulation, the density of solar salt is 1.802g/cm3, the simulated value of Wu Chenguang et al.[15] is 1.73g/cm3, with an error of 4.16%, and the experimental value of Nicole et al.[16] is 1.79g/cm3, with an error of 0.67%. The error is less than 5%, which can verify the accuracy of the simulation.

$$\overline{E_k} = \frac{1}{2}(n+1)\overline{U} \quad (1)$$

The results of molecular dynamics simulation depend on the selected potential function.According to virial theorem (equation 1), the average kinetic energy and

https://doi.org/10.1051/e3sconf/202235002004

average potential energy of the system are linearly related:

Through equilibrium molecular dynamics simulation, the self-diffusion coefficient D (m2/s) can be calculated by the Einstein method. Einstein relation[17] can be expressed as:

$$D = \lim_{t\to\infty} \frac{1}{6Nt} \sum_{i=1}^{N} \langle [r_i(t) - r_i(0)]^2 \rangle \tag{2}$$

$$\frac{G}{R_g T} - \left(\frac{G}{R_g T}\right)_{ref} = \int_{T_{ref}}^{T} -\frac{H}{R_g T^2} dT \tag{3}$$

3 Results & Discussion

3.1 Potential energy

Taking the solar salt containing 460 molecules as an example, the potential energy function of particles during heating and cooling is shown in Fig. 3(a). For pure materials, whether melting or solidification, the potential energy jumps obviously at a certain temperature, which is the phase transition temperature, and the potential energy changes linearly with the temperature before and after this jump.

The potential energy temperature curves of solar salt nanoparticles with different sizes at the cooling rate of 0.1K/ps are drawn as shown in Fig. 3(b). The potential energy of solar salt nanoparticles with different sizes has the same variation law with temperature, and there is a potential energy step between 448K-548K. At the same simulated temperature, the potential energy increases with the increase of solar salt scale, and the phase transition latent heat increases with the increase of solar salt scale. The simulated latent heat of phase transition at 2760 molecular numbers is about 117kJ/kg, which is within a reasonable error range from the experimental value of 116kJ/kg in reference[18].

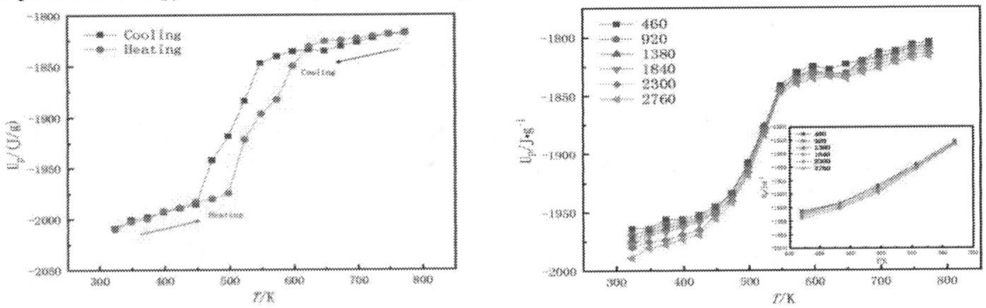

(a) (b)

Fig. 3 Potential energy temperature curve (a) 460 molecular number rise and drop temperature curve; (b) Cooling curves of nanoparticles with different scales.

3.2 Freezing point

The faster the diffusion in the binary nitrate system, the stronger the fluidity of ions in the system. Therefore, when the value of the self-diffusion coefficient changes significantly, it indicates that phase transition occurs at this temperature.

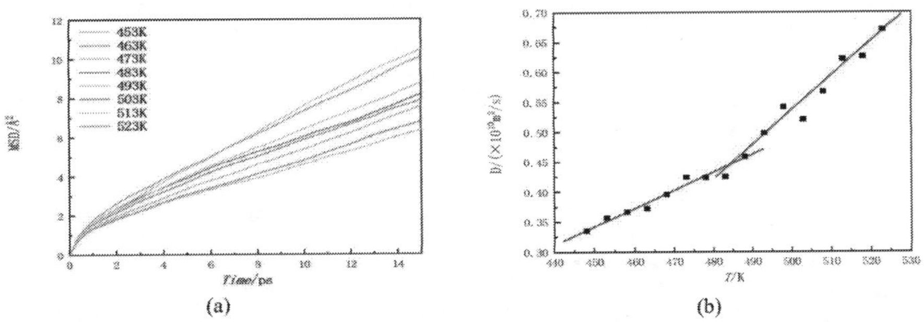

(a) (b)

Fig. 4. Mean square displacement and self diffusion coefficient of solar salt with ion number 2760: (a) mean square displacemen; (b)self diffusion coefficient.

The mean square displacement and self-diffusion coefficient of solar salt with 2760 ions in nanopores at different temperatures are shown in Fig. 4. The freezing point of solar salt at different scales is calculated. Our research group has calculated the melting point of nanoparticle solar salt at different scales, and the simulation results are summarized in Table 4. The simulation results show that the melting point and freezing point of solar salt are different at the same scale, and the melting point of particles is greater than the freezing point. This shows that the liquid-solid transition lags behind the solid-liquid transition at the nanoscale, which has been confirmed in theory and experiment[19], and also proves that it is more difficult to change from disorder (liquid) to order (solid) than from order to disorder.

$$\frac{T_c(r_{np})}{T_c(\infty)} = 1 - \frac{4V_s\gamma_{sl}r_{np}H_c(T)}{1-\frac{\delta}{r_{np}}} \quad (r_{np} < 10\text{nm}) \tag{4}$$

Where, $T_c(r_{np})$ is the solidification temperature of nanocrystals, r_{np} is the radius of nanoparticles, $T_c(\infty)$ is the solidification temperature of blocks, V_s is the molar volume of bulk crystals, γ_{sl} is the interface energy of solid-liquid phase, $H_c(T)$ is the molar solidification enthalpy, and δ is the thickness of a liquid layer at the surface.

For the liquid-solid phase transformation with uniform nucleation, the change with temperature is very small. According to formula (4), when the diameter of nanocrystals is less than 10 nm, it is equivalent to the scale, and the specific surface area has a great influence on the phase transition temperature. Since the specific surface area (surface area/mass) of solar salt is inversely proportional to the diameter, within a certain scale range, the specific surface area will increase significantly with the decrease of the number of ions. At this time, the nucleation is unstable and the formed crystal nucleus is easy to disappear. It is necessary to form a stable crystal nucleus when it is cooled to the freezing point under the condition of large undercooling.

It can be seen from table 4 that with the increase of the number of ions, the melting point increases first and then decreases, reflecting the scale effect of nitrate solidification characteristics at the nano scale.

Table 4. Phase transition temperature of solar salts at different scales

Molecular number	460	920	1380	1840	2300	2760	5520
Melting point/K	493	493	503	518	508	493	493
Freezing point/K	463	483	473	503	490	483	478

460 molecular number solar salt has a freezing point of 463K and a melting point of 493K.The reason for this phenomenon is that the proportion of surface energy of surface atoms in the total energy of particles changes.

3.3 Latent heat of phase change

The difference of system energy before and after solidification is solidification enthalpy (latent heat of solidification phase transformation). The latent heat of phase change represented by the corresponding potential energy difference of the system in Section 3.1 is summarized in Table 5. The simulation results show that the potential energy of solar salt system decreases with the decrease of temperature, and the solidification enthalpy decreases with the increase of scale.

Table 5. Phase transition latent heat of solar salts at different scales

Molecular number	460	920	1380	1840	2300	2760
Latent heat/(kJ/kg)	108.75	109.39	110.12	112.25	114.69	117.57

Similar to bulk materials, the solidification enthalpy of nanomaterials can also be expressed as the product of solidification entropy and freezing point, so formula (5) can be obtained.

$$\frac{H_c(r_{np})}{T_c(r_{np})} - \frac{H_c(\infty)}{T_c(\infty)} = \left(\frac{3k_B}{2}\right)\ln\left[1 - \frac{5r_0}{\frac{1}{\mu^2 r_{np}}}\right] \tag{5}$$

Where, ∞ represents block; r_0 is the atomic radius, nm; r_{np} is the radius of nanoparticles, nm; μ is the shape factor, the value of spherical particles is 1, H_c is the molar enthalpy of solidification, J/mol; S_c is solidification entropy, J/(mol·K); k_B is Boltzmann constant, 1.38×10^{-23}J/K.

It can be seen that when the particle size increases, the solidification enthalpy of nanoparticles increases, and when $r_{np} \to\infty$,the solidification enthalpy tends to be close to the block solidification enthalpy. Since the solidification point increases first and then decreases with the increase of scale, it is obtained from formula (5) that the solidification enthalpy increases faster with the increase of scale, which is consistent with the results obtained by simulation and the law obtained by Eryürek[20] et al.

4 Conclusions

Because it is difficult to observe the solidification process of phase change materials with nanopore size in the experiment, the solidification characteristics of solar salts with different scales are studied by equilibrium molecular dynamics simulation. The conclusions are as follows:

1) With the increase of ion number, the melting point of solar salt increases first and then decreases, indicating that the pore size of the skeleton has an optimal value.

2) The potential energy of the solar salt system decreases with the decrease of temperature, and the solidification enthalpy decreases with the increase of scale.

References

1. Z. Jiang, G. Leng, F. Ye, Z. Ge, C. Liu, L. Wang. Energy Conversion & Management, **106**(DEC.), 165-172 (2015)
2. E. Alehosseini, S. Jafari. M. Trends in Food Science & Technology, **91**, 116-128 (2019)
3. M. Umair, Y. Zhang, S. Zhang, B. Tang. Applied Energy, 235 (2019)
4. X. Chen, H. Gao, M. Yang, L. Xing, W. Dong, A. Li. Energy Storage Materials, **18**, 349-355 (2019)
5. Y. Feng, D. Feng, X. Zhang Phase transition And Heat transport properties of mesoporous composites (Beijing: Science Press) pp110-113 (in Chinese) (2019)
6. X. Chen, Z. Tang, Y. Chang, H. Gao, J. Lv. iScience **23(10)**, 101606 (2020)
7. T. Qian, J. Li, M. Xin, B. Fan. Acs Sustainable Chemistry & Engineering, **6(1)** (2017)
8. C. Xiao, A. Hg, A. Lx, A. Wd, A. Al, B. Pc. Energy Storage Materials, **18**, 280-288 (2019)
9. A. M. Goitandia, G. Beobide, E. Aranzabe, A Aranzabe. Solar Energy Materials & Solar Cells, **134**, 318-328 (2015)
10. K. Nakano, Y. Masuda, H. Daiguji. Journal of Physical Chemistry C, **119(9)**, 150205115437003 (2015)
11. Z. Y. He, Q. R. Yang, Z. Y. Li, R. Mao, L. W. Wang, C.X.J. Yan, /OL Acta Phys. Sin 1-23
12. N. Karasawa, W. Goddard Macromolecules (1992).
13. G. Pan, J. Ding, W. Wang, International Journal of Heat and Mass Transfer **103** 417 (2016)
14. G. Hu, B. Cao. Journal of Applied Physics, **114(22)**, 96 (2013)
15. C. Wu, B. Li. J/OL Journal of composites: 1-11
16. P. Nicole, B. Thomas, M. Claudia, E. Markus, W. Antje. Beilstein Journal of Nanotechnology, **6(1)**, 1487-1497 (2015)
17. D. Sheppard, R. Terrell, G. Henkelman. Journal of Chemical Physics, **128(13)**, 385-404 (2008)
18. M. K. Saranprabhu, K. S. Rajan. Renewable energy, **141**(OCT.), 451-459 (2019).
19. K. Pan, Y. Li, Q. Zhao, S. Zhang. JOM (2018)
20. M. Eryürek, Güven. Physica A: Statistical Mechanics and its Applications, **377(2)**, 514-522 (2007)

Analysis and Treatment of Flue Duct Vibration of 1000MW Generator Set with High Load

Lipeng Wang[1,*], *Jingxue* An[1], *Jinwu* Cheng[1], *Youxin* Li[1] and *Lei* Wang[1]

[1]Datang Central South Electric Power Test Research Institute, Zhengzhou 450000, China

Abstract. In view of the serious flue duct vibration problem of the outlet of a 1000MW unit with a load of 80%BRL and above, combined the vibration test with the calculation of the acoustic standing wave frequency, the cause of the flue duct vibration was analyzed and determined. A modification plan is proposed to install clapboards to change the standing wave frequency of the flue duct and keep it away from the measured resonance frequency. After the clapboards are installed, the flue duct vibration frequency of the induced draft fan outlet is dispersed from the two main frequencies of 61.3Hz and 61.4Hz before the clapboards were installed to 6.88Hz~204Hz; the maximum amplitude of the flue duct is reduced from 105.0μm to 12.6μm, and the problem of the flue duct vibration is fundamentally solved. This case provides a new idea for solving the problem of vibration of the flue duct in the large-scale power station boiler.

1 Preface

The development direction of high intelligence, large capacity and excellent parameters of power station boilers determines the continuous increase of boiler furnace, flue size and flue gas flow rate. High flue gas flow rate brings frequent occurrence of flue excitation or resonance problems. It not only produces noise pollution to the surrounding environment, but also affects the safe and efficient operation of the unit.

Wu Boyang, Li Youxin, etc.[1]~[3] analyzed the vibration mechanism of the boiler tail flue in detail by combining with the periodic shedding law of the Karman vortex when the high-speed flue gas is subjected to the hot surface tube bundle; Yan Xiaozhong, Yang Guoqi, etc.[4]~[10] analyzed the causes of flue vibration through vibration parameter retrieval and vibration test results; Zhang Huarong, etc.[11] conducted the flue vibration elimination research by optimizing the flue flow field. The researches above put forward differentiated solutions for flue vibration under different conditions, but it is rare to find the root cause of flue excitation or resonance problems by combining vibration testing and flue standing wave frequency calculation.

A HG-2913/29.3-YM2 1000MW coal-fired unit vibrates violently from the outlet of the induced draft fan to the inlet of the desulfurization when it was above 80%BRL load. Flue vibration brings problems such as fan vibration, low temperature gas saver wear and leakage, etc. It also brings threats to the safe, efficient and sustainable operation of the unit.

2 Tests and causes analysis

2.1 Vibration tests

The main design parameters of the boiler is in table 1. To find the vibration causes of the flues from the outlet of the induced draft fan to the inlet of the desulfurization, the vibration test of the flue on the right side of this section was conducted with using SK4432 vibration test analyzer when the unit brings 920MW~930MW load. The test positions is on the measuring hole flange of the flue or on the outer wall of the flue after removing the insulation layer of the flue, which is showed in detail in fig.1, and the results of the test is in the table 3 and and fig.2.

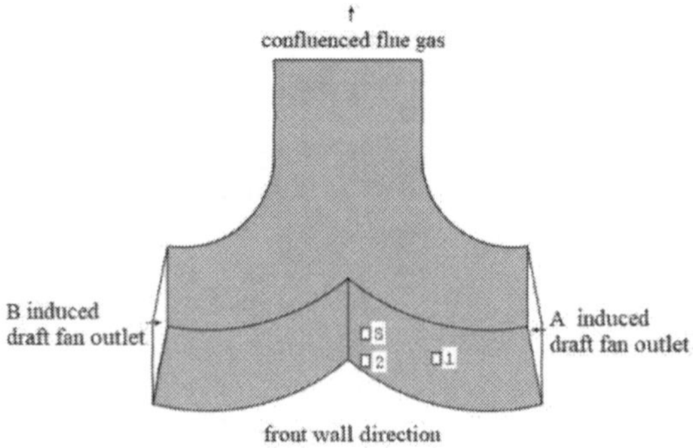

a) the induced draft fan outlet horizontal section

* Corresponding author: michealwlp@163.com

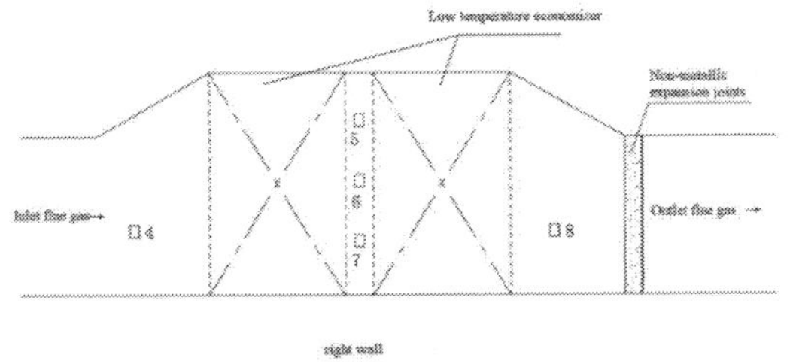

b) the low temperature economizer section

Fig. 1. The vibration test position between the outlet of the induced draft fan and the inlet of the FGD.

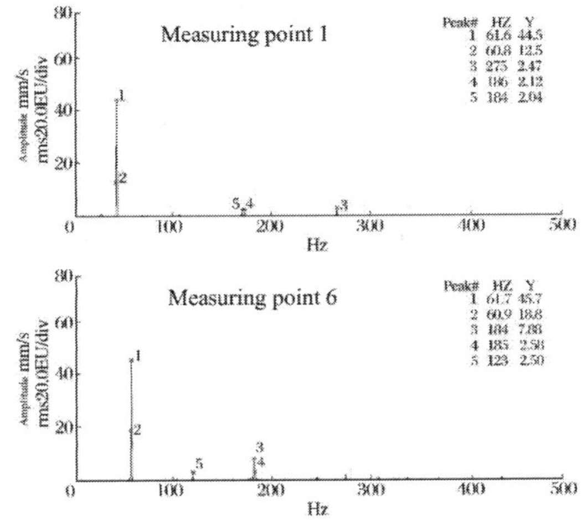

Fig.2. The typical amplitude superimpose in vibration test

Table 1. The main design parameters of the boiler

Name	Unit	BMCR	BRL
Superheated steam mass flow	t/h	2913	2828.2
Superheated steam outlet pressure	MPa(g)	29.3	29.22
Superheated steam outlet temperature	℃	605	605
Reheated steam mass flow	t/h	2384.5	2311.0
Reheated steam inlet pressure	MPa(g)	5.997	5.806
Reheated steam outlet pressure	MPa(g)	5.747	5.565
Reheated steam inlet temperature	℃	360	355.6
Reheated steam outlet temperature	℃	623	623
Feed water temperature	℃	304.9	302.7

According to the vibration amplitude of each part, it is judged that the vibration is mainly concentrated at the outlet of the induced draft fan and the body of the low temperature economizer with the 80% ECR above unit load. The flue vibration frequency falls between 61.3Hz and 61.4Hz, and the maximum amplitude at test position 7 reaches 105μm with the unit load between 920MW and 930MW.

Combined with the dynamic superposition and increase of the vibration amplitude during the test process, it is judged that the large vibration of the flue at the outlet of the induced draft fan was caused by flue resonance.

2.2 Causes analysis

The problems of flue excitation or resonance is partly caused by the integral multiple relationship between the natural acoustic frequency of the flue and the periodic shedding frequency of the Karman vortex generated by the flue gas flow subjected to the hot surface tube bundle (light pipe), the excitation of such vibrations can be attributed to unstable vortex shedding due to the lateral flow of fluid across the sides of a cylinder, which is showed in fig.3; and partly caused by the loose installation or worn of the support or diversion device of the flue.

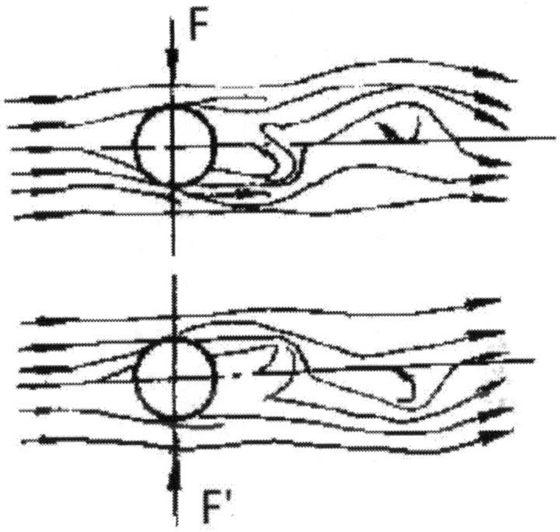

Fig. 3. The karman eddy current

To find the cause of the vibration problem of the flue at the outlet of the induced draft fan with 80% BRL above unit load, the flue natural acoustic frequency's calculation was conducted by calculation equation (1) for the flue with a significantly larger amplitude.

$$f_z = \frac{n}{2}\frac{c}{H} \tag{1}$$

The c in equation (1) is the speed of sound in flue gas medium; n is an integer, which can be 1, 2, 3...; H is the height of the flue itself

The speed of sound in the flue gas medium c at a certain temperature is calculated by equation (2):

$$c = \sqrt{KRT} \tag{2}$$

In equation (2), flue gas adiabatic index K=1.33; gas constant of flue gas R=276J/(kg·K)；T is the gas thermodynamic temperature.

Combined with the drawings of related equipment and flue, *H* of the horizontal flue at outlet of induced draft fan *is* 7000mm, *H* of the low temperature economizer flue is 12810mm. The average temperature *T* of the horizontal flue at outlet of induced draft fan is 131.25℃ when the vibration tests was conducted at 930MW unit load,which is 130.79℃ at the low temperature economizer.

The second harmonic frequency of the induced draft fan outlet flue's height direction natural acoustic frequency was calculated 55.09Hz by substituting the parameters above into equation (1) and equation (2), similarly, the fourth harmonic frequency of the low temperature economizer flue's height direction natural acoustic frequency is 60.23Hz. The ratio to the measured value 61.30Hz is respectively 0.90 and 0.98.

Research shows that the acoustic resonance of the flue occurs when the ratio k to the natural acoustic frequency's certain order harmonic is 0.8~1.2. In conclusion, the flue vibration at the outlet of the induced draft fan is acoustic resonance when the unit load is 80%BRL above.

3 The solution

To reduce the flue height H can effectively increase the natural acoustic frequency of the flue, so the clapboard installation at the induced draft fan outlet horizontal flue and low temperature economizer flue's height direction is feasible.

According to the flue design drawings of the induced draft fan outlet horizontal flue and the low temperature economizer flue, three 6mm thick and the equal flue gas deflector long clapboards are installed at each induced draft fan outlet horizontal flue as the fig.4 show, and two sets of six 6mm thick rectangular 7850mm × 1625mm clapboards are installed at low temperature economizer's upstream and downstream flue as the fig.5 show. All the clapboards' material is GB700 、 Q235-B ordinary carbon steel.

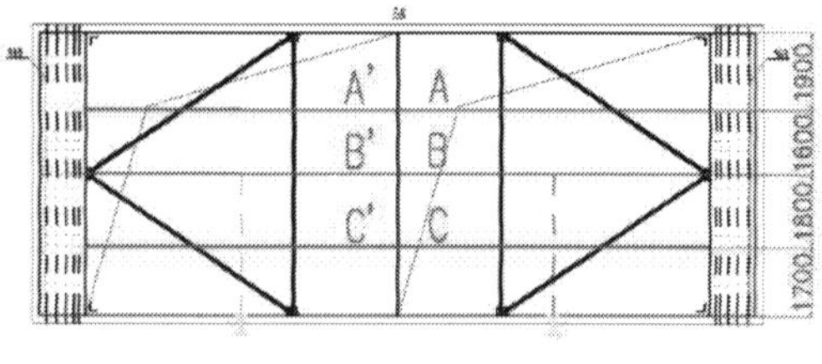

Fig.4. The clapboard installation position schematic diagram of the induced draft fan outlet horizontal fluel

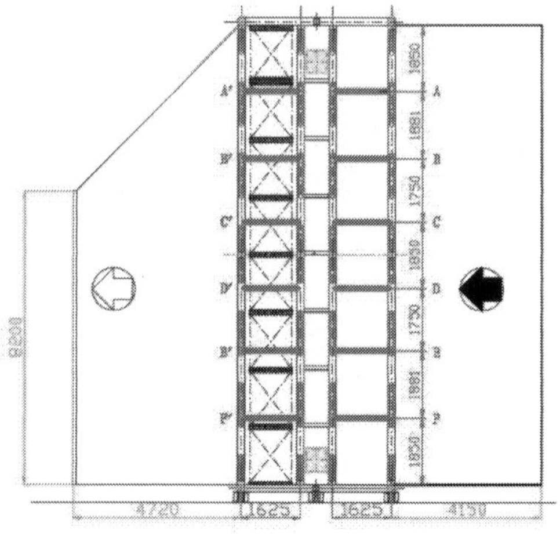

Fig.5. The clapboard installation position schematic diagram of the low temperature economizer flue

In order to avoid the coincidence of the flue natural sound frequencies, the clapboard installation on one side is set differently, which creates hindrance between the adjacent flue.

The calculation result of the each interval flue's natural acoustic frequencies after the clapboard installation is in table 2. From table 2,we can see that the each interval flue's natural acoustic frequencies is above 100Hz after the clapboard installation, which is 1.63 ratio to the measured value 61.30Hz. In part 2.2, we mentioned that the acoustic resonance of the flue occurs when the ratio k to the natural acoustic frequency's certain order harmonic is 0.8~1.2, the ratio 1.63 after the clapboard installation is far away from the acoustic resonance region, which means that the acoustic resonance of the flue will disappear.

Table 2. All the spaced acoustic standing wave theoretical calculation frequency after the clapboards' installation.

Position	Acoustic standing wave frequency(Hz)
Horizontal flue upper wall to upper clapboard	101.49
Upper clapboard to middle clapboard	120.52
Middle clapboard to lower clapboard	107.13
Lower clapboard to horizontal flue lower wall	113.43
Low temperature economizer flue upper wall to the first clapboard	104.17
The first clapboard to the second clapboard	102.46
The second clapboard to the third clapboard	110.13
The third clapboard to the fouth clapboard	104.17

 https://doi.org/10.1051/e3sconf/202235002005

Position	Acoustic standing wave frequency(Hz)
The fouth clapboard to the fifth clapboard	110.13
The fifth clapboard to the six clapboard	102.46
The six clapboard to low temperature economizer flue lower wall	104.17

4 The results before and after the clapboard installation

Considering that the width direction of the flue is too long,the actual clapboards' installation at the low temperature economizer is showed in fig.6.

Fig.6. The clapboards' installation at the low temperature economizer on site.

Adopt the solution described in 3 to install the clapboard on the horizontal section of the induced draft fan outlet and the low temperature economizer flue, the flue duct vibration frequency of the induced draft fan outlet is dispersed from the two main frequencies of 61.3Hz and 61.4Hz before the clapboards were installed to 6.88Hz~204Hz; the maximum amplitude of the flue duct is reduced from 105.0μm to 12.6μm, which is showed in table 3.

Due to the vibration frequency and vibration amplitude measurement results of the induced draft fan outlet flue before and after the installation of the integrated clapboard, the problem of the flue duct vibration is fundamentally solved.

Table 3. The contradistinction of the vibration test before and after the clapboards' installation.l

Position		Amplitude maximum (μm)		Frequency at maximum amplitude(Hz)	
		Before the clapboard installation	After the clapboard installation	Before the clapboard installation	After the clapboard installation
The horizontal section of the induced draft fan outlet	1	60.8	7.11	61.4	6.88
	2	84.1	12.6	61.3	28.8
	3	66.6	12.3	61.3/61.6	19.4
Low temperature economizer inlet	4	25.4	4.21	61.4	204
Low temperature economizer	5	34.6	4.2	61.4	10.0
	6	53.3	5.07	61.4	10.0
	7	**105.0**	2.72	61.4	10.0
Low temperature economizer outlet before expansion joint	8	44.6	9.22	61.4	9.38

5 The conclusion

In view of the serious flue duct vibration problem of the outlet of a 1000MW unit with a load of 80%BRL and above, combined the vibration test with the calculation of the acoustic standing wave frequency, the cause of the flue duct vibration was analyzed and determined. A modification plan is proposed to install clapboards to change the standing wave frequency of the flue duct and keep it away from the measured resonance frequency.

After the clapboards are installed, the flue duct vibration frequency of the induced draft fan outlet is dispersed from the two main frequencies of 61.3Hz and 61.4Hz before the clapboards were installed to 6.88Hz~204Hz; the maximum amplitude of the flue duct is reduced from 105.0μm to 12.6μm, and the problem of the flue duct vibration is fundamentally solved. This case provides a new idea for solving the problem of vibration of the flue duct in the large-scale power station boiler.

References

1. B. Wu. Acoustic Vibration Mechanism and Vibration Damping Measures for Backpass of Large Utility Boiler.Boiler technology, **(10)**: 23-25 (2000).
2. K. Dong. Influence of Kaman Vortex upon Safety of Utility Boilers and Remedial Measures Thereof.Thermal power, **(10)**:31-34 (2008).
3. Y. Li. Supercritical boiler tail flue gas duct vibration cause analysis and treatment. Huadian Technology, **39(10)**:49-51+79 (2017).
4. X. Yan, S. Yang, T. Yu, J. Lin, Y Wang. Vibration analysis and comprehensive treatment of the flue duct of a gas-fired 130 t/h boiler.Journal of Changsha University of Science and Technology (Natural Science), **(01)**:62-67 (2007).
5. G. Yang, X. Yu, P. Li, Y. Dang, P. Guo, Y. Wang. Analysis on Boiler Tail Duct Vibration in 330MW

https://doi.org/10.1051/e3sconf/202235002005

Unit.Ji Shu Jiao Liu Yu Ying Yong, **38(05)**:77-79 (2010).

6. Y. Yi. Analysis on Vibration Mechanism of Low Temperature Economizer of 350MW Coal-fired Unit and Countermeasures. GUANGDONG ELECTRIC POWER, **28(07)**:33-35+44 (2015).
7. J. Lou, J. Jiang, X. Liao. Cause Analysis and Treatment on Tail Duct Vibration of a 1990t/h Boiler.POWER EQUIPMENT, **26(05)**: 365-368 (2012).
8. C. Wang, X. Lu, W. Feng,et al. Analysis of Boiler Tail Flue Vibration and Solutions.Power System Engineering, **28(04)**:37-38+43 (2012).
9. X. Tan, S. Zhang. Vibration reasons and steps elmiinating them on site of the flue duct and airway in power plant. Heilongjiang Electric Power,**(02)**: 127-129 (2005).
10. Z. Kuang, L. Li, W. Yang, X. Yan. Vibration Analysis and Retrofit Research for a Boiler Flue System. BOILER MANUFACTURING, **(01)**:15-20 (2012).
11. H. Zhang, X. Lin, X. Ye,G. Wu. Study on flow field optimization to eliminate duct vibration. Electric Power Technology and Environmental Protection, **34(02)**:33-35 (2018).

Trend Model for Regional Energy Consumption System Based on Theory of Synergetics

Kai Chen[1], *Leiqiong* Cai[1], and *Yan* Long[1,2,*]

[1]Huazhong University of Science and Technology, China-EU Institute for Clean and Renewable Energy 430074, Wuhan, PR. China
[2]Huazhong University of Science and Technology, School of Energy and Power engineering, 430074, Wuhan, PR. China

Abstract. Trend model is a method of research on the evolution process of objects. Based on the synergetics, this paper raises a trend model for regional energy consumption systems by formula derivation and theoretical assumptions. This paper chooses Wuhan City as a case to construct the model and analyse the trend of its evolution. The results prove that the prediction effect of this model is good. The exponential smoothing method was chosen to compare with the trend model for the regional energy consumption system, and the result shows trend model raised in this paper makes good performance on the forecast.

1 Introduction

The global society, economy, environment, and politics of the 21st century are all experiencing significant changes. With the development of science and technology and also advancement of informatization, many complex and giant systems in different areas are coming into being, for example, a system like communication systems, national economic and managerial systems, business systems, and the system under study namely regional energy consumption system. The operation of the giant system can exert a significant impact on a nation and people[1]. Hence, it is necessary to research the development trend of the giant system.

In specific boundaries, a system with energy conversion is called a regional energy consumption system; for example, one place, one plant, one device, or one building can all be called a local energy consumption system[2]. The founder of general system theory, Bertalanffy, defines that a system implies a combination of various elements with mutual interaction[3]. A giant system has something in common: (1) Structure with multi-level and multi-function, lower-level part can be an element of higher-level structure; (2) Mutual connection and mutual impact between different unit thus exhibiting an extensive network; (3) The system can learn more and reorganize itself thus better its function and structure; (4) System is open and closely connected to the environment and can adapt itself in response to the development of environment. The regional energy consumption systems also show the characteristics above; besides, they share other attributes: (1) The sub-system is also complex and can cover broad areas like society, economy, energy, and environment; (2) All sub-systems are connected, and the connections are complex and not fully known.

The previous trend models mostly use the trend time series model to reflect the system change of variable time, namely, model based on curve fitting and parameter estimation[4]. These models are widely and effectively used in global economic regulation, business management, market estimation, pollution control, and weather forecasting. Nevertheless, these models are not effective for estimating giant system development due to the reasons listed here: (1) Giant system is continuous, complex, and diverse. The time series is the core of the time series model, and all the effects are related to time; thus, the mutual impacts between system variables are not accounted; for a system with many variables, variables are only analyzed with the change of time; thus the analysis results are not very satisfying; (2) Time series models do not conduct causality analysis for analysis target, and its mutual impacts complex adjustments are needed in order to improve the estimation precision, which makes the analysis more complex and arduous; (3) Time series models cannot describe the internal privity (named synergy) between sub-systems or elements.

In this investigation, optimization of the trend model is conducted by theoretic analysis and mathematical derivation combining the concepts from the theory of synergy, and this optimized model is used in regional energy consumption system. Besides, case study for Wuhan is conducted to illustrate the application of new model.

2 Basic principle of synergetics

Synergetics is a subject advanced by German physics professor H.Haken in 20th Century 70 years. He found synergetics by researching laser theory and absorbing modern cybernetics and information theory[5].

* Corresponding author: ly_hust@163.com

Synergy is the most basic concept of synergetics. In synergetics, all systems are composed of elements, links, or subsystems. These subsystems interact by the conversion of material and energy or information transmission among these subsystems. In a certain external condition, a system can form an ordered structure, in space, time, and function, through the interaction of each of the elements, links, or subsystems within the system. A system composed of many subsystems is in a state of self-organization, having a certain structure or function if the mutual cooperation between the subsystems engenders synergistic effects[6]. Therefore, the trend model of giant systems can be investigated by studying the relationship between various elements or subsystems.

Synergetics also holds that the nature of different systems is diverse, but the qualitative behavior of new structures replacing old ones has similar or even identical mechanism. The problem of evolution trends of different systems can be transformed into the analysis of the formation and development mechanism of self-organizing systems; that is, how the system changes automatically from disorder to order and from lower order to higher order under certain conditions. As an interdisciplinary science, synergitics, plays an important role in the cross-domain analysis of system evolution trends.

The order parameter is another important concept of synergetics[7]. In the evolution process of a system, there are many control parameters, divided into fast variables and slow variables (also called fast relaxation coefficient and slow relaxation coefficient). Among them, the dominant are the slow variables - the order parameter. The fast variables are dominated by the slow variables. The order parameters governing the behavior of the system can characterize the system's ordered structure and type[8]. The order parameters also characterize the combined contribution of all subsystems for the cooperative movement at the same time. The order parameter has the following characteristics: (1) The order parameter is the macroscopic parameters used to describe the behavior of the whole system; (2) The order parameter is the product of the collective motion of micro subsystems, characterizing and measuring the synergetic effect; (3) The order parameters control the subsystems' behavior and dominate the system evolution.

The selection of order parameters can ascertain the macroscopic parameters that dominate the system. By eliminating the macroscopic parameters in a subordinate position in the system, the analysis of the evolutionary trend could be greatly simplified.

3 Synergistic trend model for regional energy consumption system

3.1 Trend model theoretical basis: system synergetic equation mathematics extension and optimization

In synergetics, the Haken Model can describe the interactions of different variables and the process of structure evolution in a specific external condition[9]. We start

$$\dot{q}_1 =- \lambda_1 q_1 - a q_1 q_2 \tag{1}$$

$$\dot{q}_2 =- \lambda_2 q_2 - b q_2^2 \tag{2}$$

From a classic case of Haken Model: consider the system consists of a force and a subsystem, the state variables and force denoted respectively by q_1, q_2. The equation can be expressed as follows: in the formula, a, b, λ_1, λ_2 are control parameters[10, 11]. Assuming $\lambda_2 \gg \lambda_1$, he draws conclusions that system 2 dominates system 1 by using the method of adiabatic elimination, which can also be expressed by the following equation.

$$F(q_1, q_2) = 0 \tag{3}$$

The equation establishes a relationship between q_1 and q_2, which is called a synergetic system equation. The system synergetic equation shows that universal connections existent in the system between the macro parameters. It might as well express the equation as geometrical meaning: in the system of two macroscopic parameters, the relationship between system evolution trend and parameters can be described by a curve in the two-dimensional space.

The extension to the operation status and evolution trend of a giant complex system with s macro parameters are described: It is a curve in the m-dimensional space. The system states that every moment corresponds to a point on the curve[12].

Hence, a hypothesis must be made that at least one variation of the s macro parameters changes monotonically with time to ensure the space points on the curve move smoothly and continuously with time and that each state point corresponds to one time. Actually, for a certain system, one or more of these macroscopic parameters can always be founded under certain external conditions within the research scope of evolutionary trend. For example, in a regional energy consumption system, the unit GDP energy consumption can be controlled to decrease monotonously over time. Hence, the m-dimensional space curve can be expressed by the following m-1 equations.

$$F_1(x_1, x_2 ... x_m) = 0 \quad i = 1, 2 ... m; \tag{4}$$

In the actual case, to solve or find expression of this s-1 equations is complex, if formula (4) is converted into parametric equation group, it will lay a foundation for solving and be associated with the time series model. The parametric equations are named as an s-dimensional system synergetic equation group.

$$x_i = x_i(u) \quad i = 1, 2 ... m; \tag{5}$$

According to the order parameter principle of synergetic, the synergetic equation of m-dimensional system can be optimized. A system with m macro parameters is established, in which n macro parameters are order parameters and m-n macro parameters are dominant parameters. Hence, the m-dimensional synergetic equation group can be described.

$$x_i = x_i(u) \quad i = 1, 2 \ldots n, n+1 \ldots m; \tag{6}$$

The expression of x_{n+1} to x_m can be obtained based on order parameter principle of synergetics.

$$x_j = x_j(u) = G_j(x_1, x_2 \ldots x_n) \mathrm{j} = \mathrm{n}+1, \mathrm{n}+2 \ldots \mathrm{m}; \tag{7}$$

On the explanation, x_{n+1} to x_m can be described by the n order parameters; thus, the m-dimensional system synergetic equation can be simplified to the n-dimensional order parameter synergetic equation group.

$$x_i = x_i(u) \quad i = 1, 2 \ldots n; \tag{8}$$

3.2 Synergetics connotation of "u"

It attempts to give "u" a synergetic meaning, combining it with the time series model and make the "u" characterize the degree of self-organized conduction, and suppose:

$$u = f(t) \quad u \in (0, 1) \tag{9}$$

When $t \to 0$, $u \to 0$; when $t \to +\infty$, $u \to 1$. According to the synergetic self-organization principle, we can get $f'(t) > 0$. We might as well let "u" and "t" satisfy the relation like this:

$$u = 1 - e^{-at} \tag{10}$$

In formula (10), as a parameter, $a>0$, which is set to 0.05 by default. a can be changed according to the time span of the trend model study. When the time span is short, the larger the a value is, the more convenient the trend model analysis is. When the time span is long, a should take a smaller value.

Thus, u can be calculated through formula (10) and the relationship between the n order parameters and u can be established.

3.3 Trend model for regional energy consumption system

The general method of regional energy consumption systems trend model can be described.

(1) Identification and data collection of regional energy consumption system's order parameters. There are two identification methods, one is qualitative analysis method, namely, through the definition of the order parameter, analogy or logical reasoning to identify order parameters. The other method is quantitative analysis method, such as relaxation coefficient method, maximum entropy method, principal component analysis method and analytic hierarchy process. The specific methods of this paper will not be introduced in detail;

(2) Establishment of n-dimensional order parameter coordination equation group of regional energy consumption system, determination of $u = f(t)$ expression, and solution of the $U(u_1\, u_2 \ldots u_k)$, according to the $T(t_1, t_2 \ldots t_k)$, where k is the number of discrete values of time;

(3) Curve estimation of x_1, x_2 ... x_n, respectively, according to formula (8);

(4) Fitting of formula (8);

(5) Establishment of trend model for regional energy consumption system, after accuracy calibration.

4 Wuhan City -- a case study of trend model for regional energy consumption system

4.1 Order parameter identification and data collection

According to the order parameter definition, the order parameters of Wuhan city regional energy system order parameter are determined in Table 1.

Table 1. Order parameters of Wuhan City energy consumption system.

Subsystem	Order parameters
Social subsystem	Population density
Energy subsystem	Unit energy consumption of industrial output Coal consumption ratio of industrial Electricity consumption of the whole society Energy consumption per GDP
Economy subsystem	Per capita GDP Investment in fixed assets

Access to relevant information, the basic data of relevant indexes from the year 2002 to 2013 can be obtained, some data (like Unit energy consumption of industrial output, Coal consumption ratio of industrial) are obtained by the calculation of basic data, the order parameter values are listed in Table 2.

Table 2. Initial data for the order parameters.

parameter / Symbol /Year	Population density (People/km^2)	Energy consumption per GDP (Tec/million yuan)	Per capita GDP (yuan)	Unit energy consumption of industrial output (Tec/million yuan)	Investment in fixed assets (million yuan)	Coal consumption ratio of industrial	Electricity consumption of the whole society (kWh)
Symbol	X1	X2	X3	X4	X5	X6	X7
2002	961.69	2.26	17971	1.3312	570.43	0.4033	1587842
2003	975.82	1.96	19569	1.3114	645.06	0.4146	1788280
2004	994.65	1.85	23148	1.3630	822.2	0.4383	1855802
2005	1009.95	1.38	26548	1.2172	1055.18	0.5351	2108743
2006	1029.97	1.32	30921	1.2106	1325.29	0.4941	2308126
2007	1057.04	1.26	36347	0.9685	1732.79	0.4799	2586728
2008	1055.86	1.18	46035	0.6261	2252.05	0.4583	2864338
2009	1071.16	1.11	51144	0.5923	3001.1	0.4469	3102749
20010	1152.38	1.06	58961	0.5813	3753.17	0.4342	3536311
2011	1179.46	1.02	68315	0.5196	4255.16	0.4497	3836469
2012	1191.23	0.79	79482	0.4149	5031.25	0.4313	4032605
2013	1203	0.76	89000	0.4180	6001.96	0.3959	4372338

4.2 Establishment of u=f (t) and the value of u

The time span of this paper is short, so it takes a larger value 0.1 for " a ", then the value of u can be calculated in Table 3.

Table 3. The value of "u".

Year	2002	2003	2004	2005	2006	2007	2008	2009	2010	2011	2012	2013
t	1	2	3	4	5	6	7	8	9	10	11	12
u	0.095	0.181	0.259	0.323	0.394	0.451	0.503	0.551	0.593	0.632	0.667	0.700

4.3 The establishment of u=f (t) and the value

According to the formula (8), we analyzed the data from 2002 to 2011, using SPSS software, and did u and $x_1, x_2 ... x_7$ curve estimation respectively. Curve types include: linear, logarithmic, quadratic curve, cubic curve, compound, power, and exponential curve. The fitting curve types are determined ultimately in Table 4.

By fitting, we get the 7-dimensional order parameter of the synergetic equation group of the Wuhan City energy consumption system:

$$x_1 = 749.084u - 186.857u + 982 \quad (11)$$

$$x_2 =- 2.298u + 2.357 \quad (12)$$

$$x_3 = 217565.986u^3 - 40137.682u^2 + 21258.353u + 16072.173 \quad (13)$$

$$x_4 = 14.326u^3 - 19.308u^3 + 5.836u + 0.895 \quad (14)$$

$$x_5 = 25156.946u^3 - 9583.456u^2 + 2214.353u + 426.28 \quad (15)$$

$$x_6 = -0.813u^3 + 0.158u^2 + 0.334u + 0.364 \quad (16)$$

$$x_7 = 7268075.767u^2 - 1298640u + 1705903 \quad (17)$$

Table 4. The corresponding curve types of parameters and "u".

Parameter	x_1	x_2	x_3	x_4	x_5	x_6	x_7
Fitting curve types	quadratic curve	linear	cubic curve	cubic curve	cubic curve	cubic curve	quadratic curve

5 Test of trend model

Through the data analysis and calculation from 2002 to 2011, the 7-dimensional order parameters of the coordination equations of Wuhan energy consumption system are obtained.

Based on the equations, we predict the parameter values of each order in 2012-2013 and compared them with the actual values, as shown in Table 5.

Table 5. 2012 forecasting and real data forecasted by trend model for regional energy consumption system.

Parameter	Population density (People/km^2)	Energy consumption per GDP (Tec/million yuan)	Per capita GDP (yuan)	Unit energy consumption of industrial output (Tec/million yuan)	Investment in fixed assets (million yuan)	Coal consumption ratio of industrial	Electricity consumption of the whole society (kWh)
Predicted value	1190.90	0.8239	76988	0.4487	5107.76	0.41575	4074289
Actual value	1191.23	0.79	79482	0.4149	5031.25	0.4313	4032605
Relative error (%)	-0.03	4.3	3.1	8.1	1.5	-3.6	1.0

Table 6. 2013 forecasting and real data forecasted by trend model for regional energy consumption system.

Parameter	Population density (People/km^2)	Energy consumption per GDP (Tec/million yuan)	Per capita GDP (yuan)	Unit energy consumption of industrial output (Tec/million yuan)	Investment in fixed assets (million yuan)	Coal consumption ratio of industrial	Electricity consumption of the whole society (kWh)
Predicted value	1217.40	0.7511	85572	0.4332	5878	0.3971	4347659
Actual value	1203	0.76	89000	0.4180	6001.96	0.3959	4372338
Relative error (%)	1.2	-1.2	-3.9	3.6	-2.1	-0.3	0.5

To measure the description of the regional energy consumption system objectively, we use the exponential smoothing method of the time series forecasting model to do the same forecasting, the results in Table 7-8.

Table 7. 2012 forecasting and real data forecasted by exponential smoothing.

Parameter	Population density (People/km^2)	Energy consumption per GDP (Tec/million yuan)	Per capita GDP (yuan)	Unit energy consumption of industrial output (Tec/million yuan)	Investment in fixed assets (million yuan)	Coal consumption ratio of industrial	Electricity consumption of the whole society (kWh)
Predicted value	1228.47	1.00	67271.49	0.5540	4589.15	0.4199	3981065
Actual value	1191.23	0.79	79482	0.4149	5031.25	0.4313	4032605
Relative error (%)	3.1	26.6	-15.4	33.5	-8.8	-2.6	-1.3

Table 8. 2013 forecasting and real data forecasted by exponential smoothing.

Parameter	Population density (People/km^2)	Energy consumption per GDP (Tec/million yuan)	Per capita GDP (yuan)	Unit energy consumption of industrial output (Tec/million yuan)	Investment in fixed assets (million yuan)	Coal consumption ratio of industrial	Electricity consumption of the whole society (kWh)
Predicted value	1236.91	0.99	68186.85	0.5511	4681.07	0.4184	4030146
Actual value	1203	0.76	89000	0.4180	6001.96	0.3959	4372338
Relative error (%)	2.8	30.3	-23.4	31.8	-22	5.7	-7.8

From Table 5-8, we can see that the trend model of regional energy consumption system basis on synergetics has a better forecast effect, which can effectively describe the evolution trend of these systems.

With respect to the time series model, the trend model basis on synergetics was another notable feature, which can describe the system evolution trend with m-dimensional space curve. It means that if the future one of order parameter values can be determined, we can then determine the other macro parameter value. This formulation has a very important practical significance for energy management and energy policy.

Similar to the regional energy system of Wuhan City as an example, we will control the Energy consumption per GDP in 2015, making its control under a certain value, we can obtain other macroscopic parameters value in the synergetic trend model, such as per capita GDP, the electricity consumption of the whole society, etc. It will provide a basis for policy. Similarly, we can analyze whether the indicators of policy are expected to achieve compliance with the laws of science, according to the model.

The model makes good performance on description and prediction on the parameters, which change slowly in the trend significantly. However, from the table, we can see that the description of fluctuant macroscopic parameters, like coal consumption ratio of industrial, is not accurate enough.

6 Conclusion and Prospect

Through the construction, analysis, and comparison of trend model for Wuhan City regional energy consumption system, it proved the trend model of regional energy consumption system can effectively describe the trend of system's development and evolution, can comprehensively predict different macroscopic parameters, according to the characteristics of the self-organizing system. Eventually, a basis for energy management and energy policy can be provided by the model.

Although this paper raises a synergetic trend model for the regional energy system to one based on synergetics initially established model, there must exist a lot of space to optimize. Further study can be carried out on this model from the following aspects:

(1) The application of trend model for regional energy consumption system in energy management;

(2) Solution of the equation group mode without establishing the $u - t$ equation;

(3) The optimization of data selection and fitting curve way.

The work is supported by the National Natural Science Foundation of China (No.51106055).

References

1. N. Labanca, Â.G. Pereira, Watson, M., Krieger, K., Padovan, D., Watts, L., Moezzi, M., Wallenborn, G., Wright, R., Laes, E., Fath, B.D., Ruzzenenti, F., De Moor, T., Bauwens, T. and Mehta, L. "Transforming innovation for decarbonisation? Insights from combining complex systems and social practice perspectives." Energy Research & Social Science, **65**: 101452 (2020).
2. J. Liu, Y. Tian, K. Huang, and T. Yi, "Spatial-temporal differentiation of the coupling coordinated development of regional energy-economy-ecology system: A case study of the Yangtze River Economic Belt." Ecological Indicators, **124**: 107394 (2021).
3. P.D.V.Valentinov, "Sustainability in classical institutional economics: a systems theory view." Sustainable Production and Consumption (2021).
4. L. Göke, and M. Kendziorski, "Adequacy of time-series reduction for renewable energy systems." Energy, **238**: 121701 (2022).
5. H. Haken, "Synergetics." Physica B+C, **127**: 26-36 (1984).
6. D.C. Marinescu, Chapter 10 - Complex Systems and Self-Organization.In: Marinescu, D.C.(Eds.), Cloud Computing. Morgan Kaufmann, Boston: 301-315 (2013).
7. T.N. Zhang, and Y. J. Zhang, "Research on Enterprise Routine Search Based on Order Parameter Equation." East China Economic Management, **25**: 74-78 (2011).
8. F. Tan, and Q. Kong, "Uncovering the driving mechanism of regional synergistic development based on Haken model: case of the Bohai Rim region." Environment, Development and Sustainability, **22**: 3291-3308 (2020).
9. W. Zhong, J. Song, J. Ren, W. Yang, and S. Wang, "Revealing the nexus among energy-economy system with Haken model: Evidence from China's Beijing-Tianjin-Hebei region." Journal of Cleaner Production, **228**: 319-330 (2019).
10. Y.D. Zhu, T. Zhang, and B.J. Li, "Evolution mechanism and optimum control of regional industrial system." Journal of Manegement Sciences In China: 73-78 (2001).
11. L.P.Yang, and J.Y.Chen, "Evaluation of collaborative development level of innovation system in Yangtze River Economic Belt —— An empirical study based on Harken model." Journal of Beijing University of Posts and Telecommunications (Social Science Edition), **22**: 20-40 (2020).
12. Y. Long, S.Y. Huang, and H.C. Zhang, "Approach to synergy among material flow, energy flow and information flow." Journal of Chemical Industry Engineering: 2135-2139 (2006).

Survey on Geothermal Resources in Zhangjiakou Area

Peng Yang[1,4], *Qiang* Guo[1,4], and *Delong* Zhang[1,2,3,4,*]

[1]Beijing Institute of Exploration Engineering, Research Department, 100083 Beijing, China
[2]China University of Geosciences (Beijing), School of Engineering and Technology, 100083 Beijing, China
[3]Ministry of Natural Resources, Key Laboratory on Deep Geo-Drilling Technology, 100083 Beijing, China
[4]Ministry of Natural Resources, Technology Innovation Centre for Geothermal & Hot Dry Rock Exploration and Development, 050060 Shijiazhuang Hebei, China

Abstract. The Zhangjiakou area is rich in geothermal resources, and many counties in the region have discovered low-temperature geothermal. Structurally, Huailai County, Zhangjiakou City is located at the intersection of Zhangjiakou-Penglai Fault Zone and Shanxi Fault Basin (or Fenwei Seismic Tectonic Zone), namely the Yanqing-Huailai Basin (Yanhuai Basin). A large amount of geothermal resource investigation and research works has preliminarily defined that the area has good geothermal resource accumulation conditions. From 2019 to 2020, the China Geological Survey has organized several investigation and research in the area for in-depth geothermal resource. As one method of the survey results verification, a parameter well for geothermal resource survey was deployed as ZK02 and completed at 3000m depth. The ZK02 well was completed at a depth of 3006.9m, drilling through the Quaternary, Neogene, and Jixian strata, and entered the Archean gneiss strata. The well successfully explored high-quality artesian heat storage in the Jixian strata, obtained deep geothermal geological data and physical data, revealed the regional stratigraphic sequence and geological structure characteristics, and provided a scientific basis for regional geothermal resource potential evaluation. Based on the engineering practice of ZK02 well, this paper systematically summarizes drilling technology and experience from well structure to drilling equipment, construction technology and key technology, analyses the complex conditions and countermeasures downhole, and provide reference for the follow-up regional geothermal resources investigation, research and development and.

1 Preface

The Zhangjiakou area is rich in geothermal resources, among which, Chicheng, Huailai, Yangyuan, Yuxian and other counties have discovered low-temperature geothermal, the current development and utilization methods are mainly wellness tourism [1]. Structurally, Huailai County, Zhangjiakou City is located at the intersection of Zhangjiakou-Penglai Fault Zone and Shanxi Fault Basin (or Fenwei Seismic Tectonic Zone), namely the Yanqing-Huailai Basin (Yanhuai Basin). A large amount of geothermal resource investigation and research work has preliminarily defined that the area has good geothermal resource accumulation conditions [2][3][4]. In order to further ascertain the accumulation conditions and resource potential of deep geothermal resources in the Yanhuai Basin, from 2019 to 2020, the China Geological Survey organized investigation and research for in-depth geothermal resources in the area, and deployed high-precision gravity and magneto telluric (MT) , Controllable source audio magneto telluric (CSAMT) and seismic surveys and other comprehensive geophysical exploration work [5]; At the same time, in order to verify the survey results, a 3000m deep geothermal resource survey parameter well "ZK02" had been deployed and completed. The ZK02 well was completed at a depth of 3006.9m, drilling through the Quaternary, Neogene, and Jixian strata successively, and entered the Archean gneiss strata. The well successfully explored high-quality artesian heat storage in the Jixian strata, obtained deep geothermal geological data and physical data, revealed the regional stratigraphic sequence and geological structure characteristics, and provided a scientific basis for regional geothermal resources potential evaluation.

This article combines the engineering practice of ZK02 well and systematically summarizes the drilling construction technology and experience, which can provide reference for the follow-up regional geothermal resource investigation, research, development and utilization.

2 Geological Profile

ZK02 Well is located in the north of Donghuayuan Town, Huailai County, adjacent to Guanting Reservoir in the north, and structurally located in the southwest of Yanfan Basin. As the depth of the peripheral known boreholes is shallow, and there are very few reference

* Corresponding author: tgszdl@126.com

materials [6][7], this well is the first mid-deep geothermal parameter well in the survey area. Inferred from regional geological data analysis and geophysical prospecting results, the strata in the Donghuayuan area are mainly composed of Quaternary, Neogene, Jurassic, Jixian, Great Wall and Archean gneiss, as shown in Table 1.

Table 1. Prediction of drilling formation in ZK02 well

Stratum		Bottom depth (m)	Layer thicknes (m)	Brief description of lithology
Chronology	Symbols			
Quaternary + Neogene	Q+N	600	600	Clay, sand, gravel, Sandy gravel
Jurassic	J	1100	500	sandstone, rhyolite
Jixian System + Great Wall System	Jx+Ch	3000	1900	Dolomite

During the drilling of this well, the strata encountered were Quaternary, Neogene, Jixian and Archean from shallow to deep, no Jurassic strata were seen. As shown in Table 2.

Table 2. Summary of drilled formations in ZK02 well

Stratum		Bottom depth (m)	Layer thickness(m)	Brief description of lithology
Chronology	Symbols			
Quaternary + Neogene	Q+N	1327.9	1327.9	Clay, sand, gravel, Sandy gravel
Jixian System + Great Wall System	Jx+Ch	2960	1632.1	Dolomite, shale, Sandy gravel, mudstone, diorite porphyry
Archaean	Ar	3006.9	46.9 Unperforated	granite gneiss

The Quaternary stratigraphic lithology from shallow to deep are: brown fine gravel, mainly composed of quartz and granite; light gray clay, medium plasticity, containing silty fine sand; light gray coarse sand, mainly composed of quartz, sub-circular, sorted Etc., partly with sandy soil and fine gravel.

From shallow to deep, the Neogene stratigraphic lithology are light gray sandy mudstone, gray claystone, light gray brown sandy gravel, sandy mudstone, and sandy gravel.

The Jixian system is composed of Tieling Formation, Hongshuizhuang Formation, Wumishan Formation, Yangzhuang Formation, and Gaozizhuang Formation from shallow to deep. The lithology is dominated by dolomite, and there are well-developed fissures in some well sections.

The lithology of Archean strata is granite gneiss, gray and off-white gneiss-like structure, and the main components are feldspar, quartz, hornblende, mica and other dark minerals, which are recrystallized.

3 Well Structure

Before the ZK02 well, the stratification and burial depth in the deep part of the region were inferred based on the existing geological data and geophysical prospecting results. Therefore, in order to ensure the smooth implementation of the drilling project to achieve the geological purpose, an extra section is reserved in the well structure design. According to the drilling engineering design, the well structure has three sections with the completion diameter is Φ215.9mm bare hole. The Φ152.4mm diameter is reserved as an alternative plan for downhole complex drilling situation. Shown in Table 3.

Table 3. Structure design of ZK02 well

Section	Chronology	Well structure			
		Drill bit size (mm)	Drilling depth (m)	Casing size (mm)	Casing depth (m)
First	Quaternary + Neogene	444.5	602	339.7	600
Second	Jurassic	311.1	1102	244.5	1100
Third	Jixian System + Great Wall System	215.9	3000	177.8 *(reserve)*	
Forth *(reserve)*		*152.4*	*3000*		

In the actual drilling process, the bottom depth of the Neogene was 1327.9m, and no Jurassic stratum was seen. The Jixian system as the main reservoir has well-developed fissures and relatively fragmented partial stratum, which will cause risk of well wall collapse during the pumping test. According to the engineering design, the well structure and casing depth have been optimized and adjusted: First is increasing depth of the first and second section casing to the top of the Jixian formation; Second is insert Φ139.7mm casing + screen pipe when the third section completed, and arranged the screen pipe according to the distribution of the hot water layer. Shown in Table 4.

Table 4. Completion structure of ZK02 well

Section	Well structure			
	Drill bit size (mm)	Drilling depth (m)	Casing size (mm)	Casing depth (m)
First	444.5	1020.2	339.7	1010.86
Second	311.1	1494	244.5	1488
Third	215.9	3006.9	139.7	3006.9

The well structure design is very important to the exploratory well drilling project especially in unknown or more complex formations [8][9], and the adjustment of well structure during the drilling process is more common [10][11]. A common practice in well structure designing in unknown or more complex formations is to reserve an extra section, or to develop an alternative well structure which can effectively prevent the implementation of the drilling project from being too passive or forced to stop due to the complexity situation downhole [12][13]. In the engineering design, The ZK02 well has reserved an extra section which provides ample room for the adjustment of the well structure, during the implementation of the project, reasonable adjustments were made to the well structure according to the stratum conditions, which provided a strong guarantee for the smooth implementation of the project.

4 Drilling equipment selection and matching

The selection of drilling engineering equipment must considering about geological purpose of drilling, regional environment, construction technology, completion depth, completion diameter and other factors [14][15][16]. In recent years, with the continuous acceleration of the exploration and development of geothermal (hot dry rock) resources, the workload of geothermal drilling has increased sharply. Existing large-diameter geothermal drilling equipment mainly includes water well drilling rigs and oil drilling rigs. The advantages of water well drilling rigs are that the equipment is relatively compact, easy to relocate, and there are few construction crews; its disadvantages are mainly that the drilling depth is limited by the lifting ability, and the installation conditions of the well control equipment and solid phase control equipment are limited by the height of the first-floor platform. The installation conditions of logging sensors are poor. The oil drilling rigs are relatively mature, with a complete range of drilling depth capabilities. The structure and design of the drilling rig meet most needs of well control equipment and solids control equipment, also the ability to handle complex accidents is relatively strong. In the process of drilling equipment selection, especially exploratory wells, new types of drilling rigs with maturity and recent age should be selected first, which can provide an effective guarantee for the smooth implementation of the project.

The ZK02 well is the first exploratory well in the area. There is no reference to the well-informed well. It only has a rough estimate of the strata encountered downhole, and there is little information available for prediction of downhole conditions, and based on the principle of ensuring the safety of drilling, large capacity redundancy needs to be considered when selecting drilling rigs. Therefore, standard petroleum ZJ-30 electric drilling rigs are used, and all supporting equipment are electric, which is more friendly to the surrounding environment of the construction, and the diesel generator set is only used as an emergency backup. The main equipment list is shown in Table 5.

First, based on the drilling depth requirements, the selection of standard petroleum ZJ30 equipment can effectively ensure provide the lifting force required to drill to the designed well depth; Secondly, the supporting system according to petroleum standards can effectively ensure the compatibility of equipment, operation safety, construction efficiency and fulfil other requirements; Third, the use of all-electric equipment can greatly improve work efficiency, reduce environmental pollution, and save drilling costs. At the same time, in order to improve drilling complexity and risk response capabilities, a full set of well control equipment and blowout prevention tools are provided on site to cope with blowout risks, and diesel units are provided to cope with complex situations such as equipment failures and emergency power outages.

Table 5. List of main equipment

No.	Name	Model	Quantity
1	Drilling rig	ZJ30	1
2	Diesel unit	G12V190PZL-3/O	1
3	Auxiliary generator	CDVS-400	1
4	Mud pump	F-1000	2
5	Circulation tank	174m³	4
6	Four-level solid control		1
7	BOP	2FZ35-35	1
8	BOP control device	FKQ-480-5	1
9	Kill manifold	PFF103-35	1
10	Choke manifold	PFF65-35	1
11	Internal Blowout Prevention Tool		1

5 VDrilling Construction and Key Technologies

5.1 First section drilling

The first section drilling encountered strata are mainly Quaternary and Neogene. Quaternary stratum is mainly brown fine gravel, light gray clay, light gray coarse sand, etc., Neogene stratum is mainly light gray sandy mudstone, gray claystone, shallow Grey-brown sandy gravel, sandy mudstone, sandy gravel, etc. The drill tool combination used is Φ444.5mm drill bit + Φ203.2mm drill collar + Φ 177.8mm drill collar + Φ127 drill rod + Kelly bar. 71.5m-1020.2m use Φ444.5mm GA115G roller bit to drill, weight on bit 10-100KN, speed 20-60rpm, displacement 30L/s, average ROP 2.9m/h, maximum ROP 9.47m/h, The longest footage of a single bit is 360.59m, of which the core drilling speed is 0.65m/h.

5.2 The second section drilling

The strata encountered in the second section drilling are mainly Tertiary sandy gravel, claystone, Jixian sandy gravel, mudstone, and dolomite. The drill tool combination used is Φ311.2mm drill bit + Φ203.2mm drill collar + Φ177.8mm drill collar + Φ127 drill rod + Kelly bar. The mud specific gravity is 1.02~1.08g/cm3, the weight on bit is 20-120kN, and the displacement is 22-30L/s.

For drilling in this section, the following points should be paid attention to:

(1) Adopt pendulum drilling tool combination to prevent deflection;

(2) In High steepness well sections with easy deviating conditions should intensify inclination

measurement, master the data and take technical measures timely, and use the MWD inclinometer to prevent inclination and maintain straightness according to the actual situation;

(3) Make a record of the drilling fluid, discover formation leakage and water inrush in time, and adjust the performance of the drilling fluid and drilling parameters accordingly;

(4) Adjust the performance of the drilling fluid in time, balance the formation pressure, prevent block dropping and collapse, connect a single root to reduce the number of strokes, minimize the number of trips, and prevent the drill bit from colliding with the well wall;

(5) Ensure that the drilling equipment is operating well;

(6) The use of screw + cone bit compound drilling can effectively increase the mechanical drilling speed. However, attention should be paid to the formation conditions, and conventional drilling tool combinations should be used as far as possible in the leak-prone formations to make preparations for leaking at any time.

5.3 Third section drilling

The third section drilling adopts screw drill + roller bit compound drilling speed, speed-up drilling tool combination: Φ215.9mm drill bit +172 screw +177.8 drill collar +Φ213 centralizer +178 drill collar +127 drill rod +NC501*NC460 conversion joint + 114 drill rod + NC461*NC500 conversion joint + Kelly bar. The specific gravity of the third section mud is 1.02~1.08g/cm3, the weight on bit is 20-120kN, the speed is 20-70rpm (screw), and the displacement is 22-44L/s.

5.4 Environmentally Friendly Drilling Fluid

The selection of drilling fluid formula should consider about all the aspects of pertinence, suitability, temperature resistance, security and environmental protection.

The stratum encountered in the first drilling section was the Quaternary and Neogene. The lithology is mainly light gray clay, light gray coarse sand, sandy gravel, sandy mudstone, etc. The stratum is relatively loose and broken. The higher viscosity bentonite drilling fluid is used to quickly drill through and place the casing. Using bentonite drilling fluid, the formula is as follows:

Water+6%~8% Bentonite+0.1%~0.2% Caustic Soda+0.2%~0.4% Tackifier (GZN).

The strata encountered in the second section are mainly tertiary sandy gravel and claystone, and Jixian sandy gravel, mudstone and dolomite. The formation is relatively stable, but there are problems such as mudstone hydration and swelling and gravel loss. A low-solid phase polymer drilling fluid system with good suppression and plugging performance is selected. The formula is as follows:

Clear water + 0% ~ 3% bentonite + 0.1% ~ 0.2% caustic soda + 0.1% ~ 0.5% tackifier (GZN) + 1% ~ 2% anti-collapse plugging agent + 1% ~ 1.5% fluid loss agent (GPNH)+0.1%~0.2% coating agent (GBBJ).

The stratum encountered in the third section is granite gneiss which is stable. A solid phase free polymer drilling fluid system is adopted, and polymers are added to improve the carrying capacity of the mud and ensure the cleanness of the wellbore. If the well wall is unstable, use the mud positive circulation construction instead. When the temperature exceeds 80℃, use the temperature-resistant solid phase free polymer drilling fluid system. The formula is as follows: water + 0.1%~0.2% caustic soda + 0.3%~0.6% thickener (GZN) + 1%~2% plugging agent (GFD-1) + 1%~1.5% fluid loss agent (GPNH) +0.1%~0.2% coating agent (GBBJ).

6 Pumping test

6.1 Overview of the pumping test

The ZK02 well used air compressors to clean the wells before the pumping test from January 23 to 24, 2021 in order to remove the blockages such as mud and rock cuttings in the hole especially in the thermal reservoir section to achieve clean water without sand. Combining with the geothermal resources in this area, it is defined that the pumped water is confined water, and the pumping test adopts the three-time depressurization and steady flow test method in a single well. The pumping test started at 20 o'clock on January 25, 2021 and ended at 0 o'clock on January 30, 2021. It lasted 100 hours. the duration and stabilization time of each pumping phase shown in Table 6.

Table 6. Time Statistics Table for Each Pumping Stage

Test pumping	Beginning at 20 o'clock on January 25, 2021 and ends at 20 o'clock on January 26, lasting 24 hours, the stable time is 6 hours.
Observation of water level recovery	Beginning at 20 o'clock on January 26, 2021, and ends at 4 o'clock on January 27, the observation time is 8 hours, the stable time is 6 hours.

pumping	The first drop distance	Beginning at 4 o'clock on January 27, 2021 and ends at 4 o'clock on January 29, lasting 48 hours, the stable time is 24 hours.
	The second drop distance	Beginning at 4 o'clock on January 29, 2021, and ends at 21 o'clock on the 29th, lasting 17 hours, the stable time is 9 hours.
	The third drop distance	Beginning at 21:00 on January 29, 2021 and ends at 15:00 on January 29, lasting 17 hours, the stable time is 10 hours.
Observation of water level recovery		Beginning at 15:00 on January 29, 2021 and ends at 0:00 on January 30, lasting 9 hours, the stable time is 8 hours.

In this pumping test, the static water level is +9.8m (9.8m above the ground), and the water outlet temperature is 55°C. The dynamic water level depth of the first fall is 24.2m, the water level drop is 34.0m, and the water inflow is 160.364 m³/h. The unit water inflow is 4.717m³/hm; the depth of the dynamic water level in the second fall is 10.3m, the water level drop is 20.1m, the water inflow is 120.629m³/h, and the unit water inflow is 6.01m³/hm; the dynamic water level in the third fall is +3.6m, water level drop is 6.2m, water inflow is 53.677 m³/h, unit water inflow is 8.568m³/hm, details shown in Table 7.

Table 7. Data of pumping test

Drop distance	Still water level(m)	Buried depth of dynamic water level (m)	Water level drop (m)	Water inflow (m³/h)	Unit water inflow (m³/h.m)	Water temperatur e (℃)	Stable time (h)
S_3	+9.8	24.2	34.0	160.364	4.717	55	24
S_2		10.3	20.1	120.629	6.001		9
S_1		+3.6	6.2	53.677	8.568		10

6.2 Q=f(s) relation curve define

According to the results of the pumping test of the ZK02 well, combined with the field data, the basic data of the pumping test was defined.

According to the curvature method, the value of n is calculated according to the judgment formula to determine the Q=f(s) relationship curve type of the well, and n_1 and n_2 are calculated with three stable drawdowns:

$$n_1=(\lg S_2-\lg S_1)/(\lg Q_2-\lg Q_1) \quad (1)$$

$$n_2=(\lg S_3-\lg S_2)/(\lg Q_3-\lg Q_2) \quad (2)$$

In the formula, Q and S are the water volume and water level drop in the same pumping time.

When n=1, the curve is a straight line; when $1<n<2$, the curve is a power curve; when n=2, the curve is a parabola; when n>2, the curve is not a logarithmic curve; When n<1, there is an error in the data.

Calculated by the above formula, $n_1=1.45$, $n_2=1.85$, the Q=f(s) relationship curve of this well is a power curve type. It means that in areas where groundwater is mainly stored and the source of replenishment is poor and the water conductivity is strong, when the water level drops, the water inflow increases greatly with the drop distance; When the water level drops beyond a certain depth, the amount of water inflow increases with the drop distance very little.

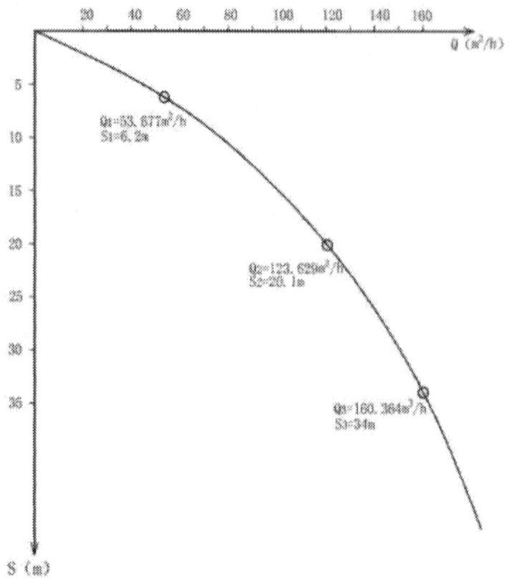

Fig. 1. Q-S curve

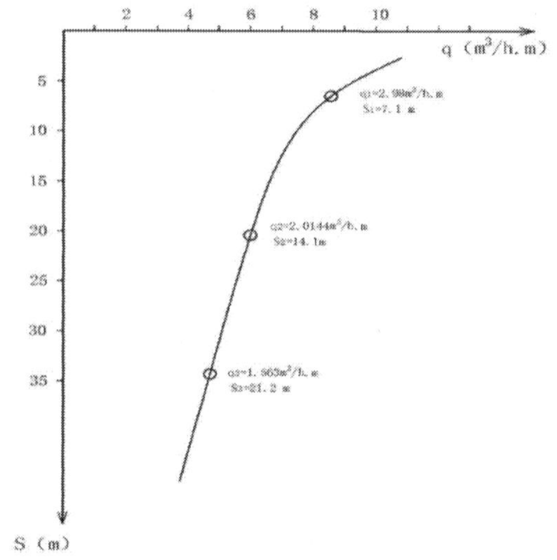

Fig. 2. q-S curve

The change range of the pumping test when the flow rate is stable is shown in Fig. 2, where q is the fluctuation range of the water inflow relative to the average value, and S is the drop distance in the same period. The data is obtained by field measurement. It can be seen from Fig. 2 that as the drop distance increases, the water inflow gradually stabilizes.

The entire change process of the pumping test can be intuitively understood Through Fig. 3. The vertical axis is marked with a one-way dual axis Q (pumping volume) and S (pumping depth), the horizontal axis is t (pumping time). It can be seen from Fig. 3 that when the pump is stopped after pumping, the water level can be quickly restored.

The entire change process of the pumping test can be intuitively understood by Fig. 3. The vertical axis is marked with a one-way dual axis Q (pumping volume) and S (pumping depth), the horizontal axis is t (pumping time). It can be seen from Fig. 3 that if the pump is stopped after pumping, the water level can be quickly restored.

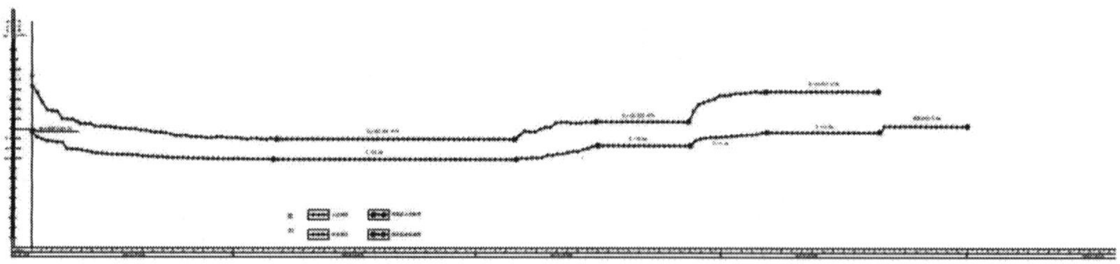

Fig. 3. Q·S-t duration curve

6.3 Calculation of hydrogeological parameters

Calculate the permeability coefficient k and the influence radius R by combining the Dupuit formula and the empirical formula of the influence radius.

Calculation formula of steady flow:

$$K = \frac{0.366Q}{MS} \lg \frac{R}{r} \quad (3)$$

$$R = 10 \cdot S \cdot \sqrt{K} \quad (4)$$

In the formula: K is the permeability coefficient (m/d)

Q is the wellbore water inflow (m^3/d)

M is the effective thickness of the thermal reservoir (m)

S is the water level drop (m)

R is influence radius (m)

r is the radius of the well (m)

$$T = KM \quad (5)$$

In the formula: T is the coefficient of hydraulic conductivity (m^2/d)

K is coefficient of permeability (m/d)

M is the effective thickness of the thermal reservoir (m)

According to the results of the pumping test of the ZK02 well, the hydrogeological parameters of the well are shown in Table 8.

The geothermal productivity test of The ZK02 well include three pumping tests with depressurization and steady flow in single well. Comprehensive analysis is carried out based on the results of this pumping test. The Q=f(s) relationship curve of this well is a power curve (exponential curve) type, and the pumping duration is analyzed at the same time. It can be seen from the curve that the water level of this well quickly stabilizes during pumping, and can recover quickly after stopping the pump, indicating that the well's thermal reservoir has good permeability and medium water richness, which is basically consistent with the hydrogeological characteristics of this area. It also proves that the pumping test is successful and the hydrogeological parameters obtained are true and reliable.

Table 8. Calculation table of hydrogeological parameters

Water inflow m^3/h	Drop down m	Influence radius m	Permeability coefficient m/d	coefficient of hydraulic conductivity m^2/d	Radius of pumping section m	effective thickness of thermal reservoir m
160.364	34.0	340	0.955	144.568	0.208	151.38
120.629	20.1	215	1.149	173.936		
53.677	6.2	74	1.425	215.717		

7 Conclusion

The ZK02 Well is an exploratory well deployed in Huailai County for deep geothermal resource exploration in Zhangjiakou area. By carefully designing the well structure, optimizing drilling equipment and technology, and controlling key parameters, the mission of 3006 meters footage was successfully completed. The geothermal productivity test through pumping test proved that the thermal reservoir of this well has good permeability and medium water richness, laying the foundation for the future development of geothermal resources in this area. Through the practice of the construction process, the systematic summary of drilling technology and experience can provide reference and support for the follow-up regional geothermal resource investigation, research, development and utilization.

This work was supported by the National Natural Science Foundation of China (No. 41802197), the project of China Geological Survey (No. DD20221681, DD20201102) and Research Foundation of Key Laboratory of Deep Geodrilling Technology, Ministry of Natural Resources (No. KF202103).

References

1. G. Wang. Research on Zhangjiakou Geothermal Resources Development and Utilization Model. Earth, **(02)**:91(2019).
2. Y. Li, S. Huang, H. Lin . Research and analysis of geothermal resources and earthquake distribution in Zhangjiakou City. Western Resources, **(05)**: 100-101 (2018).
3. J. Han, X. Li, J. Yu. The application of comprehensive geophysical methods in the survey of geothermal resources in the northern part of Zhangjiakou. Computing Techniques for Geophysical and Geochemical Exploration, **40(03)**: 372-377 (2018).
4. J. Zu, G. Wu, Y. Lian. Research on geothermal in Yanqing-Huailai Basin and its neighboring areas. Acta Seismologica Sinica, **(04)**: 107-109 (1997).
5. D. Zhang, W. Yuan, Y. Xing, et al. Application of integrated geophysical methods in geothermal exploration: a case in Zhangjiakou. IOP Conference Series: Earth and Environmental Science, **660(1)**:012108 (6pp) (2021).
6. C. Zhang, X. Song, H. Li, F. Zhang, J. Ren. Digital gas mercury observation and data analysis of Huailai 4 well. Seismological and Geomagnetic Observation and Research, **(03)**: 65-68 (2008).
7. Y. Tian, W. Wang, Y. Zhao. Research on Comprehensive Utilization of Geothermal Resources in Beijing-Tianjin-Hebei Region. Coal Geology of China, **32(02)**: 51-58 (2020).
8. D. Zhang. DZK1 geothermal exploration hole drilling technology in Julong Town, Quanzhou. Exploration Engineering (Rock and Soil Drilling and Tunneling), **45(11)**: 29-33 (2018).
9. D. Pan, J. He, T. Yang, Z. Zhang. Optimized design of wellbore structure for geothermal deep wells in karst thermal reservoirs in Niutuo Town, Xiong'an. Drilling Engineering, **48(02)**: 78-84 (2021).
10. Y. Li, J. Guo and W. Wang, Drilling construction difficulties and countermeasures of the Yunxuandi 1 well in shale gas survey in eastern Yunnan . Drilling Engineering, **48(08)**: 12-18 (2021).
11. Z. Su, S. Zhang, W. Wang, H. Cao, L. Liu, J. Xiao, H. He, J. Guo. Difficulties and countermeasures of drilling and construction of Qianshuidi 1 well. Drilling Engineering, **48(05)**: 54 -59 (2021).
12. H. Wang. Construction technology of DR9 geothermal well in Qiabuqia area of Gonghe

Basin, Qinghai . West-china Exploration Engineering, **33(06)**: 62-64 (2021).

13. K. Zhong, S. Huang. Drilling construction technology of Xiangyongdi 1 well. Exploration Engineering (Rock and Soil Drilling and Tunneling), **45(08)**: 28-33 (2018).
14. L. Fan, X. Jia, G. Wang, T. Zhang, P. Zhang, C. Lu, J. Li. Drilling practice of D03 geothermal exploration well in Xiong'an New Area. Exploration Engineering (Rock and Soil Drilling and Tunneling Engineering), **47(10)**: 13-22 (2020).
15. P. Gao, X. Dong, F. Ma, Z. Wang, X. Wang. Drilling construction technology of D15 geothermal exploration well in Xiong'an New Area. Drilling Engineering, **48(03)**: 106-112 (2021).
16. W. Liu, J. Liu, K. Guo, Z. Zhao, W. Shan. Research on Drilling Construction Technology of D14 Geothermal Well in Xiong'an New Area. West-china Exploration Engineering, **31(08)**: 102-106 (2019).

Grass Hopper Optimization Algorithm for Off-Grid Rural Electrification of an Integrated Renewable Energy System

Polamarasetty P Kumar[1,*], *Ramakrishna* S S Nuvvula [1], and *Vasupalli* Manoj[1]

[1] Department of Electrical & Electronics Engineering, GMR Institute of Technology, Rajam, India

Abstract. A suitable alternative to grid expansion has been found in renewable energy sources like wind, solar, and biomass. To put it another way, relying solely on one of the major renewable sources is both inefficient and expensive. As a result, an integrated renewable energy system is a viable option. The purpose of this article is to discuss the use of the Grasshopper Optimization Algorithm (GOA) for renewable energy sizing in the current study area. For an autonomous microgrid network, the proposed technique finds the optimum system size on the basis of Loss of Power Supply Probability (LPSP). The proposed microgrid consists of PV panels, wind turbines, biomass generator and a battery storage system. The proposed GOA algorithm's convergence efficiency in resolving the current optimization problem is investigated and compared with Particle Swarm Optimization (PSO) and Genetic Algorithm (GA) in the MATLAB software environment. The simulation results show that the GOA algorithm outperforms its counterparts, GA and PSO, in terms of system sizing.

1 Introduction

In India's rural areas, the bulk of the people still does not have access to electricity. Extending the grid power supply to such remote areas is neither feasible nor cost-effective [1]. Other drawbacks of typical diesel-powered generators include high maintenance and fuel costs, as well as excessive noise levels and, most importantly, greenhouse emissions. Renewable energy sources that are environmentally friendly can help meet the electrical load requirement. Given the unpredictability of renewable energy resources, the use of an (IRES) is preferable to improving the reliability of the off-grid system. With the IRES, users can meet the energy needs by combining two or more regionally accessible energy sources. In order to maximize efficiency and cost-effectiveness when using renewable energy resources to meet load demand, optimal models must be developed [2].

For the microgrid size problem, this article developed a novel approach named the "Grasshopper Optimization Algorithm" (GOA). 5 villages in an off-grid region in Odisha State, India, were supplied with electricity through a micro-grid approach. Solar panels, wind turbines, a biomass generator, and a battery bank make up the proposed microgrid. Zero Percent LPSP error is assumed to identify the best system configuration.The outcomes of GOA are compared to those of two other well-known optimization algorithms PSO and GA. GOA's performance is clearly seen in the comparison, which results in very quick convergence, local optima avoidance and a balance between exploration and exploitation.

2 Resources and Load Estimation

India has a plentiful supply of renewable energy resources. Five villages in the Indian state of Odisha have been recognized as possible locations for off-grid microgrid development. Fig. 1 depicts the energy demand for residential houses during the summer and winter seasons. This information comes from a door-to-door load demand survey conducted in the proposed rural off-grid villages. There are around 8.35 tonnes of biomass each year that comes from forest vegetation such as pine needles and firewood. Figs. 2 and 3 shows hourly data from a 10-year average of National Renewable Energy Laboratory data on wind speed, solar radiation, and ambient temperature (01/01/2004 to 31/12/2014).

* Corresponding author: nramkrishna231@gmail.com

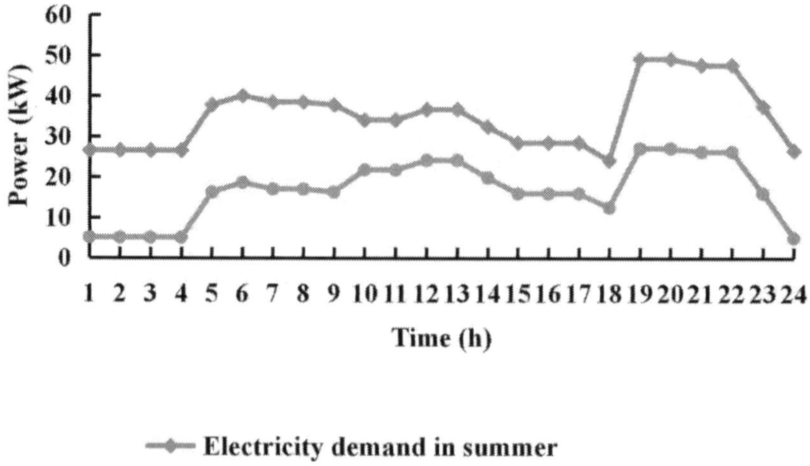

Fig. 1. Load demand in the study area

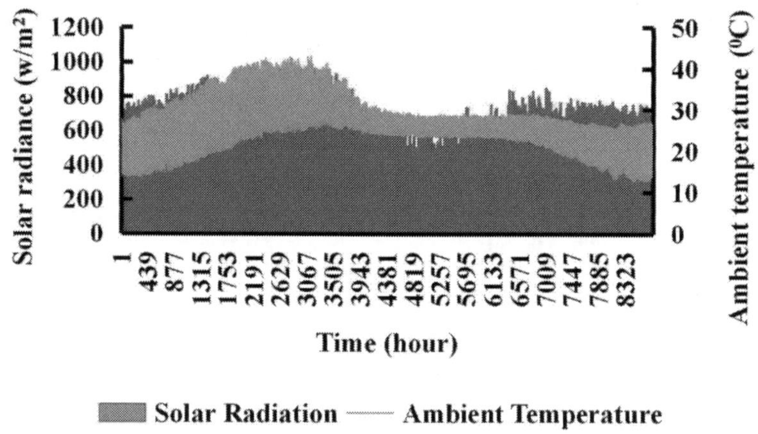

Fig. 2. study area solar radiation and ambient temperature

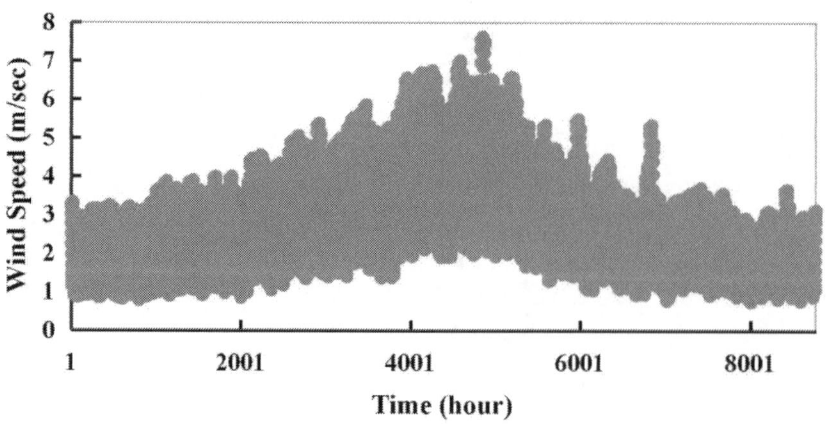

Fig. 3. Study area wind speed

3 Integrated Renewable Energy System Modelling

It is necessary to model the components prior to determining the optimal system size. Wind farms, Pv module, a biomass generator, a converter, and batteries make up the existing system. A schematic of the IRES is depicted in Fig. 4.

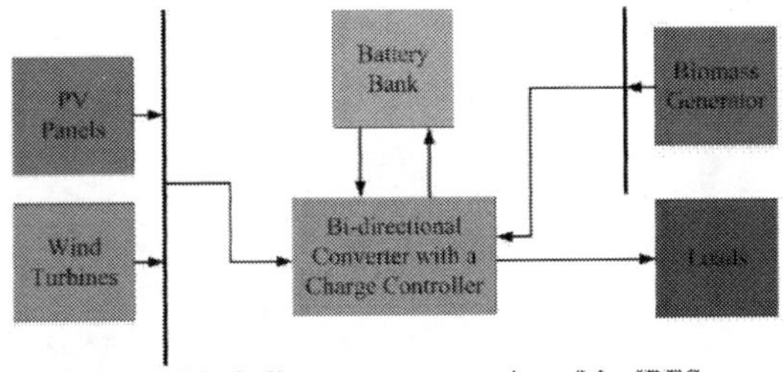

Fig. 4. Block diagram representation of the IRES

3.1 Solar energy

The PV panel's power output (P_{PV}) may be computed as follows [3]:

$$P_{pv}(t) = N_{pv} * P_{pv-rated} \times G_t(t)/1000 \times [1 + alpha_t \times (t_{amb}(t) + (0.0256 * G_t)) - T_{C\_STC})] \quad (1)$$

Here, $G_t(t)$ is hourly solar radiation, $P_{pv-rated}$ is the esteemed power of the PV panel, $alpha_t$ = -3.7×10^{-3} (1/^{0}C), $t_{amb}(t)$ is the hourly ambient temperature (^{0}C). N_{pv} is the no of PV panels, T_{C_STC} is the cell temperature.

3.2 Wind energy

This is done by calculating the wind speed and converting it to a power output for the wind turbine ($v_1(t)$) is first converted into the corresponding wind turbine hub height ($v_2(t)$) by the power rule equation.

$$v_2(t) = v_1(t) \times \left(\frac{h}{h_{ref}}\right)^g \quad (2)$$

Where g is the power law exponential.

The power generated by the WT (P_{WT}) can be calculated as [3]:

$$P_{WT}(t) = N_{wt} * \begin{cases} 0, \; v_2 < V_{cut-in}, v_2 > V_{ct} \\ v_2^3(t)\left(\frac{P_r}{V_{rated}^3 - V_{cut-in}^3}\right) - P_r\left(\frac{V_{cut-in}^3}{V_{rated}^3 - V_{cut-i}^3}\right) \\ P_r, \; V_{rated} \le v_2 \le V_{cut-} \end{cases} \quad (3)$$

Where N_{wt} is the number of wind turbines, V_{cut-in}, $V_{cut-out}$ and V_{rated} are the cut-in, cut-out and rated speed of the wind turbine respectively.

3.3 Biomass energy

The biomass generator's output power can be calculated as [4]:

$$P_{bms}(t) = \frac{q_{bm} \times \eta_{bm} \times cv_{bm} \times 1000}{doh_{bm} \times 365 \times 860} \quad (4)$$

Here, calorific value of biomass , $cv_{bm} = 4015\,\text{kcal/kg}$, q_{bm} is the volume of biomass (tons/year), doh_{bm} is the daily operating hours of the biomass generator.

3.4 Battery bank

The process of charging a battery bank may be described as follows [5]:

$$E_{bat}(t) = E_{bat}(t-1) + \left(E_g(t) - E_l(t)/\eta_{Con}\right) \times \eta_{cc} \times \eta_{chg} \quad (5)$$

The process of discharging a battery bank may be described as follows [5]:

$$E_{bat}(t) = (1-\sigma) \times E_{bat}(t-1) - \left(E_l(t)/\eta_{Con} - E_g(t)\right)/\eta_{dchg} \quad (6)$$

Here, $E_g(t)$ is hourly-generated energy, E_{bat} is the energy stored in the battery, $E_l(t)$ is hourly load demand, η_{Con} efficiency of converter, η_{cc} efficiency of charge controller, σ a battery's hourly self-discharge rate , η_{chg} charging effectiveness of the battery, η_{dchg} is discharging efficiency of the battery.

3.5 Bidirectional converter having charge controller (BDC-CC)

The rated power (kW) of the BDC-CC can be calculated as [6]:

$$P_{bdc-cc} = E_{t,max} \times 1.1 \quad (7)$$

Here,
the multiplication factor 1.1 shows here that the BDC-CC has 10% overload capacity.

4 Objective Significance and Limitation

The objective function is aimed to minimize the Life Cycle Cost (LCC) of an independent off-grid renewable energy system with an LPSP value of 0%. The system's LCC is determined by decision variables such as wind turbines, PV panels and batteries. The LCC can be calculated as:

$$min\ LCC(N_{pv}, N_{wt}, N_{bat}) = \sum_{p=pv,wt,bms,bat,BDC-CC}^{min} (LCC)_p \quad (8)$$

4.1 Constraints

Fixed-power biomass generators, such as a 5kW model, are assumed with a daily operational capacity of 4kWh of energy per hour for 5 hours. As a result, it is not constrained by any constraints. Furthermore, the following limitations are applicable on remaining sources, such as wind and solar, are explained as follows:

$$0 \leq N_{pv} \leq N_{pv-max} \quad (9)$$

$$0 \leq N_{wt} \leq N_{wt-max} \quad (10)$$

In addition, battery storage system is bound by the following constraint.

$$0 \leq N_{bat} \leq N_{bat-max} \quad (11)$$

4.2 Limits on battery bank energy storage

The following phrase limits the amount of energy kept in the battery system at any given time 't' [5]:

$$E_{bat\_min} \leq E_{bat}(t) \leq E_{bat\_max} \quad (12)$$

The E_{Bat_max} can be calculated as:

$$E_{Bat\_max} = \left(\frac{N_{bat} \times V_{bat} \times S_{bat}}{1000}\right) \times SOC_{max} \quad (13)$$

Meanwhile, E_{Bat_min} can be calculated as:

$$E_{Bat\_min} = \left(\frac{N_{bat} \times V_{bat} \times S_{bat}}{1000}\right) \times SOC_{min} \quad (14)$$

Here, SOC_{min} and SOC_{max} are the minimum maximum and state of charges of the battery.

4.3. Power reliability index

Batteries and renewable energy resources will be unable to meet load demand for a few hours per year. The power supply will then be lost in those hours. This can be calculated as follows:

$$lps(t) = \frac{E_l(t)}{\eta_{con}} - E_g(t) - [(1-\sigma) \times E_{bat}(t-1) - E_{bat\_min}] \times \eta_{dchg} \quad (15)$$

For evaluation of the system's power reliability its reliability index should be considered for various LPSP values and it can be defined as follows:

$$LPSP = \frac{\sum_{t=1}^{T} lps(t)}{\sum_{t=1}^{T} E_l(t)} \quad (16)$$

5 Grasshopper Optimization Algorithm

The Grasshopper Optimizations Algorithm (GOA) was developed by Saremi et[7] al in 2017 as a new optimization algorithm. The algorithm imitates the actions of the grasshopper in the nature in the food hunt.

Grasshopper's swarm behaviour is mathematically modelled, where the authors proposed the following equations for resolving optimization problem.

$$x_i^d = C\left(\sum_{j=1, j\neq 1}^{n} C\frac{UB_d - LB_d}{2} s(|x_j^d - x_i^d|)\frac{X_j - X_i}{D_{ij}}\right) + t_d \quad (17)$$

The component C can be determined by Eq. (18)

$$C = Cmax - L\frac{Cmax - Cmin}{l} \quad (18)$$

The current optimization problem of optimal off-grid sizing with the proposed Grass hopper optimization algorithm is described with a neat flowchart in Fig. 5.

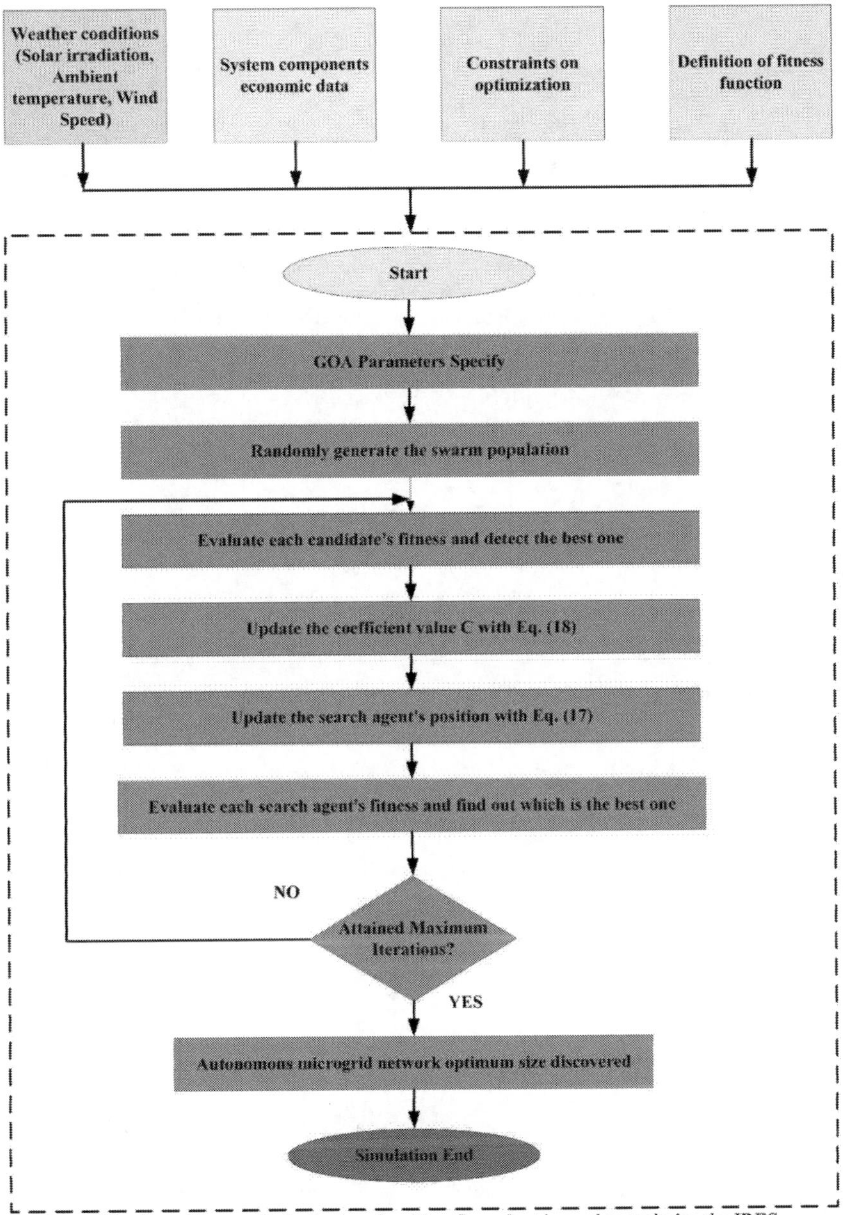

Fig. 5 The Grass Hopper Optimization algorithm flow chart is used to optimize the IRES

6 Results and Discussion

The case study examined the availability of electricity for the proposed study area. The PV (Photo Voltaic Panels)/WT (Wind Turbines)/BMS (Biomass Generator)/BAT (Batteries) configuration is tested for the proposed study area in the MATLAB environment with GA, PSO and GOA algorithms. These algorithms have been developed with their default parameters, population of 100 and maximum number of iterations as100. Table 1 shows the input values for off-grid microgrid optimization process.

Table 1. input values of the all components

Component	Life span	Rated power	Capital cost ($)	Replacement cost ($)	Annual O&M cost ($)
PV panel [6]	25 years	0.25 (kWp)	250	0	6.25
Wind turbine [6]	25 years	1 kW	2500	0	75
Biomass generator [6]	1500 hours	5 kW	4505	3153	135
Battery [6]	5 years	360 Ah	300	300	0.75

Converter [6]	10 years	55 kW	5940	5940	15

Table 2. optimization values of the IRES with proposed algorithms

configuration	component	GA	PSO	GOA
PV/WT/ BMS/BAT(LA)	N_{pv}	935	933	**932**
	N_{wt}	2	1	**0**
	N_{bat}	308	306	**304**
	Converter (kW)	55	55	**55**
	LCC ($)	8,23,975	8,15,969	**8,08,374**

The optimal configuration of IRES is achieved by simulating hourly load demand at an LPSP value of 0%, Table 2 shows the corresponding optimization results with three optimization algorithms: GA, PSO and GOA. The optimal off-grid microgrid includes 932 number of PV panels (N_{pv}), 304 number of LA batteries (N_{bat}), 0 wind turbines (N_{wt}) and one 55 kW converter with Life Cycle Cost (LCC) of $8,08,374. Table 2 shows that GOA has the least number of components and the lowest life-cycle cost when compared to other well-known GA and PSO algorithms. Fig. 6 shows a detailed review of the GA, PSO, and GOA algorithms convergence processes to find the optimum configuration in the off-grid microgrid framework.

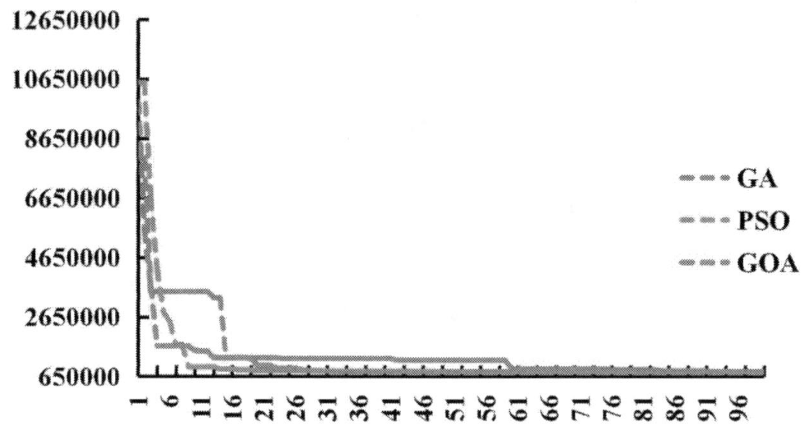

Fig. 6. Convergence curves of all proposed algorithms for IRES optimization

7 Conclusions

A framework for rural electrification utilising renewable resources such as wind, solar and biomass is presented in this paper. Five remote villages in India's Odisha state have been identified as a study area. A model of interconnected renewable sources has been Using the GOA algorithm, the proposed Integrated Renewable Energy System (IRES) model is optimised in the MATLAB environment to minimize the system Life Cycle Cost (LCC). Furthermore, the LCC obtained with the Grass Hopper Optimization algorithm has been provided at the lowest cost in comparison to its counterparts, GA and PSO algorithms. For developers of off-grid projects the proposed rural electricity IRES model may be useful.

References

1. P. P Kumar and R. P. Saini, "Optimization of an off-grid integrated hybrid renewable energy system with various energy storage technologies using different dispatch strategies," Energy Sources, Part A Recover. Util. Environ. Eff., vol. 00, no. 00, pp. 1–30, doi: 10.1080/15567036.2020.1824035. (2020)
2. P. P. Kumar and R. P. Saini, "Optimization of an off-grid integrated hybrid renewable energy system with different battery technologies for rural electrification in India," J. Energy Storage, vol. 32, no. September, p. 101912, doi: 10.1016/j.est.2020.101912. (2020)
3. A. L. Bukar, "Optimal sizing of an autonomous photovoltaic/wind/battery/diesel generator microgrid using grasshopper optimization algorithm," vol. 188, no. May, pp. 685–696 (2019)
4. A. Chauhan and R. P. Saini, "Techno-economic optimization based approach for energy management of a stand-alone integrated renewable energy system for remote areas of India," Energy, vol. 94, pp. 138–156, doi: 10.1016/j.energy.2015.10.136. (2016)
5. A. Chauhan and R. P. Saini, "Discrete harmony search based size optimization of Integrated Renewable Energy System for remote rural areas

of Uttarakhand state in India," Renew. Energy, vol. 94, pp. 587–604, doi: 10.1016/j.renene.2016.03.079. (2016)

6. A. M. Patel and S. K. Singal, "Optimal component selection of integrated renewable energy system for power generation in stand-alone applications," Energy, vol. 175, pp. 481–504, doi: 10.1016/j.energy.2019.03.055. (2019)
7. S. Saremi, S. Mirjalili, and A. Lewis, "Grasshopper Optimisation Algorithm: Theory and application," Adv. Eng. Softw., vol. 105, pp. 30–47, doi: 10.1016/j.advengsoft.2017.01.004. (2017)

https://doi.org/10.1051/e3sconf/202235002009

Modified Fuzzy Inference Method for Heat Flux Inversion of Geothermal Reservoir Heat Source

Chongxin Xu[1], *Chunwei* Zhou[1], *Gang* Liu[1], and *Shengming* Liao[1,*]

[1]School of Energy Science and Engineering, Central South University, Changsha, Hunan 410083, China

Abstract. The key to determine the performance of fuzzy inference inversion is to select a reasonable domain. However, there is no universal method for selecting domain at present. According to the characteristics of heat flux of geothermal heat source and the research of fuzzy inference inversion process, this paper modified the fuzzy inference method from two aspects of domain setting and iteration termination condition. The recommended domain and selection scheme for solving the problem of geothermal heat flux are given, and the modified fuzzy inference inversion method is applied to Rucheng geothermal field to verify the method. The results showed that the modified fuzzy inference inversion method could overcome the problem that the solution of the traditional method fell into a cycle, and the relative error of the verification term was less than 5%. Compared with the traditional method, the modified method greatly improved the computational efficiency, and the number of iterations was reduced to only 7. This method has a good application prospect for geothermal heat source inversion and resource evaluation.

1 Introduction

Obtaining geothermal source information accurately, especially the flux is crucial for the efficient evaluation and rational utilization of geothermal energy. Currently, the geophysics inversion methods [1-2] are mainly applied, which always need huge drilling and geological information [3]. However, these methods often require lots of drilling to obtain data, which not only needs a lot of cost, but also the obtained geological abnormal area cannot be directly confirmed as a geothermal heat source. Although thermal methods and remote sensing methods can obtain temperature distribution intuitively, due to the self-sealing characteristics of geothermal, it is often only possible to obtain temperature information in shallow areas [4,5]. For geothermal exploration, thermal method is easier and more intuitive than others. Based on inverse heat transfer problem, applying a few data can invert expected parameters, which is more economical and efficient [6].

However, due to the complex structure of geothermal reservoir and the highly ill-posed characteristic of inverse problem, the researches on inverse geothermal heat transfer problem are extremely rare. At present, the methods to solve the inverse heat transfer problems are divided into three major categories: Tikhonov regularization method [7], gradient-based methods (steepest descent method, conjugate gradient method, Levenberg - Marquardt method, etc.) [8], and evolutionary- and artificial intelligence-based methods (genetic algorithm, neural network method, particle swarm optimization algorithm, fuzzy inference method, etc.) [9]. The regularization method can overcome the illposed problem to some extent, but its calculation speed and accuracy are greatly affected by the selection of regularization parameters [10]. There is still no common method of parameter selection. Gradient-based method is often fast and has strong anti-ill-posed ability [11]. However, they are greatly influenced by selection of the initial value and are easily fall into the local optimal solution. In addition, the inversion results are significantly deteriorated when the information is incomplete or the error is large. When the situation is complex or there is a large amount of information, the calculation efficiency will be significantly affected, and even the inversion results may not be calculated. Although there are many researches and methods on the inverse heat transfer problem, the ill-posed problem still cannot be completely overcome, and most of the research is also limited to low-dimensional and simpler problems [12]. Geothermal reservoir is complex, with discontinuous physical parameters, non-uniform and discontinuous anisotropic media, and the existence of seepage. Related studies have shown that the fuzzy inference method has good calculation efficiency and robustness[13]. Compared with the low computational efficiency of genetic algorithms and the need for huge data to train of neural networks, the fuzzy inversion method has a brighter application prospect for solving the inverse geothermal heat sources problem [14].

A modified fuzzy inference inversion method for geothermal heat flux inversion is presented. The application of the modified method to Rucheng geothermal field case verifies the effectiveness and accuracy of the method.

2 Modified fuzzy inference method

*Corresponding author: smliao@csu.edu.cn

Fuzzy inference method is mainly based on fuzzy set theory, which simulates human thinking for uncertain reasoning, decision-making and calculation [15]. The establishment of fuzzy rules determines the efficiency and accuracy of calculations [16], and the judgment conditions also affect the accuracy of the results and the efficiency of calculations. The main determinant of fuzzy rules is the selection of the domain. Based on expert knowledge, it is still a large human factor and uncertainty factor, and there is still no common rule; the judgment condition is currently generally determined by the sum of squares of errors. Various factors are considered more comprehensively, but according to different situations, there is still improvement for the judgment condition. According to the actual situation of geothermal reservoir heat source heat flux inversion, this paper studied the influence of different domain selections on the efficiency and accuracy of fuzzy inversion.

2.1 Fuzzy inversion for geothermal source

There is still some controversy about the definition of geothermal heat source. The main point is that geothermal heat sources include magma packages, radioactive material decay, fracture friction, and chemical reaction heat. For hydrothermal geothermal system, especially in areas near volcanoes, the magma packet theory is generally accepted [17]. When studying geothermal utilization, more attention is paid to the areas where the heat flow is concentrated and strong in the geothermal reservoir in order to arrange the location of wells reasonably. Therefore, it is assumed that a certain area at the bottom of the reservoir is a heat concentrated area to provide energy to the reservoir.

According to the general characteristics of geothermal reservoirs, a simplified three-dimensional geothermal reservoirs model is established. The simplified model is divided into three layers. The uppermost layer is the surface area, and its main heat transfer way is heat conduction; the second layer is a thermal storage layer, and its main heat transfer ways are heat conduction and heat convection; the third layer is the area below the thermal storage layer , The main heat exchange way is heat conduction. According to the reservoir characteristics and the existence of seepage, each reservoir is set as a porous medium, and the heat exchange mode is controlled by adjusting its parameters. The energy equation of a three-dimensional geothermal reservoir is:

$$\left(\rho C_p\right)_{eff}\frac{\partial T}{\partial \tau} + \rho_f C_f q \nabla T - \nabla q_T = -\Phi_T \quad (1)$$

Where T indicates the reservoir temperature (K), $(\rho C_p)_{eff}$ represents the effective volumetric capacity, defined as $(\rho C_p)_{eff} = \varphi\rho_f C_f + (1-\varphi)\rho_s C_s$, where ρ_s indicates the solid density (kg/m^3), C_f and C_s are fluid and solid heat capacities (J/kg·K) respectively. q_T can be described as follow:

$$q_T = \lambda_{eff}\nabla T \quad (2)$$

Where λ_{eff} represents the effective thermal conductivity, defined as $\lambda_{eff} = \varphi\times\lambda_f + (1-\varphi)\times\lambda_s$, where λ_f and λ_s are fluid and solid thermal conductivities (W/m·K) respectively.

Because the geothermal heat source heat flux is relatively small, after establishing the reservoir model, twelve kinds of collocations and four initial values are set up to study the influence of the domain on fuzzy inversion. The input domain is set to three cases: PA: [-75,75], PB: [-90,90], PC: [-105,105]. The output universe is set to four cases: OA: [-1.5,1.5], OB: [-3,3], OC: [-4.5,4.5], OD: [-6,6]. The twelve combinations are: Case 1 (PA-OA), Case 2 (PA-OB), Case 3 (PA-OC), Case 4 (PA-OD), Case 5 (PB-OA), Case 6 (PB-OB), Case 7 (PB-OC), Case 8 (PB-OD), Case 9 (PC-OA), Case 10 (PC-OB), Case 11 (PC-OC) and Case 12 (PC-OD). After the direct problem is solved, six temperature measurement points are selected to obtain the data, and the actual measurement point temperature under the true value of the heat flux density is obtained, and the six measurement points are used to calculate the four cases with the initial value of 0.25, 1, 3, and 5 respectively.

According to the actual situation, we select the part in Fig.1 [18] as the research area and select ten wells for research. Among the ten wells, ZK4 and ZK8 are production wells, and the rest are testing wells. Their distribution is shown in Fig.1. The data of six wells for calculation (ZK3, ZK4, ZK5, ZK6, ZK7, ZK9) is selected, and the data of the remaining wells are used to verify the accuracy of the method.

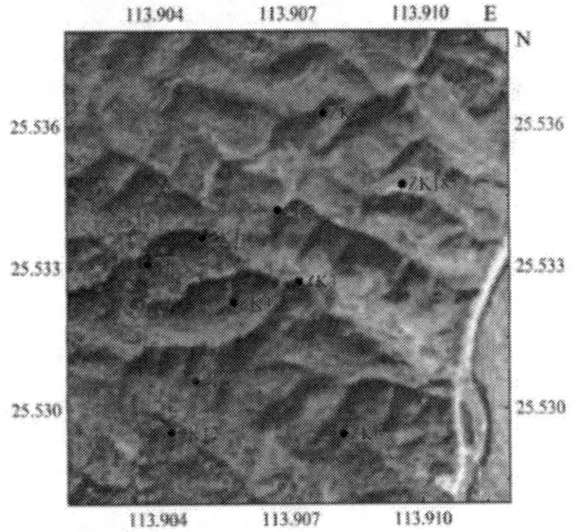

Fig. 1. Location of boreholes.

2.2 Results and discussion

2.2.1 Initial value is 0.25

When the initial value is set to 0.25, the inversion results of the 12 cases are shown in Fig.2 and Fig.3. The first number in the brackets in the figure represents the unilateral length of the input domain, and the second number represents the unilateral length of the output domain. According to the results, it is not difficult to find that the entire inversion process presents two

 https://doi.org/10.1051/e3sconf/202235002009

characteristics. One is that a trend is gradually approaching the real value, as shown in Fig.2; the other is that it oscillates at both ends of the real value, as shown in Fig.3. The ideal inversion process is the form shown in Fig.2. The results of Case 6 are the most ideal in various situations in Fig.2, with high calculation efficiency and accurate calculation results. When the input domain is the same and the output domain increases, the number of iterations decreases. However, when the output domain is too large, the adjustment value will increase. In some cases, results can be obtained, such as cases 2, 7, and 11. The domain value matches well. When the ratio of input and output domain values is further changed, accurate inversion results cannot be obtained, which then manifests as equal amplitude oscillations on both sides of the real value. When the output domain is the same, the input domain and the inversion performance do not show a correlation, and it is more important to select the appropriate value. The smaller the input domain, the larger the output domain, the greater the oscillation amplitude, and the larger the output domain of the same input domain, the slower it will reach the same amplitude oscillation. When the input domain is large, when the output domain is small, the second situation is less likely to occur. When the second situation occurs, if the final oscillation occurs, the average of the two repeated values is exactly the result of the inversion. The input-output domain ratio in Fig.2 is over 30, and each case in Fig.3 is less than 30. Of course, 30 is not necessarily a critical value, but it can be explained that the larger the input domain, the better. However, the increase in the ratio will increase the number of iterations, and affect the calculation efficiency. So, selecting a reasonable value and ratio is more appropriate.

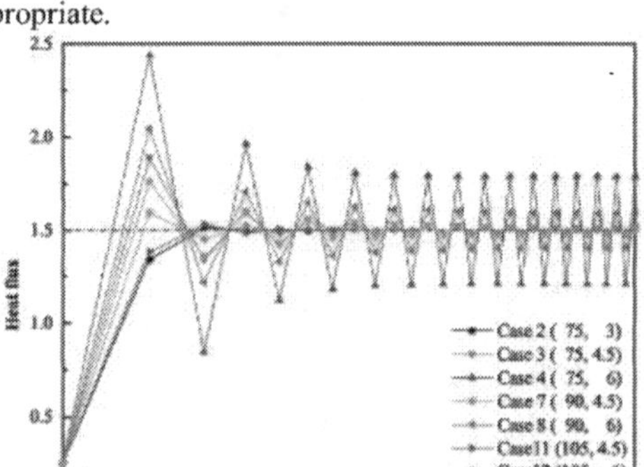

Fig. 2. Inversion of type 1 when initial value is 0.25.

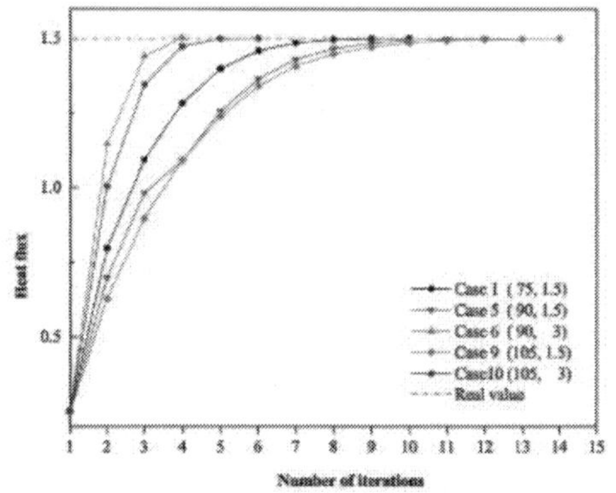

Fig. 3. Inversion of type 2 when initial value is 0.25.

2.2.2 Initial value is 1

When the initial value is 1, the inversion results are shown in Fig.4 and Fig.5. When the initial value is 1, the inversion performance and rules of various cases are basically the same. The set of case 6 is still the best performing group, and the results are still shown in two characteristics. When the inversion process is as in the first situation, the number of iterations is reduced to varying degrees, but the change is not large, and the number of iterations is reduced at most 2. In some cases, the number of iterations does not change.

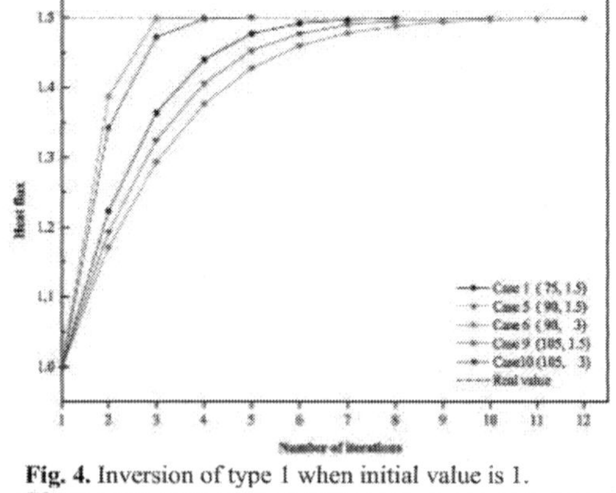

Fig. 4. Inversion of type 1 when initial value is 1.

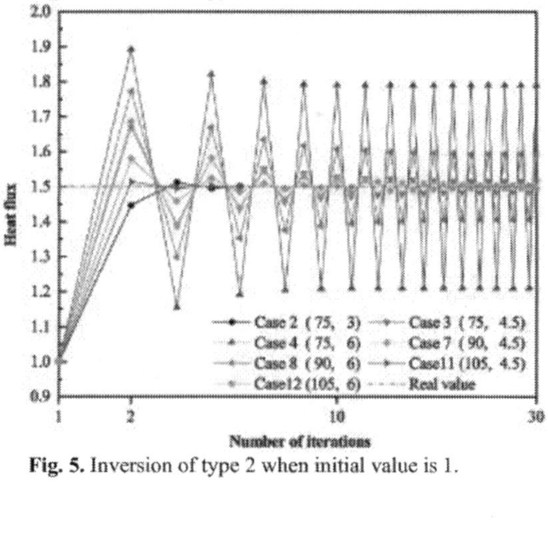

Fig. 5. Inversion of type 2 when initial value is 1.

2.2.3 Initial value is 3

When the initial value is 3, the inversion results are shown in Fig.6 and Fig.7. The performance and law of the inversion process are basically the same as the above two cases, indicating that this method can solve the inversion problem regardless of whether the initial value is greater or less than the real value. At this point, the best-performing case is also case 6.

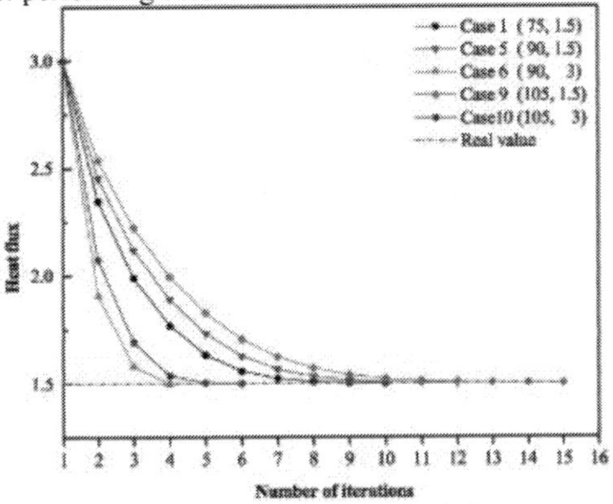

Fig. 6. Inversion of type 1when initial value is 3.

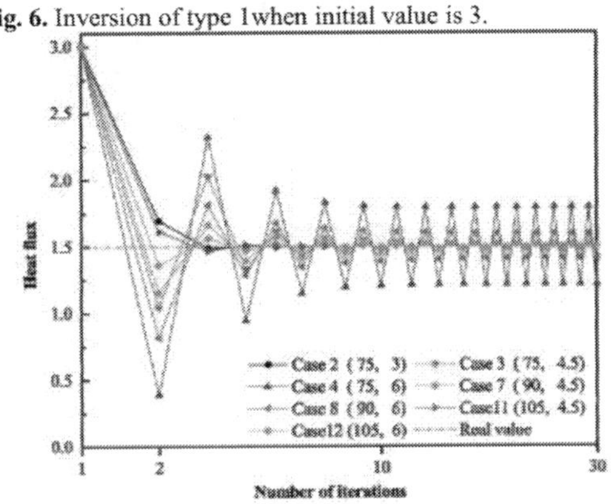

Fig. 7. Inversion of type 2 when initial value is 3.

2.2.4 Initial value is 5

When the initial value is 5, the inversion results are shown in Fig.8 and Fig.9. Through the comparison of the above four cases, the fuzzy inversion is not very sensitive to the selection of the initial value. Although it also shows the characteristics that the closer to the real value, the faster the inversion, but the number of iterations in different cases is almost the same, the maximum difference is 2, the inversion speed is not much different, and the final inversion results are all the same. The performance and rules of the inversion process of each case are consistent. Among the 12 cases set, the performance of case 6 is the best. It is not difficult to find that when the second inversion value and the first inversion value are on the same side of the real value, the inversion result can usually be obtained. When they are different, the situation of equal amplitude oscillation on both sides of the real value will often appear.

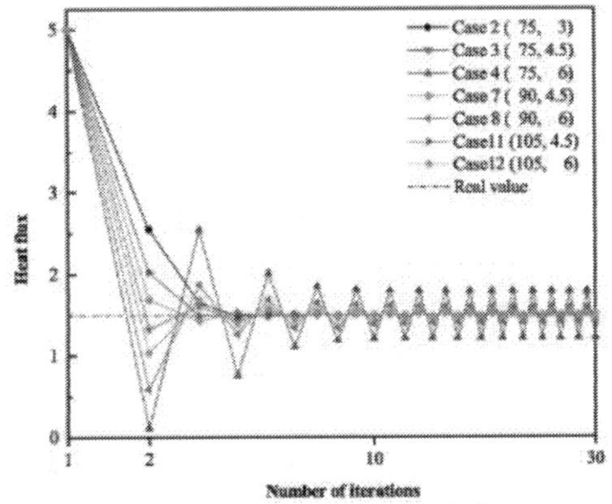

Fig. 8. Inversion of type 1 when initial value is 5.

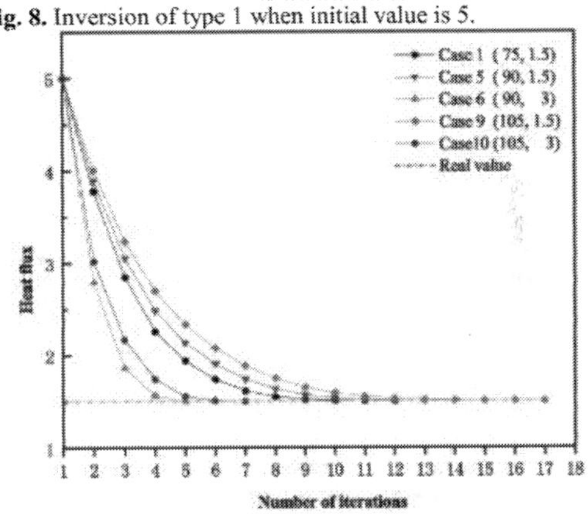

Fig. 9. Inversion of type 2 when initial value is 5.

2.3 Introduction of modified method

According to the characteristics of geothermal and the above research, a modified fuzzy inversion method is proposed. This method is modified in two aspects.

2.3.1 Selection of input and output domains of fuzzy rules

In view of the characteristics of geothermal and the performance of the above-mentioned cases, it is recommended to choose Case 6 (input domain [-90,90], output domain [-3,3]) as the domain choice for inverting geothermal heat flux. It can also adopt another scheme. Select the input and output domains according to the actual situation, take a larger value for the input domain, and set the input-output domain ratio to be greater than 30.

2.3.2 Selection and improvement of iteration termination conditions

According to the results, although the above selection of the domain has been optimized, there may still be unavoidable oscillations. The accurate inversion can also be obtained by selecting and improving the iteration termination conditions. The iterative termination condition generally chooses the method in which the sum of squares of errors is less than a small value.

Although this method can consider the impact more comprehensively, it cannot converge to obtain the inversion result after the oscillation occurs, and it will continue to fall into the loop. Therefore, additional judgment conditions must be added. Because the establishment of the direct problem model has many artificial factors that cannot fully simulate the geological situation. For geothermal problems, if the limit value of the error square is often selected too large, a satisfactory result may not be obtained. If the selection is too small, the termination condition will never be met when applied to the actual situation. When the same value is found consecutively, the calculation is stopped, the result is output and the manual inspection is prompted. When it is determined that there are inversion values with the same or very similar intervals, the average value of the two end points of the oscillation is taken as the input of the next operation until the requirement of the sum of squares of the error is satisfied to obtain the inversion result.

When the fuzzy rules are selected reasonably, the inversion result can be obtained quickly. In addition, when the iteration termination condition is further restricted, the phenomenon of non-convergence can be effectively solved and the inversion result can be obtained. Two improvements to fuzzy inversion can effectively solve the problem of heat flux inversion in geothermal reservoirs.

3 Rucheng verification case

To verify the reliability and superiority of the modified fuzzy inversion method, the modified method is applied to the inversion of heat flux in Rucheng hydrothermal geothermal field. The Rucheng geothermal field is located on the north side of the middle section of the Nanling Mountains, surrounded by a depression basin of high and low in the middle, which belongs to the terrain of middle and low mountains. The geomorphic structure is divided into three types: structural erosion, structural denudation and erosion accumulation type. The geothermal field area belongs to a warm and humid subtropical climate, with an annual precipitation of 1051.9-2303.6 mm, a multiyear average temperature of 289.75 K, and an annual average absolute humidity of 16.8%. In the geothermal area, the width of the hot water riverbed is 20-70 m, and there are deposits of sand and gravel ranging from 0.5-20 m. The highest flood water level of the Reshui River is 342.734 m, the lowest water level is 339.879 m, and the maximum flow is about 300 m^3/s. The exposed stratum in this area is simple, mainly composed of Quaternary alluvium, Sinian sandy slate, shallow metamorphic sandstone and early granite rock mass of Yanshan. Then, compare it with the results of traditional fuzzy inversion. The inversion results are shown in Fig.10. The modeling of the Rucheng geothermal reservoir is also simplified to the general geothermal reservoir model mentioned above. The parameters of each part are set as shown in Table 1 according to the actual exploration results, and the shape and location of the heat source are artificially assumed. The area at the bottom except for the region of the heat source is in an adiabatic condition, and the surrounding area is also in an adiabatic condition with certain initial value. The top is in robin condition. Since the actual situation is natural convection heat exchange with air in the large space, the convective heat transfer coefficient is 5 $W/(m^2 \cdot K)$. The locations of selected measurement points are shown in Table 2.

Using the traditional fuzzy inversion method and the improved fuzzy inversion method, the same heat flux result (1.361 W/m^2) was obtained, which is shown in Fig.10. It can be seen that when only the termination condition is improved, the number of iterations is reduced from the original infinite loop at the same value (19) to 15, which improves the calculation efficiency and prevents getting into the loop. If the choice of domain is also improved, the fuzzy inversion does not show an oscillating form, but smoothly inverts to the final value, and greatly reduces the number of iterations to 7. Since there is no accurate result of heat flux obtained from actual exploration, the remaining drilling data is selected and is compared with the data obtained from the inversion value to verify the accuracy of the inversion results. The results are shown in Table 3. The measured temperature id obtained from actual drilling data at selected points. Inversion temperature is calculated by substituting inversion parameters into the established simulation model.

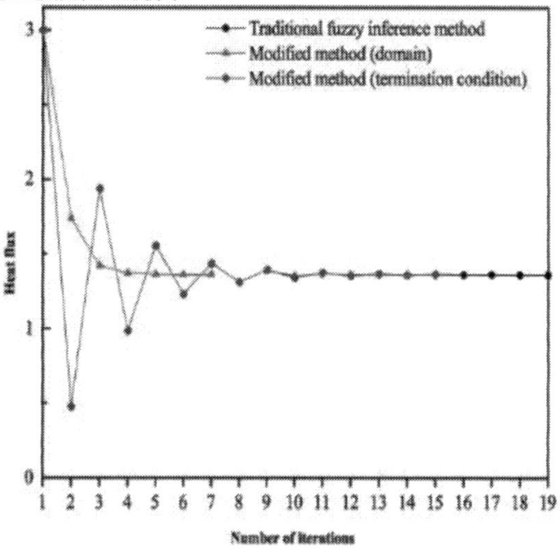

Fig. 10. Inversion results f Rucheng geothermal field.

Table 1. Parameters of the simplified model of Rucheng geothermal field.

Layer	Length	Width	Height	Density
	m	m	m	kg/m^3
1	800	800	40	1730
2	800	800	60	2000
3	800	800	400	2600
Layer	Thermal conductivity		Heat capacity	
	W/(m·K)		J/(kg·K)	
1	0.9		840	
2	2.0		950	
3	2.6		837	

Table 2. Locations of selected measurement points.

Number	Length	Width	Depth
	m	m	m
ZK3	550.54	453.33	100
ZK4	430.11	408.89	100
ZK5	344.09	275.56	100
ZK6	627.96	173.33	100
ZK7	275.27	480	100
ZK8	490.32	591.11	100
ZK9	576.34	777.78	100
ZK10	275.27	533.33	95
ZK12	309.68	177.78	100
ZK18	722.58	635.56	100

The relative errors between the temperature field data obtained by the inversion and the actual measured temperature data are within 5%, which verifies the accuracy of the fuzzy inversion method. Compared with the traditional fuzzy inversion method, the improved fuzzy inversion method greatly improves the calculation efficiency and the problem of loop.

Table 3. Verification of the inversion results of Rucheng geothermal field.

Number of drilling wells	Measured temperature	Inverse temperature	Relative error
	K	K	%
ZK8	318.75	316.31	-0.77
ZK10	312.85	320.55	2.46
ZK12	302.15	313.36	3.71
ZK18	300.95	308.37	2.47

4 Conclusion

By analyzing the characteristics of the fuzzy inversion process, combined with the characteristics of the researched object, the fuzzy inference method is modified in terms of the domain setting and the iterative termination condition. A recommended domain and selection method are given, and the reliability of the modified method and its superiority compared with the traditional fuzzy inversion method are verified through the Rucheng geothermal case. The specific conclusions are summarized as follows:

(1) The results of the fuzzy inversion process under different initial values and different domain settings show that fuzzy inversion can efficiently obtain accurate inversion results. The fuzzy inversion process presents two characteristics: one-sided approximation to the true value and final oscillation with equal amplitude around the real value on both sides.

(2) In view of the characteristics of the traditional fuzzy inversion process, improvements were made in two aspects: the domain setting and the selection of the iterative termination condition. The recommended domain of geothermal heat source heat flux (input domain [-90,90], output domain [-3,3]) and selection methods are presented, and modified iteration termination conditions are proposed.

(3) The modified fuzzy inversion method was applied to Rucheng geothermal reservoir case, and the inversion results were verified based on other drilling data. The results showed that the relative errors of the inversion results were all within 5%, indicating the method's effectiveness and accuracy. Compared with the traditional method, the modified method overcomes the shortcomings (falling into an oscillation cycle) of the traditional method. At the same time, the number of iterations is greatly reduced to 7.

Acknowledgements

This work was supported by the National Natural Science Foundation of China [52076218]; the Foundation of National Sustainable Development Agenda [2019sfq02]; the Fundamental Research Funds for the Central Universities of Central South University [2021zzts0665].

References

1. E. Cabrera-Navarrete, G. Ronquillo-Jarillo and A. Markov, Wavelet analysis for spectral inversion of seismic reflection data, Journal of Applied Geophysics **177**, 104034 (2020)
2. R. Li, H. Zhang, Q. Zhuang, R. Li and Y. Chen, Bp neural network and improved differential evolution for transient electromagnetic inversion, Computers and Geosciences **137**, 104434 (2020)
3. J. Domra Kana, N. Djongyang, D. Ra¨ıdandi, P.N. Nouck and A. Dadj´e, A review of geophysical methods for geothermal exploration, Renewable and Sustainable Energy Reviews **44**, 87-95 (2015)
4. A.A. Minissale, A simple geochemical prospecting method for geothermal resources in flat areas, Geothermics **72**, 258-267 (2018)
5. M. Li, L. Zhang, G. Liu, Step-wise algorithm for estimating multi-parameter of the ground and geothermal heat exchangers from thermal response tests, Renewable Energy 150 (2020) 435–442.
6. G. Liu, B. Zhou and S. Liao, Inverting methods for thermal reservoir evaluation of enhanced geothermal system, Renewable and Sustainable Energy Reviews **82**, 471-476 (2018)
7. H. Najafi, K.A. Woodbury and J.V. Beck, Real time solution for inverse heat conduction problems in a twodimensional plate with multiple heat fluxes at the surface, International Journal of Heat and Mass Transfer **91**, 1148-1156 (2015)
8. M. Cui, K. Yang, X.L. Xu, S.D. Wang and X.W. Gao, A modified levenberg-marquardt algorithm for simultaneous estimation of multi-parameters of boundary heat flux by solving transient nonlinear inverse heat conduction problems, International Journal of Heat and Mass Transfer **97**, 908-916 (2016)
9. K.H. Lee, Application of repulsive particle swarm optimization for inverse heat conduction problem - parameter estimations of unknown plane heat source, International Journal of Heat and Mass Transfer **137**, 268-279 (2019)

10. D. Ma, W. Tan, Z. Zhang and J. Hu, Parameter identification for continuous point emission source based on tikhonov regularization method coupled with particle swarm optimization algorithm, Journal of Hazardous Materials **325**, 239-250 (2017)
11. X. Luo, Z. Yang, A new approach for estimation of total heat exchange factor in reheating furnace by solving an inverse heat conduction problem, International Journal of Heat and Mass Transfer **112**, 1062-1071 (2017)
12. H. Li, J. Lei and Q. Liu, An inversion approach for the inverse heat conduction problems, International Journal of Heat and Mass Transfer **55(15)**, 4442-4452 (2012)
13. K. Wang, G. Wang, H. Chen, S. Wan and C. Lv, Quantitative identification of three-dimensional subsurface defect based on the fuzzy inference of thermal process, International Journal of Heat and Mass Transfer **133**, 903-911 (2019)
14. G. Wang, S. Wan, H. Chen, C. Lv and D. Zhang, A double decentralized fuzzy inference method for estimating the time and space-dependent thermal boundary condition, International Journal of Heat and Mass Transfer **109**, 302–311 (2017)
15. A. Hocine, N. Kouaissah, S. Bettahar and M. Benbouziane, Optimizing renewable energy portfolios under uncertainty: A multi-segment fuzzy goal programming approach, Renewable Energy **129**, 540–552 (2018)
16. P.A. Adedeji, S.A. Akinlabi, N. Madushele and O. O. Olatunji, Neuro-fuzzy resource forecast in site suitability assessment for wind and solar energy: A mini review, Journal of Cleaner Production **269**, 122104 (2020)
17. J. Tian, Z. Pang, Q. Guo, Y. Wang, J. Li, T. Huang and Y. Kong, Geochemistry of geothermal fluids with implications on the sources of water and heat recharge to the rekeng high-temperature geothermal system inthe eastern himalayan syntax, Geothermics **74**, 92–105 (2018)
18. X. Long, R. Yuan, X. Deng, F. Li, J. Pi and H. Li, Hot dry rock geothermal resources in ru county (chinese), Science & technology review **33**, 68-73 (2015)

Preparation and Application of Nano-structured Lead Dioxide from Waste Lead Slag

Xi Wang[1], *Dandan* Wu[1,2], *Du* Yuan[1], and *Xu* Wu[1,*]

[1]School of Environmental Science and Engineering, Huazhong University of Science and Technology, Wuhan, 430074, China
[2]Hubei HuaDeLai (HDL) Co. Ltd, Wuhan, 430079, China

Abstract. To develop an efficient and green method to recycling lead slag, a novel strategy to fabricate nano-lead dioxide from lead slag was applied by the hydro-electrometallurgy. In leaching system, the optimum condition of leaching time 100 min, 80 °C, stirring rate 500 rpm, liquid/solid ratio 20 and 1 mol/L methanesulfonic acid resulted in lead recovery of 89% and residue obtained without toxicity. The kinetic study revealed that the methanesulfonic acid leaching of lead slag shows good agreement with a diffusion-controlled shrinking-particle model. Additionally, the apparent activation energy of MSA leaching of lead slag was determined using Arrhenius model as 13.621 kJ/mol. Furthermore, Nano-PbO_2 recovered from waste lead paste was prepared for the electrolysis of water to produce ozone. The experimental results showed that the ozone production of nano-PbO_2 recovered from waste lead slag was not significantly different from the nano-PbO_2 prepared by pure reagents. In conclusion, waste lead slag can be recycled for the fabrication of nano-PbO_2, which has a significant advantage on realizing the recycling of lead resources.

1 Introduction

As the major solid waste generated during the smelting process of primary and secondary lead, lead slag has gained extensive focus in the environmental engineering field due to its enormous amount and high migratory toxic elements containing. Nowadays, lead slag is mainly disposed by landfill and stockpiling, which occupies extensive area [1]. However, there is a potential risk of severe pollution of surrounding environment and critical waste of resources. The high concentrations of precious metal in lead slag reconciled with the uncertain environmental stability inducing further research relating to the efficient recovery of metal lead from lead slags. Thus, growing attention has been drawn to recovery of lead slags.

Various methods for recovery of metals and decrease of the lead dispersion from lead slag are proposed. The current recycling process of lead slag includes mainly pyrometallurgical smelting, hydrometallurgical leaching and bioleaching technology [2-4]. The traditional pyrometallurgical smelting method is a widely adopted to recovering the secondary lead. But there are higher requirements for equipment and SO_2 emission and lead dust in pyrometallurgical technology can cause a hazardous threat to the environment. For this reason, many researchers draw attention to the recovery of lead via hydrometallurgy technology [5, 6]. Hydrometallurgical technology can separate and extract valuable elements from lead slag effectively and selectively. Forte [7] reported the lead recovery from lead slag via acid leaching, and this method is applicable to the recovery of metallic lead from lead residue. Nitrate-based leaching systems include HNO_3, H_2O_2-HNO_3 and Fe-HNO_3 etc. Kim [8] proposed an 2nd leaching step using HNO_3-Fe(III), about 90% of lead was leaching. However, the waste acid, waste salt, the residue and other by-products would be generated in the nitrate-based leaching systems.

As a strong organic acid, methanesulfonic acid (MSA) has many properties such as good stability, high conductivity, low toxicity, biodegradability and high solubility of metallic salts [9]. It is a polar molecule with a strong complexing ability to Pb^{2+}. In recent years, methanesulfonic acid has been used widely in Sn, Pb and Cu plating as electrolyte methanesulfonic acid leaching. In the meanwhile, methanesulfonic acid has a broad application in recovering precious metals such as Pb, Zn, Cu and Sn as lixiviant. methanesulfonic acid was a common leaching reagent to recover Pb from cerussite, galena, brass, etc. [10-11]. Therefore, methanesulfonic acid was selected and studied in the experiments.

PbO_2, as an excellent electrode material, is widely used in wastewater treatment, electrochemical measurement, and electrochemical ozone production (EOP) [12]. According to the research of Chen [13], nano-PbO_2 can improve the current efficiency of ozone production effectively, thereby reducing energy consumption. Nano-structured PbO_2 has a larger specific surface area, which increases the specific surface area of electrocatalytic activity. To realize the recycling of lead

* Corresponding author: profxuwu@hust.edu.cn.

resources, the fabrication of nano-structured lead dioxide from lead slag is proposed.

2 Materials and methods

2.1 Preparation and materials

The spent lead slag sample was collected from traditional furnaces at a lead smelting factory in Hubei Xiangyang, China. The sample was dried at 120 °C for 24 h under vacuum after crushing. Then the sample was put into a planetary high-energy ball-milling device (XQM-4) with the sealed zirconia ball-milling pot until 300 mesh.

2.2 Stability evaluation

The short-term stability of solid samples obtained from the spent lead slag and samples after leaching was assessed by the Toxicity Characteristic Leaching Procedure (TCLP), using an extraction fluid of pH 4.93±0.05. The 1000 mL extractant was made up of 5.7 mL of CH_3CH_2COOH and 64.3 mL of 1 M NaOH and deionized water. The solid samples and extractant were added to 10 mL centrifuge tubes in the solid-to-liquid ratio of 1:20, and continuously oscillated at 200 rpm for 18 ± 2 hours in a tumble shaker. Suspensions were finally filtered using 0.22 μm membrane filters.

2.3 Leaching experimental procedure

All leaching experiments of lead slag were initiated in a 200 mL beaker, which was fixed with a mechanical stirrer. First, the lead slag was added to a mixed solution of H_2SO_4 and H_2O_2, which converted lead element into lead sulphate. Then the lead sulphate was transformed into lead carbonate by ammonium carbonate. Leaching agent, methanesulfonic acid, with a desired concentration was put into the beaker and heated to a predetermined temperature before the lead carbonate was added to the beaker with agitation. After the target temperature was reached, sieved lead slag concentrate sample was added to the system.

2.4 Diaphragm electrolysis

The main part of the electrolytic cell was made of organic glass (polymethylmethacrylate, PMMA) A 2.0 cm×2.0 cm hollow was set in the opposite direction of the cathode chamber and the anode chamber (3.0 cm×4.0 cm×4.0 cm). The exchange membranes were permeated in 1 mol·L^{-1} NaCl solution for 12 hours and then cut into a size of 3.0 cm×3.0 cm. Then the two sides were fixed with fluorine rubber pads, the anode and cathode chamber and ion exchange membrane were pressed together with bolts assembly. The cathode and anode plates were respectively fixed by the electrode plate holders, which were made by 3D printing technology. There were slits in the middle to ensure that the electrode plates were always upright. The electrode plate holders can slide on the tank and the distance between the cathode and anode plates can be adjusted freely between 2.0 cm and 10.0 cm.

An HMTECH-A4080DT ion exchange membrane (Hangzhou Huamo Technology Co. Ltd., China) was used to separate an electrolysis cell. The ion exchange membrane was mainly synthesized from polyvinylidene fluoride, polyacrylonitrile and styrene. The important parameters for the ion of < 2.7 Ω·cm^2, and exchange capacity of 2.2 meq/mg. The anode plate is a fluorine-doped tin oxide (FTO) conductive glasses, and the cathode is a titanium electrode of 2.8 cm×5.0 cm. The width of the anode plate is slightly smaller than the width of the anode chamber to ensure that there is no obstruction for convection of electrolyte.

2.5 Characterization and analyzation

After all leaching experiment, the obtained filtrate was analysed by inductively coupled plasma atomic emission spectroscopy (ICP-AES, Thermo Elemental) to determine the amount of metal extraction. The original sample and obtained solid residues were analysed by X-ray diffraction (XRD, Philips) in order to determine the mineralogical composition. Scanning electron microscopy (SEM, JSM-IT300LA) was employed to characterize the morphology of precipitations.

3 Results and discussion

3.1 Characterization of lead slag

In the process of pre-treatment of the lead slag, the sample was ground into 300 mesh particles with a ball mill, and then pass them through a sieve. X-ray diffraction analysis (XRD) analysis was used for mineralogy of the slag sample with a particle size of less than 74 μm. In Fig.1, the main mineral of the Pb sample was $Pb_4(SO_4)(CO_3)_2(OH)_2$, $PbCO_3$ and SiO_2.

 https://doi.org/10.1051/e3sconf/202235003001

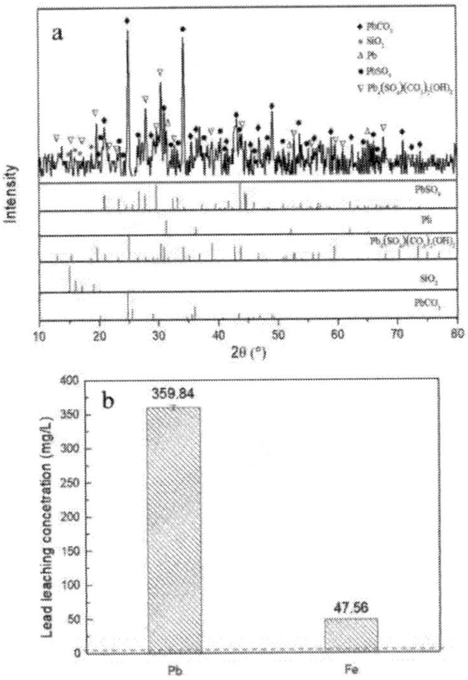

Fig. 1. XRD pattern (a) and TCLP result of lead slag (b).

To be classified as non-hazardous waste, the leaching concentration of Pb should be lower than 5 mg/L, while there is no limit of Fe [15, 16]. The results obtained with the TCLP in Fig.1 showed that the original sample of lead slag was significantly higher than the standard. High leachability indicates greater potential harm to the environment. Although elements combined in lead slag are not contact with leaching solutions directly, there are potential mobility of heavy mobility and it should not be considered safe for environment if landfilled or processed for building materials.

3.2 Batch experiments

To optimize the efficiency of lead leaching in lead slag, the effect of operating parameters such as liquid/solid ratio, stirring rate, temperature, methanesulfonic acid concentration on recovery of lead was examined. The experiment of stirring rate was investigated in the range of 200 to 600 rpm. The determination of the mass transfer and diffusion is the key to influence the speed in the leaching process [17]. The stirring intensity play an important role in the reaction rate during the solid-liquid reaction, which increased the leaching rate by 40%–70% [18]. However, the stirring rate didn't influent leaching rate in Fig.3, which suggest the mass transfer is not affected by fluid film surrounding the solid particles, and the liquid film diffusion is not a control step of this process. The result of the effect of liquid-solid ratio between 5:1 and 40:1 was shown that the larger the liquid-solid ratio has a higher the leaching rate of lead. When the liquid-solid ratio is large, the viscosity and diffusion resistance of solutions could be reduced to improve the leaching rate [19]. The effect of methanesulfonic acid concentration ranging from 0.5 to 5 mol/L was studied. Fig.2 shows that the leaching efficiency is increasing with an increasing concentration, which indicate that the accelerate dissolution of lead slag.

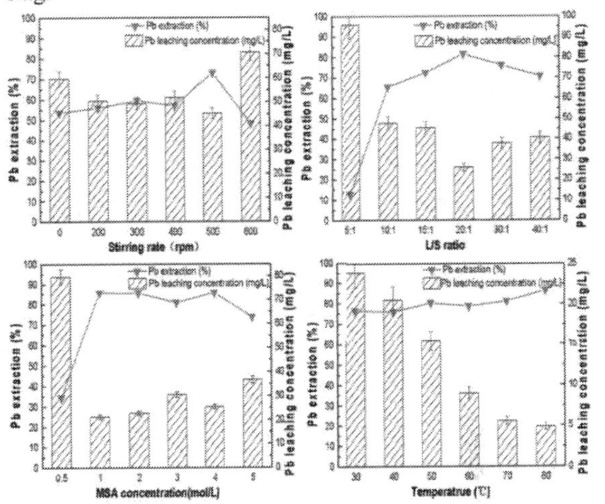

Fig. 2. Pb extraction and TCLP result of Pb leaching concentration of the residue at various string rates, L/S ratios, MSA concentrations and temperatures.

3.3 Kinetic analysis

After optimizing the parameters, then the kinetic study was performed in the optimized condition. The leaching results obtained at different temperature (40-80 °C) in Fig.3. The temperature influences the dissolution of lead, which has a significant effect on the complexation of methanesulfonic acid.

The application of shirking core models as Eq. (1), (2) and (3) was used to determine the lead leaching kinetics. In the leaching experiment, samples are taken regularly to analyse the leaching efficiency. Eq. (1) is a film diffusion-controlled process, while Eq. (2) is applicable to chemical reaction-controlled process. When the leaching reaction is in a diffusion-controlled process of stationary film, the kinetic reaction equation is according to Eq. (3), where α is the leaching efficiency of lead, k is the apparent chemical reaction rate, t is the leaching time.

$$\alpha = kt \tag{1}$$

$$1-(1-\alpha)^{\frac{1}{3}} = kt \tag{2}$$

$$1-\frac{2}{3}\alpha-(1-\alpha)^{\frac{2}{3}} = kt \tag{3}$$

Based on Eq. (1), (2), (3) and obtained data, α, $1-(1-\alpha)^{\frac{1}{3}}$, $1-\frac{2}{3}\alpha-(1-\alpha)^{\frac{2}{3}}$ value versus time were plotted in Fig.3. The experimental data was linearly fitted with a diffusion-controlled shrinking-particle model. It could be seen that Fig.3 showed a good linear relationship in the leaching experiment, indicating that the dynamic data of leaching reaction fitted the diffusion-controlled shrinking-particle model well. The kinetic equation at each temperature was obtained from Fig.3 as Eq. (4), (5) and (6), where α is the leaching efficiency of lead, t is the leaching time.

$$40\ ℃: 1-\frac{2}{3}\alpha-(1-\alpha)^{\frac{2}{3}}=0.001145t-0.005101 \quad (4)$$

$$60\ ℃: 1-\frac{2}{3}\alpha-(1-\alpha)^{\frac{2}{3}}=0.001443t-0.003537 \quad (5)$$

$$80\ ℃: 1-\frac{2}{3}\alpha-(1-\alpha)^{\frac{2}{3}}=0.001606t-0.002197 \quad (6)$$

The correlation coefficients value corresponding to kinetic equations (4), (5) and (6) are 0.9753, 0.9694 and 0.9383 respectively, indicating that the measured data is in good agreement with the fitting equation.

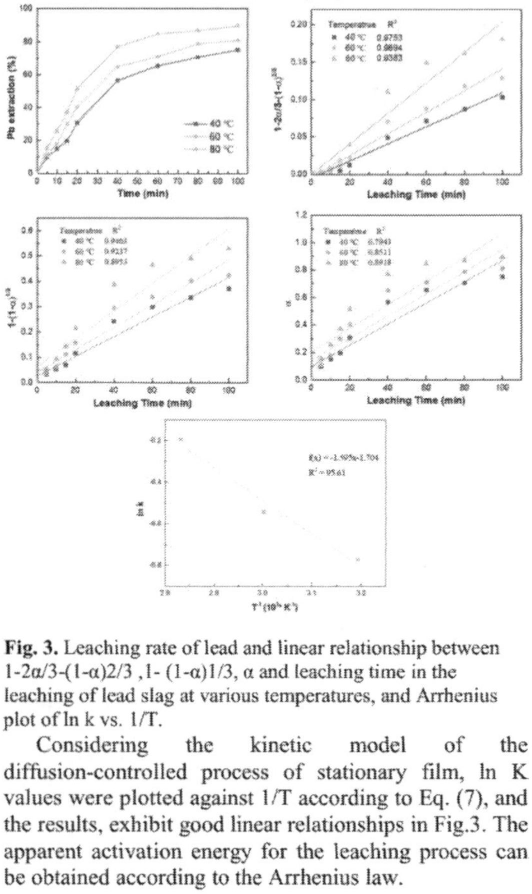

Fig. 3. Leaching rate of lead and linear relationship between 1-2α/3-(1-α)2/3 ,1- (1-α)1/3, α and leaching time in the leaching of lead slag at various temperatures, and Arrhenius plot of ln k vs. 1/T.

Considering the kinetic model of the diffusion-controlled process of stationary film, ln K values were plotted against 1/T according to Eq. (7), and the results, exhibit good linear relationships in Fig.3. The apparent activation energy for the leaching process can be obtained according to the Arrhenius law.

$$\ln K = -\frac{E}{RT} + B \quad (7)$$

The estimated apparent activation energy from the slope of fitted curve calculated as 13.621 kJ/mol according to Eq. (7) and Fig.3. The activation energy is within the range of internal diffusion control between 8 and 30 kJ/mol [20, 21], which further proves that the leaching reaction is a diffusion-controlled process of stationary film.

3.4 A proposed process for preparation of nano-structured PbO_2

In this study, a diaphragm electrodeposition device for fabricating PbO_2 was used to avoid the deposition of lead that made a cause of material loss and energy consumption caused on the anode in Fig.4. To realize the recycling of acid liquid, the anion exchange membrane was used to enrich the acid solution in the anode chamber, and the collected acid solution can be used in the leaching stage. The two products of electrolysis are separated and collected in two electrode chambers, which simplifies the operation and avoids product contamination.

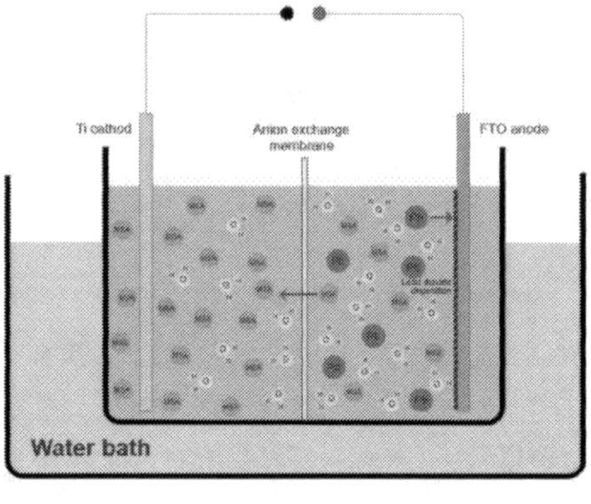

Fig. 4. Schematic of the diaphragm electrolysis cell.

Fig.5 shows the process of recovering and preparing nano-PbO_2 from waste lead slag. Lead slag is pre-treated as the above-mentioned leaching experiment, the solid sample was transformed into lead-containing methanesulfonic acid solution prepared for the next stage. Then PbO_2 was collected at the anode electrode by diaphragm electrolysis. The optimum temperature for calcining Pb_3O_4 was determined to be 400 °C, and the reaction time reached 10 h to obtain pure Pb_3O_4 from PbO_2. Pb_3O_4 obtained after calcination was added to acetic acid to obtain the precursor $Pb(CH_3COO)_4$. Pure nano-PbO_2 was obtained with the hydrolysis method as reported before. And the nanoparticles were range from 10-30 nm, which shows pure β-PbO_2 crystallization.

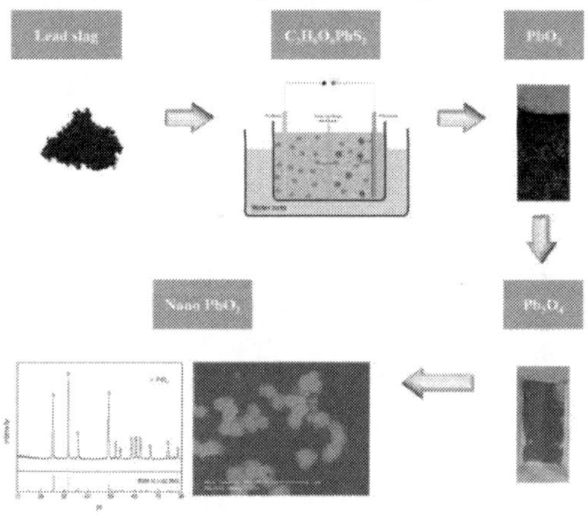

Fig. 5. Preparation of the nano-structured lead dioxide.

3.5 Application of nano-structured PbO_2

The obtained nano-PbO_2 was prepared according to the method as reported before [22] to prepare membrane electrode assembly for electrolyzing water to generate ozone, and its ozone-producing performance was measured as Fig.6.

Under the same current density, the cell voltage of obtained PbO_2 is basically the same as the cell voltage of nano-PbO_2 prepared with $Pb(CH_3COO)_4$ reagent. The output, current efficiency and energy consumption are also similar with the increase of current density. At 0.9 $A \cdot cm^{-2}$, the lowest energy consumption is about 182 wh $(gO_3)^{-1}$. The ozone-production of nano-PbO_2 recovered from waste lead slag and the nano-PbO_2 prepared by pure reagent is equivalent. Therefore, waste lead slag be employed to the preparation of nano-PbO_2 to realize the recycling of lead resources.

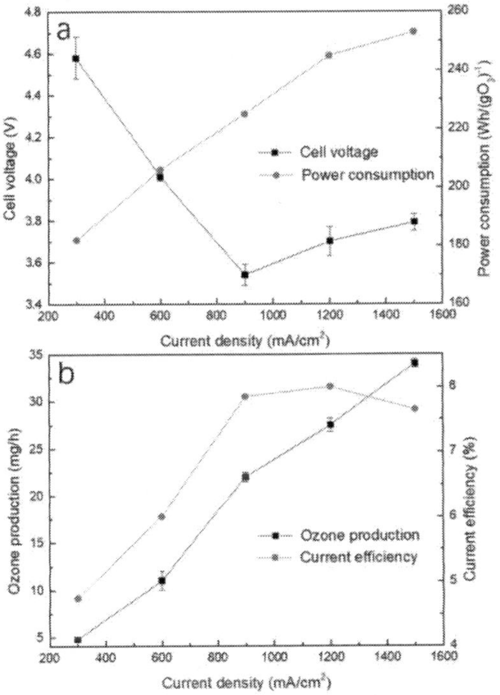

Fig. 6. The cell voltage and energy consumption (a), ozone production and current efficiency (b) with different current density.

4 Conclusion

In this research, the preparation and application of nano lead dioxide from lead slag is studied. In leaching system, the factor of stirring speed, temperature, methanesulfonic acid concentration, and liquid/solid ratio on the leaching performance of lead slag were investigated. The optimum condition of stirring rate 500 rpm, L/S ratio 20, 1 mol/L methanesulfonic acid, and 80 °C resulted in lead extraction of 89% and residue obtained without toxicity. The kinetics of leaching process was linearly fitted with a diffusion-controlled shrinking-particle model. Additionally, the Arrhenius diagram was plotted for leaching reaction and the apparent activation energy was obtained as 13.621 kJ/mol. Nano-PbO_2 recovered from waste lead paste was prepared for the electrolysis of water to produce ozone. The experimental results showed that the ozone production of nano-PbO_2 recovered from waste lead slag was not significantly different from the nano-PbO_2 prepared by pure reagents. In conclusion, waste lead slag can be employed to the fabrication of nano-PbO_2, which has a significant advantage on realizing the recycling of lead resources.

References

1. M. Alwaeli, Investigation of gamma radiation shielding and compressive strength properties of concrete containing scale and granulated lead-zinc slag wastes - ScienceDirect. Journal of Cleaner Production, **166**, 157-162 (2017)
2. F. Dai, et al., Recovery of high purity lead from spent lead paste via direct electrolysis and process evaluation. Separation and Purification Technology, **224**, 237-246 (2019)
3. J. Zhang, et al., A new pre-desulphurization process of damped lead battery paste with sodium carbonate based on a "surface update" concept. Hydrometallurgy, 2016. 160: p. 123-128.
4. A. Potysz, E.D. van Hullebusch and J. Kierczak, Perspectives regarding the use of metallurgical slags as secondary metal resources - A review of bioleaching approaches. Journal of Environment Management, **219**, 138-152 (2018)
5. C.S. Chen, Y.J. Shih and Y.H. Huang, Recovery of lead from smelting fly ash of waste lead-acid battery by leaching and electrowinning. Waste Management, **52**, 212-220 (2016)
6. J. Tang and B.M. Steenari, Leaching optimization of municipal solid waste incineration ash for resource recovery: A case study of Cu, Zn, Pb and Cd. Waste Management, **48(FEB.)**, 315-322 (2016)
7. F. Forte, et al., Closed-loop solvometallurgical process for recovery of lead from iron-rich secondary lead smelter residues. Rsc Advances, **7(79)** (2017)
8. E. Kim, et al., Process development for hydrometallurgical recovery of valuable metals from sulfide-rich residue generated in a secondary lead smelter. Hydrometallurgy, **169**, 589-598 (2017)
9. H. Wang, et al., Direct oxidative pressure leaching of bismuth sulfide concentrate in methanesulfonic acid medium. Hydrometallurgy, **194**, 105347 (2020)
10. T.J. O'brien, Method for measuring the quantity of lead on the surface of a brass component. US5612224 A (2017)
11. Z. Wu, et al., Fundamental study of lead recovery from cerussite concentrate with methanesulfonic acid (MSA). Hydrometallurgy, **142**, 23-35 (2014)
12. B.P. Chaplin, Critical review of electrochemical advanced oxidation processes for water treatment

applications. Environmental Science: Processes and Impacts, **16(6)**, 1182-1203 (2014)
13. T. Chen, et al., Electrochemical sensing of glucose by carbon cloth-supported Co3O4/PbO2 core-shell nanorod arrays. Biosens Bioelectron, **53**, 200-206 (2014)
14. S.C. Stark, et al., Assessment of metal contamination using X-ray fluorescence spectrometry and the toxicity characteristic leaching procedure (TCLP) during remediation of a waste disposal site in Antarctica. Journal of Environmental Monitoring, **10** (2010)
15. G.M.F. Gomes, T.F. Mendes and K. Wada, Reduction in toxicity and generation of slag in secondary lead process. Journal of Cleaner Production, **19(9-10)**, 1096-1103 (2011)
16. A. Smaniotto, et al., Qualitative lead extraction from recycled lead-acid batteries slag. Journal of Hazardous Materials, **172(2-3)**, 1677-80 (2009)
17. X. Deng, X., et al., Hydrothermal desulfurization of spent lead paste based on comproportionation reaction. Separation and Purification Technology, **259**, 118115 (2021)
18. X. Zhu, X. Liu, and Z. Zhao, Leaching kinetics of scheelite with sodium phytate. Hydrometallurgy, 186, 83-90 (2019)
19. W. Liu, et al., Preparation of calcium stannate from lead refining slag by alkaline leaching-purification-causticization process. Separation and Purification Technology, **212**, 119-125 (2019)
20. J.S.J.V. Deventer, Kinetics of metallurgical processes. Minerals Engineering, **13(3)**, 329-330 (2020)
21. G. Shi,et al., Kinetics of copper extraction from copper smelting slag by pressure oxidative leaching with sulfuric acid. Separation and Purification Technology, **241**, 116699 (2020)
22. X. Wu, Facile Synthesis of β-PbO2 Nanoparticles Range from 10–30 nm and their Application for Ozone Generation. Journal of The Electrochemical Society, **168(12)**, 123504 (9pp) (2021)

A Study of Phosphate Solubilizing Capacity by *Penicillium Aurantiogriseum* under Different Carbon and Nitrogen Resources

Jun Hu[1,2], *Liyan* Wang[1,2], *Liangliang* Zhang[1,2], *Hongjian* Gao[1,2], and *Da* Tian[1,2,*]

[1]Anhui Province Key Laboratory of Farmland Conservation and Pollution Prevention, School of Resources and Environment, Anhui Agricultural University, Hefei 230036, China
[2]Research Centre of Phosphorus Efficient Utilization and Water Environment Protection along the Yangtze River Economic Belt, Anhui Agricultural University, Hefei 230036, China

Abstract. Phosphate-solubilizing fungi have been successfully applied to the release of phosphorus from insoluble tricalcium phosphate. A phosphate-solubilizing fungi Penicillium aurantiogriseum was isolated and investigated the phosphorus solubilizing capacity under different carbon and nitrogen resources. The highest released phosphorus content reached 1000 mg/L in ammonium and sucrose conditions. Carbon resources did not limit the release of phosphorus by Penicillium aurantiogriseum. However, nitrate and urea significantly reduced the phosphorus release, which had a low phosphorus content (~780 mg/L). Glucose and ammonium were more efficient for fungal growth and organic acid secretion. Oxalic acid secreted by Penicillium aurantiogriseum dominated the release of phosphorus. The formed calcium oxalate promoted the dissolution of tricalcium phosphate. This study indicated that the effective utilization of Penicillium aurantiogriseum to dissolve tricalcium phosphate need to support more ammonium nitrogen.

1 MIntroduction

Phosphorus (P) is a necessary element for crop growth and function in chlorophyll and proteinsynthesis[1]. However, the P uptake by crops is limited due to the low mobility of P in soil[2]. Most P in the soil exists in the form of insoluble inorganic phosphate, such as calcium, ferric and aluminium phosphate[3]. These phosphates usually have a low K*sp* value (10^{-19}~10^{-33}), which is difficult to be directly absorbed and utilized by plants[4-6]. Although chemical phosphate fertilizer supply can significantly increase the crop yield, more than 85% P was adsorbed or precipitated in soil via the formation of insoluble phosphate[7]. The amount of P stored in the soil would be continuously used by crops for 50 years[8]. In addition, the increased input of chemical phosphate fertilizer would also accelerate the consumption of phosphate ores. As a non-renewable resource, the global phosphate ores would be depleted in the next 50-100 years[9]. Therefore, improving the utilization of soil insoluble phosphate is an effective way to reduce the consumption of phosphate ore and sustainable agricultural development.

The acidic conditions can significantly promote the release of P from insoluble phosphate in the soil[10]. In nature, the release of P is mainly driven by plant roots and microorganisms[10]. On the one hand, the CO_2 produced by roots and microorganisms via respiration can form a lower pH micro-environment, which promotes the release of phosphorus[10]. On the other hand, the roots and microorganisms can also secrete organic acid to promote the release of P via acidity and chelation[10, 11]. Compared with chemical acid, the organic acid is more efficient in the dissolution of insoluble phosphate. For example, the average P released from rock phosphate by oxalic acid was 21 mg, while sulfuric acid only solubilized 14 mg[12].

Phosphate solubilizing fungi (PSF) are able to produce lots of organic acids, especially for oxalic acid. Compared with bacteria, PSF has a stronger ability of organic acid secretion. The solubilized P by PSF is ~10 times compared with bacteria[13]. Cations such as calcium and iron promote the release of phosphorus in phosphate minerals[14]. More importantly, PSF also has a cell wall and mycelium structure, which is not only improving environmental resistance, but also easy to capture more P via the extension of hyphae [13, 15]. In addition, PSF can also maintain the phosphate-solubilizing ability after continuous incubation[13]. Therefore, the PSF is usually considered as the primary candidate for the dissolution of insoluble phosphate in the soil.

The oxalic acid produced by PSF usually dominates the release of P from insoluble phosphate. Compared with other organics (e.g., citric, malic, and tartaric acids), oxalic acid has the highest acidity constant (pK_{a1} = 1.25 and pK_{a2} = 4.27)[14]. The oxalic acid secreted by PSF can significantly reduce the pH to ~2 after seven days of incubation[13]. In addition, oxalic acid also shows a high

* Corresponding author: tianda@ahau.edu.cn

ability to chelate metal cations such as Ca^{2+}, Fe^{3+}, Al^{3+}, etc.

However, the secretion of oxalic acid by PSF are usually influenced by different environmental factors, including pH, carbon and nitrogen resources, metal cation, etc[14]. In the cause of pH, the optimal oxalic acid secreted by PSF occurred at pH 5-6 and was inhibited at pH<3[14]. A metal cation like Mn^{2+} can promote oxalic acid production[14]. The existence of nitrate promotes the secretion of oxalic acid by PSF, while ammonium inhibits, and then affecting the phosphate solubilizing capacity[14]. Hence, choosing a suitable environment is important for the secretion of oxalic acid by PSF, especially for the solubilization of P from insoluble phosphate.

This study aims to investigate the P solubilizing capacity by PSF-*Penicillium aurantiogriseum* (*P. aurantiogriseum*) under different carbon and nitrogen resources. The organic acid secreted by *P. aurantiogriseum* was also investigated in this research. The organic acid was analysed by high-performance liquid chromatography (HPLC). The soluble P content was analysed by inductively coupled plasma-optical emission spectrometry (ICP-OES). The mineralogical of precipitation was analysed by X-ray powder diffraction (XRD).

2 Materials and methods

2.1 Isolation and identification of fungi

2.1.1 Screening phosphate solubilizing fungi

The phosphate-solubilizing fungus was screened from the wheat rhizosphere soil located at Suzhou city, Anhui province, China. The soil sample was collected and prepared for a suspension with sterile water at a 1:9 ratio. After shaking for 30 min at 25 °C, the suspension was diluted to 10^{-3}, 10^{-4}, 10^{-5}, 10^{-6} and then coated on Pikovskaya (PVK) solid medium with 0.1mL, respectively. After five days of incubation (28 °C), the fungi with obvious phosphate-solubilizing circles were collected and purified. Then, the purified fungi were punched and inoculated to PVK solid medium for the preliminary identification of P solubilizing capacity. The size of colony diameter (d, cm) and phosphate-solubilizing circle (D, cm) were observed and recorded.

2.1.2 Molecular biology identification

Genomic DNA of phosphate-solubilizing fungi with high P solubilizing capacity was extracted using a fungal DNA kit (GBCBIO Technologies). The semi-nested PCR protocol was performed to amplify the internal transcribed spacer (ITS) region. Universal primers ITS1 (5′ - TCCGTAGGTGAACCTGCGG-3′) and ITS4 (5′ - TCCTCCGCTTATTGATATGC-3′) was applied for the amplification round. The ITS rRNA gene was sequenced and searched in GenBank. All the selected sequences were aligned using the Clustal W. The neighbour-joining phylogenetic tree was constructed by the MEGA 7.0.

2.2 Carbon and Nitrogen preparation

Glucose, sucrose and starch were used as the three different carbon sources in the experiment. Ammonium-nitrogen ($(NH_4)_2SO_4$), nitrate-nitrogen (KNO_3) and urea were selected as the three different nitrogen sources. The carbon source and nitrogen source reagents used in this experiment were supplied by China Pharmaceutical Group Chemical Reagent Co., Ltd. (Shanghai, China).

2.3 Fungal incubation under different carbon and nitrogen

The P solubilizing capacity by *P. aurantiogriseum* was performed with six treatments, i.e., glucose, sucrose, starch, ammonium, nitrate and urea. Tricalcium phosphate ($Ca_3(PO_4)_2$) was used as the initial insoluble phosphate. The initial PVK liquid medium contains 10 g glucose, 0.5 g $(NH_4)_2SO_4$, 0.2 g NaCl, 0.2 g KCl, 0.25 g $MgSO_4 \cdot 7H_2O$, 0.03 g $MnSO_4 \cdot 4H_2O$, and 0.03 g $FeSO_4 \cdot 7H_2O$ in 1 L deionized water. The reagents in this study were supplied by China Pharmaceutical Group Chemical Reagent Co., Ltd, Shanghai, China. Before incubation, the PVK liquid medium with different carbon and nitrogen conditions were sterilized (121 °C, 20 min). Then, 0.5 g $Ca_3(PO_4)_2$ and 1 mL fungal spores (10^7 CFU/mL) were added into a 250 mL Erlenmeyer flask with 100 mL PVK medium. Finally, the flask was sealed with a sealed membrane (BS-QM-003, Biosharp) containing 0.22 μm air-permeable membrane. All the tests were conducted under sterile conditions. After seven days of incubation, the PVK medium with different carbon and nitrogen was filtered through a 0.45 μm polyethersulfone (PES) membrane. The filtrate was collected for the analysis of organic acids and soluble P. The filter was collected for the analysis of biomass and mineralogical. All the treatment was performed with three replicates.

2.4 Statistical analysis

The P solubilizing experiments under different carbon and nitrogen by PSF were performed with triplicates. Both the means and standard deviations were calculated and presented for each treatment. The Tukey's honestly significant difference test ($p < 0.05$) via one-way ANOVA was identified in each treatment by using SPSS 16.0.

2.5 Instrumentation

The pH value in the medium was measured by the FE20 pH meter (Mettler Toledo, Int. Inc.). The soluble P concentration in the medium was analyzed by ICP-OES (PerkinElmer Avio 200). The P calibration curves (1, 5, 10, 20, 50, 100 mg/L) were prepared before the test, which had a 0.999 R square of the external standard curve.

The organic acid concentration was analyzed by HPLC (Agilent 1200). The temperature of the chromatographic column was 30 °C. The standard curves of citric and oxalic acids were performed to 50, 100, 200, 500, 1000, and 2000 mg/L. The R square value of the standard curve was 0.999. The methanol (CH_3OH) and 2.5 wt‰ potassium phosphate monobasic (KH_2PO_4) consisted of the mobile phase with a ratio of 1:99. The mobile phase pH was adjusted to 2.8 by using phosphoric acid [16]. The flow rate of the mobile phase was 1 mL/min.

The precipitate was analyzed by X-ray diffraction (D/Max-2500, Rigaku Corporation) (Cu-Kα; 36 kV; 20 mA; scanned from 5° to 60° at a speed of 4°/sec). The XRD pattern was identified by MDI Jade 6.5 software.

3 Results & Discussion

3.1 Identification and preliminary analysis of phosphate solubilizing ability

A phosphate-solubilizing fungi strain of AHF-4 was screened from wheat rhizosphere soil in Wanbei area Anhui province. After purification, the strain was cultured on a glucose potato agar medium (PDA) to observe the colony morphology and microscopic characteristics. The formed flat colony had a regular edge and was distributed with white-yellow and blue-white circles (Fig. 1A). The slightly flocculent occurred in the centre of the colony (Fig. 1A). Microscopic observation showed that the strain has broom-shaped conidia (Fig. 1A). Combined with colony characteristics, The strain was preliminarily identified as *Penicillium*. After homologous sequences blast in NCBI (Fig. 1C). The ITS sequence of AHF-4 is highly similar to *Penicillium aurantiogriseum* in Genbank (Fig. 1C). The results of the phylogenetic tree showed that the strain of AHF-4 belongs to *Penicillium aurantiogriseum* (Fig. 1C). After one, three, five and seven days of incubation, the phosphate-solubilizing circle by *Penicillium aurantiogriseum* increased to 2.13, 2.72, 3.05 and 3.74 cm, respectively (Fig. 2).

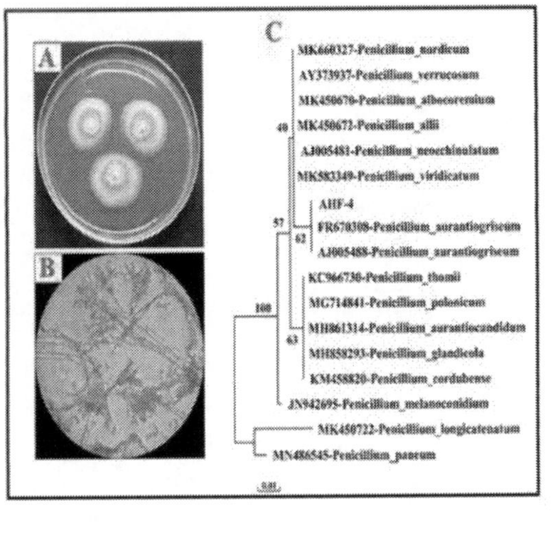

Fig. 1. The colony image (A), microscopic image (B) and phylogenetic tree analysis (C) of *Penicillium aurantiogriseum*.

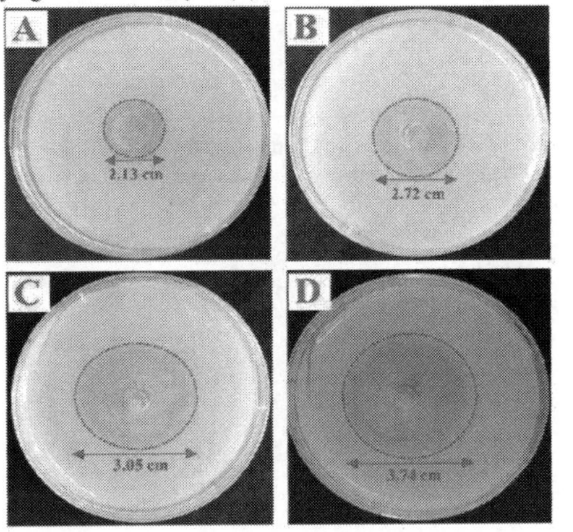

Fig. 2. The formed phosphate-solubilizing circles by *Penicillium aurantiogriseum* on the PVK solid medium after one (A), three (B), five (C) and seven (D) days of incubation.

3.2 The pH value in the medium under different C and N resources by *P. aurantiogriseum*

The initial pH of the PVK medium was 6.5. The pH values in glucose, sucrose and starch treatments decreased to 2.6, 3.3 and 2.7 after seven days of incubation, respectively (Fig. 3). Under different nitrogen resources, the medium pH decreased to 2.6, 3.7 and 3.5 in ammonium, nitrate and urea after seven days of incubation, respectively (Fig. 3). These results indicated that glucose, starch and ammonium nitrogen was more efficient for the secretion of organic acids by P. *aurantiogriseum*.

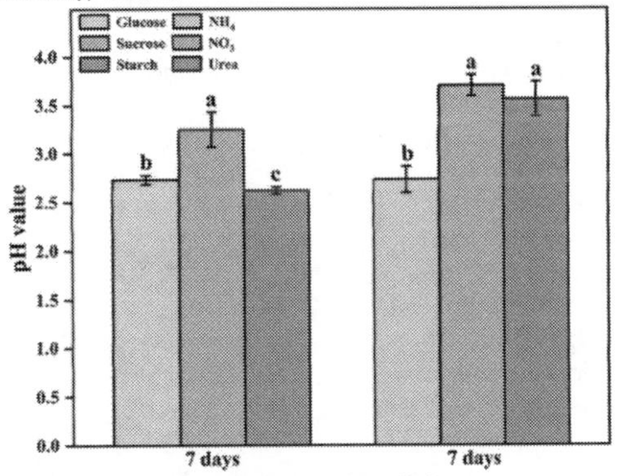

Fig. 3. The pH values in the PVK medium after seven days of incubation under different carbon and nitrogen resources. The error bars in each column represent the standard deviations of three replicates.

3.3 The dry biomass of fungi under different carbon and nitrogen resources

The dry biomass of fungi has no significant difference among the three types of carbon resources, which ranged from 0.31 to 0.32 g (Fig. 4). However, the urea limited the growth of fungi. The dry biomass was 0.28 g, significantly lower than ammonium and nitrate, i.e., 0.32 and 0.34 g (Fig. 4). The growth of fungi did not influence by the types of carbon, but rather by nitrogen. Unlike other PSF, *P. aurantiogriseum* was favourable in both utilization of ammonium and nitrate. However, the existence of carbonate would limit the growth of PSF[17]. The adsorption of urea by fungi would release the CO_2, hence the formed carbonate would limit the growth of *P. aurantiogriseum*.

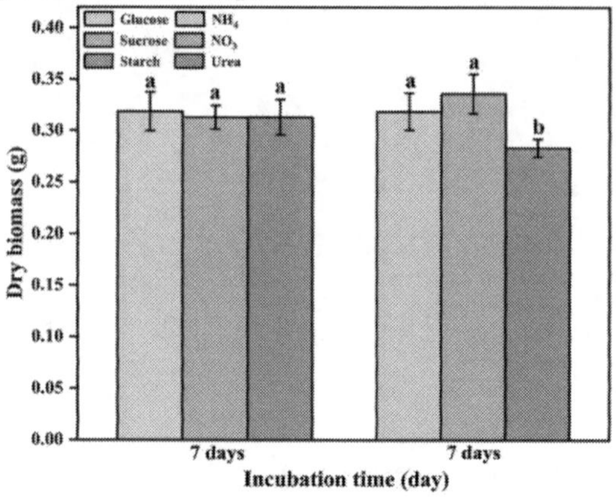

Fig. 4. The dry biomass values after seven days of incubation under different carbon and nitrogen resources. The error bars in each column represent the standard deviations of three replicates.

3.4 The release of P under different carbon and nitrogen resources

The P content in three types of carbons had no significant difference after seven days of incubation, more than 950 mg/L P was released from tricalcium phosphate (Fig. 5). However, the P content in nitrate and urea was significantly lower than ammonium after seven days of incubation, i.e., 779.5 vs. 950 mg/L (Fig. 5). This result indicated that *P. aurantiogriseum* can promote the dissolution of insoluble phosphate under different carbon and nitrogen resources (Fig. 5). Compared with other PSF(e.g., *Aspergillus niger* and *Penicillium oxalicum*), *P. aurantiogriseum* showed a higher P solubilizing capacity in calcium phosphate[13, 18, 19].

The dissolution of calcium phosphate would be more efficient in the lower pH value. However, the pH in nitrate and urea almost reached 4, much higher than 2.5 in ammonium (Fig. 3). In addition, the H^+ produced by the assimilation process of ammonium would also contribute to the decrease of pH value[14]. Hence the release of P from calcium phosphate by *P. aurantiogriseum* in ammonium was much higher than other nitrogen. In other words, the utilization of *P. aurantiogriseum* to dissolve insoluble phosphate in soil should increase the input of ammonium nitrogen.

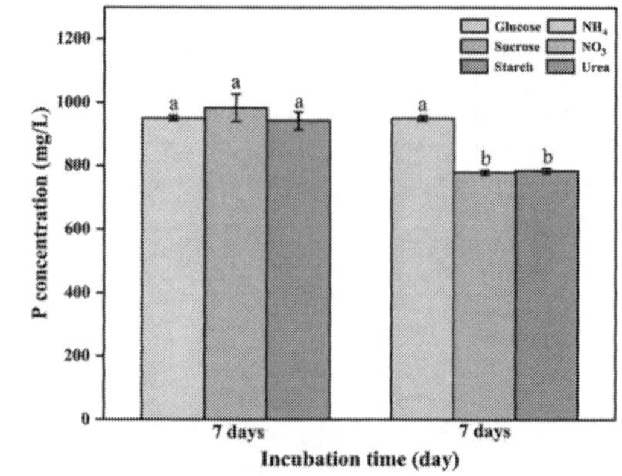

Fig. 5. The P content after seven days of incubation under different carbon and nitrogen resources. The error bars in each column represent the standard deviations of three replicates.

3.5 The organic acid secretion by fungi under different carbon and nitrogen resources

Oxalic and citric acids were the primary organic acids secreted by *P. aurantiogriseum* (Fig. 6). However, both the oxalic and citric acids content were lower than 300 mg/L. The highest oxalic acid secretion occurred in glucose and ammonium resources (Fig. 6A). While sucrose and urea can stimulate *P. aurantiogriseum to* secrete more citric acid (Fig. 6B). The primary organic acid (oxalic and citric acid) secreted by *P. aurantiogriseum* was similar to other PSF but had a lower content[18, 19]. Especially for oxalic acid, the secretion content was much lower than *Aspergillus niger* and *Penicillium oxalicum*, i.e., 205 mg/L vs. ~1000 mg/L[18, 19]. However, *P. aurantiogriseum* showed lower biomass compared with the above two PSF (0.3 vs. 0.6 g). That is to say, *P. aurantiogriseum* was more efficient in the release of P from calcium phosphate. In addition, oxalic acid also had a strong combination ability with calcium cations. The formed calcium oxalate would also contribute to the low content of oxalic acid.

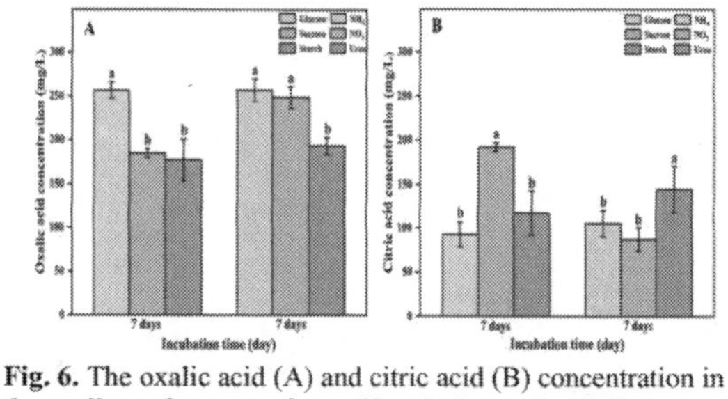

Fig. 6. The oxalic acid (A) and citric acid (B) concentration in the medium after seven days of incubation under different carbon and nitrogen resources. The error bars in each column represent the standard deviations of three replicates.

3.6 XRD analysis

The peaks at 12.1°, 25.8°, 29.5° and 32.2° stand for the mineral of calcium phosphate ($Ca_3(PO_4)_2$) (Fig. 7)[3]. The newly formed mineral of calcium oxalate (CaC_2O_4) located at 13.7° and 23.7° occurred in each treatment

https://doi.org/10.1051/e3sconf/202235003002

(Fig. 7)[3]. This result also indicated that the oxalic acid secreted by *P. aurantiogriseum* dominated the release of P. The different carbon and nitrogen did not influence the mineralization in the process of P dissolution. This XRD result was also confirmed by the other PSF.

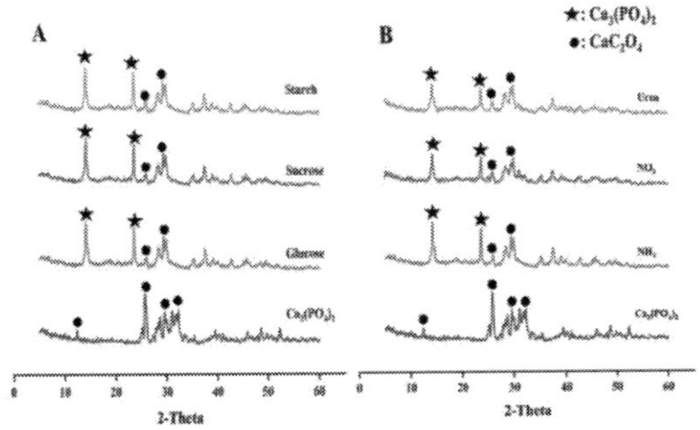

Fig. 7. The XRD patterns of precipitation in carbon (A) and nitrogen (B) resources.

4 Conclusion

The screened PSF *P. aurantiogriseum* shows a high P solubilizing capacity under different carbon and nitrogen resources. Compared with carbon, the types of nitrogen significantly influenced the release of P by fungi. Ammonium nitrogen is more efficient for fungal growth and organic acid secretion, especially for oxalic acid. Both the carbon and nitrogen did not influence the formation of calcium oxalate. The utilization of *P. aurantiogriseum* to dissolve insoluble phosphate need to increase ammonium nitrogen input.

Acknowledgements

This work was supported by the National Natural Science Foundation of China (NO. 42007030) and the Natural Science Foundation of Anhui Province (NO. 2008085QD187), the program at Anhui Agricultural University (NO. yj2019-20).

References

1. C. Bisson, N.B.P. Adams, B. Stevenson, A.A. Brindley, D. Polyviou, T.S. Bibby, P.J. Baker, C.N. Hunter C.N and A. Hitchcock, The molecular basis of phosphite and hypophosphite recognition by ABC-transporters. Nat. Commun., **8**, 1746 (2017)
2. D.L. Jones and E. Oburger, Solubilization of phosphorus by soil microorganisms. In: Bünemann E, Oberson A, Frossard E. (Eds.), Phosphorus in Action: Biological Processes in Soil Phosphorus Cycling. Springer Publisiing Inc., Berlin, Heidelberg 169-198 (2011)
3. D. Tian, L. Wang, J. Hu, L. Zhang, N. Zhou, J. Xia, M. Xu, K.K. Yusef, S. Wang, Z. Li and H. Gao, A study of P release from Fe-P and Ca-P via the organic acids secreted by Aspergillus niger. J. Microbiol., **59**, 819-826 (2021)
4. X. Wang, Y. Wang, J. Tian, B.L. Lim, X. Yan and H. Liao, Overexpressing AtPAP15 enhances phosphorus efficiency in soybean. Plant. Physiol., **151** 233-240 (2009)
5. G.M. Filippelli The global phosphorus cycle. Rev. Mineral. Geochem., **48**, 391-425 (2002)
6. Y. Jiang, J. Tian and F. Ge, New insight into carboxylic acid metabolisms and pH regulations during insoluble phosphate solubilisation process by Penicillium oxalicum PSF-4. Curr. Microbiol., **77**, 4095-4103 (2020)
7. D. Tian, Z. Li, D. O'Connor, Z. Shen and D. Hou, The need to prioritize sustainable phosphate-based fertilizers. Soil. Use. Manage., **36**, 351-354 (2020)
8. J. Zhu, M. Li and M. Whelan, Phosphorus activators contribute to legacy phosphorus availability in agricultural soils: A review. Sci. Total. Environ., **612**, 522-537 (2018)
9. D. Cordell, J.O. Drangert and S. White, The story of phosphorus: Global food security and food for thought. Glob. Environ. Change., **19**, 292-305 (2009)
10. D. Tian, M. Su, X. Zou, L. Zhang, L. Tang, Y. Geng, J. Qiu, S. Wang, H. Gao and Z. Li, Influences of phosphate addition on fungal weathering of carbonate in the red soil from karst region. Sci. Total. Environ., **755**, 142570 (2021)
11. Z. Li, M. Su, D. Tian, L. Tang, L. Zhang, Y. Zheng and S. Hu, Effects of elevated atmospheric CO2 on dissolution of geological fluorapatite in water and soil. Sci. Total. Environ., **599-600**, 1382-1387 (2017)
12. G.O. Mendes, H.M. Murta, R.V. Valadares, W.B. Silveira, I.R. Silva and M.D. Costa, Oxalic acid is more efficient than sulfuric acid for rock phosphate solubilization. Miner. Eng., **155**, 106458 (2020)
13. Z. Li, T. Bai, L. Dai, F. Wang, J. Tao, S. Meng, Y. Hu, S. Wang and S. Hu, A study of organic acid production in contrasts between two phosphate solubilizing fungi: Penicillium oxalicum and Aspergillus niger. Sci. Rep., **6**, 25313 (2016)
14. F. Palmier, A. Estoppey, G.L. House, A. Lohberger, S. Bindschedler,P.S.G. Chain and P. Junier, Chapter Two-Oxalic acid, a molecule at the crossroads of bacterial-fungal interactions. In: Gadd G.M, Sariaslani S. (Eds.) Advances in Applied Microbiology. Academic Press, Elsevier publishing Inc, **106**, 49-77 (2019)
15. F.P. Coutinho, W.P. Felix and A.M. Yano-Melo, Solubilization of phosphates in vitro by Aspergillus spp. and Penicillium spp. Ecol. Eng., **42**, 85-89 (2012)
16. D. Tian, Z. Jiang, L. Jiang, M. Su, Z. Feng, L. Zhang, S. Wang, Z. Li and S. Hu, A new insight into lead (II) tolerance of environmental fungi based on a study of Aspergillus niger and Penicillium oxalicum. Environ. Microbiol., **21**, 471-479 (2019)
17. M. Su, L. Meng, L. Zhao, Y. Tang, J. Qiu, D. Tian, and Z. Li, Phosphorus deficiency in soils with red color: Insights from the interactions between

minerals and microorganisms. Geoderma., 2021, **404** 115311 (2021)

18. D. Tian, Z. Lai, X. Zou, C. Guo, L. Tang, M. Su, Z. Li and S. Hu, A contrast of lead immobilization via bioapatite under elevated CO2 between acidic and alkaline soils. Soil. Use. Manage., **34**, 542-544 (2018)
19. Z. Li, F. Wang, T. Bai, J. Tao, J. Guo, M. Yang, S. Wang and S. Hu, Lead immobilization by geological fluorapatite and fungus Aspergillus niger. J. Hazard. Mater., **320**, 386-392 (2016)

Oxalic Acid Modification Improves the Removal Effect of Cr(VI) by Biochar through Three Important Means: Adsorption, Reduction and Mineralization

Fangfang Min[1], *Haoming* Chen[1,*] and *Dehua* Ma[1]

[1]Nanjing University of Science and Technology, School of Environmental and Biological Engineering, 210094 200 Xiaolingwei Street, Xuanwu District, Nanjing City, Jiangsu Province, China.

Abstract. Oxalic acid, one of the common organic acids secreted by microorganisms, was used to modify the biochar of sludge (SB) and rice husk (RB), and the adsorption function of biochar was combined with the reductive property of oxalic acid to repair Cr(VI). The biochar materials were characterized by scanning electron microscopy (SEM), X-ray diffraction (XRD), X-ray photoelectron spectroscopy (XPS) and Fourier transform attenuated total reflection infrared spectroscopy (ATR-IR), and the mechanism of chromium hexavalent removal was studied. Adsorption experiments showed that the removal rate of Cr(VI) with the modified SB and RB was 1.68 and 2.19 mg/g within 4h, which was 7.4 and 33 times of that before modification. Oxalic acid modification can enhance the repair of Cr(VI) by biochar in three aspects: 1. increases the BET surface area (109.34-174.13%) and decrease the average pore width (86.3-93.7%) of biochar, which increases the adsorption sites. 2. the addition of reducing oxygen-containing functional groups improves biocahr reduction performance. 3. improve reduced and unreduced Cr ion forms stable minerals on the surface of biochar. Therefore, oxalic acid can not only improve the adsorption and reduction of Cr(VI) by biochar, but also further improve the fixation/passivation of Cr ion after reduction and reduce the migration of Cr ion

1 Introduction

With the rapid development of industry, the discharge and pollution of industrial wastewater has become a major problem for environmental protection. Among them, Cr contained in industrial wastewater has certain potential risks to the ecosystem and human health (Ma et al., 2021). There are two valence states of Cr in the environment: Cr(III) and Cr(VI). Cr(VI) has higher water solubility and fluidity, and is easily absorbed by plants and penetrated into groundwater, causing great harm to human body, animals and plants (Barrera-Díaz et al., 2012; Johan et al., 2020). The toxicity of Cr(III) is less than that of Cr(VI) (Fu et al., 2011), so it is safer to change Cr(VI) to a low valence state in the repair process.

At present, the methods used to treat excessive Cr ions in water can be roughly divided into three categories: physical, chemical and biological. Traditional treatment in water generally adopts chemical methods (Femina Carolin et al., 2017), but most chemical treatment has high operating cost and inconvenient operation, and it is easy to cause secondary pollution due to overdosage (Mohan et al., 2011). Although the biological method is cost-effective and environmental friendly, the microorganism is vulnerable to the stress of heavy metals, and it is difficult to play the biological repair function (Chen et al., 2020). Physical adsorption method uses substances with high specific surface area or highly developed pore structure as adsorbents to remove heavy metals from water. Physical adsorption method is simple and easy to obtain raw materials, but a range of environmental factors may affect the properties of adsorbent over the long term, resulting in low adsorption capacity. Therefore, it is one of the research hotspots to improve the repair function of adsorbent by means of modification.

Biochar, an excellent adsorbent, is mainly produced by pyrolysis of waste organic biomass. It has the advantages of stable structure, well-developed pores, large specific surface area, abundant surface functional groups and cheap and easy access to raw materials (Tan et al., 2016). Studies have shown that the removal rate of Cr(VI) from apple wood biochar is greater than 99.9% at pH=2 (Liu et al., 2020). There are many modification methods of biochar, such as surface oxidation and reduction, magnetization and biochar functionalization, which can improve the removal efficiency of pollutants (Manik Mian et al., 2018). However, for Cr with variable valence, reduction modification is more conducive to further improve the significance of biochar for Cr repair.

Reducing organic acids have the ability to reduce variable valence heavy metals in the environment. For the treatment of Cr-polluted wastewater and soil, low molecular weight organic acids such as citric acid and oxalic acid have been found to be effective reducing

[*] Corresponding author: chenhaoming89@hotmail.com

agents (Qin et al., 2020). In this experiment, the physical adsorption of biochar was combined with the chemical reduction ability of oxalic acid, and the modified material was used to repair Cr(VI). We compared and analyzed the Cr(VI) removal effect and related mechanism before and after biochar modification by XRD, ATR and XPS, so as to provide better reference value for the practical application of biochar.

2 Materials and methods

2.1 Preparation of modified biochar

5g sludge biochar (SB) and rice husk biochar (RB) were mixed with 100 mL oxalic acid (0.5 mol/L, $H_2C_2O_4$, 99.8% purity, Macklin) solution, and shook in a shaker at 180 r/min, 28℃ for 12 h. Then the solids after extraction and filtration are fully dried in an oven at 60℃ (48 h). The modified sludge biochar was denoted as HSB and the modified rice husk biochar as HRB.

2.2 Adsorption experiment method

The pH of 100 mL potassium dichromate ($K_2Cr_2O_7$, 99% Purity, Aladdin) solution with a concentration of 50 mg/L was adjusted to 3. 2 g of biochar before and after modification were added to chromium solution, and the four treatments were placed on a magnetic stirrer of 300 r/min for adsorption experiments. Samples were taken at 30, 60, 90, 120, 180 and 240 min, respectively. After adsorption experiment, the four treated solids were centrifuged and collected, which were denoted as SB-Cr, RB-Cr, HSB-Cr and HRB-Cr, respectively. Total chromium in the solution was determined by ICP-OES (iCAP PRO ICP-OES, Germany). Determination of Cr(VI) in solution by ultraviolet spectrophotometer at 540 nm (752N, China). Amount of chromium reduction (Cr(III)) is the difference between total Chromium and Cr(VI).

2.3 Material characterization methods

The surface morphology and structure of biochar were determined by scanning electron microscope (SEM) equipped with X-ray energy dispersive spectrometer (EDS) (ZEISS Supra 55, Germany) and automatic specific surface and porosity analyzer (BET) (ASAP 2460, USA). Decaying total reflection infrared spectrometer (ATR-IR) (Nicolet iS5, USA) was used to analyze the functional groups on the surface of biochar before and after modification. The chemical composition of the surface crystal structure of biochar was analyzed by X-ray photoelectron spectroscopy (XPS) (K-Alpha, USA) and X-ray diffrotometry (XRD) (D8ADVANCE, Germany).

3 Results and discussion

3.1 Adsorption experiment results

Adsorption experiment showed that (Figure 1), the removal rate HRB (82.62%) > HSB (63.18%) > SB (8.51%) > RB (2.43%), HRB has the largest Cr(VI) removal capacity within 4 h, reaching 2.19 mg/g, which is 33 times higher than RB. The removal amount of Cr(VI) by HSB was also significantly increased from 0.23 mg/g to 1.68 mg/g. From the reduction effect, the reduction rate after modification is significantly increased, the maximum reduction amount of HRB is 0.56 mg/g, and the reduction effect of HSB is better than HRB, reaching 0.65 mg/g.

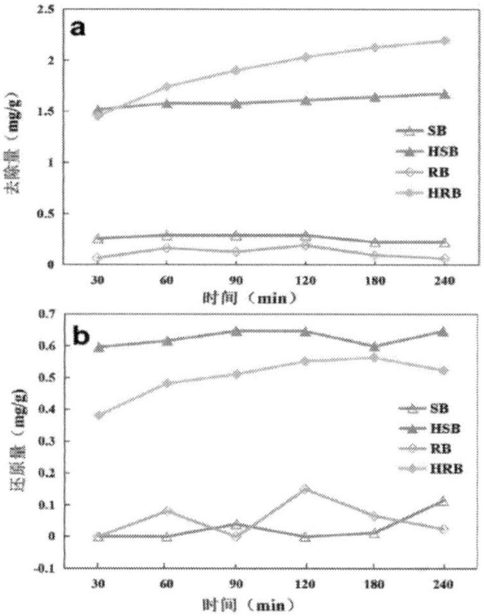

Fig. 1. Chromium adsorption curves of different biochar treatments. (a)The amount of hexavalent chromium removed. (b)The amount of hexavalent chromium reduction. SB=sludge biochar, HSB=oxalic acid modified sludge biochar, RB= rice husk biochar, HRB= oxalic acid modified rice husk biochar.

3.2 Material structure analysis

Before oxalic acid modification, the BET surface area of SB and RB was 33.67 m^2/g and 43.66 m^2/g, respectively. After modification, the BET surface area of HSB and HRB was 92.30 m^2/g and 91.40 m^2/g, which increased significantly (Table 1). The average pore width decreased, which may be caused by the structural collapse of medium and large pores of biochar (Liu et al., 2021). The total pore volume of the two materials changed little before and after modification. The increase of BET surface area provides more adsorption sites for materials, thus increasing the adsorption/binding of Cr(VI) on the surface of biochar. BET surface area of HSB was larger than that of HRB, while removal rate of Cr(VI) of HRB was 1.31 times higher than that of HSB, which indicating that physical characteristics such as BET surface are not the main factors determining the adsorption capacity of biochar for Cr(VI).

Table 1. BET surface area and pore size of biochar before and after modification

	BET surface area (m^2/g)	Average pore width (nm)	Total pore volume (cm^3/g)
SB	33.67	11.892	0.079
RB	43.66	5.352	0.036
HSB	92.30	0.744	0.040
HRB	91.40	0.731	0.041

From the SEM image (Figure 3), the surface of SB has a large number of continuous gullies and deep holes, while RB has a more developed pore structure, in the form of flake. After modification, the surfaces of HSB and HRB become loose and broken to varying degrees, exposing more pores. Therefore, combined with the change of BET surface area, oxalic acid modification can significantly increase the contact area between biochar and pollutants.

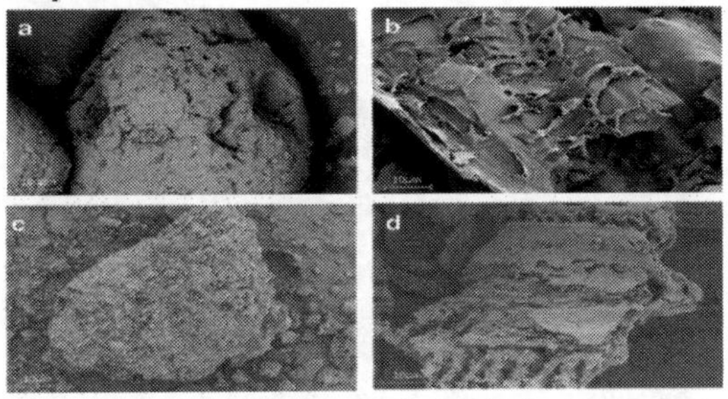

Fig. 2. SEM image of biochar. (a)sludge biochar, (b)rice husk biochar, (c)oxalic acid modified sludge biochar and (d)oxalic acid modified rice husk biochar.

3.3 Functional group analysis

ATR-IR spectra showed that HSB and HRB had more abundant functional groups than SB and RB (Figure 3). The peak of HSB was more obvious at 1614 cm^{-1}, and a new peak appeared at 1315 cm^{-1} and 776 cm^{-1}. The peak at 1585 cm^{-1} of HRB shifted to 1614 cm^{-1} and the peak intensity increased, and a new peak also appeared at 1315 cm^{-1}. The characteristic peak of 1585 cm^{-1} and 1614 cm^{-1} are generated by C=O double bond or C=C vibration on aromatic structure (Liu et al., 2021). 1315-1065 cm^{-1} belongs to C-H in-plane bending vibration and C-O stretching vibration. 1000-776 cm^{-1} is produced by out-of-plane bending vibration of aromatic compounds C-H. These peak changes are likely related to the oxidation-reduction of Cr, especially oxygen-containing functional groups that can promote the adsorption and reduction of Cr(VI) on the surface of biochar to Cr(III) (Ahmad et al., 2014). After adsorption of Cr, the signal of reducing oxygen-containing functional groups on the modified biochar (HSB-Cr and HRB-Cr) decreased significantly, indicating that the reducing functional groups on the modified biochar participated in the reduction of Cr, which is the second main mechanism of Cr(VI) removal by modified biochar. In addition, the vibration peak of HSB-Cr at 536 cm^{-1} is related to the vibration peak of phosphoric acid group (Zhu et al., 2016), indicating that HSB may form related mineral on the surface after contacting with heavy metals.

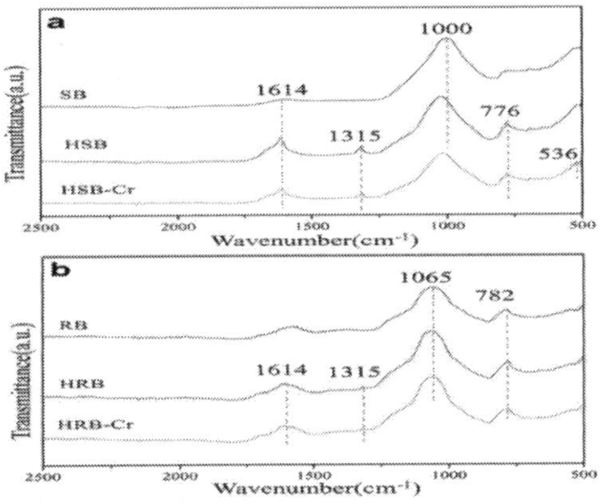

Fig. 3. ATR-IR spectra of biochar. (a)HSB-Cr=oxalic acid modified sludge biochar after reaction with Cr(VI), (b)HRB-Cr= oxalic acid modified rice husk biochar after reaction with Cr(VI).

3.4 Oxidation-reduction performance analysis

The XPS spectrum of biochar after reaction with Cr(VI) is shown in Figure 4, and the double peaks corresponding to Cr $2P_{3/2}$ and Cr $2P_{1/2}$ orbits can be observed respectively. Two Peaks of Cr(III) are located at 577.60 and 586.76 eV, and two peaks of 579.70 and 588.86 eV correspond to Cr(VI). For HSB-Cr and HRB-Cr, the peak intensity of Cr(III) is much higher than that of Cr(VI), indicating that most of Cr(VI) is reduced to Cr(III). Compared with SB-Cr and RB-Cr, the reduction performance of modified biochar was significantly improved, in which HSB-Cr(III) (90.98%) > SB-Cr(III) (48.96%) and HRB-Cr(III) (97.31%) > RB-Cr(III) (45.61%), indicating that reduction is one of the main mechanisms of Cr(VI) removal.

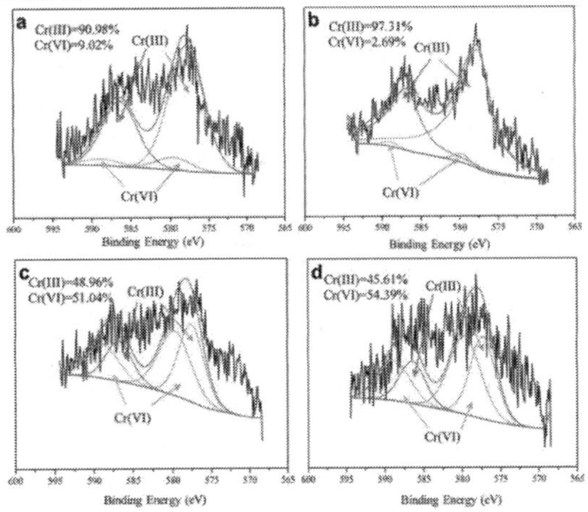

Fig. 4. The Cr XPS spectra of biochar. (a) HSB-Cr=oxalic acid modified sludge biochar after reaction with Cr(VI), (b) HRB-Cr= oxalic acid modified rice husk biochar after reaction with Cr(VI), (c) SB-Cr= sludge biochar after reaction with Cr(VI), (d) RB-Cr= rice husk biochar after reaction with Cr(VI).

3.5 Mineral analysis

SEM images of modified biochar (Figure 5. a and b) showed that crystal particles existed on the surface and pores of the HSB-Cr and HRB-Cr. EDS scanning results showed that these crystals contained a large amount of chromium, which proved that the crystals were chromium-containing minerals. In addition, the XRD results (Figure 5. c and d) showed that there were typical characteristic peaks 2θ=24.61°, 36.36°, 54.85° and 65.64° on HSB-Cr and HRB-Cr, this peaks may be the typical characteristic peaks of Cr_2O_3 (Cr mineral standard cards). It is also found that there are chromium phosphate minerals on HSB-Cr, 2θ=20.63°, 25.28° 33.61° and 35.99°. SEM and XRD results further confirmed that biochar adsorbed Cr(Ⅵ) in the solution and reduced it with anionic elements or reductive functional groups on the surface, thus forming a more stable Cr mineral form. In this study, the reaction background pH is 3, so carbonate may not exist stably in a strong acidic environment. Phosphate is more stable than carbonate in acidic environment, so phosphorus provides a new way for biochar passivation and fixation of Cr. Heavy metal mineralization can effectively avoid the desorption of adsorbed heavy metals back to the environment and cause secondary pollution, which is a very effective means to remove heavy metal pollution.

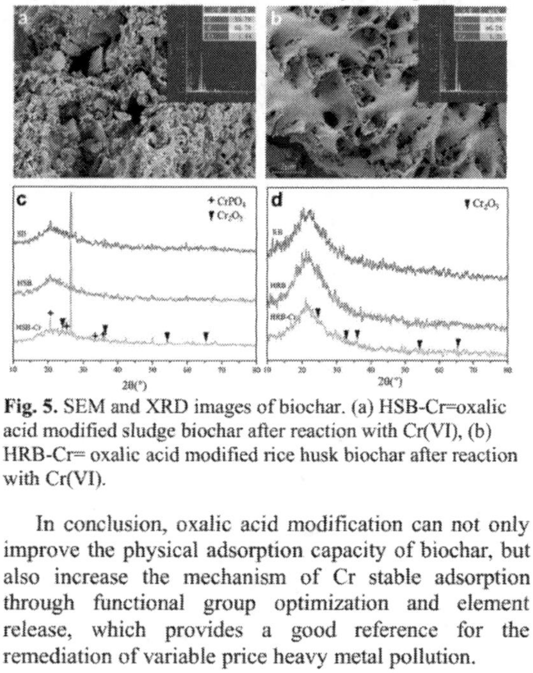

Fig. 5. SEM and XRD images of biochar. (a) HSB-Cr=oxalic acid modified sludge biochar after reaction with Cr(VI), (b) HRB-Cr= oxalic acid modified rice husk biochar after reaction with Cr(VI).

In conclusion, oxalic acid modification can not only improve the physical adsorption capacity of biochar, but also increase the mechanism of Cr stable adsorption through functional group optimization and element release, which provides a good reference for the remediation of variable price heavy metal pollution.

References

1. Y. Ma, S.W. Li, S.Z. Ye, S.X. Tang, D. Hu, L. Wei and F. Xiao, Hexavalent chromium inhibits the formation of neutrophil extracellular traps and promotes the apoptosis of neutrophils via AMPK signaling pathway. Ecotoxicol. Environ. Saf., 223 (2021)
2. C. Barrera-Díaz, V. Lugo-Lugo and B. Bilyeu, A review of chemical, electrochemical and biological methods for aqueous Cr(VI) reduction. J. Hazard. Mater. 223-224 (2012)
3. J.C. Johan, B. Neetu and M.N.C. Evans, Chromium in Environment, Its Toxic Effect from Chromite-Mining and Ferrochrome Industries, and Its Possible Bioremediation. Exposure Health.**12(1)** (2020)
4. F.L. Fu and Q. Wang, Removal of heavy metal ions from wastewaters: A review. J. Environ. Manage. **92(3)** (2011)
5. C.F. Carolin, P.S. Kumar, A. Saravanan, G. Janet Joshiba and M. Naushad, Efficient techniques for the removal of toxic heavy metals from aquatic environment: A review. J. Environ. Chem. Eng. **5(3)** (2017)
6. D. Mohan, S. Rajput, V. Singh, P. Steele and C. Pittman, Modeling and evaluation of chromium remediation from water using low cost bio-char, a green adsorbent. J. Hazard. Mater. **188(1)** (2011)
7. H.M. Chen, L.Y. Tang, Z.J. Wang, M. Su, D. Tian, L. Zhang and Z. Li, Evaluating the protection of bacteria from extreme Cd (II) stress by P-enriched biochar. Environ. Pollut. **263(PA)** (2020)
8. G.C. Tan, W.L. Sun, Y.R. Xu, H.Y. Wang and N. Xu, Sorption of mercury (II) and atrazine by biochar, modified biochars and biochar based activated carbon in aqueous solution. Bioresour. Technol..**211** (2016)
9. N. Liu, X.T. Zhang, C. Xu, P. Liu, J. Lv, Y.Y. Liu and Q.Y. Wang, Removal mechanisms of aqueous Cr(VI) using apple wood biochar: a spectroscopic study. J. Hazard. Mater. **384** (2020)
10. M.M. Mian, G.J. Liu, B. Yousaf, B. Fu, H. Ullah, M.U. Ali and Q. Abbas, Simultaneous functionalization and magnetization of biochar via NH 3 ambiance pyrolysis for efficient removal of Cr (VI). Chemosphere. **208** (2018)
11. J.H. Qin, Q.W. Li, Y.Q. Liu, A.Y. Niu and C.X. Lin, Biochar-driven reduction of As(V) and Cr(VI): Effects of pyrolysis temperature and low-molecular-weight organic acids. Ecotoxicol. Environ. Saf. 201 (2020)
12. J. Liu, X.Y. Yang, H.H. Liu, X.P. Jia and Y.C. Bao, Mixed biochar obtained by the co-pyrolysis of shrimp shell with corn straw: Co-pyrolysis characteristics and its adsorption capability. Chemosphere. 282 (2021)
13. M. Ahmad, A.U. Rajapaksha, J.E. Lim and M. Zhang, Biochar as a sorbent for contaminant management in soil and water: A review. Chemosphere. 99 (2014)
14. N.Y. Zhu, T.M. Yan, J. Qiao and H.L. Cao, Adsorption of arsenic, phosphorus and chromium by bismuth impregnated biochar: Adsorption mechanism and depleted adsorbent utilization. Chemosphere. 164 (2016)

WISCO's Low-carbon Transformation Based On LEAP And Scenario Analysis

Yongping Huang [1], *Ben* Wang [1], *Wu* Yang [1], *Lushi* Sun [1,*]

[1] State Key Lab of Coal Combustion, Hua Zhong University of Science and Technology, 430074, Wuhan, China

Abstract. The paper takes Wuhan Iron and Steel Co., Ltd. (hereinafter referred to as " WISCO ") energy consumption and carbon emissions as the research object, and analyzes WISCO's carbon emissions. Through the construction of WISCO's Long range Energy Alternatives Planning system (hereinafter referred to as LEAP) prediction model for energy consumption and carbon emissions, three scenarios have been set up: a baseline scenario, an energy efficiency improvement scenario, and an energy structure change scenario to measure WISCO's carbon emissions. Based on the results of the LEAP model, this article puts forward the following recommendations for WISCO's low-carbon transformation. WISCO must strictly control steel output. WISCO should introduce electric furnaces to increase the proportion of electric furnace steel. While ensuring the supply of scrap steel resources, WISCO ought to use scrap steel efficiently. In addition, introduction of hydrogen direct reduction iron technology is also crucial to WISCO in a planned way.

1 Introduction

Statistics from the World Iron and Steel Association show that, on average, 1.8 tons of carbon dioxide are emitted per ton of steel produced in the world. The carbon dioxide emissions of the steel industry account for about 6% of the total global carbon dioxide emissions. As a major steel country, the carbon emissions of the steel industry account for about 15% of the country[1]. The study of WISCO's low-carbon transformation path is of great significance to a low-carbon pilot city like Wuhan, and at the same time provides experience for the low-carbon transformation of China's steel industry.

Many countries in the world have begun research and development of low-carbon technologies in the steel industry. Japan's most famous low-carbon metallurgical project is the COURSE50 project[2], which focuses on the research and development of blast furnace gas separation and carbon dioxide recovery technology based on hydrogen reduction. It plans to reduce carbon dioxide emissions by 30% in 2030 and to use and promote the technology in 2050. South Korea's POSCO's hydrogen-rich blast furnace ironmaking technology aims to reduce carbon dioxide emissions by 10%[3]. The low-carbon technology paths of European countries represented by Sweden and Austria are mainly green hydrogen direct reduction and electric furnaces. The Swedish HYBRIT project[4] is a breakthrough ironmaking technology that uses hydrogen to replace coking coal, coke or natural gas. If this project is put into use, Sweden's greenhouse gas emissions will be reduced by 10%[5]. The German low-carbon technology path takes into account both blast furnace hydrogen-rich smelting and direct reduction of green hydrogen[6]. In terms of long processes, the project of the German ThyssenKrupp Group converts waste gas into raw materials for chemical products, and carbon dioxide will no longer be emitted into the air. In terms of short processes, the SALCOS project in Germany aims to produce steel through the use of green hydrogen and biomethane[7].

2 Research methods

For WISCO's carbon emissions calculations, the total greenhouse gas emissions of steel production enterprises should be equal to the emissions from fossil fuel combustion, industrial production processes, and the net purchases of electricity and heat from all production systems within the boundaries of the company. The sum should also be deducted from the emissions implied by carbon sequestration products. Both the consumption of fossil fuels and the conversion coefficient required to calculate the average low calorific value of fossil fuels can be directly obtained from Wuhan Iron and Steel, while the carbon content and carbon oxidation rate per unit of calorific value are calculated using the "Guidelines for Accounting Methods and Reporting of Greenhouse Gas Emissions of Chinese Steel Production Enterprises (Trial)."

* Corresponding author: sunlushi@hust.edu.cn

This paper chooses the LEAP model as a research tool.

The LEAP model (The Long range Energy Alternatives Planning system) is a software widely used in energy policy analysis and climate change assessment jointly researched and developed by the Stockholm Environmental Research Institute and Boston University in the United States[8]. It has been used by hundreds of countries in 150 countries around the world. Used by thousands of organizations, including government agencies, non-governmental organizations, academic institutions, energy units, and consulting companies[9]. LEAP has now become an effective tool for many countries to implement energy development planning and greenhouse gas emission assessment. The model takes energy supply, energy consumption and environmental impact as the research object, according to energy demand and future social and economic development trends, according to different policies and technology choices, set up different scenarios, and use mathematical models to predict the energy of various sectors demand and environmental impact, detailed economic benefit analysis of various energy options. This model realizes the simulation of the energy consumption system and is usually referred to as the "terminal energy consumption model". This article uses the scenario analysis method, sets the scenario analysis time interval to 2019-2050, and designs different emission reduction scenarios to explore reasonable emission reduction measures.The scenarios are described as follows:

2.1 Scene settings

(1) The baseline scenario is the reference scenario. The relevant parameters of steel production in 2019 are used as the baseline parameters, and other emission reduction approaches or methods are optimized schemes relative to this scenario. Under this scenario, based on the development status of WISCO during the 13th Five-Year Plan period, combined with existing policy plans, predict WISCO's future development in accordance with current development trends. The scenario is set as the slow process of structural adjustment, green development and independent innovation; elimination of outdated production capacity; the impact of market recovery and slower implementation of policies, occasional fluctuations in crude steel production, and slower decline; promotion of energy conservation and emission reduction, but unit product energy consumption declines slowly; WISCO's electric furnace steel ratio remains zero without an increase; there is a large gap between the iron-to-steel ratio and the domestic advanced level, the penetration rate of non-blast furnace ironmaking technology has grown slowly; the utilization rate of scrap steel has increased slowly. WISCO's energy consumption structure and energy technology level remain basically unchanged in 2019.

(2) Under the energy efficiency improvement scenario, we will increase efforts to eliminate outdated production capacity and strictly prohibit new steel production capacity. WISCO's crude steel production will gradually decline after reaching its peak in 2018; deepening energy conservation and emission reduction, and the overall energy consumption per ton of steel will drop rapidly; the iron-steel ratio has begun to slowly decline, but there is still a large gap with the domestic advanced level. Non-blast furnace ironmaking technology has accelerated popularization; scrap steel utilization has been greatly improved.

(3) The energy structure change scenario is based on the energy efficiency improvement scenario. Under this scenario, the key energy saving and emission reduction technologies are fully popularized, and advanced energy saving and emission reduction technologies are gradually promoted; the overcapacity of crude steel is contained; the product structure is optimized and the output is stable decrease; long-process steelmaking is gradually replaced by short-process steelmaking; Wuhan Iron and Steel's unit product energy consumption has dropped significantly, reaching the international advanced level; it is set that the proportion of electric furnace steelmaking in Wuhan's iron and steel industry will continue to increase from 2019 to 2050, and non-blast furnace ironmaking technology popularization; the utilization rate of scrap steel has been greatly increased, and the ratio of iron to steel has been declining; the consumption of fossil energy has dropped significantly.

2.2 parameter settings

The influencing factors of this study mainly include steel product output, energy intensity, equipment structure, production structure, electric furnace ratio, blast furnace ratio, iron-steel ratio, and energy-saving low-carbon technology penetration rate. The specific settings of the parameters are based on five time nodes: 2020, 2025, 2030, 2035, and 2050.

(1) Crude steel output

This article is based on the policy opinions issued by the National Development and Reform Commission in the "Key Points for Resolving Excess Capacity in Iron and Steel in 2019" and the "Implementation Measures for Capacity Replacement in the Iron and Steel Industry" issued by the Ministry of Industry and Information Technology, combined with the actual situation of WISCO. During 2019-2050, WISCO's crude steel output showed a slow downward trend. After parameter adjustments, the crude steel output declines at different rates in the three different scenarios. Due to the decline in crude steel production, the consumption of various fossil energy sources has decreased.

(2) Energy intensity

In 2020, the comprehensive energy consumption per ton of steel of WISCO is 569 kg of standard coal. Considering that the electric furnace ratio between foreign steel and WISCO and the scrap utilization gap are too large, the comprehensive energy consumption per ton of steel of WISCO is set at 540 kg in the energy efficiency improvement scenario. Coal/ton of steel declines at different rates under different scenarios.

(3) Equipment structure

In the baseline scenario and the energy efficiency improvement scenario, the energy structure is not changed, and the electric furnace ratio remains zero. In the energy structure change scenario, the electric furnace ratio parameter setting refers to the increase in the rate of change of the electric furnace ratio in developed countries. Electric furnace steel accounts for about 30% in the world, while the United States accounts for 60%[10]. From 1960 to 1985, the ratio of electric furnaces in the United States is rising, and the long-term trend is to gradually increase the proportion of electric furnace steel. With reference to the above data and the actual situation in my country, the electric furnace ratio parameters of WISCO continue to rise.

(4) Emission factors

The emission factors of fossil fuels remain basically unchanged, set to constant values, using WISCO's statistical monitoring values or the default values in the "Guidelines for Accounting Methods and Reporting of Greenhouse Gas Emissions for Chinese Steel Production Enterprises (Trial)" [11].The default value of IPCC in the LEAP model TED template and the power carbon emission factor are adopted by the central China regional power grid carbon emission factor of the State Grid.

(5) Scrap ratio

According to calculations by the Metallurgical Industry Economic Development Research Center in 2015, my country's steel reserves have reached 8 billion tons. In 2020, my country's steel reserves will reach 11.5 billion tons, and will reach 15 billion tons by 2025. By 2030, domestic steel reserves will be basically completed. After industrialization, it reached 17 billion tons. In the future, China's scrap supply will be sufficient to meet the growth of domestic scrap demand. In the medium and long term, the increase in the amount of scrap steel will cause the price of scrap steel to fall, which in turn will drive the increase in demand for scrap steel. The parameter setting refers to the forecast results of my country's total scrap steel resources from 2015 to 2030 in the "Strategic Research Report on a Powerful Country in Ferrous Metal Mineral Resources" compiled by the special group of "Strategic Research on a Powerful Country in Ferrous Metal Mineral Resources".

Table 1. Parameter settings under the baseline scenario.

Year	2020	2025	2030	2035	2050
Crude steel production (10,000 tons)	1556	1493	1440	1410	1300
Clean coal (10,000 tons)	756	704	672	639	576
Coke (10,000 tons)	56	71	79	87	97
Anthracite (10,000 tons)	220	210	204	199	190
Bituminous coal (10,000 tons)	104	94	90	85	75
Natural gas (10,000 cubic meters)	4149	4200	4250	4300	4350
Electricity (million kWh)	5000	4860	4780	4690	4620
Iron to steel ratio (%)	93.03	85	82	80	78
Electric furnace ratio (%)	0	0	0	0	0
Scrap ratio (%)	16.7	17.4	18	18.4	20

Table 2. Parameter setting under the energy efficiency improvement scenario.

Year	2020	2025	2030	2035	2050
Crude steel production (10,000 tons)	1556	1361	1259	1164	1000
Clean coal (10,000 tons)	744	679	628	588	500
Coke (10,000 tons)	47	40	38	35	30
Anthracite (10,000 tons)	216	204	485	174	150
Bituminous coal (10,000 tons)	102	85	76	67	45
Natural gas (10,000 cubic meters)	4149	4115	4089	4064	4000
Electricity (million kWh)	5000	4730	4570	4430	4220
Iron to steel ratio (%)	86	82	78	74	68
Electric furnace ratio (%)	0	0	0	0	0
Scrap ratio (%)	17	20.5	23.2	26.2	32.3

Table 3. Parameter setting under the energy structure change scenario.

Year	2020	2025	2030	2035	2050
Crude steel production (10,000 tons)	1500	1200	1055	950	700
Clean coal (10,000 tons)	720	594	500	450	20
Coke (10,000 tons)	46	38	34	30	19
Anthracite (10,000 tons)	210	136	113	83	20
Bituminous coal (10,000 tons)	61	20	7	0	0
Natural gas (10,000 cubic meters)	4149	3851	3206	1422	300
Electricity (million kWh)	5000	4530	4060	3500	1020
Iron to steel ratio (%)	80	75	72	69	64
Electric furnace ratio (%)	0	10	20	30	50
Scrap ratio (%)	19	23.5	26.5	30.2	36.5

3 Results and Discussion

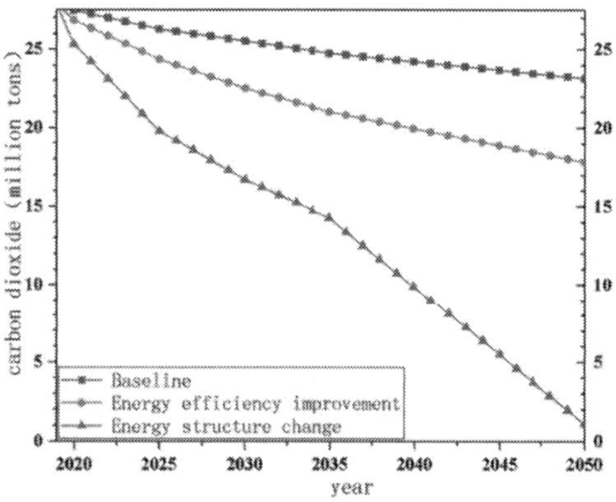

Fig. 1. Comparison of carbon emission forecast changes in three scenarios.

The above figure is a comparison chart of carbon emission forecast changes in the three scenarios. According to the baseline scenario, WISCO completed the "carbon peak" target in 2018, and the subsequent carbon emissions have been slowly decreasing year by year. By 2035, it will not be able to achieve the 35% carbon reduction target, and by 2050, it will be far from achieving the "carbon neutral" target.

According to the energy efficiency improvement scenario, WISCO completed the "carbon peak" in 2018. Carbon emissions are declining year by year. By 2035, carbon emissions will drop by about 5 million tons, which is about 20% carbon reduction. However, the target of 35% carbon reduction in 2035 and the target of "carbon neutrality" in 2050 cannot be achieved.

According to the energy structure change scenario, WISCO completed the "carbon peak" in 2018. Carbon emissions are declining rapidly year by year. By 2035, about 15 million tons of carbon will be reduced, which will reduce carbon by more than 50%, exceeding the target of 35% carbon reduction by 2035, and basically completing the "carbon neutral" target by 2050. The following focuses on the analysis of energy structure change scenarios.

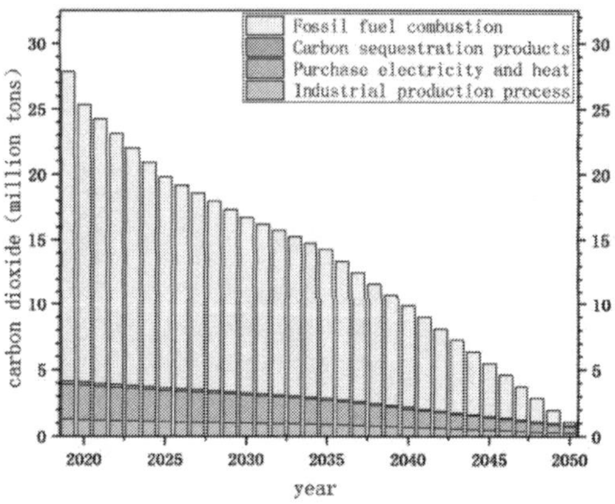

Fig.2. Carbon Emission Prediction for Energy Structure Change Scenarios

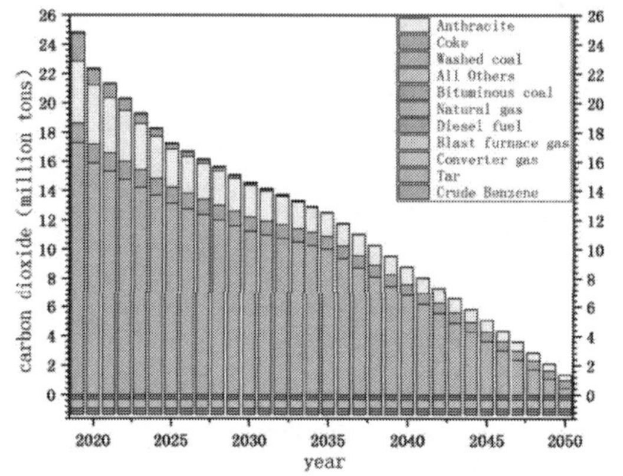

Fig. 3. Fossil fuel combustion carbon emission prediction in the energy structure change scenario

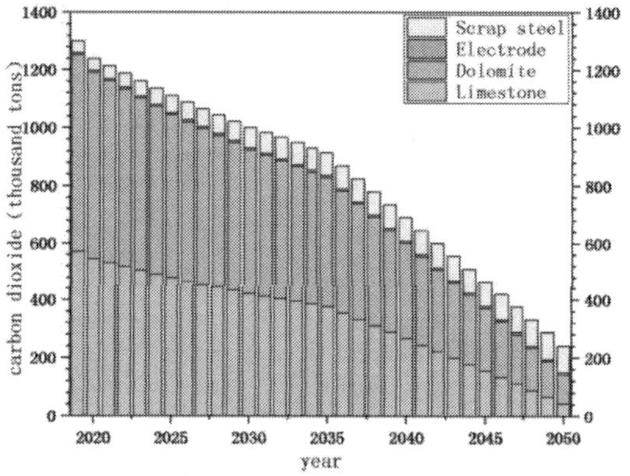

Fig. 4. Carbon Emission Prediction of Industrial Production Process in the energy structure change scenario

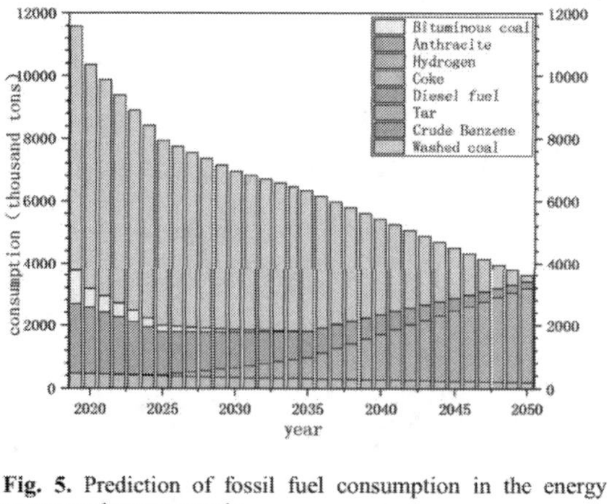

Fig. 5. Prediction of fossil fuel consumption in the energy structure change scenario

WISCO continued to develop in accordance with the energy structure change scenario. While further implementing the promotion of energy saving and emission reduction technologies, WISCO introduced new iron-making and steel-making technologies. Short-process steel-making gradually replaced traditional long-process steel-making, and the burning of fossil fuels produced a large amount of carbon emissions. After 2035, bituminous coal will be completely abandoned by Wuhan Iron and Steel, and the consumption of clean coal, coke and anthracite is also declining year by year.

It can be seen from **Fig.5** that hydrogen has begun to be used as an energy source in 2025. 2035 is an important time node. The hydrogen smelting technology is mature and the amount of hydrogen used has begun to increase significantly. Under the structural change scenario, it is assumed that the experiment of hydrogen metallurgy technology will start in 2025, and the technology will mature and be put into use in large quantities by 2035. Therefore, after 2035, fossil energy such as clean coal and anthracite will withdraw from the stage of steel smelting at a faster rate. From 2020 to 2050, the net consumption of carbon dioxide emissions from fossil fuel combustion activities continues to drop significantly. By 2050, the carbon dioxide emissions from fossil fuel combustion activities have been controlled below 500,000 tons.

The carbon dioxide emissions produced in the industrial production process are mainly due to the amount of carbon dioxide produced by the consumption of dolomite and limestone as solvents in the ironmaking process. Under the scenario of changes in the energy structure, energy-saving and emission-reduction technologies are added. Under the change of ironmaking and steelmaking technologies, Before 2035, the consumption of dolomite and limestone will decrease with the output of steel; after 2035, the consumption of dolomite and limestone will drop significantly. Due to the new steelmaking technology, hydrogen energy steelmaking will replace the traditional long-process steelmaking process, dolomite and limestone no longer act as solvents in the sintering and ironmaking processes, so the carbon dioxide emissions generated by the industrial production process will also approach zero in 2050.

Since the carbon dioxide emissions implied by WISCO's carbon fixation products account for a small proportion of WISCO's carbon dioxide emissions, the amount of change is also very small under the scenario of changes in the energy structure, which is not the focus of WISCO's carbon emission reduction work.

4 Conclusions

4.1 Control steel production

Controlling steel production is the main means to reduce carbon emissions in the steel industry. According to the crude steel output under the energy structure change scenario in the model calculation results, the specific crude steel output is shown in **Table 4**. The crude steel output in the simulation forecast does not include the scrap-electric arc furnace steelmaking process, and the crude steel of the non-scrap-electric furnace steel route. The output is strictly controlled and continues to decline.

According to the forecast result, the crude steel output of WISCO will fall to 9.5 million tons by 2035, an average annual decrease of about 3.0%. It will be reduced to 7 million tons by 2050, an average annual decrease of 2.0%. At the same time, the proportion of WISCO's scrap steel needs to be continuously increased. In 2035, the proportion of scrap-electric furnace steel will reach 30%, and the proportion of scrap-electric furnace steel will reach 50% in 2050.Crude steel production continues to decline, and the amount of scrap steel continues to increase. Finally, the growth rate of total steel production generally shows a downward trend, before 2035 The average decrease was 0.67%, and then the decrease increased to 1.43%.

Table 4. Steel production control plan.

Year	2020	2025	2030	2035	2050
Crude steel production (ten thousand tons)	1500	1200	1055	950	700
Percentage of scrap steel (%)	/	23.5	26	30	36
Total output (ten thousand tons)	/	1568	1425	1357	1093

4.2 Change the energy structure

The change of energy structure is the most important emission reduction measure. According to the forecast results, three suggestions for green transformation are given. (1) WISCO should introduce electric furnaces to increase the proportion of electric furnace steel. (2) While ensuring the supply of scrap steel resources, WISCO should use scrap steel efficiently. (3) WISCO should introduce hydrogen direct reduction iron technology in a planned way.

The above suggestions are given from the three perspectives of electric arc furnace, scrap steel, and direct reduced iron. The three suggestions are integrated in terms of energy structure changes. Use hydrogen to smelt direct reduced iron to make steel in an electric arc furnace and use scrap steel to make steel in an electric arc furnace. The basis of the two approaches is the introduction of electric arc furnaces by Wuhan Iron and Steel.

Combining the results of simulation prediction and the status quo of my country's scrap steel industry, the specific supporting implementation plan is as follows: WISCO will begin to introduce electric arc furnaces in 2022 and gradually increase the proportion of electric arc furnaces. According to the results of the forecast model, the proportion of electric arc furnaces will reach 10% in 2025, the proportion of electric arc furnaces will reach 30% in 2035, and the proportion of electric arc furnaces will reach 50% in 2050. With the growth of my country's steel reserves and the decline in domestic scrap prices, the cost of using scrap for electric arc furnace steelmaking will gradually decrease. Before 2025, due to the cost of electric arc furnace steelmaking, the main method of putting scrap steel into the converter is still used to reduce carbon emissions. In 2025, scrap steel will be used to make steel in the electric arc furnace, and the amount of scrap steel will be increased year by year. By 2035, the use amount of scrap steel should reach 30%, and by 2050, the use amount will reach 36.5%. WISCO will start the direct hydrogen smelting direct reduced iron technology in 2025, and it will be put into use on a large scale in 2035. In 2035, the proportion of electric arc furnaces will reach 30%. WISCO can use direct reduction iron electric arc furnaces to make steel, and at the same time use scrap electric arc furnaces to make steel. It can be seen from **Fig.5** , after 2035, the use of hydrogen will increase significantly, and the output of direct reduced iron will continue to rise. Direct reduced iron will be put into electric arc furnace steelmaking. The time nodes are shown in the following table.

Table 5. Energy structure change time.

Item	2025	2035	2050
Proportion of electric arc furnaces	10%	30%	50%
Scrap steel	start	30%	36.5%
Hydrogen direct reduction of iron technology	Carry out technical research	Industrial application	

References

1. W. Xu, Y. Li, T. Zhu, W. Cao, CO2 Emission in Iron and Steel Making Industry and Its Reduction Prospect, The Chinese Journal of Process Engineering,13:175-180. (2013)
2. S. Watakabe, K. Miyagawa, S. Matsuzaki, et al. Operation Trial of Hydrogenous Gas Injection of COURSE50 Project at an Experimental Blast Furnace, ISIJ International, 53:2065-2071.(2013)
3. X.S. Wang, D. Li, X.Y. Wang, Analysis on Development Strategy of German Iron and Steel Industry, Metallurgical Economics and Management,4: 22-26.(2017)
4. M. Pei, M. Petäjäniemi, A. Regnell, O. Wijk, Toward a Fossil Free Future with HYBRIT: Development of Iron and Steelmaking Technology in Sweden and Finland,Metals, 10(7):972 (2020)
5. Q. He, Progress of HYBRIT Fossil-free Smelting Technology in Swedish Iron and Steel Industry, Metallurgical Economics and Management,01: 52-56. (2021)
6. L. Zhang, H. Li, L. Cheng, B. Li,Overview of the Low Carbon Technology Development in the Foreign Steel Industry, Metallurgical Economics and Management, 5:30-33 (2018)
7. G. Zuo, A. Hirsch, The trial of the top gas recycling blast furnace at LKAB's EBF and scale-up,Revue de Métallurgie,9:387-392 (2009)
8. H. Song, S. Lee, S. Maken, et al. Environmental and economic assessment of the chemical absorption process in Korea using the LEAP model,Energy Policy,29(12):2109-2124 (2007)
9. K. Wang, C. Wang, X.D. Lv, et al. Abatement potential of CO2 emissions from China's iron and steel industry based on LEAP, Journal of Tsinghua University ,46(12): 1982-1986 (2006)
10. Г, Я, Klimo, B, И, Sizov, Chang Zheng, Status and development trend of foreign electric furnace steelmaking production, Hebei Metallurgy, 02: 62-64. (1991)
11. General Office of the National Development and Reform Commission Guidelines for Accounting Methods and Reporting of Greenhouse Gas Emissions for Chinese Steel Production Enterprises(Trial) https://www.ndrc.gov.cn/xxgk/zcfb/tz/201311/t20131101_963960.html?code=&state=123 (2013)

Life Cycle Assessment of Biochar Preparation of Chinese Traditional Medicine Residue by Low-temperature Pyrolysis

Qian Deng[1], *Aijun* Li[1,*], *Yangwei* Wu[1], and *Lushi* Sun[2]

[1] Huazhong University of Science and Technology, School of Energy and Power Engineering, 430074 Wuhan, China
[2] Huazhong University of Science and Technology, School of Energy and Power Engineering, State Key Laboratory of Coal Combustion, 430074 Wuhan, China

Abstract. This study aimed to evaluate the environmental impacts of the biochar production process through low-temperature pyrolysis of Chinese medicine residue via life cycle assessment (LCA). An LCA model consisting of biomass pretreatment, pyrolysis, separation and cooling was developed. The results indicated that the low-temperature pyrolysis process has the highest environmental impact on AP and GWP. The consumption and direct emission of pyrolysis process were the primary sources of environmental impact. In addition, by comparing to the traditional landfill process of Chinese medicine residue, it turns out that the low-temperature pyrolysis process has improved environmental protection performance.

1 Introduction

With the rapid development of Chinese medicine and the extension of the related resource based industrial chains, the yield of Chinese medicinal residue is increasing year by year. The annual discharge of Chinese medicinal residue in China is up to 30 million tons in 2015, among which the production of Chinese patent medicine takes up the largest part, accounting for about 70% of the total[1]. Chinese medicine residue contains a certain amount of active ingredients and a large amount of cellulose, hemicellulose, lignin, protein and other rich organic ingredients. Recently, it has been reported that traditional Chinese medicine residue is used for edible fungus, feed, pyrolysis and gasification[2-4], partly realizing its resource-oriented utilization. However, due to the long fermentation cycle, complex composition, difficult separation, and some residues may contain harmful substances, the utilization of the residue as feed and fertilizer is still under restriction[5]. Agricultural straw and garden waste have been widely used in the preparation of biochar, showing broad application prospects in soil improvement and environmental protection [6,7]. Chinese medicine residue is highly similar to agricultural and forestry wastes, which are characterized by high carbon content and easy to collect, and also can be used to prepare biochar. However, the existing literature has mostly focused on the performance and application of biochar, the environmental impact caused by its production process, and the quantitative assessment of its environmental performance are required to be identified systematically and comprehensively.

So far, several LCA studies have evaluated the production of activated carbon or bio-oil from biomass[8-13]. As a typical biomass, some researchers have attempted to investigate pyrolysis gasification from Chinese medicine residue[14,15], but there is still a lack of LCA investigation on biochar production from Chinese medicine residue through low-temperature pyrolysis.

The purpose of this study is to analyze the environmental impacts on low-temperature pyrolysis of Chinese medicinal residue to prepare biochar. This article indicates the influence of low-temperature pyrolysis on various environmental impact categories by comparing with the landfill process, so as to provide more intuitive and reliable scientific basis for its environmental protection performance.

2 Data source and methods

2.1 Data source

The Chinese medicine residue analyzed in this article is based on the research of Guo[16] from Henan Wanxi Pharmaceutical Co., Ltd. Table 1 represents the elemental and proximate analysis of Chinese medicine residue, respectively.

* Corresponding author: hust_laj@mail.hust.edu.cn

Table 1. Proximate and elemental analysis and LHV of Chinese medicine residue[16]

Chinese medicine residue	Proximate analysis (wt% wet basis)				Elemental analysis (wt% dry basis)					
Project	M	FC	V	A	C	H	O	N	S	LHV（MJ/kg）
Content	12.5	12.41	72.62	2.47	42.4	6.2	47.39	1.06	0.15	14.90

2.2 Process description

The pyrolysis system of Chinese medicine residue consists of four units: biomass pretreatment, pyrolysis, separation and cooling. The residue was first broken by crusher until the particle size is between 40-60 mu[17], and then dried until the moisture content is less than 10%. Once they meet the requirements of pyrolysis, low-temperature pyrolysis reaction can be carried out in the pyrolysis furnace to generate biochar and gas products. The temperature of pyrolysis is 550 ℃. Then biochar shall be rapidly cooled down to about 80 ℃ in a short time through an air-cooled and water-cooled two-stage heat exchanger. Gas products reacted with air in the combustion furnace to provide heat for drying and pyrolysis, resulting in exhaust gas such as CO2.

2.3 Life cycle assessment framework

2.3.1 Research methods and tools

This study intends to examine the gate-to-gate life cycle impacts of biochar derived from Chinese medicine residue with low-temperature pyrolysis techniques. LCA was carried out with eBalance software which is the first general LCA software with independent intellectual property rights developed by IKE Environmental Technology Co., LTD. (IKE). It provides high-quality database support in China and the world, which is suitable for LCA analysis of various products. Life cycle impact assessment (LCIA) with the eBalance software can provides comprehensive quantitative analysis results. The impact categories considered in this study include acidification (AP), Chinese resource depletion potential (CADP), primary energy demand (PED), eutrophication (EP), global warming potential (GWP) and respiratory inorganics (RI).

2.3.2 Goal and scope

The goal of this study was to evaluate the environmental performance of biochar preparation by low-temperature pyrolysis. The environmental footprints of all input processes of the entire life cycle from raw material (Chinese medicine residue) pretreatment to the final product (biochar) were included in this study. Besides, by comparing with landfill process, the degree of impact of low-temperature pyrolysis process of Chinese medicine residue on various environmental categories was determined. Additionally, further suggestions for optimizing production and reducing pollution were put forward. The functional unit was the dispose of 10 tons Chinese medicine residue. According to Fig. 1, the subsystems included biomass pretreatment unit (process 1), pyrolysis unit (process 2), separation unit (process 3), and cooling unit (process 4). All inputs and outputs involved within the system boundary were based on the processing of 10 tons of Chinese medicine residue. The use and abandonment of the product, and the manufacturing and maintenance of the equipment were not considered here to ensure the accuracy and operability of the model.

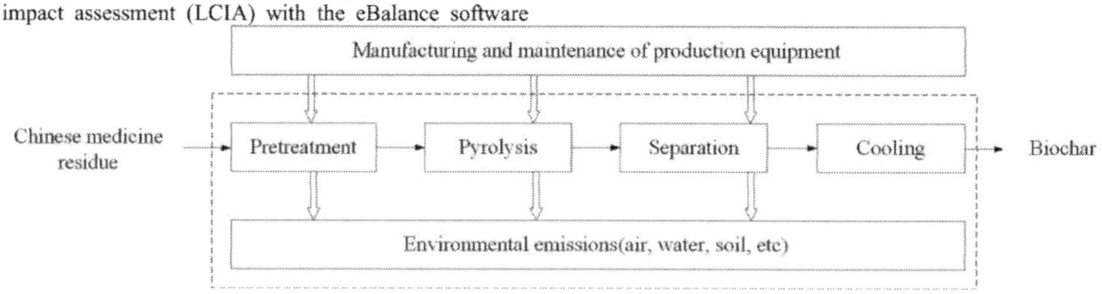

Fig. 1. System boundary of biochar production from Chinese medicine residue.

2.3.3 Life cycle inventory (LCI)

The Life Cycle Inventory (LCI) is the second phase of an LCA study, in which all inputs and outputs data to or from the system boundary are collected. The data for inventory analysis is based on the material input, output, resource and energy consumption and environmental emissions in all sectors within the defined system boundaries. Energy and water consumption for operation, biochar production, and the composition of the exhaust gas are given by previous simulation results. The CLCD Public, Ecoinvent Public and ELCD databases of the eBalance software were used for the background data (production of materials and energy sources). Tables 2 and 3 showed the details on input and output data.

Table 2. The input data used for four processes in biochar production.

Process	Substance	Unit	Input
Process 1	Chinese medicine residue	t	10
	Electricity	kWh	1480
	Air	kg	50000
	Heat	GJ	2.746
Process 2	Dry medicinal residue	t	9.722
	Heat	GJ	23
	Nitrogen	kg	96.3
Process 3	Air	kg	50000
	H2	kg	186.68
	CH4	kg	768.56
	CO	kg	312.46
	H2O	kg	2418.57
	N2	kg	92.75
	CO2	kg	3690.97
Process 4	Biochar	kg	1904.14
	Air	kg	15800
	Electricity	MJ	1494.77
	Water	kg	2495

Table 3. The output data used for four processes in biochar production.

Process	Substance	Unit	Output
Process 1	Dry medicinal residue	t	9.722
	Wastewater	t	0.278
Process 2	Biochar	kg	1904.14
	Ash	kg	245
	H2	kg	186.68
	CH4	kg	768.56
	CO	kg	312.46
	H2O	kg	2418.57
	N2	kg	92.75
	CO2	kg	3690.97
Process 3	H2O	kg	5820.37
	N2	kg	38332.70
	O2	kg	7014.30
	CO2	kg	6290.32
	SO2	kg	26.22
	NO2	kg	0.0263
Process 4	Biochar	kg	1904.14
	Wastewater	kg	2495

3 Results and discussion

3.1 Life cycle impact assessment (LCIA)

Fig. 2 showed the life cycle impact assessment results of low-temperature pyrolysis process of Chinese medicine residue. The order of impact categories was as following AP > GWP > RI > EP > CADP > PED. To dispose of 10 tons Chinese medicine residue, the highest amounts of impact categories were found to be for AP, GWP, and RI. The greatest indicator was related to acidification. This problem is due to the release of acid gases such as SO2, N O2, etc. resulting from biogas combustion. The second environmental problem was correlated to the global warming. Indeed, greenhouse gas emissions such as CO2 are directly involved with power consumption. The third main indicator of biochar production was RI.

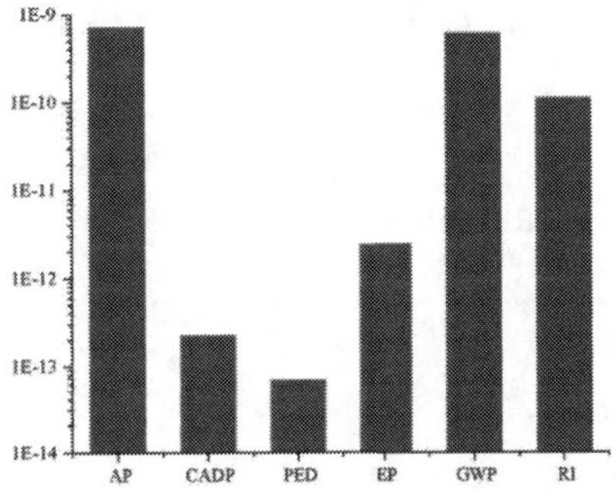

Fig. 2. Life cycle impact assessment results of low-temperature pyrolysis process of Chinese medicine residue

3.2 Analysis of process contribution

Analysis of process contribution was conducted for the above six environmental impact categories to identify the major pollution processes for effective improvement measures. Fig. 3 showed the analysis results of process contribution for key environmental impact categories. It's worth noting that pollution from upstream is also taken into account in eBalance software. Therefore, substances in process contribution assessment include pollution sources from upstream raw material production. Results presented that in the process of low-temperature pyrolysis of Chinese medicine residue, the separation unit has a relatively large impact on AP, GWP and RI, indicating that the consumption and direct emission in pyrolysis process were the main sources of environmental impact. For CADP and PED, the pyrolysis and cooling unit have great influence, accounting for more than 30 %. Moreover, the pretreatment unit also has a certain influence. For EP, the separation process has the largest environmental impact, reaching more than 40 %. Therefore, it can be concluded that the pyrolysis unit needs to be optimized urgently.

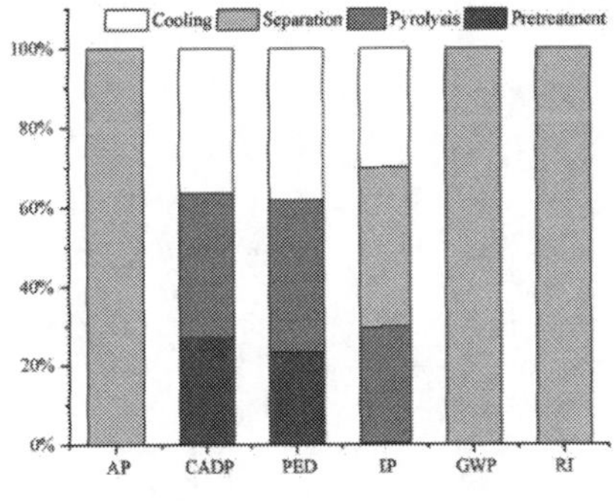

Fig. 3. Analysis of process contribution

3.3 Comparison with landfill process of traditional Chinese medicine residue

By comparing the low temperature pyrolysis with the landfill process of 10 tons Chinese medicine residue, we can further quantitatively analyze the environmental protection performance of the low-temperature pyrolysis process of Chinese medicine residue. The results of the comparison of environmental impact analysis between low-temperature pyrolysis and landfill process of Chinese medicine residue were shown in table 3.

Table 4. Comparison of environmental impact analysis results between low-temperature pyrolysis and landfill process

Impact category	Unit	Low-temperature pyrolysis	Landfill
AP	kg SO2 eq	26.2	52.5
CADP	kg Coal-R eq	3.46	4.72
PED	MJ	5.82	8.05
EP	kg PO43-eq	8.87×10-3	1.51×10-2
GWP	kg CO2eq	6290	12600
RI	kg PM2.5 eq	2.05	4.10
ECER-125-CN-2010	-	8.01×10-9	1.60×10-8

The results indicated that for the above six environmental impact categories, low-temperature pyrolysis had lower values in comparison to landfill process, respectively. Especially for GWP, AP and RI, low-temperature pyrolysis of Chinese medicine residue showed obvious advantages. The main reasons can be attributed into two aspects: On the one hand, the low-temperature pyrolysis process recycles the residue and reduces the discharge of solid waste and harmful substances; On the other hand, Chinese medicinal residue directly buried in the landfill, resulting in a large amount of land occupation, and will affect the mining of minerals. Additionally, Chinese medicinal residue decay in the natural environment, bringing about a large number of harmful substances which cannot be ignored.

In general, the low-temperature pyrolysis process of Chinese medicine residue showed better environmental performance than the traditional landfill process, which can better reflect the characteristics of environmental friendliness in all aspects.

4 Conclusions

This study assessed the environmental impacts of biochar production from Chinese medicine residue through low-temperature pyrolysis, and compared low-temperature pyrolysis with landfill process of Chinese medicine residue. The main conclusions are as following:

(1) Normalization results showed that low-temperature pyrolysis of Chinese medicine residue has the highest impact on acidification and global warming potential, followed by respiratory inorganics, eutrophication, Chinese resource depletion potential and primary energy demand.

(2) Analysis of process contribution revealed that the separation unit had a relatively larger impact on AP, GWP and RI, indicating that the energy consumption and direct emission of pyrolysis process were the main sources of environmental impact. The pyrolysis unit is in urgent need of optimization.

(3) From the perspective of life cycle, the low-temperature pyrolysis process of Chinese medicine residue showed a better environmental protection performance in various environmental impact categories than traditional landfill process. It's not only because the residue is recovered in the low-temperature pyrolysis process, which reduces the discharge of solid waste and harmful substances, but also because the Chinese medicine residue is directly landfilled, resulting in a large number of land occupation and harmful substances.

In short, the low-temperature pyrolysis system showed evident advantages than landfill system, but still needs to be optimized. This study laid a foundation for the application and popularization of the low-temperature pyrolysis of Chinese traditional medicine residue.

Acknowledgments

This work was supported by "National Key R&D Program of China (2019UFC1906600)".

References

1. M. L. Na, J Chen, W. Z. Wei, et al. Research progress of biological processing of traditional chinese medicine residues. Lishizhen Medicine and Materia Medica Research. 01: 194-196. (2016)
2. J. Z Chen., Y. Z.Tan, X. H. Wang, et al. Effects of bushen yishou capsule residues on culture of pleurotus abalonus han,k.m.chen et.s.cheng and nutrition composition analysis. Southwest China Journal of Agricultural Sciences. 02: 740-742. (2012)
3. J. A. Duan, S. L. Su, G Sheng, et al. Research practices of conversion efficiency of resources utilization model of castoff from chinese material medica industrialization. China Journal of Chinese Materia Medica. 38: 3991-3996. (2013)
4. E. H. Wang, X. F. Cai and F. L. Zhang Effects of Degraded Astragalus Residue on Growth Performance and Flavor of Broilers. Southwest China Journal of Agricultural Sciences.(10): 69-72.(2020)
5. S. Guo, J. A. Duan, X. J. Lu, et al. Research and Practice on Utilization of Chinese Medicinal Materials Solid Waste by Pyrolysis and Carbonization. Modern Chinese Medicine. 19(12): 1665-1671. (2017)
6. W. F. Chen,, W. M. Zhang and J. Meng. Advances and prospects in research of biochar utilization in agriculture. entia Agricultura Sinica. 46(016): 3324-3333. (2013)
7. L. Qian, W. Zhang, J. Yan, et al. Effective removal of heavy metal by biochar colloids under different pyrolysis temperatures. Bioresource Technology. 206: 217-224. (2016)
8. K. Hjaila., R. Baccar, M. Sarra, et al. Environmental impact associated with activated carbon preparation from olive-waste cake via life cycle assessment. Journal of Environmental Management. 130: 242-247. (2013)

9. J. F. Peters, D Iribarren and J Dufour, Simulation and life cycle assessment of biofuel production via fast pyrolysis and hydroupgrading. Fuel. 139: 441-456. (2015)
10. L. Heng, H. Zhang, J. Xiao, et al. Life Cycle Assessment of Polyol Fuel from Corn Stover via Fast Pyrolysis and Upgrading. Acs Sustainable Chemistry & Engineering. 6(2): 2733-2740. (2018)
11. S. Suganya, P. S. Kumar, Evaluation of environmental aspects of brew waste-based carbon production and its disposal scenario. Journal of Cleaner Production. 202: 244-252. (2018)
12. X. Yang, D. Han, Y. Zhao, et al. Environmental evaluation of a distributed-centralized biomass pyrolysis system: A case study in Shandong, China. Science of the Total Environment. 716.(2020)
13. X. Zhang, L. L. Zhang, R. Li, et al. Life cycle assessment of straw fast pyrolysis based on energy integration. CIESC Journal. 72(5): 2792-2800. (2021)
14. P. Wang, H.B. Yu, X. F. Xue, et al. Study on characteristics and kinetics of pyrolysis of herb residue. Chinese Journal of Environmental Engineering. 04(9): 2115-2119. (2010)
15. P. F. Fan, J. D. Li, Y. T. Liu, et al. Experimental study of gasification of herb residues of Ganmaoqingre granules in pilot-scale dual-loop circulating fluidized bed. Chemical Industry and Engineering Progress.(8): 1979-1985,1991. (2014)
16. F. Q. Guo Study on pyrolysis gasification and tar oxidation reforming process of traditional Chinese medicine residue biomass. (Doctoral dissertation, Shandong University).(2013)
17. P. Badger, S. Badger, M. Puettmann, et al. Techno-economic analysis: Preliminary assessment of pyrolysis oil production costs and material energy balance associated with a transportable fast pyrolysis system. Bioresources. 6(1): 34-47. (2011)

Adsorption of Ciprofloxacin and Tetracycline by Organically Modified Magnetic Bentonite

Rui Tang[1], *Hanbing* Zhang[2], *Caimei* Lu[2], *Kun* Liu[1], *Sishan* Yu[1] and *Zhangfa* Tong[1,3,*]

[1]Guangxi Colleges and Universities Key Laboratory of New Technology and Application in Resource Chemical Engineering, School of Chemistry and Chemical Engineering, Guangxi University, Nanning 530004, Guangxi, China
[2]College of Resources, Environment and Materials, Guangxi University, Nanning 530004, Guangxi, China
[3]Guangxi Engineering Academy of Calcium Carbonate Industrialization, Guangxi University, Nanning 530004, Guangxi, China

Abstract. In order to endow magnetic bentonite (MB) with better magnetic separation and magnetic stability, obtain stronger adsorption of antibiotics, carboxymethylcellulose-chitosan/magnetic bentonite composite (MB/CC) was prepared by modifying previously fabricated MB with carboxymethylcellulose (CMC) and chitosan (CS). The characterization of MB/CC and the adsorption properties towards ciprofloxacin (CIP) and tetracycline (TC) was investigated. The characterization results revealed that CC improves the Fe_3O_4 stability and removal capability of CIP and TC for the final product. Adsorption results indicated that MB/CC exhibited superior adsorption performance for CIP (182 $mg \cdot g^{-1}$) and TC (189 $mg \cdot g^{-1}$). Adsorptive removal of CIP and TC by MB/CC was more than 90% even after 5 cycles of adsorption-desorption process. The adsorption process fitted well with the Langmuir model and the pseudo-second order kinetic model. The main steps during the successful adsorption process of MB/CC for CIP and TC includes pore diffusion, ion exchange and electrostatic interaction. MB/CC has advantages of simplified synthesis, convenient operation and high-efficiency, which could be deemed as a promising alternative adsorbent for antibiotics removal from wastewater.

1 Introduction

Ciprofloxacin (CIP) belongs to fluoroquinolone antibiotics, and tetracyclines (TC) belongs to tetracycline antibiotics. CIP and TC are two kinds of antibiotics produced and used in large quantities in China, which are widely used in clinical medical treatment and livestock aquaculture [1-2]. CIP and TC after use are discharged into the environment in different ways and exist in various forms. Because they have certain biological toxicity, they cause great harm to the environment and hinder the healthy development of human body.

Magnetic bentonite (MB) is a kind of adsorption material developed by our research group in the early stage. It is found that MB has a certain adsorption effect on antibiotics, dyes and heavy metal ions [3]. However, the properties of Fe_3O_4 are not stable, it is easy to oxidize and agglomerate contacting with air. Moreover, some absorption points of bentonite will be occupied after Fe_3O_4 added to bentonite, thus, the adsorption capacity of MB would be reduced, and also affect the solid-liquid separation ability of MB.

Chitosan (CS) is a kind of environmental friendly material, which has remarkable adsorption capacity for organic pollutants, antibiotics and metal ions. The structure contains a large number of amino and hydroxyl groups[4]. CMC is cheap, well soluble and biodegradable polyanionic cellulose[5]. Owing to the electrostatic interaction between CS and CMC, a non-toxic, high adsorption capacity and economical CC was prepared.

2 Experimental

2.1 Materials

Bentonite (RB) was obtained from Tianjin Guangfu Fine Chemical Research Institute, chemical purity. Chitosan $(C_6H_{11}NO_4)_n$ (deacetylation degree is more than 95%, viscosity is 100-200 MPa·s), Sodium carboxymethylcellulose, $[C_6H_7O_2(OH)_2CH_2COONa]_n$ (viscosity 800-1200 MPa·s) were purchased from Aladdin reagent (Shanghai) Co., Ltd., chemical purity. $FeCl_2 \cdot 4H_2O$, $FeCl_3 \cdot 6H_2O$, $NH_3 \cdot H_2O$, C_2H_5OH, HNO_3, CIP ($C_{17}H_{18}FN_3O_3$) and TC ($C_{22}H_{24}N_2O_8$) were purchased from Guangdong Guanghua Chemical Reagent Co., Ltd., analytical purity.

2.2 Adsorption studies

0.2 g MB and MB/CC were added into CIP and TC solution (200 ml, 10 $mg \cdot L^{-1}$) respectively. Under the rotating speed of 150 $r \cdot min^{-1}$, the CIP and TC supernatants were adsorbed in a constant temperature water bath oscillator for 2 h, and then removed from the oscillator according to the time within 0 ~ 120 min. The absorbance values of CIP and TC supernatants were measured by ultraviolet spectrophotometer at 270 nm and

* Corresponding author: zftong@gxu.edu.cn

360 nm, respectively, and the kinetic process was analyzed. A series of CIP and TC solutions with different concentration gradients ranging from 10 to 80 mg·L^{-1} were prepared, and 1.0 g·L^{-1} MB and MB/CC were added respectively. The adsorption was oscillated for 2 h at the rotating speed of 150 r·min^{-1}. And the effect of initial concentration of solution was explored.

The calculation formulas of adsorption removal of CIP and TC by MB and MB/CC are shown in (1) and (2):

$$q = (C_0 - C_t)/m \tag{1}$$

$$R = (C_0 - C_t)/C_0 \times 100\% \tag{2}$$

Where C_0 is the concentration of CIP and TC before adsorption (mg·L^{-1}). C_t is the concentration of the supernatant of CIP and TC were adsorbed at t (min) (mg·L^{-1}). m is the dosage of MB and MB/CC (g·L^{-1}). q is the adsorption capacity of MB and MB/CC for CIP and TC (mg·g^{-1}). R is the adsorption efficiency (%) of adsorbates on MB and MB/CC.

3 Results and discussion

3.1 Effect of pH and initial concentration of CIP and TC

The adsorption efficiency of antibiotics on MB and MB/CC were changed significantly at pH were 2 ~ 6 and 7 ~ 10. The solution was acidic at pH 2 ~ 6, CIP and TC existed in the form of cation after hydrolysis, and reacted with —OH, —NH_2 on the surface of the MB/CC, thus the adsorption percentage of CIP and TC on MB/CC was higher. With the increase of pH, OH^- in the solution increased, CIP and TC gradually deprotonated, the combining ability with — OH, — NH_2 gradually weakened. Therefore, the removal efficiency of CIP and TC on MB and MB/CC were decreased[6].

The adsorption efficiency of CIP and TC first increased and then reached the adsorption equilibrium gradually[7]. This was due to the more adsorption sites on MB and MB/CC to be occupied at the initial concentration of CIP TC is low, thus the adsorption efficiency of CIP and TC on MB and MB/CC was faster. When the initial concentration increased, the adsorption process has been reached adsorption saturation and the adsorption sites was limited, hence, the adsorption capacity of CIP and TC on MB and MB/CC was not increased.

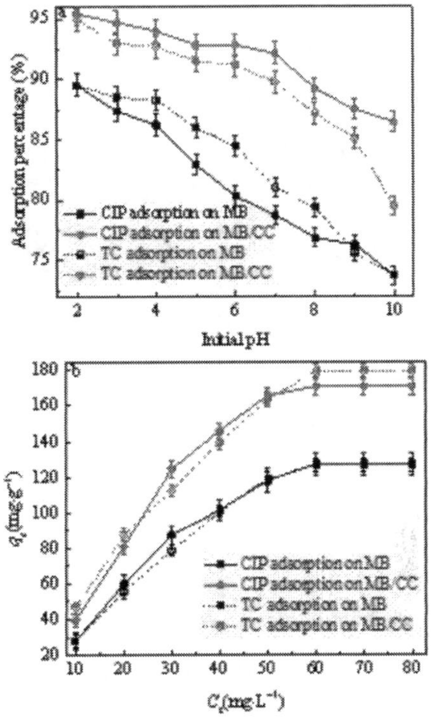

Fig. 1. Effect of (a) pH and (b) initial concentration on the adsorption of CIP and TC.

3.2 Recycling performance

After 5 adsorption-desorption cycling experiments, the adsorption efficiency of CIP and TC on MB decreased from 82% and 61% to 61% and 42% respectively, while the adsorption efficiency of CIP and TC by MB/CC decreased from 98% and 96% to 90% and 91% respectively, which indicated that MB/CC has good recycling performance.

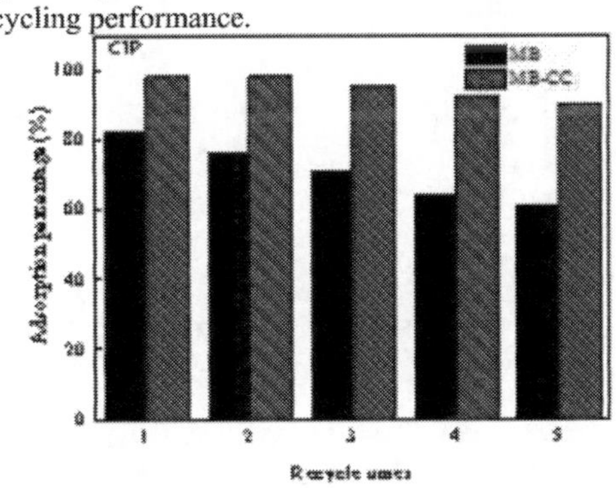

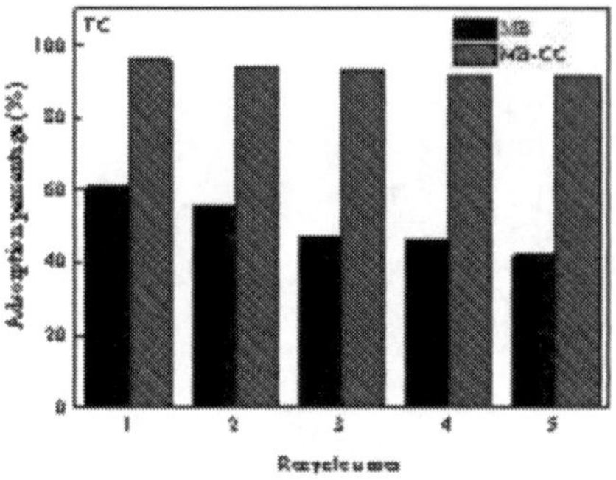

Fig. 2. Recycling performance of CIP and TC on MB and MB/CC.

3.3 Adsorption kinetics and isotherms studies

The R^2 were 0.999 (Langmuir adsorption isotherm model), the Langmuir adsorption capacities of MB and MB/CC for CIP were 147 mg·g^{-1} and 182 mg·g^{-1}, and for TC were 136 mg·g^{-1} and 189 mg·g^{-1}, respectively. The adsorption process was more consistent with Langmuir model, and belongs to monolayer adsorption [7]. The K_L values of CIP and TC adsorbed by MB/CC were much higher than those on MB, which indicated that MB/CC had stronger adsorption capacity for CIP and TC.

Table 1. Fitting parameters of CIP and TC adsorption isotherm models of MB and MB/CC.

	Langmuir model			Freundlich model			$q_{m,exp}$/mg·g^{-1}
	K_L/L·mg^{-1}	$q_{m,cal}$/mg·g^{-1}	R^2	K_f/mg·g^{-1}	1/n	R^2	
MB/CC(CIP)	1.27	182	0.999	51.3	0.41	0.939	170
MB/CC(TC)	1.31	189	0.999	68.2	0.43	0.955	179
MB(CIP)	0.28	147	0.999	36.1	0.37	0.937	127
MB(TC)	0.55	136	0.999	45.7	0.39	0.958	126

The maximum adsorption capacities ($q_{e,cal}$) of CIP and TC on MB/CC were 181 mg·g^{-1} and 191 mg·g^{-1}, and on MB were 102 mg·g^{-1} and 119 mg·g^{-1}, respectively, which are closer to the actual adsorption capacities ($q_{e,exp}$) measured by experiments. In conclusion, the adsorption process of CIP and TC on MB/CC was more consistent with the quasi second-order kinetic model, which indicated that the adsorption process was a process of fast initially, then slowed down n and finally reached equilibrium [8].

Table 2. Fitting parameters of CIP and TC adsorption kinetic models of MB and MB/CC.

	quasi first-order kinetic model			quasi second-order kinetic model			
	k_1×0.01/ min^{-1}	$q_{e,cal}$/ mg·g^{-1}	R^2	k_2×0.01/ g·mg^{-1}·min^{-1}	$q_{e,cal}$/ mg·g^{-1}	R^2	$q_{e,exp}$/ mg·g^{-1}
MB/CC(CIP)	0.003	92	0.466	0.36	181	0.999	170
MB/CC(TC)	0.006	103	0.588	0.41	191	0.999	179
MB(CIP)	0.002	68	0.582	0.32	102	0.999	127
MB(TC)	0.005	77	0.673	0.35	119	0.999	126

3.4 Adsorption mechanism

(1) Pore diffusion: MB/CC was equipped with abundant mesopores which can accommodate the adsorption process of CIP and TC on MB/CC.
(2) Ion exchange: CIP and TC exist in the form of cation at the solution is acidic, thus K^+, Ca^{2+}, Na^+ and Mg^{2+} in bentonite can adsorption CIP and TC through ion exchange.
(3) At acidic, CIP and TC exist in the form of cation. Hence CIP and TC combine with MB/CC (negative charge) through the electrostatic attraction.

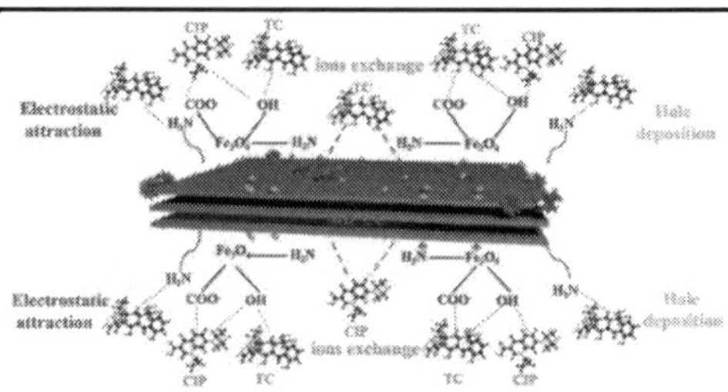

Fig. 3. Proposed mechanism for CIP and TC adsorption on MACB-CC.

4 Conclusions

CMC and CS were used as organic modifiers, and MB/CC was prepared by microwave solid-liquid phase synthesis. The removal efficiency of CIP and TC on MB were above 80%, whileon MB/CC were above 90% at pH 2 ~ 6. The adsorption process of CIP and TC on MB/CC was more consistent with Langumir isotherm model and quasi second-order kinetic model. The adsorption process of CIP and TC on MB/CC belongs to monolayer adsorption, and the adsorption process was a process of fast initially, slowed down then and finally reached equilibrium.

Acknowledgements

The authors acknowledge the financial support by the National Natural Science Foundation of China (Grant number 21576055) and Petrochemical Resources Processing and Process Reinforcement Technology Key Laboratory Project of Guangxi province (Grant numbers 2020Z008).

References

1. S.J. Olusegun and N.D.S. Mohallem, Comparative adsorption mechanism of doxycycline and Congo red using synthesized kaolinite supported CoFe2O4 nanoparticles. Environ. Pollut. **260**, 114019 (2020)
2. Y. Youngbeom, H.C. Yang, C. Jaewon, C. Soohoon and K.M. Sung, Multi-barrier approach for removing organic micropollutants using mobile water treatment systems. Sci. Total. Environ. **639**, 331-338 (2018)
3. C. Cazzorla, G. Bensi, G. Biasucci, V. Leuzzi, F. Manti, A. Musumeci, F. Papadia, V. Stoppioni, A. Tummolo, M. Vendemiale, G. Polo and A. Burlina, Living with phenylketonuria in adulthood: The pku attitude study. Mol. Genet. Metab. Rep. **16**, 39-45 (2018)
4. H. Wang, L. Wang and S. Ye, Construction of Bi2WO6-TiO2/strach nanocomposite films for visible-light catalytic degradation of ethylene. Food. Hydrocolloid. **39**, 75-90 (2018)
5. L. Xu, X. Xu, G. Cao, S. Liu, Z.Y. Duan, S.M. Song and M.J. Zhang, Optimization and assessment of Fe-electrocoagulation for the removal of potentially toxic metals from real smelting wastewater. J. Environ. Management. **218**, 129-138 (2018)
6. J. Zheng, D.J. Luo and Y.H. Qiao, Surface complexation modeling of U(VI) sorption on GMZ bentonite in the presence of fulvic acid. Radiochim. Acta. **105**, 33-41 (2016)
7. F.L. Fu and Q. Wang, Removal of heavy metal ions from wastewaters: a review. J. Environ. Management. **92**, 407-418 (2011)
8. D. Colangelo, F. Torchio, D.M.D. Faveri and M. Lambri, The use of chitosan as alternative to bentonite for wine fining: Effects on heat-stability, proteins, organic acids, colour, and volatile compounds in an aromatic white wine. Food. Chem. **264**, 301 (2018)

The Life Cycle Assessment on Mixed Incineration of High Heating Value Industrial Solid Waste with Municipal Solid Waste

Aijun Li[1], *Yuhao* Liu[1,2,*], *Hongyun* Hu[2] and *Hong* Yao[2]

[1]School of Energy and Power Engineering, Huazhong University of Science and Technology, Wuhan 430074, China
[2] State Key Laboratory of Coal Combustion, School of Energy and Power Engineering, Huazhong University of Science and Technology, Wuhan 430074, China

Abstract. The industrialization process of China has achieved remarkable economic and social development in recent decades, which companies with excessive production of industrial solid waste (ISW). It took up a large amount of land and led to potential environmental problems such as air, water, soil pollution. With the increasing awareness of environmental protection and the better vision to build a pattern of sustainable economy, the comprehensive utilization of ISW is becoming a challenging and promising research field. In this work, a waste incineration power plant in Guangdong Province, China was selected as the research object. Its novel incineration scheme by mix ISW and MSW was evaluated, which aims to improve power generation efficiency. The comprehensive environmental impact assessment of this refuse incineration-power generation technology was also carried out by life cycle assessment. Results show that for waste incineration power generation system, the numerical order of all normalization indicators is Waste Solids>EP>NOx>ADP>AP>CADP>RI>CADP (fossil fuel) >CO_2> SO_2>GWP>PED>COD>IWU> Water Use > NH_3-N. It shows that the ash reduction and denigration system are the most worthy of optimization. And the contribution of flue gas purification to ECER index reaches 38.98%, which is the most influential link and significantly ahead of other links, as well as the most needed optimization link.

1 Introduction

In recent years, both developed and developing countries are faced with serious waste discharge and disposal problems caused by the high degree of industrialization, rapid economic development and population growth, and the severity of environmental problems are becoming increasingly prominent [1]. With the increasing awareness of governments around the world on environmental pollution and resource shortage, people have paid more and more attention to environmental protection and sustainable development [2]. To achieve this ambitious goal, countries around the world by making international regulations and provisions to strengthen international cooperation and build a community of human destiny, for example, the *Kyoto protocol* has been replaced by the *Paris agreement*, this is very important not only to solve the problem of carbon emissions, and for the sustainable development of economy

China's decades of industrialization have produced enormous economic and social development. But at the same time, it also causes serious environmental pollution, such as waste pollution, dust pollution, water pollution, etc. [5]. Especially, the excessive production of industrial solid waste (ISW) occupies a large amount of land and leads to ecological, environmental problems such as air, water and soil pollution, which has become an important cause of environmental pollution in China [6].

ISW is characterized by large output, wide source and large fluctuation of components, so it is difficult to achieve large-scale treatment with a unified method. At present, landfill treatment is the most widely used method. However, it has a low degree of resource recovery and occupies too many land resources, which is not in line with the trend of environmental protection [8]. Despite a great deal of research in recent years, the comprehensive utilization of ISW is still a serious problem in the sustainable development of developing countries. Some researchers have applied ISW to building materials such as cement, baking-free bricks and concrete [10-12], as well as the preparation of porous materials and sintered glass-ceramics [13-15], but these studies are still at the laboratory level, and it is difficult to treat ISW on a large scale and realize the industrial application. Municipal solid waste (MSW) which has a similar situation with it has achieved large-scale incineration power generation treatment, and the effect is very significant, which is a treatment mode worthy of reference [16-17]. The feasibility of using ISW for incineration power generation needs to be evaluated. Different components of ISW may have different impacts on pollutant emission, energy consumption and power generation in the process, and the life cycle assessment (LCA) can include all the -

* Corresponding author: huserlyh94@hust.edu.cn

 https://doi.org/10.1051/e3sconf/202235003007

above indicators, which is a suitable assessment method for the process.

This work attempts to analogize the treatment mode of MSW. The relevant experiments on ISW are made in a waste incineration power plant, which aims to explore the possibility of using large quantities of ISW with the complex components for incineration power generation, Then the comprehensive indicators of life cycle assessment are estimated to evaluate from the perspective of environmental impact.

2 LCA modeling and results

2.1 Introduction to waste incineration power generation system

This work selects a large waste incineration power plant in Guangdong Province, China as the research object. The plant consists of three branch factories and a total of 8 incinerators. The mature technology based on grate furnace is adopted, which has an excellent representation. It incinerates about 4500t mixed solid waste (ISW and MSW) every day and generates about 450kWh of electricity per ton of solid waste. We select the second factory for ISW mixed with the MSW incineration power generation experiment, the whole process is made up of waste feed system, incinerator system, steam turbine power generation system, flue gas purification system and fly ash solidification system, besides raw materials and industrial by-product ash transportation system. The mixed waste is dumped into the waste storehouse by the transport truck and then captured by the waste crane into the incinerator for full combustion. After that, the steam turbine converts the heat energy generated by the combustion into electricity, and the high-temperature flue gas produced by the incinerator enters the flue gas purification system after denitrification in the incinerator. Then Using deacidification tower, denitration system outside the incinerator, cloth bag dust collector and activated carbon injection system for removal of the harmful material in flue gas, and the purified flue gas is discharged by reaching the national standard. The removed fly ash will form bottom ash by chelating stable solidification system, which will be transported out the factory area together with the slag produced by incinerators and safely landfill.

2.2 Goal and scope definition

The research scope of the life cycle assessment should be according to the purpose of the research content, this work used eBalance software for the operation of the waste incineration power plant in the environmental impact assessment, and the scope of evaluation objects is set to from waste collection of transportation process to the all forms waste emissions, namely the main body of waste incineration power plant operation, and the specific scope boundaries are shown in Figure 1. Due to the immeasurable nature of equipment manufacturing and plant construction and the small degree of environmental impact, they are not considered in this work.

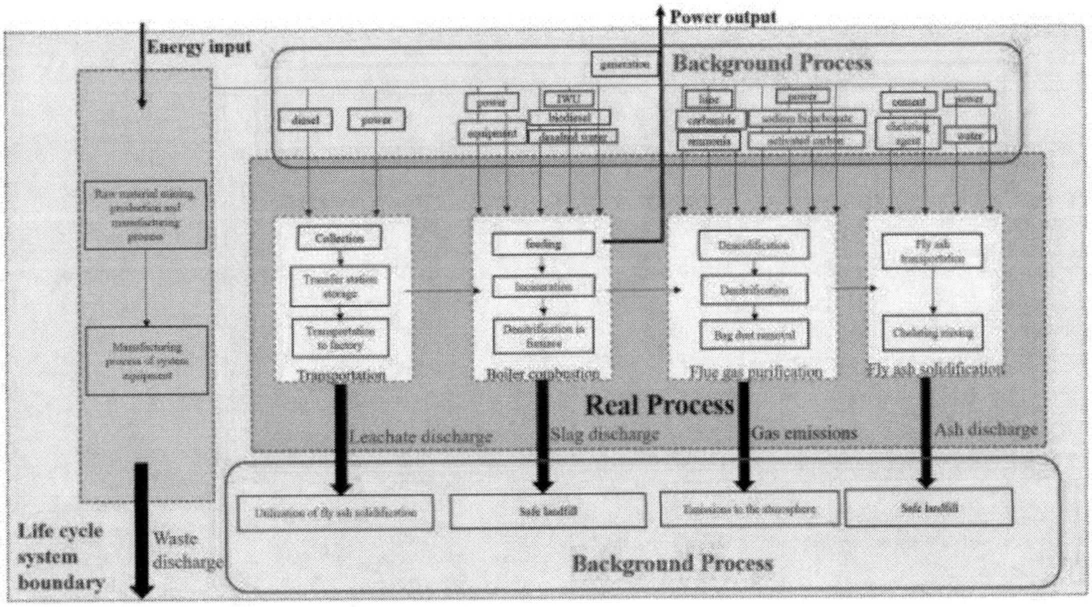

Fig. 1. The boundary of LCA in waste incineration power generation system

To facilitate the subsequent quantitative analysis, it is necessary to establish an appropriate calculation benchmark in the LCA evaluation model and calculate the input and output data in the modeling process. This work sets up a dual function unit reference flow from the perspective of waste treatment and power generation. (1) With waste treatment as the benchmark flow, the functional unit is set as the system processing

1t of mixed waste. (2) With power generation as the reference flow, the function unit is set as 1kWh electric energy produced by the system.

According to the system boundary, the real process and background process are divided. The data of the real process mainly come from the actual investigation of the factory and the results of large-scale experiments, while the data of the background process are selected from the database in the eBalance software and related papers. The database used in this model includes the built-in database of eBalance software: Chinese life cycle database (CLCD) and Ecoinvent, which follows the principle of giving priority to the local database in China.

2.3 Life cycle inventory

The waste incineration power generation process of this plant is divided into six major links: transportation and feeding - boiler combustion - steam turbine power generation - flue gas purification - fly ash solidification – ash and slag landfill. The plant operation data and the experimental data are selected to establish a material list database for each major link. All the power consumption and water consumption of the plant come from the real-time meter reading data, and all the substances added in the process also come from the accurate data records.

2.4 Environmental impact assessment

2.4.1 Classification

This work selects two types, namely resource consumption type and material emission type, with a total of 15 environmental impact indicators to comprehensively evaluate the energy consumption and environmental pollution performance of the system in low-carbon path planning.

2.4.2 Characterization results

Since different pollutants have different contributions to the environmental impact of their classification and do not meet the same dimension, they cannot be directly accumulated. Therefore, it is necessary to introduce the characteristic factor representing the impact degree with the reference pollutant as the coordinate. The characterization results can directly reflect the quantitative values of different environmental impact indicators, and facilitate the comparison of multiple indicators among different process systems.

2.4.3 Normalization results

The characteristic results of different dimensions of environmental impact indicators cannot be directly compared, but the normalization of the characteristic results can be realized. In this work, eBalance built-in normalization scheme -- China 2010 Normalized Total (CN2010) was used. The utilization background of this scheme is to collect the proportion of the total emission or consumption of the main functional substances of various environmental impact indicators in China in 2010, which is the normalized base value to illustrate the contribution. Table 1 shows the normalization results of various environmental impact indicators in the life cycle of the waste incineration power generation system for consuming 1t mixed waste.

Table 1. Normalization results of waste incineration power generation system

Indicator	Normalization result
ADP	3.47E-11
AP	3.43E-11
CADP (fossil fuel)	1.23E-11
CADP	2.30E-11
PED	6.79E-12
CO_2	8.27E-12
COD	5.48E-12
EP	5.61E-11
GWP	6.87E-12
IWU	5.30E-12
NH_3-N	1.13E-12
NOx	4.53E-11
RI	1.68E-11
SO_2	7.35E-12
Waste Solids	1.91E-09
Water Use	1.28E-12

According to the results in Table 1, the normalization value of Waste Solids ranks first among all indexes, and its magnitude is much higher than other indexes. This is because the incineration of waste still produces about 25% slag and bottom ash for landfill treatment, which is still a huge number, so it is a very necessary and meaningful technical breakthrough direction for ash reduction. The second and third indexes are EP and NOx, which fully illustrate the importance of denitration in the waste incineration power generation system. How to reduce the amount of ammonia, urea and other substances while maintaining or even reducing NOx emissions is a technical difficulty that needs to be overcome. The order of all indexes is Waste Solids>EP>NOx>ADP>AP>CADP>RI> CADP (fossil fuel) > CO_2> SO_2>GWP>PED> COD>IWU> Water Use > NH_3-N.

2.4.4 Comprehensive evaluation index calculation

Normalization results of environmental impact indicators do not combine indicators and reduce the number of indicators. When there are multiple indicators, they often conflict with each other, which makes the analysis results of LCA inconclusive. To further consolidate the indicators, in this work, two integrated evaluation calculation schemes built into eBalance are used for calculation. The first scheme is the Index of sustainable consumption and production (ISCP), which includes eight environmental impact indicators (GWP, EP, AP, COD, RI, CADP, Water Use and Waste Solids). The weight is calculated by the expert scoring method and the analytic hierarchy process. The second scheme is the energy conservation and emission reduction comprehensive index (ECER), which includes seven environmental impact indexes (PED, SO2, NOx, COD, CO2, NH3-N and IWU). ECER is a policy indicator, which is selected according to the policy target clause "reduction rate per unit GDP" in China's "14th Five-year Plan" for National Economic and Social Development. Therefore, ECER plays a strong role in process comparison, enterprise evaluation and policy suggestions. The Pareto diagram analysis of the ECER index of the waste incineration power generation system is shown in Figure 2.

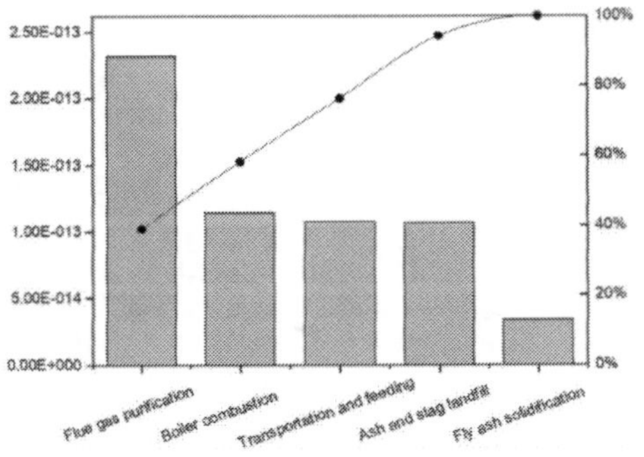

Fig. 2. Pareto chart in different links of ECER indicator of waste incineration power generation system

As can be seen from Figure 2, the contribution of flue gas purification to ECER index reaches 38.98%, which is the most influential link and significantly ahead of other links. The contribution ratio of boiler combustion, transportation and feeding, ash and slag landfill is nearly the same, and the contribution ratio of fly ash solidification is the smallest. The steam turbine power generation link is not included in the statistics because it is the link that produces income to the index. The flue gas purification link has a huge impact and needs to be further analyzed. Therefore, the ECER contribution ratio of related products in this link is statistically analyzed as shown in Figure 3. Products with a contribution ratio of less than 0.1% are not included in the statistics.

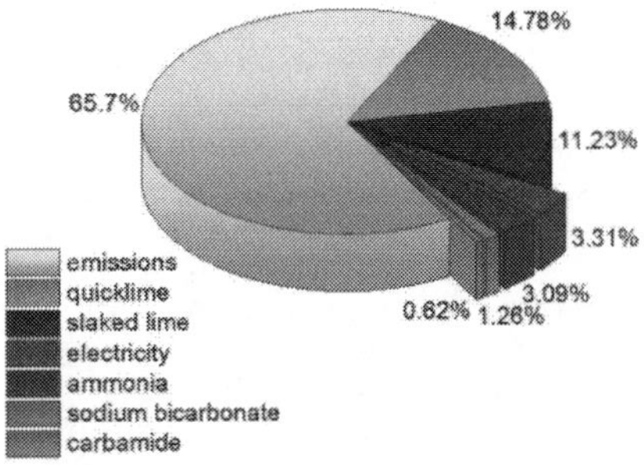

Fig. 3. Proportion of ECER indicator contribution of related products in flue gas purification

According to the contribution ratio of related products in the flue gas purification link in Figure 3, it can be seen that the flue gas directly emitted after treatment accounts for the largest contribution ratio, reaching 65.7%, which is much higher than the contribution of other products. Therefore, the direct emission of flue gas is the key factor affecting the overall environment. However, in the waste incineration power generation system, all flue gas can be discharged only after being strictly treated to meet the requirements of the national standard, and the impact on the environment is already relatively small, thus indicating that the environmental impact caused by other links of the system is smaller, and the specific numerical comparison of the environmental impact of different processes in the next part can further illustrate the superiority of the waste incineration power generation system.

3 Discussion and suggestions

The results of LCA show that, from the comparison of environmental impact indicators, the value of EP indicators is very prominent, far exceeding other indicators, and it should be optimized most. Incomplete denitration will cause a certain amount of NOx to be discharged into the atmosphere in the form of gas. Meanwhile, a large amount of ammonia water, urea and other substances are added in the denitration process, which are the contribution sources of EP indicators. Therefore, the processing system should establish a more stringent emission control system, optimize denitration technology, and carry out the double handle from the input side of additives and NOx emission side to reduce the environmental load.

4 Conclusions

This study was conducted at a large waste incineration power plant in Guangdong Province, China as the studied. And it aims to explore the possibility of using a large proportion of ISW for incineration power generation from the perspective of environmental impact assessment. Firstly, the waste mixed with ISW and MSW is taken as the research object, and the LCA model with the boundary from waste collection and transportation process to all kinds of waste discharge process is established by selecting a variety of environmental impact indicators, and the calculation results of the system are characterized, normalized and weighted comprehensively calculated. Results show that for waste incineration power generation system, the numerical order of all normalization indicators from largest to smallest is Waste Solids, EP, NOx, ADP, AP, CADP, RI, CADP (fossil fuel), CO_2, SO_2, GWP, PED, COD, IWU, Water Use, NH_3-N. It shows that the ash reduction and denitration system is most worthy of optimization, which can be carried out by the ash recycling system. And the contribution of flue gas purification to ECER index reaches 38.98%, which is the most influential link and significantly ahead of other links, as well as the most needed optimization link.

This paper is one of the phased achievements of the National Natural Science Foundation of China (general project), "Study on the Mechanism of Low-Temperature Cascade Pyrolysis of Medical Waste Based on Interaction and Secondary Reaction Enhancement" (52076092).

References

1. Global Waste Management Outlook. UNEP, Nairobi(2015).
2. J. Malinauskaite, H. Jouhara, D. Czajczyńska, P. Stanchev, E. Katsou, P. Rostkowski, R.J. Thorne, J. Colón, S. Ponsá, F. Al-Mansour, L. Anguilano, R. Krzyżyńska, I.C. López, A. Vlasopoulos, N. Spencer. Municipal solid waste management and waste-to-energy in the context of a circular economy and energy recycling in Europe. Energy, **141**, 2013-2044 (2017).
3. H. Fekete, K. Riahi, et al. Paris Agreement climate proposals need a boost to keep warming well below 2 degrees C. Nature (2016).
4. China's Ecological and Environmental Protection Plan in the 13th Five-Year Plan (2016).
5. Y. Guan, G. Huang, L. Liu, M. Zhai, B. Zheng. Dynamic analysis of industrial solid waste metabolism at aggregated and disaggregated levels. J. Clean. Prod., **221**, 817-827 (2019).
6. L. Liu, G. Huang, B. Baetz, C.Z. Huang, K. Zhang. A factorial ecologically-extended input-output model for analyzing urban GHG emissions metabolism system. J. Clean. Prod., **200**, 922-933 (2018).
7. D. Zhang. The standing committee of the National People's Congress law enforcement inspection group about the inspection of the People's Republic of China on the prevention and control of environmental pollution by solid waste report on the implementation of the - on November 1, 2017 at the 30th session of the standing committee of the 12th national people's congress. Bulletin of the Standing Committee of the National People's Congress of the People's Republic of China, 1001-1010, (2017).
8. S. Rahimi, A. Hafezalkotob, S. M. Monavari, et al. Sustainable landfill site selection for municipal solid waste based on a hybrid decision-making approach: Fuzzy group BWM-MULTIMOORA-GIS. Journal of Cleaner Production, 248(Mar.1):119186.1-119186.15 (2020).
9. Z. Tang, W. Li, G. Ke, J.L. Zhou, V.W. Tam. Sulfate attack resistance of sustainable concrete incorporating various industrial solid wastes. J. Clean. Prod., **218**, 810-822 (2019).
10. Y.F. Gao, Z.F. Li, J. Zhang, et al.Synergistic use of industrial solid wastes to prepare belite-rich sulphoaluminate cement and its feasibility use in repairing materials. Conster Build Mater, **264**, 1-13 (2020).
11. J. Zhou, H. Gao, Z. Shu, Y. Wang, C. Yan.Utilization of waste phosphogypsum to prepare non-fired bricks by a novel Hydration–Recrystallization process. Constr Build Mater, **34**, 114-119 (2012).
12. Z. Tang, W. Li, G. Ke, J.L. Zhou, V.W.Y. Tam. Sulfate attack resistance of sustainable concrete incorporating various industrial solid wastes. J Clean Prod, **218**, 810-822 (2019).
13. R. Liu, F. Huang, R. Du, C. Zhao, Y. Li, H. Yu. Recycling and utilisation of industrial solid waste: an explorative study on gold deposit tailings of ductile shear zone type in China. Waste Manag Res, 33, 570-577 (2015).
14. H. Chen, B. Li, M. Zhao, X. Zhang, Y. Du, Y. Shi, J.S. McCloy. Lanthanum modification of crystalline phases and residual glass in augite glass ceramics produced with industrial solid wastes. J. Non-Cryst. Solids, **524**, 1-10 (2019).
15. H.A. Abdel Gawwad, M.S. Mohammed, M. Heikal. Ultra-lightweight porous materials fabrication and hazardous lead-stabilization through alkali-activation/sintering of different industrial solid wastes. J Clean Prod, **244**, 1-12 (2020).
16. Y. Ding, J. Zhao, J.W. Liu, et al. A review of China's municipal solid waste (MSW) and comparison with international regions: Management and technologies in treatment and resource utilization. Journal of Cleaner Production, **293(10)**: 126144 (2021).
17. M.M. Mian, X. Zeng, A.a.N.B. Nasry, S.M.Z.F. Al-Hamadani. Municipal solid waste management in China: a comparative analysis. J. Mater. Cycles Waste Manag., **19 (3)**, 1127-1135 (2016)

Adsorption of Tetracycline by Ammonium Molybdate Modified Chinese Fir Biochars

Caimei Lu[1], *Hanbing* Zhang[1,*], *Rui* Tang[2], *Sishan* Yu[1] and *Zehua* Li[1]

[1]College of Resources, Environment and Materials, Guangxi University, Nanning 530004, Guangxi, China;
[2]Guangxi Colleges and Universities Key Laboratory of New Technology and Application in Resource Chemical Engineering, School of Chemistry and Chemical Engineering, Guangxi University, Nanning 530004, Guangxi, China

Abstract. Tetracycline (TC) adsorption amount on original wood biochar was finitely, so modification is imperative to improve its adsorption capacity. This study mixed chinese fir sawdust with ammonium molybdate to prepare a modified chinese fir biochar (MCB) via pyrolysis at 800 ℃. Meanwhile, pristine chinese fir biochar (CB) was produced as control. The effects of pH, adsorption time, initial concentration and adsorbent dosage on the adsorption process were also studied. Langmuir model and quasi second-order kinetic model were used describe the characteristics of adsorption process. The optimally capacity of TC adsorption on MCB reached 36 mg/g, which was 6 – fold increase to CB. Therefore, modification by ammonium molybdate can be used as a potential way of chinese fir sawdust biochar for increasing adsorption capacity of TC from antibiotic sewage water.

1 Introduction

Till now, antibiotics have been diffusely used worldwide, especially in China. Antibiotics are generally used in not only for human therapies as an important class pharmaceuticals, but also for aquaculture and animal husbandry as the feed additive to enhance the growth efficiency of animals. But in the meanwhile, it is inevitable to discharge into aquatic environments for the overuse as well as the slow metabolism of antibiotics, resulting in refractory pollution sources and potential harm to human health. Among all of the antibiotics, tetracycline (TC) is one of the most widely used antibiotics in both animal and human treatment. However, what is noteworthy is that human beings and animals could not break down tetracycline basically by normal metabolism, and a great deal of tetracycline was released into the environment with the discharge of sewage. Misuse of TC will increased antimicrobial resistance of microorganism lead to increase potential threats to ecosystem's food chain, and consequently to human health. Therefore, it is necessary to not only avoid the abuse of antibiotics but also to effectively remove them from water [1].

Up to now, many technologies have been served for removing TC from the aquatic environment, such as biodegradation, advanced oxidation, photolysis and adsorption [2]. Among them, adsorption treatment is considered as one of the most effective methods and is expected to be widely used in practical applications, because of its relatively cheap, easy operation and environmental friendliness. The selection of adsorbent is the key of adsorption technology. In recent years, carbon-based materials have been discovered by researchers to exhibit good performance in adsorbing antibiotics from aqueous solution. However, the high cost and complex preparation process of most carbon materials limit their wide application [3].

Of carbon-based adsorption materials, biochar is obtained by pyrolysis of waste or inexpensive biomass under an inert atmosphere. As an environmentally friendly adsorbent, biochar has gained importance due to wide raw material sources, low-cost preparation, having diverse functional groups and a large surface area. There have been some scholar used biochar as a sorbent to treat antibiotic wastewater [4]. However, the surface area, pore size and surface active group of the original biochar are limited, so further activations or modification are required to increase their adsorptive property [5]. Acid and alkali activation are common methods for the production of porous biochar, but it is easy to corrode the equipment during the preparation. Therefore, looking for eco-friendly raw material and modifier or activator is the mainstream of the present study.

Chinese fir is a unique tree species in China and is widely used in papermaking, construction, furniture and shipbuilding. At the same time, in a large number of chinese fir sawdust was generated during production. To be environmentally friendly, chinese fir sawdust can be used as raw materials for biochar preparation instead of discard. In addition to their large output, chinese fir sawdust are rich in lignin, cellulose and hemicellulose, which make it a promising raw material of biochar adsorbent [6]. Currently, few reports have investigated the pyrolysis treatment and modification of chinese fir sawdust. Daniela et al[7] prepared zinc chloride-activated

* Corresponding author: coldicezhang0771@163.com

chinese fir carbon as an adsorbent for p-nitrophenol removal from wastewater. However, there is no report on the preparation of chinese fir biochar for antibiotic wastewater treatment. Hence, it is important to investigate how to produce high value-added chinese fir biochar adsorbent by modification and realize low-cost treatment for antibiotic removal from wastewater.

In this study, choosing eco-friendly inorganic ammonium molybdate as a modifier, chinese fir sawdust was mixed with ammonium molybdate to prepare an ammonium molybdate modified chinese fir sawdust biochar (MCB) at 800 °C. Meanwhile, chinese fir biochar (CB) was also prepared as control. The adsorption performance of TC on the MCB is studied. The effects of adsorption time, pollutant concentration, adsorbent dosage and pH on the adsorption performance are then discussed. The adsorption kinetics, isotherms of TC on biochars were further studied, and the adsorption mechanism of TC on MCB was also revealed. This research can provide reference for the exploitation of chinese fir sawdust resources and the removal of TC from antibiotic sewage water.

2 Materials and methods

2.1 Raw materials

The chinese fir sawdust were collected from Guangxi, China. Chinese fir sawdust was washed 3 times by water and then dried in oven at 60℃ for 24 h, stored in the desiccator. Ammonium molybdate was procured from Guangdong Guanghua Sci-Tech Co., Ltd. Tetracycline (>95% purity) was obtained from Shanghai Macklin Biochemical Co., Ltd. HCl (36%) was purchased from Xilong chemical Co., Ltd. The solutions were prepared using ultrapure water (Milli-Q, Germany).

2.2 Preparation of biochar

Initially, 5 g chinese fir sawdust was mixed with 5 g ammonium molybdate in 100 mL ultrapure water stirred for 4 h, and then the product was rinsed with ultrapure water 3 times and dried at 80℃ for 12 h. Secondly, the solid materials were pyrolyzed in the pipe furnace at 800℃ for 2 h (under nitrogen atmosphere and heating rate was 10 °C/min). The obtained carbonaceous material was washed by 2 M HCl in order to eliminate inorganic components and then washed with ultrapure water until neutral pH. Finally, the obtained biochar products were dried at 80 °C for 12 h, ground, sieved through a 200-mesh sieve. The resulting samples were named MCB, where M stands for modification. By contrast, biochar without modification was prepared by Chinese fir sawdust pyrolysis in the pipe furnace at 800℃ for 2 h. Thereafter, the biochar was washed, dried, ground followed the above method, and noted CB.

2.3 Batch adsorption studies

Stock solutions of TC (100 mg/L) was prepared, from which working solutions were prepared in sequence during follow-up experiment.

Batch adsorption experiments were performed in conical flasks with wobbling (140 rpm) in a water bath shaker at room temperature (25±1°C). Experimental parameters including contact time (0 ~ 240 min), initial TC concentration (10 ~ 50mg/L), adsorbent dosage (0.6 ~ 1.4 g/L) and solution pH (3 ~ 11) were studied in details. The initial pH of the TC solution was transformed with 0.2 M HCl and NaOH solution. Each batch test was carried out triplicates to ensure repeatability. After adsorption, the solid-liquid mixtures were filtered by 0.45 μm membrane and then measured with UV-2550 at a wavelength of 356 nm [8].

2.4 Adsorption isotherms

The amount of TC sorbed onto adsorbents was calculated by the mass balance equation as below:

$$q_e=\frac{C_0-C_e}{m} \tag{1}$$

$$R=\frac{(C_0\text{-}C_t)}{C_0\times 100\%} \tag{2}$$

C_0 is the initial concentration of TC (mg/L). C_t is the concentration of the supernatant of TC (mg/L) was adsorbed at t (min). m is the dosage of biochar (g/L). q_e is the adsorption capacity of biochar for TC (mg/L). R is the removal efficiency (%) of TC on biochar.

To understand the characteristics of the adsorption process, adsorption kinetics included quasi first-order kinetic model (Eq. (3)) and quasi second-order kinetic model (Eq. (4) were performed with an initial TC concentration of 30 mg/L and a biochar dosage of 1 g/L.

$$\ln(q_e-q_t)=\ln q_e-k_1 t \tag{3}$$

$$\frac{t}{q_t}=\frac{1}{k_2 q_e^2}+\frac{t}{q_e} \tag{4}$$

q_e and q_t are the adsorption capacity of TC (mg/L) at equilibrium and t (min) respectively; k_1 (min^{-1})and k_2 (g/mg/min)are the degradation rate constant of quasi -first-order and quasi second-order kinetic model respectively.

Besides, the isotherm models studies of TC adsorption was performed by stirring 50 mL TC (10–50 mg/L) in 100 mL conical flasks for 180 min until reaching equilibrium, and the adsorption date was fitted via adsorption Freundlich (Eq. (5)) and Langmuir (Eq.(6)) equations.

$$\ln q_e=\ln K_f+\frac{1}{n}\ln C_e \tag{5}$$

$$\frac{C_e}{q_e}=\frac{C_e}{q_m}+\frac{1}{K_L q_m} \tag{6}$$

q_e is the adsorption capacity of TC (mg/L) at equilibrium; C_e is the concentration of TC (mg/L) at equilibrium; K_f (mg/L) is the Freundlich model adsorption capacity (mg/g); n is the heterogeneity factor; q_m is the maximum adsorption capacity of biochar (mg/g); and K_L (L/mg) is the Langmuir constant.

3 Results and discussion

3.1 Characteristics of the adsorbent

The XRD patterns of the samples after TC adsorption was showed in Fig. 1. Two broad peaks at 22.63 and 43.54 which corresponding to amorphous carbon (002) and graphite structure (100) planes, respectively [9]. Comparing the XRD patterns of CB, MCB give distinct crystal peaks corresponding to Mo_2C (JCPDS card No. 71–0242). This result indicated that the Mo ion was successfully loaded in the MCB as Mo_2C. While after TC adsorption, the typical peaks of MoO_2 could be observed in the XRD spectra of spent MCB that may be due to the reaction between Mo_2C and TC in the adsorption process or oxidation during the spent MCB recovery.

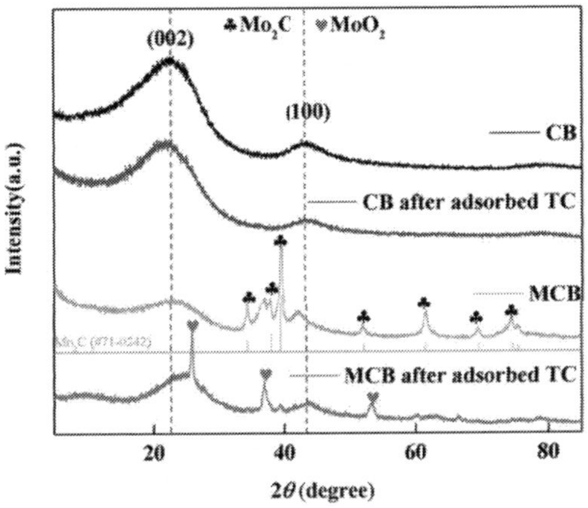

Fig. 1. XRD spectra of various biochar samples before and after TC adsorption

The surface functional groups in CB and MCB before and after TC adsorption were showed by FTIR spectroscopy (Fig. 2), which confirmed the existence of characteristic peaks of sawdust biochar including 1648 cm^{-1} assigned to the stretching vibration of C=O for carbonyl and carboxyl groups, the band at 1426 cm^{-1} ascribed as the aromatic rings of C=C of the unsaturated carbon structure, and 1133 cm^{-1} ascribed to C-O [10]. A broad band at about 3200-3600 cm^{-1} in samples was observed and corresponded to the stretching vibration of the -OH from hydroxyl. Upon treatment with ammonium molybdate, the intensity of the peaks at 2370 cm^{-1} ($-CH_3$) was increased and the new peaks at 680 cm^{-1} was observed and corresponded to the Mo-C, that is result of the generation of Mo_2C. After the adsorption of TC on MCB, the intensity of the peaks at 2370 cm^{-1} ($-CH_3$) was decreased may be due to Mo_2C was oxidized to MoO_2 during the adsorption process. This result from FTIR spectra coincide with the XRD.

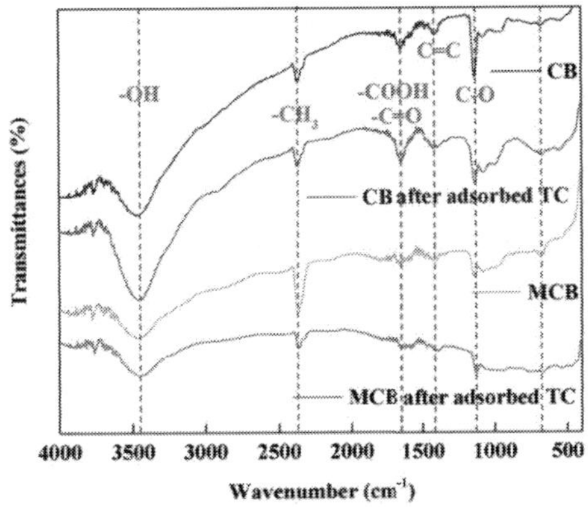

Fig. 2. FTIR spectra of various biochar samples before and after TC adsorption

3.2 Effect of contact time

From Fig. 3, it is observed that the effect of contact time on TC (30 mg/L) removal by MCB and CB. TC adsorption on MCB and CB were rapid initially and then relatively slowed down, finally attains equilibrium. For MCB, the TC adsorption efficiency reached approximately 35% within 5 min, equilibrium was achieved in 180min and the maximum adsorption efficiency attained more than 85%. Unlike MCB, the maximum TC adsorption efficiency of CB is only about 20%. The possible reason is that higher surface area and more available binding sites existed in biochar after modification [11].

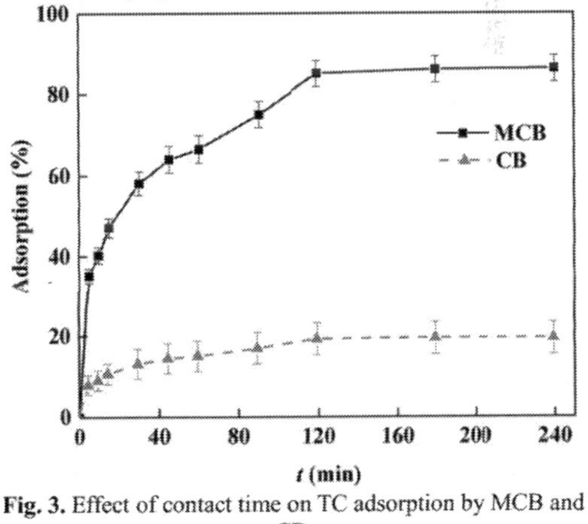

Fig. 3. Effect of contact time on TC adsorption by MCB and CB.

3.3 Effect of initial concentration

The influence of initial concentration of TC was studied by increasing the amounts of TC concentration from 10 to 50 mg/L. As we can see from Fig. 4, the TC adsorption efficiency decreased as the initial concentration increased. That is result of a limited amount of adsorption sites on the surface of biochar. In the concentration extent of 10 to 50 mg/L, the adsorption efficiency of MCB reached 70 to 100 %, while that of CB only reached about 10 to

 https://doi.org/10.1051/e3sconf/202235003008

20 %, which indicates the modification can effectively improve the adsorption capacity of biochar for TC.

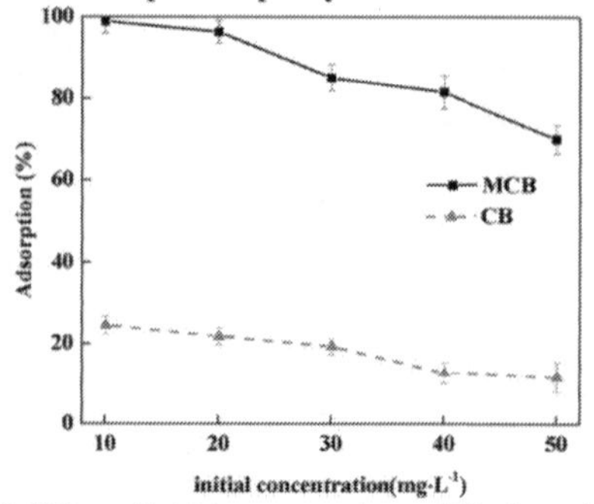

Fig. 4. Effect of initial concentration on TC adsorption by MCB and CB.

3.4 Effect of adsorbent dosage

Fig. 5 shows a similar regularity of MCB and CB adsorbed TC about the effect of adsorbent dosage. With increasing the biochar dosage, the adsorption efficiency of biochar improved to a certain extent. However, excessive dosage could decrease the adsorption capacity due to the aggregation effect of biochar [12]. Suitable dosages that can be used in TC removal is 1.0 g/L.

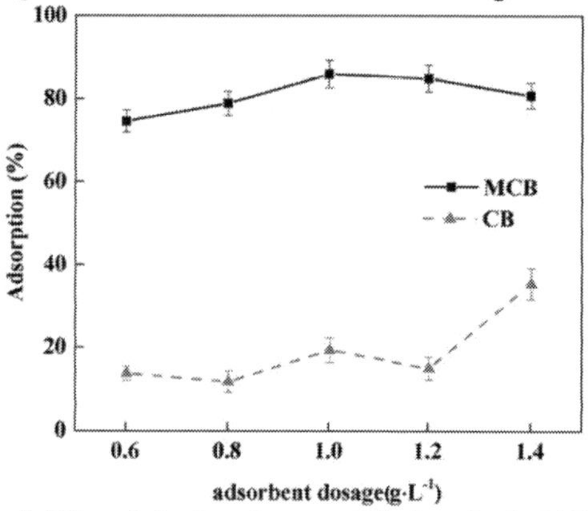

Fig. 5. Effect of adsorbent dosage on TC adsorption by MCB and CB.

3.5 Effect of pH

The TC solution initial pH obviously influences the adsorption consequence by changing surface characteristics of both adsorbent and adsorbate, because of solution pH not only affects surface charge of biochar but also change the existence form of TC. The adsorption efficiency of TC as a function of pH was presented in Fig. 6. The TC initial pH is 6.5, and TC adsorption quantity increased at initially in the pH range of 3.0 to 5.0, then significant drop with the pH increased from 7.0 to 11.0. As a result of TC is an amphipathic molecule and exists as TC^{+} at a pH value of less than 8.0, TC^{-} and TC^{2-} at pH higher than 8.0, that resulting in electrostatic attraction between TC and biochar, and an increase in the adsorption capacity of biochar [13]. On the contrary, significantly decreased adsorption ability due to electrostatic repulsion between adsorbent and adsorbate under pH 9.0 to11.0.

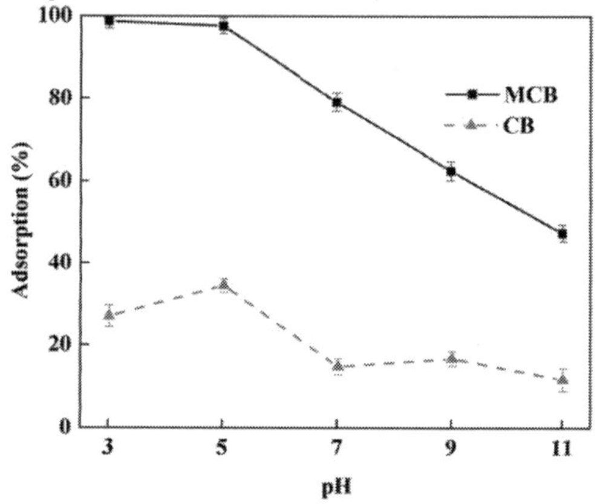

Fig. 6. Effect of pH on TC adsorption by MCB and CB.

3.6 Adsorption isotherms

The constants of isothermal adsorption for TC is shown in Table 1. TC adsorption on MCB and CB fit Langmuir model well with R^2 0.996 and 0.979, and the Langmuir adsorption capacities of MCB and CB for TC were 36.0 mg/g and 7.5 mg/g, respectively. The theoretical q_m of MCB is very close to the experimental value and it demonstates that this adsorption process probably was monolayer adsorption that related to chemisorption [14]. The K_L values of TC adsorbed by MCB was much higher than those on CB, indicating that MCB had higher affinity and stronger adsorption capacity of TC.

Table 1. Fitting parameters of TC adsorption isotherm models of MCB and CB.

	Langmuir model			Freundlich model			$q_{m,exp}$/mg·g^{-1}
	K_L/L·mg^{-1}	$q_{m,cal}$/mg·g^{-1}	R^2	K_f/mg·g^{-1}	1/n	R^2	
MCB	1.19	36.0	0.996	18.37	0.46	0.958	35.02
CB	0.07	7.5	0.979	1.07	0.23	0.915	6.01

The parameters of kinetic models for adsorption of TC on the biochar are shown in Table 2. The data for the TC showed a better adjustment to the quasi second-order kinetic model as supported by the higher R^2. Hence, the adsorption of TC on MCB and CB occurs through chemisorption mainly and is a process of fast initially, slowed down then and finally reached equilibrium [15]. For MCB, adsorption of TC is enhanced as the chemical interactions are further improved because of the high surface area and more functional groups imparted on the biochar by modifier. Table 3 shows that MCB has been compared with other reported adsorbents for TC adsorption.

Table 2. Fitting parameters of TC adsorption kinetic models of MCB and CB.

	quasi first-order kinetic model			quasi second-order kinetic model			
	k_1×0.01/ min^{-1}	$q_{e,cal}$/ $mg \cdot g^{-1}$	R^2	k_2×0.01/ $g \cdot mg^{-1} \cdot min^{-1}$	$q_{e,cal}$/ $mg \cdot g^{-1}$	R^2	$q_{e,exp}$/ $mg \cdot g^{-1}$
MCB	0.68	23.2	0.903	8.19	37	0.992	35.02
CB	0.55	4.6	0.912	1.39	6.12	0.984	6.01

Table 3. Comparison of adsorption performance of various adsorbents for TC removal.

Targeted pollutant	Adsorbent	Adsorption conditions	q_{max} (mg/g)	Ref
TC	Zinc (II) modified hydroxyapatites	20 °C, pH 5, dosage 25 g/L, concentration 0.6 mmol/L	16.5	[16]
	Magnetic sludge bio char	25 °C, pH 5, dosage 0.2 g/L, concentration 200–500 mg/L	145	[17]
	Shell powder that was coated with ZnO nanoparticles	30 °C, pH 5, dosage 1 g/L, concentration 100 mg/L	99	[18]
	Grapefruit peel biochar	25 °C, pH 7, dosage 1 g/L, concentration 50 mg/L	32.5	[19]
	Cow manure biochar		11.7	[20]
	MCB	25 °C, pH 5, dosage 1 g/L, concentration 10–50 mg/L	36	This study

4 Conclusions

MCB was prepared via a one-pot pyrolysis method at 800 ℃. Modification increased the hydrophilicity and surface area of biochar. The optimal adsorption capacity of TC on MCB is 36 mg/g, which was attributed to pore filling and electrostatic interaction occurred on surface of MCB. Acidic conditions benefited TC adsorption on MCB and the adsorption process fit Langmuir model and quasi second-order kinetic model well. Furthermore, the practical application of MCB should be performed in the future studies.

Acknowledgment

The authors acknowledge the financial support by the National Natural Science Foundation of China (Grant number 22168004), Petrochemical Resources Processing and Process Reinforcement Technology Key Laboratory Project of Guangxi province (Grant numbers 2020Z008) and National Natural Science Foundation of Guangxi province (Grant number 2020GXNSFAA297036).

References

1. J. Zhao, G.W. Liang, X.L. Zhang, et al. Coating magnetic biochar with humic acid for high efficient removal of fluoroquinolone antibiotics in water. Sci. Total. Environ. **688**,1205-1215 (2019)
2. N. Cheng, B. Wang, P. Wu et al. Adsorption of emerging contaminants from water and wastewater by modified biochar: A review. Environ Pollut. **273**, 116448 (2021)
3. C.E.R. Barquilha, M.C.B. Braga. Adsorption of organic and inorganic pollutants onto biochars: Challenges, operating conditions, and mechanisms. Bioresour Technol. **15**, 100728 (2021)
4. Y.F. Ma, M. Lia, P. Li et al. Hydrothermal synthesis of magnetic sludge biochar for tetracycline and ciprofloxacin adsorptive removal. Bioresour Technol. **319**, 124199 (2021)
5. X.G. Zheng, X.Y. He, H. Peng et al. Efficient adsorption of ciprofloxacin using Ga2S3/S-modified biochar via the high-temperature sulfurization. Bioresour Technol. **334**, 125238 (2021
6. C.Y. Chen, W.H. Chen, S. Lim, et al. Synergistic interaction and biochar improvement over co-torrefaction of intermediate waste epoxy resins and fir. Environ Technol Inno. **21**, 101218 (2021)
7. D.R. Popovici, M. Neagu. C.M. Dusescu-Vasile. et al. Adsorption of *p*- nitrophenol onto activated carbon prepared from fir sawdust: isotherm studies and error analysis. J. React Kinet Mech Cat. **133**, 483-500 (2021)
8. K. Liu, Z.F. Tong, Y. Muhammad, et al. Synthesis of sodium dodecyl sulfate modified BiOBr/magnetic bentonite photocatalyst with Three-dimensional parterre like structure for the enhanced photodegradation of tetracycline and ciprofloxacin. Chem. Eng. J., **388**, 124374 (2020)
9. J. He, Y. Xiao, J.C. Tang, et al. Persulfate activation with sawdust biochar in aqueous solution by

enhanced electron donor-transfer effect. Sci. Total. Environ. **690**, 768-777 (2019)

10. Y.Z Zhang, M.Q Xu, S.X. Liang, et al. Mechanism of persulfate activation by biochar for the catalytic degradation of antibiotics: Synergistic effects of environmentally persistent free radicals and the defective structure of biochar. Sci. Total. Environ. **794**, 148707 (2021)
11. C.Y. Li, X.X. Zhu, H.L. He, et al. Adsorption of two antibiotics on biochar prepared in air-containing atmosphere: Influence of biochar porosity and molecular size of antibiotics. J. Mol. Liq. **274**, 353-361 (2019)
12. P. Verma, T. Watanabe, Y. Honda, et al. Microwave-assisted pretreatment of woody biomass with ammonium molybdate activated by H_2O_2. Bioresour Technol. **102**, 3941-3945 (2011)
13. Z.K. Wang, Y. Muhamm, R. Tang, et al. Dually organic modified bentonite with enhanced adsorption and desorption of tetracycline and ciprofloxacine. Sep. Purif Technol. **274**, 119059 (2021)
14. F.X. Dong, L. Yan, X.H. Zhou, et al. Simultaneous adsorption of Cr(VI) and phenol by biochar-based iron oxide composites in water: Performance, kinetics and mechanism. J. Hazard Mater. **416**, 125930 (2021)
15. R. Tang, Z. Wang, Y. Muhammad, et al. Fabrication of carboxymethyl cellulose and chitosan modified Magnetic alkaline Ca-bentonite for the adsorption of hazardous doxycycline. Colloids Surf. A, **610**, 12573 (2021)
16. C. Oliveira, A.L.M. de Oliveira, L. Chantelle, et al. Zinc(II) modified hydroxyapatites for tetracycline removal: Zn(II) doping or ZnO deposition and their influence in the adsorption, Polyhedron, **194**, 114879 (2021)
17. Y. Ma, M. Li, P. Li, et al. Hydrothermal synthesis of magnetic sludge biochar for tetracycline and ciprofloxacin adsorptive removal, Bioresour Technol, **319**, 124199 (2020)
18. A. A. Mohammed, T. J. Al–Musawi, S. L. Kareem, et al. Simultaneous adsorption of tetracycline, amoxicillin, and ciprofloxacin by pistachio shell powder coated with zinc oxide nanoparticles, Arab J Chem, **13**, 4629–4643 (2020)
19. D.W. Zhang, Q.Q. He, X.L. Hu, et al Activation of grapefruit derived biochar by its peel extracts and its performance for tetracycline removal, Bioresour. Technol., **316**, 123971 (2021)
20. P. Zhang, Y. Li, Y. Cao, et al. Characteristics of tetracycline adsorption by cow manure biochar prepared at different pyrolysis temperatures, Bioresour. Technol., **285**, 121348 (2019)

Acid Activated Bentonite-La Crosslinked Sesbania Gum Beads for Efficient Congo Red Adsorption

Sishan Yu[1], *Caimei* Lu, *Rui* Tang[2], *Qiuyu* Luo[1], *Mingyong* He[1,*] and *Hanbing* Zhang[1]

[1]College of Resources, Environment and Materials, Guangxi University, Nanning 530004, China
[2]Guangxi Colleges and Universities Key Laboratory of New Technology and Application in Resource Chemical Engineering, School of Chemistry and Chemical Engineering, Guangxi University, Nanning 530004, Guangxi, China

Abstract. In this study, to recycle acid activated bentonite (AAB), novel hybrid beads were prepared from AAB and sesbania gum (SG), which might be an efficient adsorbent for removing an anionic dye Congo red (CR). To gain the greatest adsorption conditions of AAB-SG-La beads, experiments were carried out to evaluate the effects of initial pH value, AAB-SG-La beads dosage on CR removal. The optimal pH value of AAB-SG-La beads for CR elimination was in the range 4.0 ~ 8.0, which can broaden the applications of beads. In addition, the adsorption isotherm datas were consistent with the Langmuir adsorption model and adsorption kinetics conformed to the pseudo-second-order model. Furthermore, the adsorption capacity of CR was 588 mg/g, which revealed a superior adsorption capacity to other materials used as sorbents for CR. This work accredits that AAB-SG-La beads have the ability to remove anionic dye as an economical adsorbent.

1 Introduction

Due to the development of industry, some industries involved in cosmetics, medicine, and textiles emission an abundance of dye wastewater. In particular, In particular, Congo red is a representative anionic dye. This dye was harmful to human health, for instance causing eyes damaging, gastrointestinal malaise and cancer [1].

Various techniques are being used to eliminate dyes from aqueous waste, for instance, biological degradation, membrane filtration, photocatalytic oxidation process, electrochemical method, ion exchange, and adsorption. Above all, adsorption is considered one of the most potential technologies. It has great superiority in its high efficiency, low cost, and easy operation.

It's known that clay minerals such as bentonite have a great adsorption effect on dyes, which are excellent adsorbents, mainly due to their physicochemical properties, high surface area, and ion-exchange process. However, bentonite exist negative charge, which is more beneficial to adsorb basic dyes than acid dyes. Surface modification of bentonite via acid activation enhances the adsorption efficiency of acid dye[2]. Nonetheless, when bentonite needs to be used in practical application, the main obstacle is to deal with solid-liquid separation[3].

Many researchers have studied recovery of acid activated bentonite (AAB) using to sodium alginate,which is the natural biopolymer[4]. Newly, natural polymers have the focus of interest in the field of aqueous treatment as a result of their distinctive advantages: numerous active groups for instance hydroxyls, biodegradability, and nontoxic nature to the environment. Sesbania gum (SG) regard as a natural polymer obtained from the endosperm of sesbania seeds. Its main structure involves D-galactose and D-mannose, which are composed of 1:2. and it has biocompatibility, biodegradability and water solubility. SG can be acted as a novel adsorbent for diverse toxic and harmful compounds due to the functional groups like plenty of hydroxyls[5]. Acid activated bentonite composition with Sesbania gum beads predigests solid-liquid separation disposal compared with the use of fine clay materials.

Herein, we invested objectives of this study that is to prepare an easy and green of composite beads from the combination of acid activated bentonite-sesbania gum beads via a simple fabrication method; to observed the effect of the removal of CR on adsorbent dosage, pH; moreover, to explore the adsorption performance toward CR containing adsorption kinetics as well as adsorption isotherm.

2 Materials and Methods

2.1 Materials

Bentonite was got from Sinopharm Chemical Reagent Co., Ltd, sesbania gum (SG), Congo red (CR), lanthanum nitrate were obtained from Macklin. The reagents used in this study were of analytical grade, and all solutions were made up with deionization water.

2.2 Preparation of AAB-SG-La beads

* Corresponding authors: mingyong0771@163.com.

AAB was prepared according to our previous research [6]. Briefly,1 g of AAB was added to 30mL of 2%(w/v) sesbania gum solution, then vigorous stirring for 4 h until homogeneous. The mixture was decant into the syringe, then dropped into 100 mL 4%(w/v) lanthanum nitrate solution to form beads at stewing 24 h. The bead surface had remain lanthanum ions, which was removed by rinsing with deionized water. Dried, stored in a clean bottle for further use, and named the sample as AAB-SG-La beads. The fabrication process of AAB-SG-La beads is given in Fig.1.

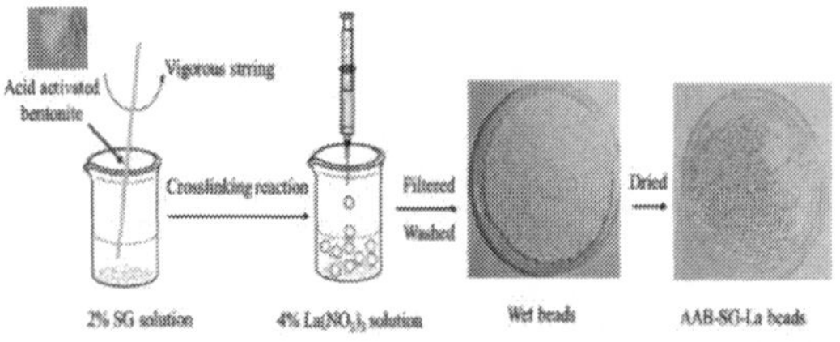

Fig. 1. The fabrication process of AAB-SG-La beads.

2.3 Characterization

The functional groups of the adsorbents were using KBr pellet technique and determined by Fourier transform infrared spectrometer (Nicolet Nexus 470 apparatus, Nicolet, United States) were recorded in a scanning range of 4000 ~ 400 cm^{-1}.

2.4 Adsorption experiments

Batch adsorption experiments on the removal of CR by AAB-SG-La beads dosage, which performed with 100 mL of CR solutions of predetermined concentration (20 ~ 800 mg/L) in a series of 250 mL flasks, solution pH values (4~10), AAB-SG-La beads dosage (0.1~0.1 g/L) and reaction times (0~4 h). In this study, 0.1 mol/L HCl and NaOH solutions were used to adjust the pH. After each adsorption experiment, a certain volume (approximately 3 mL) of the aqueous solution was filtered through the 0.45μm filtering membrane, and the concentration of non-adsorbed CR was measured on the UV-visible spectrophotometer (UV-2550) with the maximum absorbance wavelengths of 497 nm. The percentage of dyes removal and The adsorption equilibrium capacity qe (mg/g) was measured by using the following Eq.(1) and (2), respectively:

$$R = (C_0\text{-}C_e)/C_0 \times 100\% \quad (1)$$

$$q = (C_0\text{-}C_e)V/m \quad (2)$$

Where, C_0(mg/L) and C_e(mg/L) are the initial concentrations of CR and the remaining CR concentration in the solution after adsorption. Each experiment was repeated three times and mean values were considered in the fitting process.

3 Results and Discussion

3.1 Effect of AAB-SG-La beads dosage

The influence of adsorbent doses of AAB-SG-La beads on the adsorption of CR was indicated in Fig.2. It was observed that the effect of adsorbent dose on CR removal from aqueous solution using CR concentration 300 mg/L, contact time 240 min, and initial pH of CR solution 7 at 25 °C. The removal of CR from 15.56 up to 99.57% with an increase of adsorbent dose from 0.1 to 0.1 g/L. It is an ordinary appearance and is shown in this investigation[7].

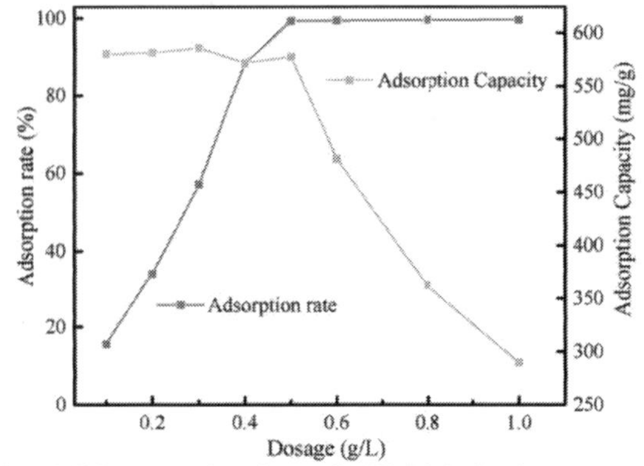

Fig. 2. Effect on various doses of AAB-SG-La beads on adsorption of CR.

3.2 Effect of adsorption on solution pH

Along with the surface properties of adsorbent in solutions will be effect by the initial pH value of CR. Fig. 3 described the role of pH on the adsorption of CR dye. For this, when the pH value on CR solution was grown from 4 up to 10, elimination of CR dropped to 99.65% from 19.45%. At high pH values, AAB-SG-La beads surface may be negatively charged, which lessened the adsorption of CR because of an electrostatic repulsion. On the contrary, AAB-SG-La beads surface may be positively charged in the acid region, making AAB-SG-La beads could be favorable to the adsorption of CR own to the electrostatic attraction. A similar report that holds this phenomenon in the literature[8].

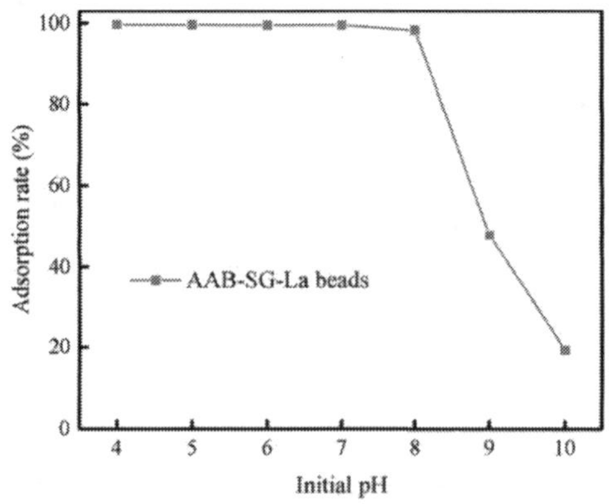

Fig. 3. Effect on the initial pH value on adsorption of CR on AAB-SG-La beads.

https://doi.org/10.1051/e3sconf/202235003009

3.3 Adsorption isotherm studies

The Langmuir and Freundlich isotherm models are regarded as explaining the CR adsorption isotherm parameters onto AAB-SG-La beads as shown in Table 1. According to the correlation coefficient (R^2) and the adsorption capacity ($q_{m,cal}$ = 588 mg/g), CR adsorption onto prepared AAB-SG-La beads is fitting by the Langmuir model. It is recommend that the adsorption of AAB-SG-La beads are monomolecular layer and active sites are uniform.

3.4. Adsorption kinetics

As show in Table 2, the R^2 of the pseudo-second order kinetic model is higher than the pseudo-first-order as well as the adsorption capacity ($q_{m,cal}$ = 625 mg/g) derive from the pseudo-econd order kinetic model was similar to that gain experimentally ($q_{m,exp}$ = 597 mg/g). Therefore, the adsorption process is suitable to fit the pseudo-second order kinetic model. It is described that chemisorption may be predominantly command the CR adsorption rate by the AAB-SG-La beads.

The practical utilization of AAB-SG-La beads removes CR from to study. The maximum adsorption capacity of AAB-SG-La beads for CR is acquired to be 588 mg/g which shows that the adsorbent competes favorably with other related adsorbents as shown in Table 3.

Table 1. Adsorption isotherms parameters for CR dyes onto AAB-SG-La beads.

Adsorbent	Langmuir model			Freundlich model			$q_{m,exp}$/ mg·g^{-1}
	K_L/L·mg^{-1}	$q_{m,cal}$/ mg·g^{-1}	R^2	K_f/mg·g^{-1}	1/n	R^2	
AAB-SG-La beads	0.39	588	0.998	189.10	0.24	0.505	597

Table 2. Adsorption kinetic parameters for CR adsorption onto AAB-SG-La beads.

Adsorbent	Pseudo-first-order model			Pseudo-second-order model		
	k_1 ×0.01 /min^{-1}	$q_{m,cal}$/ mg·g^{-1}	R^2	k_2×0.001/g·mg^{-1}·min^{-1}	$q_{m,cal}$/ mg·g^{-1}	R^2
AAB-SG-La beads	1.24	138	0.660	0.17	625	0.997

Table 3. Contrast with the maximum adsorption capacities of several different sorbents for CR removal.

Adsorbent	Q_{CR}/ mg·g^{-1}	References
gelatin/bentonite beads	123	[9]
PAB hydrogel	42	[10]
Chitosan-Fe(OH)$_3$ beads	445	[11]
ZnO-clay-alginate beads	547	[12]
Alginate/natural bentonite beads	111	[13]
AAB-SG-La beads	**588**	**This study**

3.5. Adsorption mechanism

Fig. 4 showed the FTIR spectra of AAB-SG-La beads before and after CR adsorption. The band in spectra with wavenumber range 3410 cm^{-1} correspond to vibration stretching of O−H, peak at 2919 cm^{-1} which could be related to aliphatic C−H vibrations. The peaks at 1629 and 1379 cm^{-1} could be assigned to the stretching vibration of −OH, and the symmetric stretching of −COO−, respectively[14]. The corresponding peaks of −OH groups were also shifted from 3410 cm^{-1} to 3435 cm^{-1} for CR uptake. Moreover, the peak at 1379 cm^{-1} disappeared and a new peak at 1421 cm^{-1} onto AAB-SG-La beads after CR adsorption. These changes confirmed that the −OH groups and −COO− groups were involved in the adsorption of CR. The similar report that holds this result in the literature[15].

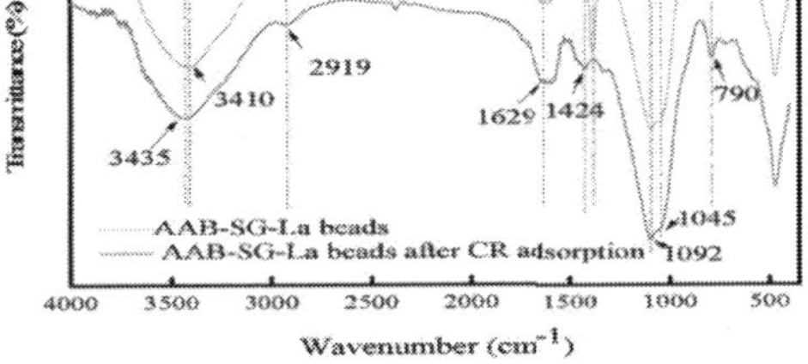

Fig. 4. FTIR spectra of AAB-SG-La beads before and after

CR adsorption.

4 Conclusions

We presented an inexpensive adsorbent based on composite AAB-SG-La beads via lanthanum nitrate solution crosslinking method, which was prepared from acid activated bentonite(AAB) and sesbania gum(SG). The optimum pH value for CR elimination by AAB-SG-La beads was broaden the applications of the beads. In addition, the adsorption isotherm data complied with the Langmuir adsorption model, the adsorption kinetics followed in the pseudo-second-order model. The maximum adsorption capacity was observed to be 588 mg/g. AAB-SG-La beads were used to anionic dye the adsorption capacity was comparably high compared with other synthetic materials. The results of FTIR spectral characterization of AAB-SG-La beads before and after CR adsorption indicated that the adsorption was mainly completed by hydrogen bonding. This work may provide new insight into the recovery of acid activated bentonite for environmental remediation, which would not cause secondary pollution.

Acknowledgments

The authors appreciatively acknowledge for their financial support such as the National Natural Science Foundation of China (22168004), Guangxi Natural Science Foundation (2020GXNSFAA297036), the Project of Guangxi Key Laboratory of Petrochemical Resource Processing and Process Intensification Technology (No. 2020Z008), Guangxi Cooperative Innovation Centre for Calcium-based Materials (GJZX2020–4).

References

1. M.A. Adebayo, J.I. Adebomi, T.O. Abe, et al. Removal of aqueous Congo red and malachite green using ackee apple seed–bentonite composite. Colloids Surf. A Physicochem. Eng. Asp., **38**, 100311 (2020)
2. N. Belachew, G. Bekele, Synergy of Magnetite Intercalated Bentonite for Enhanced Adsorption of Congo Red Dye. Silicon., **12**, 603-612 (2019)
3. X.Y. Xu, B. Wang, H. Tang, et al. Removal of phosphate from wastewater by modified bentonite entrapped in Ca-alginate beads. J Environ Manage., **260**, 110130 (2020)
4. R.R. Pawar, P.G. Ingole, et al. Use of activated bentonite-alginate composite beads for efficient removal of toxic Cu2+ and Pb2+ ions from aquatic environment. Int J Biol Macromol., **164**, 3145-3154 (2020)
5. S. Lan, Z.H. Leng, N. Guo, et al. Sesbania gum-based magnetic carbonaceous nanocomposites: Facile fabrication and adsorption behavior. Colloids Surf. A Physicochem. Eng. Asp., **446**, 163-171 (2014)
6. H.B. Zhang, Z.F. Tong, T.Y. Wei, et alSorption characteristics of Pb(II) on alkaline Ca-bentonite. Appl Clay Sci., **65**, 21-23 (2012)
7. A. Nasrullah, A.H. Bhat, A. Naeem, High surface area mesoporous activated carbon-alginate beads for efficient removal of methylene blue. Int J Biol Macromol., **107**, 1792-1799 (2018)
8. M.A. Adebayo, J.I. Adebomi, T.O. Abe, et al. Removal of aqueous Congo red and malachite green using ackee apple seed–bentonite composite. Colloids Surf. A Physicochem. Eng. Asp., **38**, 100311 (2020)
9. W.Y. Li, Q.L. Ma, Y.S. Bai, et al. Facile fabrication of gelatin/bentonite composite beads for tunable removal of anionic and cationic dyes. Chem Eng Res Des., **134**, 336-346 (2018)
10. A. Mokhtar, S. Abdelkrim, A. Sardi, et al. Preparation and Characterization of Anionic Composite Hydrogel for Dyes Adsorption and Filtration: Non-linear Isotherm and Kinetics Modeling. J Polym Environ., **28**, 1710-1723 (2020)
11. X.X. Yang, Y.M. Li, H.M. Gao, et al. One-step fabrication of chitosan-Fe(OH)3 beads for efficient adsorption of anionic dyes. Int J Biol Macromol., **117**, 30-41 (2018)
12. S. Vahidhabanu, D. Karuppasamy, A.I. Adeogun, et al. Impregnation of zinc oxide modified clay over alginate beads: a novel material for the effective removal of congo red from wastewater. RSC Advances, 7: 5669-5678 (2017)
13. A. Oussalah, A. Boukerroui, A. Aichour, et al. Cationic and anionic dyes removal by low-cost hybrid alginate/natural bentonite composite beads: Adsorption and reusability studies. Int J Biol Macromol, **124**, 854-862 (2019)
14. Y.M. Zhao, L.X. Gai, H. Liu,et al. Network interior and surface engineering of alginate-based beads using sorption affinity component for enhanced phosphate capture. Int J Biol Macromol, **162**, 301-309 (2020)
15. B.G. Li, H.Y. Yin, Excellent biosorption performance of novel alginate-based hydrogel beads crosslinked by lanthanum(III) for anionic azo-dyes from water. J Disper Sci Technol., **42**, 1830-1842 (2020)

E3S Web of Conferences **350**, 03010 (2022) https://doi.org/10.1051/e3sconf/202235003010
EREGCE 2022

Synthetic Effects and Economic Accounting of Low-temperature Pyrolysis Process of Typical Medical Waste based on Kinetics Analysis

Yangwei Wu [1], *Aijun* Li[1,*], *Su* Lei[1,2], *Qian* Deng[1], and *Hong* Yao[1,2]

[1] Huazhong University of Science and Technology, School of Energy and Power Engineering, 430074 Wuhan, China
[2] State Key Laboratory of Coal Combustion Huazhong University of Science and Technology, 430074 Wuhan, China

Abstract. The thermogravimetric (TG) analysis was adopted to study the pyrolysis characteristics of two typical medical wastes at different heating rates (10, 20, and 50 ℃/min), and the kinetic parameters were analyzed by the Coats-Redfern method. Finally, the economic evaluation of the technology was carried out. The results showed that the two typical medical wastes both have only one weight loss peak. The syringes and bamboo sticks pyrolysis ranged from 264.80 to 554.07 ℃ and from 205.66 to 487.05 ℃, respectively. With the increase of the heating rate, the initial, final and peak temperatures of the two raw materials shifted to higher values. The pyrolysis reaction mechanisms of the syringes and bamboo sticks were summarized as the nucleation growth model (A2.5) and the second-order model (F2). The total investment of medical waste pyrolysis project is 59 million yuan, and the annual net gain is 11.85 million yuan. The investment recovery period of the project takes five years.

1 Introduction

With the improvement of medical enterprise, a large quantity of medical wastes have been generated. According to statistics, there were about 843,000 tons of medical wastes produced in China in 2019, and the production increased by 4-8% annually [1]. Due to the high infectivity and pathogenicity of medical waste, rapid harmless treatment has attracted the attention of the whole society.

Medical waste contains more than 80% organic fractions [2]. Concretely speaking, it includes 39.30-50.00% plastics, 14.32-45.10% biomass and 3.40-6.60% rubber [3]. And the calorific value is higher than 4000 kcal/kg [4-5], so it has the potential to be converted into energy through thermal treatment. Pyrolysis can break medical wastes into high value products such as coke, combustible gas and tar in an oxygen-free or anoxia environment [6]. Compared with traditional incineration technology, pyrolysis technology has been considered to have obvious advantages in producing energy, high value products and in reducing pollution [6-8].

The prerequisite and key to the development and design of the pyrolysis process lies in the thorough understanding of the pyrolysis characteristics of medical waste. The complexity and variability of medical waste result in the extremely complex pyrolysis process. Therefore, in order to understand the overall pyrolysis characteristics of medical waste, it is necessary to grasp the pyrolysis behavior of single component. Thermogravimetric analysis has been widely applied in researching the thermal analysis and kinetic parameters of organic matter [9], and the Coats-Redfern method is a general method for the thermal kinetic analysis of coal, biomass and solid wastes [10]. The Coats-Redfern method can simultaneously calculate the activation energy, pre-exponential factor and reaction mechanism function [11]. Therefore, the Coats-Redfern method can be used to obtain kinetic parameters and reaction mechanisms.

In this study, syringes and bamboo sticks were selected as typical components of plastic and biomass in medical wastes. The pyrolysis characteristics of raw materials with different heating rates were studied by TG. Based on thermogravimetric data, the Coats-Redfern method was used to calculate the kinetic parameters and determine the most appropriate reaction mechanism. Pyrolysis behaviors and kinetic analysis grasped in this paper are helpful to the selection and optimization of the reactors as well as the setting of operation parameters in the industrial pyrolysis process.

2 Materials and methods

2.1 Materials

Two typical medical waste components, syringes and

* Corresponding author: hust_laj@mail.hust.edu.cn

bamboo sticks were selected in this study, which were purchased from Jiangsu Zhiyu Medical Instrument Co., Ltd. and Qingdao Hainuo Biological Engineering Co., Ltd. respectively. The samples were crushed to powder with a particle size of less than 0.5 mm, and then dried in an oven at 105 ℃ for 3 h. The proximate analysis was performed according to the Chinese National Standard GB/T212-2008, and the ultimate analysis was tested by elemental analyzers (Elementar Vario Micro cube, Germany and Euro Vector EA3000, Italy). The test results are shown in Table 1.

Table 1. Proximate analysis and ultimate analysis of samples

Sample	Proximate analysis/wt.%			Ultimate analysis/wt.%				
	Ad	Vd	FCd	Cd	Hd	Nd	Sd	Od
Syringes	0.02	99.10	0.88	85.90	13.53	0.00	0.00	0.55
Bamboo sticks	8.64	83.33	8.03	44.58	5.73	0.28	0.19	40.58

2.2 Experiment methods

The pyrolysis experiment was carried out on a thermogravimetric analyzer (STA 449 F3 Jupiter, NETZSCH Company, Germany). About 10 mg of samples were placed in an alumina crucible and heated from 25 to 600 ℃ at heating rates of 10, 20 and 50 ℃/min. A nitrogen flow rate of 20 ml/min was applied during the entire heating process. For the sake of reducing the system error, the blank experiment of each heating rate was carried out before the actual experiments.

2.3 Coats-Redfern kinetic method

The Coats-Redfern (CR) method [12] was used for kinetic analysis in this study, whose equation was as follows:

$$\ln\frac{G(\alpha)}{T^2}=\ln\frac{AR}{\beta E}-\frac{E}{RT} \quad (1)$$

Therefore, the values of activation energies and pre-exponential factors can be obtained by the slope and intercept of the fitted regression equation. The most suitable reaction mechanism was determined by comparing activation energies with model-free method.

3 Results and discussion

3.1 Thermogravimetric analysis

The TG and DTG curves of samples at different heating rates (10, 20, 50 ℃/min) are displayed in Fig. 1. Both of two kinds of samples had only one weight loss peak, indicating that both of them had one-step thermal degradation characteristics. The initial pyrolysis temperature (Ti), maximum weight loss rate pyrolysis temperature (Tp), final pyrolysis temperature (Tf) and mass residue (mf) were summarized in Table 2. From Fig. 1 and Table 2, syringes had a rapid reaction rate and a narrow temperature range. The pyrolysis interval of syringes with different heating rates was 364.80-554.07 ℃, and the peak temperature of DTG was 459.41 ℃ for the heating rate of 10 ℃/min, 467.94 ℃ for the heating rate of 20 ℃/min and for the heating rate of 50 ℃/min, respectively. Syringes produced almost no

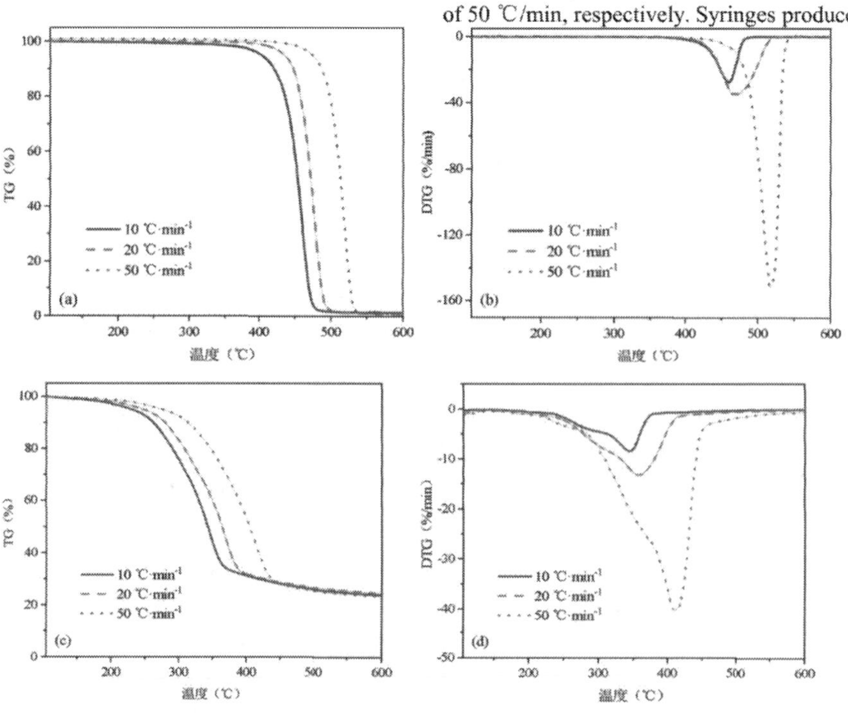

Fig. 1. TG/DTG curves of two samples: (a)(b) Syringes; (c)(d) Bamboo sticks

coke after pyrolysis, and the mass residues at three heating rates were less than 1.3 wt.%. Compared with syringes, bamboo sticks appeared to have a slower weight loss rate and a wider pyrolysis temperature range. The pyrolysis process of bamboo sticks started at 205.66-220.83 ℃, and finished at 410.18-487.03 ℃. It can be found that the DTG curves of bamboo sticks were asymmetrical, and there was an obvious shoulder on the left half of every DTG curve, which may be caused by the pyrolysis of different components in bamboo sticks.

Table 2. Pyrolysis characteristics of two samples

Sample	β/℃·min-1	To/℃	Tp/℃	Tf/℃	mf/%
Syringes	10	364.80	459.41	496.44	1.25
	20	374.50	467.94	529.87	0.73
	50	394.06	518.95	554.07	0.98
Bamboo sticks	10	205.66	344.53	410.18	23.92
	20	210.94	358.43	439.97	24.72
	50	220.83	410.76	487.03	23.52

Noteworthily, the initial, final and peak temperatures of the two raw materials shifted to higher temperatures with the increase of heating rate from 10 to 50 ℃/min. From the DTG curves, it was not difficult to find that the maximum weight loss rates under high heating rates were higher than those under low heating rates, indicating that the higher the heating rate was, the more intense the pyrolysis reaction was.

3.2 Kinetic analysis

The determination of pyrolysis reaction models and kinetic parameters is very important to study the pyrolysis mechanisms of materials. The common solid reaction mechanism functions were substituted into the Coats-Redfern method for calculation, and activation energies E, pre-exponential factors A and correlation coefficients R2 of syringes and bamboo sticks were obtained and summarized in Table 3 and 4, respectively. As shown in Table 3 and 4, the correlation coefficients of all reaction mechanism functions exceeded 0.9, which demonstrated that the kinetic parameters calculated by the Coats-Redfern method were credible. Meanwhile, the kinetic parameters of different reaction mechanism function models fluctuated greatly. For example, The values of activation energy were distributed in 78.51-629.71 KJ/mol and 12.71-122.48 KJ/mol for syringes and bamboo sticks, respectively. It was obvious that the difference between the maximum and minimum was nearly 9 times, which showed that not all mechanism function models were applicable to the selected samples.

The activation energies of syringes calculated by the A2.5 mechanism at three heating rates were 96.60-135.95 KJ/mol, which were close to the values calculated by the FWO, KAS and Friedman method. Thus, it can be inferred that the pyrolysis behavior of syringes follows the nucleation growth model (A2.5), and the integral form of the mechanism function is G(α)=[-ln(1-α)]2/5. It shows that during the pyrolysis process of syringes, several active sites for the decomposition reaction were first formed at the reaction interface. The decomposition products formed the nuclei of the product around these active sites, and the surrounding materials then reacted on the nuclei. As the pyrolysis reaction proceeded, the reaction nuclei continued to increase and grow [13] until the pyrolysis reaction was completed. For bamboo sticks, the values of activation energy with three heating rates obtained by the F2 model ranged from 87.49 to 91.18 KJ/mol, similar to the values estimated by model-free models. Therefore, the second-order model (F2) may be the most suitable mechanism of the pyrolysis behavior of bamboo sticks, whose integral form is G(α) =(1-α)-1-1.

Table 3. The kinetic parameters of syringes at different heating rates by CR method

Model	β=10 ℃·min-1			β=20 ℃·min-1			β=50 ℃·min-1			Average
	E/KJ·mol-1	A/min-1	R2	E/KJ·mol-1	A/min-1	R2	E/KJ·mol-1	A/min-1	R2	E/KJ·mol-1
F1	259.41	1.92×1018	0.9887	343.36	1.23×1024	0.9995	359.30	1.93×1024	0.9943	320.69
F2	347.56	9.40×1024	0.9560	468.02	1.54×1033	0.9828	481.30	5.56×1032	0.9670	432.29
F3	454.76	1.15×1033	0.9165	620.60	1.87×1044	0.9537	629.71	9.77×1042	0.9308	568.36
D1	397.99	8.53×1027	0.9981	513.87	5.07×1035	0.9887	547.88	3.17×1036	0.9973	486.58
D2	435.65	3.05×1030	0.9981	565.91	1.59×1039	0.9950	599.97	6.44×1039	0.9992	533.84
D3	481.71	2.05×1033	0.9950	630.26	1.70×1043	0.9992	663.70	3.62×1043	0.9984	591.89
D4	450.85	9.55×1030	0.9974	587.13	1.24×1040	0.9969	621.00	4.07×1040	0.9993	552.99
A2	123.74	1.89×108	0.9876	165.53	2.42×1011	0.9995	173.18	4.65×1011	0.9939	154.15
A2.5	96.60	1.71×106	0.9871	129.96	6.31×108	0.9994	135.95	1.26×108	0.9937	120.84
A3	78.51	7.10×104	0.9865	106.25	1.15×107	0.9994	111.13	2.37×107	0.9935	98.63
R1	193.03	1.59×1013	0.9980	250.78	1.92×1017	0.9881	267.47	7.45×1017	0.9972	237.09
R2	223.55	1.75×1015	0.9967	293.14	1.27×1020	0.9976	309.68	3.36×1020	0.9992	275.46
R3	234.89	8.60×1015	0.9948	308.98	1.24×1021	0.9991	325.38	2.79×1021	0.9984	289.75

Table 4. The kinetic parameters of bamboo sticks at different heating rates by CR method

Model	β=10 ℃·min-1			β=20 ℃·min-1			β=50 ℃·min-1			Average
	E/KJ·mol-1	A/min-1	R2	E/KJ·mol-1	A/min-1	R2	E/KJ·mol-1	A/min-1	R2	E/KJ·mol-1
F1	57.82	1.51×104	0.9900	63.17	6.50×104	0.9938	65.51	1.06×105	0.9919	62.17
F2	87.49	1.59×107	0.9791	90.91	3.61×107	0.9590	91.18	2.70×107	0.9542	89.86
F3	124.62	7.93×1010	0.9545	125.14	7.55×1010	0.9165	122.48	2.04×1010	0.9097	124.08
D1	84.02	1.22×106	0.9655	96.12	1.87×107	0.9909	103.22	5.20×107	0.9973	94.46
D2	95.20	8.48×106	0.9780	107.25	1.18×108	0.9966	114.06	2.67×108	0.9997	105.51
D3	109.71	5.46×107	0.9878	121.27	6.14×108	0.9981	127.39	1.01×109	0.9982	119.46
D4	99.97	5.70×106	0.9821	111.87	7.41×107	0.9978	118.46	1.51×108	0.9997	110.10
A2	23.98	8.33×100	0.9850	26.59	2.55×101	0.9919	27.47	4.94×101	0.9895	26.01
A2.5	17.01	1.50×100	0.9985	17.82	3.13×100	0.9970	19.86	9.17×100	0.9877	18.23
A3	12.71	4.85×10-1	0.9756	14.39	1.35×100	0.9887	14.79	2.76×100	0.9853	13.96
R1	37.09	9.61×101	0.9546	43.06	5.69×102	0.9881	46.32	1.47×103	0.9963	42.16
R2	46.36	4.77×102	0.9799	52.19	2.50×103	0.9977	55.12	5.33×103	0.9995	51.22
R3	49.93	7.57×102	0.9848	55.63	3.75×103	0.9978	58.41	7.36×103	0.9981	54.66

3.3 Economic analysis

Based on the kinetic calculations and experimental results, taking the medical waste pyrolysis project with a treatment capacity of 100 t/d as an example, the rotary kiln pyrolysis gasification device was selected. A minimum of 1,500 m3 gas can be generated per ton of medical waste, and the calorific value of the gas is about 6.3 MJ/Nm3. The heat consumption rate of the unit is calculated as 10 MJ/kW·h. Eight 600 GFZ1 automatic shut-type high-pressure gas generator sets were selected for the power plant, whose continuous output was about 4000 kW. The total investment of the power plant was 59 million yuan, of which the medical waste pyrolysis gasification system was about 37 million yuan and the power plant system was about 22 million yuan.

(1) Electricity selling profit:

According to the above settings, the annual output power can be calculated as:

4000 kW×8400 h (350 D) /10000=33.6 million kW·h

The average self-consumed rate of the power plant was calculated by 6%, the total power consumption of the medical waste pyrolysis gasification device was about 225 kW, and the on-grid price was budgeted by 0.65 yuan/(kW·h). Therefore, the electricity selling profit per year can be computed as:

0.65×[33.6×(1-6%)-0.225×8.4]=19.3 million yuan

(2) Metal recovery:

On the basis of investigation results, the metal content in medical waste was about 0.5%, and the metal recovery rate was estimated as 0.4%. Thus, metal recovery from waste disposal is 140 tons per year. Considering the market price of 940 yuan/ton of scrap iron, the annual income is 131,600 yuan.

(3) Saving transportation cost:

35,000 tons of medical waste disposed in situ per year can save 6,000 times transportation. Pressing 50 yuan per time can save a total of 300,000 yuan. The fuel charge can save up to 1.35 million yuan, which was calculated by 300,000 km of transportation distance and 5.64 yuan per liter of diesel prices. As a whole, the transportation cost can be saved by about 1.65 million yuan.

(4) Operating cost:

The gas making equipment lifetime is approximately 10 years, and the repair cycle is about 1 year. 0.5 million and 2 million yuan are required per year for the system maintenance and operation respectively. For the operation of the system and the management of the power plant, 25 people are needed, and the annual salary and social insurance for each of them is 50,000 yuan. Together with management cost of 300,000 yuan per year, the annual operating cost totals 4.05 million yuan.

(5) Depreciation cost:

The depreciation of houses and buildings is calculated based on 25 years, while machinery equipment and other fixed assets are depreciated for 10 years. The ratio of the remaining value is 5%, and the calculated depreciation rate is 9.5%. The depreciation cost is about 5.18 million yuan per year.

According to the above economic benefit analysis, the total investment of the medical waste pyrolysis gasification project is 59 million yuan, and the total annual expenditure is 9.23 million yuan. Without considering government subsidization, the income is 21.08 million yuan and the project investment payback period takes five years. If the government grants a medical waste treatment subsidy of 83 yuan/ton, the project investment recovery period will be four years.

4 Conclusion

The pyrolysis behaviors of two typical components of medical wastes (syringes and bamboo sticks) were investigated at three heating rates by TG. According to TG and DTG curves, the Coats-Redfern method was employed to determine the suitable reaction mechanisms and kinetic parameters. The main conclusions are as follows:

(1) It can be observed from TG and DTG curves that the pyrolysis process of two kinds of samples was only one stage. The pyrolysis reaction rate of the syringes was relatively fast, and the pyrolysis mainly occurred at 264.80-554.07 ℃. Bamboo sticks covered a decomposition temperature that ranged from

205.66-487.05 ℃, and there was an obvious shoulder on the left half of every DTG curve, which may be the result of the pyrolysis of different components in bamboo sticks. The initial, final and peak temperatures of the two raw materials shifted to higher temperatures as the heating rate increased.

(2) It is found that the activation energies of syringes calculated by the A2.5 model (96.60-135.95 KJ/mol) and bamboo sticks obtained from the F2 mechanism function (87.49-91.18 KJ/mol) were similar to values estimated by model-free models. Therefore, the pyrolysis kinetic reaction mechanisms of syringes and bamboo sticks followed as nucleation growth model (A2.5) and second-order model (F2), respectively.

(3) The total investment of 100 t/d medical waste pyrolysis gasification project is 59 million yuan, and the total annual expenditure is 9.23 million yuan. Without considering government subsidization, the income is 21.08 million yuan, and the investment recovery period of the project needs five years. If the government grants 83 yuan per ton of medical waste treatment subsidies, the project investment recovery period is four years.

Acknowledgments

This work was financially supported by the Science and Technology Planning Project of Hubei Province, China (No.2018ACA151) and the National Nature Science Foundation of China (52076092).

References

1. Ministry of Ecology and Environment. Nation annual report on prevention and control of solid waste pollution in large and medium sized cities. http://www.mee.gov.cn/hjzl/sthjzk/gtfwwrfz/. (2021)
2. S. Fang, L. Jiang, P. Li, et al. Study on pyrolysis products characteristics of medical waste and fractional condensation of the pyrolysis oil. Energy, 195: 116969.(2020)
3. G. Su, H.C Ong, S Ibrahim, et al. Valorisation of medical waste through pyrolysis for a cleaner environment: Progress and challenges. Environmental Pollution, 279: 116934. (2021)
4. N. Deng, W. Cui, W. Wang, et al. Experimental study on co-pyrolysis characteristics of typical medical waste compositions. Journal of Central South University, 21: 4613-4622. (2014)
5. A. Dash, S. Kumar, R. Singh, R. Thermolysis of medical waste (waste syringe) to liquid fuel using semi batch reactor. Waste and Biomass Valorization, 6: 507-514. (2015)
6. C.W., Purnomo, W. Kurniawan, M. Aziz, Technological review on thermochemical conversion of COVID-19-related medical wastes. Resources, Conservation and Recycling, 167: 105429.(2021)
7. J. Hong, S. Zhan, Z. Yu, et al. Life-cycle environmental and economic assessment of medical waste treatment. Journal of Cleaner Production, 174: 65-73. (2018)
8. L.C., Malav, K.K., Yadav, N Gupta, et al. A review on municipal solid waste as a renewable source for waste-to-energy project in India: Current practices, challenges, and future opportunities. Journal of Cleaner Production, 277: 123227. (2020)
9. T. Cheng, R. Guo, H. Jo, et al. Pyrolysis characteristics and kinetic studies of horse manure using thermogravimetric analysis. Energy Conversion and Management, 180: 1260-1267. (2019)
10. H. Cai, J. Liu, W. Xie, et al. Pyrolytic kinetics, reaction mechanisms and products of waste tea via TG- FTIR and Py-GC/MS. Energy Conversion & Management, 184: 436-447. (2019)
11. Z. Epelioullar, H. Haykr-Ama, S. Yaman, Kinetic modelling of RDF pyrolysis: Model-fitting and model-free approaches. Waste Management, 48: 275-284. (2016)
12. A.W. Coats, J.P. Redfern, Kinetic Parameters from Thermogravimetric Data. Nature, 201: 68–9.(1964)
13. Y Liang, H Peng, Z Yun, et al. Relationship between polymerization kinetics and microstructure in reactive polymer blends: An Avrami-Erofeev study. European Polymer Journal, 106: 72-78. (2018)

Transient Simulation of 1,4-Dioxane Fate and Transport During Enhanced Soil Vapor Extraction Treatment

Chang Zhang[1], *Libin* Zhang[1], and *Yuanming* Guo[1,*]

[1]Jiangsu Key Laboratory of Chemical Pollution Control and Resources Reuse, School of Environmental and Biological Engineering, Nanjing University of Science and Technology, Nanjing 210094, China

Abstract. 1,4-Dioxane is an emerging contaminant a that is fully miscible in water and resistant to biodegradation. The presence of 1,4-Dioxane in vadose zone creates prolonged environmental issues and risk human health. To date, treatment technologies for 1,4-Dioxane vadose zone remediation are still being developed. In this work, a numeric simulation tool was developed to investigate the fate and transport of 1,4-Dioxane in the treatment zone of enhanced soil vapor extraction (XSVE) which utilizes heated air injection to improve contaminant removal efficacy. The results suggested identify source zone is critical for XSVE treatment, inaccurate heat injection point would prolong treatment duration.

1 Introduction

1,4-Dioxane is widely used as a stabilizer for chlorinated solvents such as 1,1,1-trichloroethane and trichloroethylene, and as a solvent for resins, vegetable oils, mineral oils and waxes[1, 2]. It is reported that 1,4-Dioxane exists in surface water and groundwater in the United States, Japan, Germany, the Netherlands, the United Kingdom and Canada[3]. It is considered to be a potential organic pollutant causing human disease[3]. Studies have shown[3-5] that the presence of 1,4-Dioxane will affect the secretion of sex hormones to a certain extent, damage the liver and cause renal failure. The international agency for research on cancer classified 1,4-Dioxane as class 2B carcinogen, which may have carcinogenic effect on humans, and it is reported that it has genotoxic liver cancer in rats[5]. Due to the existence of two oxygen atoms in the chemical structure of dioxane, it has strong hydrophilicity, good solubility and stability in water[6, 7]. Therefore, 1,4-Dioxane can transfer rapidly in the environment. It can evaporate from dry soil and spread rapidly to even low permeability soils, such as silt and clay[8]. At the same time, dioxane will quickly enter groundwater through soil and dissolve almost completely[9]. The physical and chemical properties of dioxane create challenges for removing this compound.

At present, most 1,4-Dioxane remediation technologies aimed at groundwater, including advanced oxidation, activated carbon adsorption and bioremediation[10-14]. For soils containing chlorinated solvent VOCs and 1,4-Dioxane, SVE system can easily remove chlorinated solvent VOCs, but usually leaves a large amount of 1,4-Dioxane in aerated pore water, which becomes a long-term source of groundwater pollution. Enhanced soil vapor extraction (XSVE)[15-17] is recently reported technology that utilizes heated air injection and high flow rate extraction to achieve 1,4-Dioxane removal from vadose zone.

Although great successes were made in XSVE field experiments, the fate and transport of 1,4-Dioxane within the treatment zone are not well understood. There are very few studies on this topic. Previous studies have limitations. A previous modeling work only simulate 1,4-Dioxane in bulk[18]. HypeVent XSVE assumed to be isolated from exchange of 1,4-Dioxane[15,16]. But the impact of processes like heat loss of soil, non-equilibrium distribution of temperature and humidity on XSVE performance is still unknown. This study aims to gain better understanding of the effect of such processes on the fate and transport of 1,4-Dioxane in soil.

2 Model development and validation

A numerical code was developed to simulate 1-dimenstional energy and mass transport during XSVE process[17]. Building on the excel modeling tool Hypervent[19], this model offers additional information (e.g. real-time 1,4-Dioxane and temperature profile) and also provide more accurate transient heat and 1,4-Dioxane transport simulation in vadose zone. A series of equations and boundary conditions are solved using a finite-difference numerical method. Gauss-Seidel iterative method was applied to obtain explicit solutions. The governing equations used in this work are summarized in Table 1.

3 Simulation scenarios

The simulation conceptual model was comprised of three soil columns with a bottom area of 180 square centimeters. The schematic of the simulation conceptual

* Corresponding author:yuanming.guo@njust.edu.cn

Table 1. Enhanced SVE Equations

Overall energy balance:

$$\frac{d}{dt}\left\{m_{soil}C_{p,soil}(T_{soil}-T_{ref})+m_{w,T}C_{p,wL}(T_{soil}-T_{ref})+m_{w,v}\Delta H_{w,vap}+m_{air}C_{p,air}(T_{soil}-T_{ref})\right\}=$$
$$\left\{\dot{m}_{air,in}C_{p,air}(T_{in}-T_{ref})+\dot{m}_{w,in}[C_{p,wL}(T_{in}-T_{ref})+\Delta H_{w,vap}]\right\}-\left\{\dot{m}_{air,out}C_{p,air}(T_{soil}-T_{ref})+\dot{m}_{w,out}[C_{p,wL}(T_{soil}-T_{ref})+\Delta H_{w,vap}]\right\}+R_{loss}$$

$$R_{loss} = \int_0^l r \times \left(T_{soil}-T_{ref}\right) dl$$

Temperature dependent partitioning coefficients[1]:

$$H_{14D}(T_{soil}+273.15)=\frac{\exp\left(37.3025-24.3009\frac{298.15}{(T_{soil}+273.15)}+4.8084\ln\left(\frac{(T_{soil}+273.15)}{298.15}\right)-9.7034\left(\frac{(T_{soil}+273.15)}{298.15}\right)\right)}{(101.325)(0.0821)(T_{soil}+273.15)(55.55)}$$

$$P_{v,w}(T_{soil}+273.15)=\frac{10^{\left(8.07131-\frac{1730}{(T_{soil}+233.426)}\right)}}{760}$$

Mass balances on 1,4 dioxane, water, and air:

$$\frac{dm_{14D}}{dt}=-\dot{m}_{14D,out}$$

$$\dot{m}_{14D,out}=Q_{air,STP}\left(\frac{T_{soil}+273.15}{273.15}\right)H_{14D}(T_{soil}+273.15)\left(\frac{m_{14D}}{m_{w,soil}}\right)$$

$$\frac{dm_{w,T}}{dt}=\left\{\dot{m}_{w,in}-\dot{m}_{w,out}\right\}$$

$$\dot{m}_{w,in}=Q_{air,STP}\left(\frac{T_{ambient}+273.15}{273.15}\right)\left(\frac{\%RH_{ambient}}{100}\right)\left(\frac{P_{v,w}(T_{ambient}+273.15)M_{w,H2O}}{R(T_{ambient}+273.15)}\right)$$

$$\dot{m}_{w,out}=Q_{air,STP}\left(\frac{T_{soil}+273.15}{273.15}\right)\left(\frac{P_{v,w}(T_{soil}+273.15)M_{w,H2O}}{R(T_{soil}+273.15)}\right)$$

$$\dot{m}_{air,in}=\dot{m}_{air,out}=Q_{air,STP}\left(\frac{PM_{w,air}}{R(273.15)}\right)$$

Table 2. Nomenclature for HypeVent XSVE equations presented

$C_{p,air}$ = heat capacity of air [kJ/kg-C]
$C_{p,soil}$ = heat capacity of soil [kJ/kg-C]
$C_{p,wL}$ = heat capacity of liquid water [kJ/kg-C]
$C_{p,wv}$ = heat capacity of water vapor [kJ/kg-C]
$\Delta H_{w,vap}$ = specific enthalpy of water vaporization [kJ/kg]
H_{14D} = 1,4 dioxane Henry's Law Constant [L-water/L-vapor]

[1] Henry's Law Constant for 1,4 Dioxane predicted by Ondo and Dohnal equation from: *Temperature dependence of limiting activity coefficients and Henry's law constants of cyclic and open-chain ethers in water*. Fluid Phase Equilibria 262 (2007) 121–136.

Water vapor pressure equation from: *To Estimate Vapor Pressure Easily. Antoine Coefficients Relate Vapor Pressure to Temperature for Almost 700 Major Organic Compounds*, Yaws C. L., Yang H.-C., 1989, Hydrocarbon Processing, 68(10), 65–68, 1989.

$m_{w,14D}$ = mass of 1,4 dioxane in treatment zone [kg]

m_{air} = mass of air in treatment zone [kg]

m_{soil} = mass of soil in treatment zone [kg]

$m_{w,soil}$ = mass of liquid water in treatment zone [kg]

$m_{w,T}$ = total mass of (liquid + vapor) in treatment zone [kg]

$m_{w,v}$ = mass of water vapor in treatment zone [kg]

$M_{w,air}$ = molecular weight of air [kg/mole]

$M_{w,H2O}$ = molecular weight of water [kg/mole]

$Q_{air,STP}$ = air flow rate through the treatment zone as measured on a flowmeter calibrated to standard conditions (0 C and 1 atm) [L/min]

P = atmospheric pressure [atm]

$P_{v,w}$ = vapor pressure of water [atm]

R = gas constant [0.0821 L-atm/mole-K]

R_{loss}=heat loss constant over soil column [kJ/C/m^2/m]

t = time [min]

$T_{ambient}$ = average ambient air temperature [C]

$T_{initial}$ = initial temperature of soil in the treatment zone [C]

T_{ref} = reference temperature for energy calculations [0 C]

T_{soil} = temperature of soil in the treatment zone [C]

$\%RH_{ambient}$ = percent relative humidity in ambient air [%]

model is shown in Figure 1. Heated ambient air injection well and vapor extraction well were placed at each endof soil columns. 3 treatment zone conditions were investigated: a) 30 cm contaminant source near injection well followed by 30 cm clean soil; b) uniform source zone between two wells; and c) 30 cm clean soil near injection well followed by 30 cm contaminant source. Source zone 1,4-Dioxane concentration was 50 mg-1,4-Dioxane/g-soil. Ambient air was heated to 120°C and then was then injected into treatment zone at 2L/min.

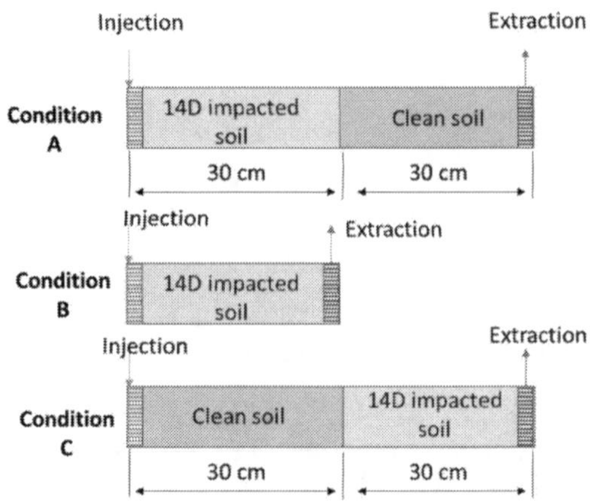

Fig. 1. Simulation conceptual model

4 Results

1,4-Dioxane mass removal rates vs time and % removal vs from different XSVE simulations are shown in Figure 2 and 3. Air injection locations played an important role on XSVE treatment duration. For the initial 700 min, XSVE hardly removed any 1,4-Dioxane in Condition A simulation. Then 1,4-Dioxane removal rates started to increase and reached the maximum value of 0.79 mg /min at about 1450 minutes. The total treatment period of Condition A was around 2100 minutes. On the other hand, treatment durations for XSVE to complete Condition B and Condition C remediation were 1100 and 1600 minutes, respectively. And the maximum removal rates were 740 mg /min for Condition B and 0.38mg/min for Condition C.

These results suggested that source zone identification is critical for a successful XSVE design. For example, none 1,4-Dioxane was removed during the first 700 min treatment in Condition A. During this period, heated air drove 1,4-Dioxane vaporize in the source zone and migrate towards extraction point, however, once it reached clean, cool soil, vaporized 1,4-Dioxane would re-partition with soil water and re-participate. It was not until all treatment zone was heated that 1,4-Dioxane started to be removed. Whereas the most effective treatment was found when injection and extraction wells perfectly bracketed source zone. The

https://doi.org/10.1051/e3sconf/202235003011

treatment time for Condition B was only ½ of Condition A.

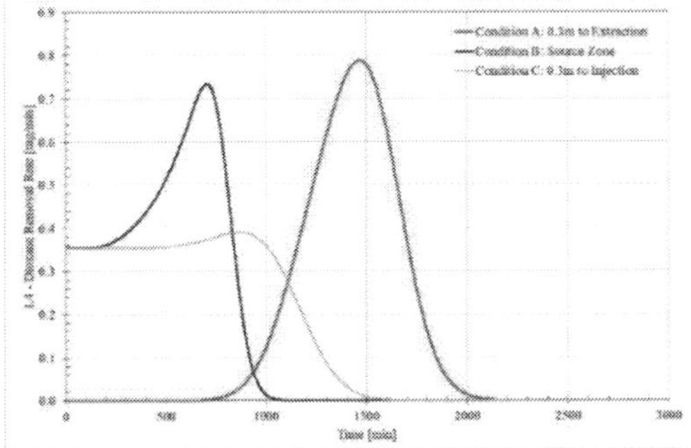

Fig. 2. 1,4-Dioxane removal rates from different XSVE simulations

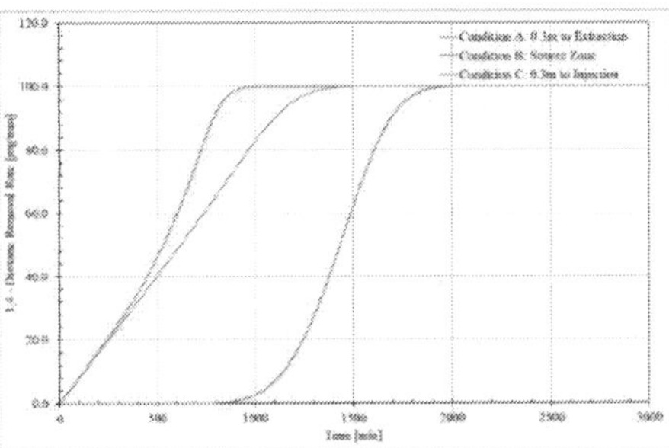

Fig. 3. % 1,4-dioxane Removal run for different XSVE simulations

5 Conclusions

In this paper, a screening-level model was established based on mass and energy balance calculation, and it was used to evaluate remediation of 1,4-Dioxane in the vadose zone using enhanced SVE with heated air injection. The results suggested source zone identification is an important factor affecting XSVE treatment efficacy. Accurate source zone delineation would reduce XSVE treatment duration and energy cost significantly.

References

1. K. Rathfelder, W.W.G. Yeh and D. Mackay, Mathematical simulation of soil vapor extraction systems: model development and numerical examples. Journal of Contaminant Hydrology, **8**, 263-297 (1991)
2. M.J. Zenker, R.C. Borden and M.A. Barlaz, Occurrence and Treatment of 1,4-Dioxane in Aqueous Environments. Environmental Engineering Science, **20**, 423-432 (2003)
3. A.C. McElroy, M.R. Hyman and D.R.U. Knappe, 1,4-Dioxane in drinking water: emerging for 40 years and still unregulated. Current Opinion in Environmental Science & Health, **7**, 117-125 (2019)
4. L.D.V. Abreu and P.C. Johnson, Effect of vapor source-building separation and building construction on soil vapor intrusion as studied with a three-dimensional numerical model. ENVIRONMENTAL SCIENCE & TECHNOLOGY, **39**, 4550-4561 (2005)
5. M. Gi, M. Fujioka, A. Kakehashi, T. Okuno, K. Masumura, T. Nohmi, M. Matsumoto, M. Omori, H. Wanibuchi and S. Fukushima In vivo positive mutagenicity of 1,4-dioxane and quantitative analysis of its mutagenicity and carcinogenicity in rats. Arch Toxicol, **92**, 3207-3221 (2018)
6. H. Nakagawa, S. Takagi and J. Maekawa, Fered-Fenton process for the degradation of 1,4-dioxane with an activated carbon electrode: A kinetic model including active radicals. Chemical Engineering Journal, **296**, 398-405 (2016)
7. H. Barndõk, D. Hermosilla, C. Han, D.D. Dionysiou, C. Negro and Á. Blanco, Degradation of 1,4-dioxane from industrial wastewater by solar photocatalysis using immobilized NF-TiO2 composite with monodisperse TiO2 nanoparticles. Applied Catalysis B: Environmental, **180**, 44-52 (2016)
8. H. Dang and A.M. Cupples, Identification of the phylotypes involved in cis-dichloroethene and 1,4-dioxane biodegradation in soil microcosms. Sci Total Environ, **794**, 148690 (2021)
9. R.H. Anderson, J.K. Anderson and P.A. Bower, Co-occurrence of 1,4-dioxane with trichloroethylene in chlorinated solvent groundwater plumes at US Air Force installations: Fact or fiction. Integrated Environmental Assessment and Management, **8**, 731-737 (2012)
10. G. Andaluri and R. Suri, Removal of 1,4-Dioxane and Volatile Organic Compounds from Groundwater Using Ozone-Based Advanced Oxidation Process. Ozone: Science & Engineering, **39**, 423-434 (2017)
11. D.T. Adamson, R.H. Anderson, S. Mahendra and C.J. Newell, Evidence of 1,4-dioxane attenuation at groundwater sites contaminated with chlorinated solvents and 1,4-dioxane. Environmental Science & Technology, **49**, 6510-6518 (2015)
12. H. Barndok, L. Cortijo, D. Hermosilla, C. Negro and A. Blanco, Removal of 1,4-dioxane from industrial wastewaters: routes of decomposition under different operational conditions to determine the ozone oxidation capacity. Journal of Hazardous Materials, **280**, 340-347 (2014)
13. N. Merayo, D. Hermosilla, L. Cortijo and A. Blanco, Optimization of the Fenton treatment of 1,4-dioxane and on-line FTIR monitoring of the reaction. Journal of Hazardous Materials, **268**, 102-109, (2014)
14. A. Tawfik, Degradation pathways of 1,4-dioxan e in biological and advanced oxidation processes. Desalination and Water Treatment, **178**, 360-386 (2020)
15. D.R. Burris, P.R. Dahlen and R.E. Hinchee, Soil Gas Sampling for 1,4-Dioxane during Heated Soil Vapor

Extraction. Groundwater Monitoring & Remediation, **38**, 85-89 (2018)

16. R.E. Hinchee, P.R. Dahlen, P.C. Johnson and D.R. Burris, 1,4-Dioxane Soil Remediation Using Enhanced Soil Vapor Extraction: I. Field Demonstration. Groundwater Monitoring & Remediation, **38**, 40-48 (2018)
17. D.R. Burris, P.C. Johnson, R.E. Hinchee and P.R. Dahlen, 1,4-Dioxane Soil Remediation Using Enhanced Soil Vapor Extraction (XSVE): II. Modeling. Groundwater Monitoring & Remediation, **38**, 49-56 (2018)
18. J.F. Horst, C.H. Bell, A. Lorenz, M. Heintz, Y. Miao, J. Saling, D. Favero and S. Mahendra, Bioremediation of 1,4-Dioxane: Successful Demonstration of In Situ and Ex Situ Approaches. Ground Water Monitoring and Remediation, **39**, 15-24 (2019)
19. P.C. Johnson, C.C. Stanley, M.W. Kemblowsik, J.D. Cothart and D.L. Byers, A practical approach to the design, operation, and monitoring of soil venting systems. Groundwater Monitoring Review, 159-178 (1990)

High Uranium Extraction by Polydopamine Functionalized MXene from Aqueous Solutions

Pengcheng Gu[1,*], *Dichen* Xia[1], *Yan* Liu[1], *Quan* Chen[1], and *Lingling* Wang[2,3]

[1]School of Resources and Environment, Anhui Agricultural University, Hefei 230036, China
[2]School of Earth and Space Sciences ,University of Science and Technology of China, Hefei, China
[3]Anhui Academy of Environmental Science Research, Hefei, China

Abstract. MXenes were outstanding materials for aqueous environment remediation, MXenes with a high capacity for radionuclides uranium (U(VI)) remains a challenge. In this study, a noval polydopamine functionalized MXene (defined as MXene@PDA) was successfully synthesised and evaluated for the aggregation towards U(VI) from wastewater. Versatile PDA owing functional groups provided more sites to capture contaminant ions. The mechanism research with MXene@PDA was explored by batch experiments toghther with the XPS analyses. The results revealed MXene@PDA with abundant functional groups exhibited superior elimination ability (90.4 mg/g) at pH = 5.0. The mechanism of U(VI) on MXene@PDA was primarily ascribed to the surface complexation force between the UO_2^{2+} and -OH、NH_2 groups. The PDA modification of MXene materials are proved to be excellent materials for the extraction of radionuclides in the aqueous solution.

1 Introduction

With the depletion of resources, nuclear power has been deemed as a key resource in energy system and become an alternative to the traditional energy[1,2]. However, various radioactive contaminants, such as 235U and 154Eu poses a fatal threat to environmental protection and human health. As the main fuel for current commercial reactors, uranium with high internal radiation and long half-life, and even at very low concentrations can cause damage to the environment and organisms[3].

To date, various two-dimensional materials have been widely used in the dispose of radioactive wastewater. Layered double hydroxides (LDHs), metal organic frameworks (MOFs). Unfortunately, most of the above materials inevitably endure slow adsorption speed, limited adsorption capacities and poor selectivity, making them unsuitable for practical applications[4,5].

Recently, an energying class of two-dimensional transition metal carbide (MXene), has emerged in the environmental fields because of the unique physical and chemical properties. In particular, multilayered MXenes are promising candidates for clean-up of several radionuclide cations (238U and $107Pd^{2+}$) through the coordination interation, ion exchange and reduction immobilization[6]. For example, Wang et al reported V_2CT_x showed the remarkable removal capacity towards U(VI)[7]. Additionally, anthother $Ti_3C_2T_x$ with more efficient adsorption properties for U(VI) was further reported by Wang' s groups. Futrhermore, the interaction mechanism of adsorption–reduction in the elimilation process of U(VI) was revealed[8]. It can't be ignored that the absence of abundant functional groups greatly hinder its inherent sewage purification capacity. To address this problem, various investigations have been devoted to the design of functionalized MXene material for enhancing its removal performances. All in all, limitations of U(VI) include low removal capacity and poor selectivity, and novel functionalized MXene materials with high sequestration capacity for U(VI) remain a goal.

Polydopamine (PDA), an excellent biopolymer, has attracted multidisciplinary interests benefit from its outstanding biocompatibility, strong adhesive force and abundant functional groups[9,10]. The PDA layer can be easily decorated on the surface of various substrates by spontaneous polymerization of dopamine. Because of these advantages, as an important coating material PDA provided an excellent platform for binding metals through covalent and non-covalent interactions incuding hydrogen bonds, electrostatic interactions etc. In the past decade, sereval researchers devotes lots of energy to utilize PDA for surface modification of several materials. As expected, the PDA-coated naocomposites exhibited the outstanding performance to sereval pollutant ions. Hence, the PDA with the versatile merits could be considered as potential material to capturing U(VI) from aqueous solution.

Herein, a noval polydopamine functionalized $Ti_3C_2T_x$ (defined as $Ti_3C_2T_x$ @PDA) was synthesised and evaluated for the uptake of U(VI) from wastewater. The interaction mechanism was further researched by batch experiments and XPS analyses. The batch experiments revealed that MXene@PDA with abundant functional groups exhibited superior removal capacity (90.4 mg/g)

* Corresponding author: gupc518@163.com

at T = 293K and pH = 5.0. Furthermore, the XPS analyses exhited that the interaction mechanism of U(VI) on the $Ti_3C_2T_x$ @PDA was primarily ascribed to the complexation force between UO_2^{2+} and rich functional groups (hydroxyl and amine groups). PDA functionalized MXene composites are proved to be exceelent materials for the extraction of radioactive nuclear wastewater.

2 Materials and Operation Methods

2.1 Prepartion of $Ti_3C_2T_x$@PDA

Typically, 100 mg Ti_3AlC_2 powders (purchased from Forsman Technology Company) was added to 21 mL of solution including 1 M LiF and 6 M HCl group by group. And then, aftermagnetic stirring for 48 h at 45 °C for 48h, the obtained outcome was centrifuged for 30 min at 3500 rpm and washed 8 times. The supernatant is collected with water until the pH drops to 6, it was dired in oven at 80 °C for ~12 hours. The final product was $Ti_3C_2T_x$. And then, 220 mg $Ti_3C_2T_x$ was dissolved in 120 mL Tris buffer, followed by ultrasonic treatment for 15 min, adding 220 mg dopamine, and mechanical agitation for overnight. The obtained residues was washed six times using ultrapure water and ethanol and desiccated for ~12 h in an oven at 80℃. The grey powderw was named as $Ti_3C_2T_x$ @PDA.

2.2 Batch Adsorption

Firstly, a certain amount of of $UO_2(NO_3)_2$. $6H_2O$ was dissolved in deionized water, the origin solution of U (VI) was prepared (200 mg L^{-1}). Various concentrations of U(VI) were carried out on $Ti_3C_2T_x$ at 25 °C. Adsorption isotherm studies explored adsorption experiments at pH = 5.0 and a series of concentrations. The adsorption rate (%) and adsorbability (Qe) are counted by the formula below [11]:

$$q_e = \frac{V\times(C_0-C_e)}{m} \quad (1)$$

$$Adsorption\% = \frac{(C_0-C_e)}{C_0}\times 100 \quad (2)$$

Here, C_0 (mg·L^{-1}) is the original concentration, Ce (mg·L-1) is the equilibrium concentration, m (mg) and V (mL) represent the dosage of MXene@PDA toghther with the total volume of the suspension.

2.3 Instrumentation

In this study, the morphologies of MXene samples was determined by S-4800 (SEM). The X-ray diffraction (XRD) patterns was characteristed by A Bruker D8 advance diffractometer (over a sweep speed of 10/min at the region from 5°C to 80°C. The XPS dates was equipped with the Al X-ray source,which operated at 10 kV. with an Al Kα source at 1486.8 eV.

3 Results & Discussion

3.1 Characterization

The structures of original MXene ($Ti_3C_2T_x$) toghther with MXene@PDA was characterized in virtue of SEM. As shown in Fig. 1A, The Al layers were removed and a representative exfoliated morphology of layered $Ti_3C_2T_x$ (layered organ-like) was successfully obtained.

After modification with dopamine, Fig. 1B clearly shows irregular outer edges and thickened edges of $Ti_3C_2T_x$, which distinctly indicates that the shuck of PDA is evenly wrapped in $Ti_3C_2T_x$. The morphologic evolution of the original organ-like $Ti_3C_2T_x$ confirms the successful surface modification of PDA.

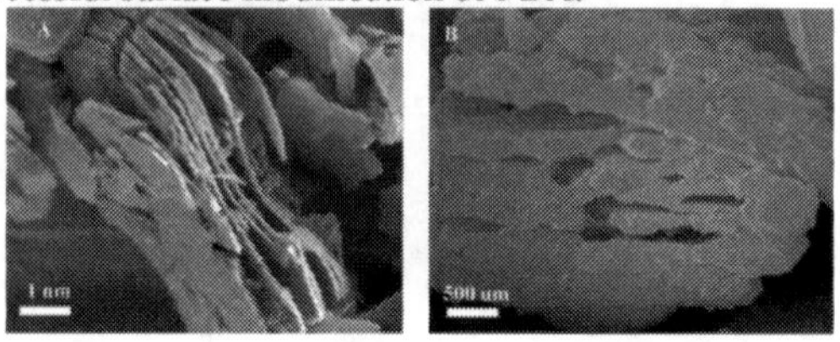

Fig. 1. (A) SEM images of $Ti_3C_2T_x$; (B) SEM images of $Ti_3C_2T_x$ @PDA.

The crystal structure transformation of samples were further measureed by XRD. As presented in Fig. 2A, The as-synthesized $Ti_3C_2T_x$ shows a featured (002),(004) peaks at 7.84° and 15.6°, which were consistent with those obtained by predecessors [12]. The intensity of the typical peak representing $Ti_3C_2T_x$ was significantly reduced and flattened after dopamine modificated. The XPS spectra in Fig. 2b revealed that Ti, O and C are the mainly constituents on the surface of $Ti_3C_2T_x$. Obviouly, an appearance of distinct N 1s sharp peak was captured from the $Ti_3C_2T_x$ @PDA samples compared with the pristine $Ti_3C_2T_x$, demonstrating that the successful surface modification PDA[9].

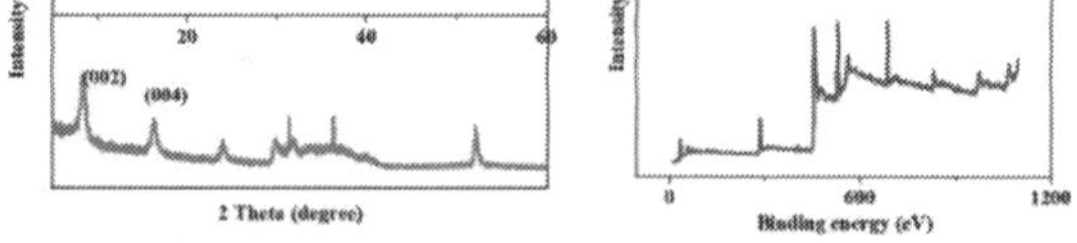

Fig. 2. (A) XRD patterns of Ti3C2Tx and Ti3C2Tx @PDA; (B) XPS spectra of Ti3C2Tx and Ti3C2Tx @PDA.

3.2 Thermodynamic studies

The results of U(VI) uptake by $Ti_3C_2T_x$@PDA are presented in Fig. 3. Fig. 3A shows a U(VI) adsorption isotherms for $Ti_3C_2T_x$@PDA under diffenert temperatures (298 K-328 K). Obviously, with the temperature increasing, the maximum value (Qe) intends to rise, manifesting that the higher temperature will accelerate the elimination of U(VI). Furthermore, the following models are used to simulate the isotherms results, refers as[13]:

Langmuir model:

$$q_e = \frac{K_L Qe_{max}}{1+K_L C_e} \quad (3)$$

Freundlich model:

$$Q_e = K_f C_e^{\frac{1}{n}} \quad (4)$$

Qe represents the mass of U(VI),which is adsorbed on the $Ti_3C_2T_x$@PDA, Ce means the equilibrium concentration, Q_{max} signifies the saturated elimination performance, and B (L mol^{-1}) represents the constant correlated to the heat of adsorption. K_f is the elimination ability, which the equilibrium concentration was close to 1; the n represents the adsorption dependence degree.

The relevant fitting parameters are shown in Table 1. Comparsion to the regression coefficients (R^2), the immobilization isotherms was more in line with the Langmuir model. Furthermore, the maximum fixation ability of $Ti_3C_2T_x$@PDA theoretically derived to be 90.4 mg/g at 313 K. This obtained results suggest the excellent adsorption capacity was may be caused by the abundant oxygen-containing function group and amine functionalization, which synergistically accelerated the uptake of U(VI).

Table 1. The fitted parameters calculated by two models

Parameters	K_L(L/mg)	$Q_{e,max}$ (mg/g)	R^2
Langmuir	0.95	90.4	0.93
Freundlich	0.93	-	0.87

To study the thermodynamic feasibility and intrinsic characteristics of adsorption process, the thermodynamic parameters (ΔG^0, ΔH^0 and ΔS^0) are calculated from the adsorption isotherms varying with temperature. The positive ΔH^0 values of 9.78 kJ mol^{-1} for $Ti_3C_2T_x$ @PDA indicated that the feature endothermic of adsorption process. Meanwhile, the positive ΔS^0 value further affirms the traits of high affinity of U(VI) for $Ti_3C_2T_x$ @PDA. Addtionsally, The value of ΔG^0 is negative, further also indicating the adsorption process was spontaneous. All in all, the thermodynamic parameters reflect that $Ti_3C_2T_x$ @PDA exhibit outstanding adsorption potential for U(VI) in aqueous solution.

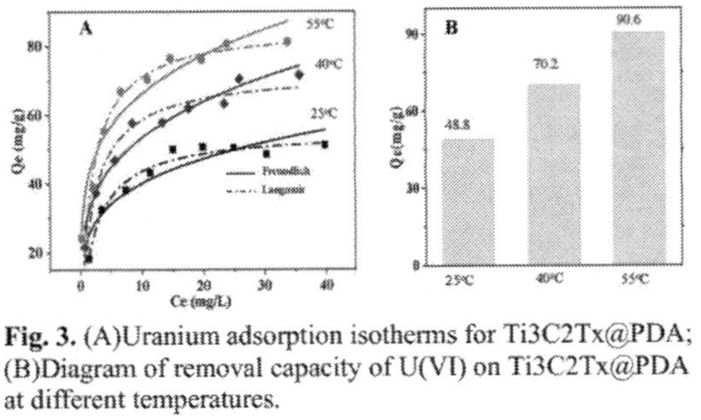

Fig. 3. (A)Uranium adsorption isotherms for Ti3C2Tx@PDA; (B)Diagram of removal capacity of U(VI) on Ti3C2Tx@PDA at different temperatures.

3.3. Effect of pH

In general, the pH plays a leading part in the migration of pollutant, especially in the aqueous solution. Therefore, the adsorption mechanism is explored through exporing the influence of pH and ionic strength[14]. As presented by Figure 4, the solution pH presents a evident impact on the adsorption of $Ti_3C_2T_x$@PDA on U(VI), which was divided into two regions. The uptake ability of $Ti_3C_2T_x$@PDA increases with the pH increasing from 1.5 to 6.5 by degrees. Subsequently, the adsorption percentage declined at 6.5 < pH <11.0. According to previous studies, the point of zero charge (pH_{PZC}) of $Ti_3C_2T_x$@PDA were approximately 4.0, noting the surface of $Ti_3C_2T_x$@PDA is negatively charged over a wide pH range[8]. when pH lower to the pH_{PZC}, the fierce electrostatic repulsion together with competition towards H^+ are the primaryly limiting factors for the uptake of U(VI). Under this pH condition .U(VI) is mainly in the form of positive charge $UO_2(OH)^{2+}$, $UO_2(OH)^+$ and UO_2^{2+}[15]. As pH increasing from pH_{PZC} to 6.5, the formation precipitation (UO_2CO_3) is the main removal mechanism. As result, the removal rate is significantly improved. However, accompanied by the increase of pH, the uptake percentage was significant reduction. In the range of 6.5-11.0, the forming species of $UO_3(CO_3)_3^{4-}$and $UO_3(CO_3)_2^{2-}$ were difficult to be adsorbed on $Ti_3C_2T_x$@PDA,which was negatively charged. The above mentions were the main reason for the reduced removal rate.

The ionic strength of U(VI) adsorption on $Ti_3C_2T_x$@PDA was performed in different concentrations $NaNO_3$ solutions (0.001M, 0.01M and 0.1M), respectively (Fig. 4). The extraction of U(VI) on $Ti_3C_2T_x$@PDA is significantly correlated with the ionic strength. In the whole pH range, it was found that the lower the salt concentration, the stronger the absorption efficiency of U(VI). This phenomenon revealed that the elimination process may be dominated by surface complexation.

With the decrease of ionic strength, the competition between Na^+ and U(VI) weakens, and the effective sites of U(VI) was increased[16]. It can be explained as follows: one hand, the increase of ionic strength presents a negative effect on the activity coefficient of U(VI), limiting the movement of U(VI) from solution to surface. On the other hand, the solid particles tend to accumulate in high ionic concentrations, which greatly reduces the available combining sites for ion exchange and thus reduces the elimination of U(VI) on the $Ti_3C_2T_x$@PDA. Finally, the increase of the thickness of the electric double layer inhibits the electrostatic attraction, thus further reducing the uptake of U(VI) by $Ti_3C_2T_x$@PDA at high salt concentration. Base on the above discussion, the interaction mechanism can be described by surface complexation of the outer sphere[17].

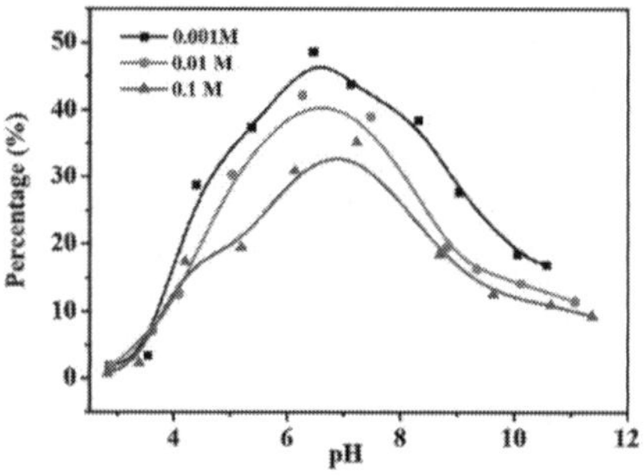

Fig. 4. Effect of pH and ionic strength on the uranium extraction efficiency

3.4. Mechanism Investigation

The interaction mechanism was further confirmed by XPS technique. The samples before ($Ti_3C_2T_x$@PDA) and after extraction ($Ti_3C_2T_x$@PDA- U(VI)) were presented in the Fig. 5 and Table 2. It could be seen in Fig. 5(A) that the O 1s of $Ti_3C_2T_x$@PDA spectrum was divided into three component peaks located at 532.8, 531.2, and 533.3 eV, which were ascribed to the groups C=O, C-O and O-H[10]. Compared to pristine $Ti_3C_2T_x$@PDA, the peak position of C-O and -OH were shifted to 532.3 eV and 533.0 eV, respectively. The shifted of position indicated that a new bond was formed between O and U(VI)[14,17]. Furthermore, the quantity of -OH increased from 40.4% to 29.5% after adsorption. The change of peaks position and quantity showed that the oxygen-containing functional groups (such as –OH) plays an key part in the adsorption process. For the N 1s spectrum (Figure 5B and Table 2), the N 1s spectrum of $Ti_3C_2T_x$@PDA located at 400.0 eV and 402.5 eV, which were ascribed to the groups −N= and the $-NH_2$. The $-NH_2$ group is present because of the introduction of PDA layer. It is clear that no discernible difference was observed for −N = species (400.00 eV). However, the peak of $-NH_2$ (402.5 eV) changed to a lower binding energy (401.5 eV)[9]. This phenomenon indicates that the $-NH_2$ functional group may combine with U(VI), which is one of the fundamental reasons for the efficient extraction of U(VI). Therefore, benefited from hydroxyl groups and amine groups from PDA, the interaction mechanism between $Ti_3C_2T_x$@PDA and U(VI) was majorly ascribed to the surface complexation .

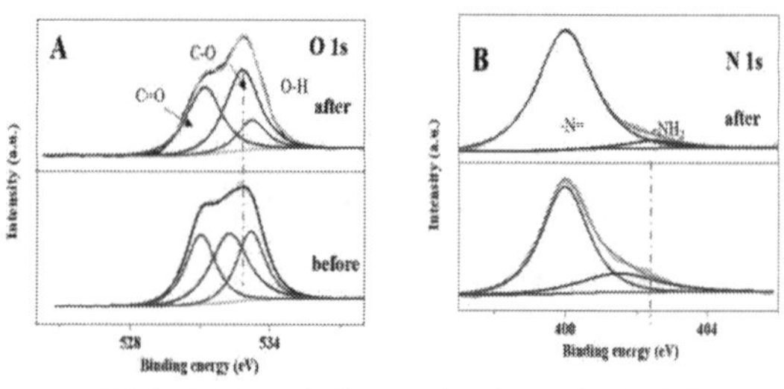

Fig. 5. XPS spectra of $Ti_3C_2T_x$@PDA of U(VI) uptake. (A,B) high-resolution O 1s and N 1s region.

Table 2. Binding energies of $Ti_3C_2T_x$@PDA: O and N before and after U(VI) adsorption.

Peaks E_B (eV)	O 1s (eV)			N 1s (eV)	
	C-O	-OH	C=O	-N=	$-NH_2$
$Ti_3C_2T_x$@PDA	532.8	533.3	531.2	400.0	402.5
$Ti_3C_2T_x$@PDA-U	532.3	533.0	531.2	400.0	401.5

4 Conclusion

As consequence, this work reported a noval polydopamine functionalized $Ti_3C_2T_x$ (defined as $Ti_3C_2T_x$ @PDA) and applied as U(VI) scavenger from. Owing abundant functional groups, The PDA could provide multifarious sites to fix contaminant ions. Based on the XPS analyses O 1s and N 1s, the excellent affinity of MXene@PDA towards U(VI) could be mainly ascibed to their coordination interaction with hydroxy groups and amine functional groups. The batch experiments revealed that MXene @PDA with abundant functional groups exhibited superior elimination performance (90.4 mg/g) . This work provides guidance for the development of MXene adsorbents with excellent uranium extraction ability .

Acknowledgements

This work was supported by the Natural Science Foundation of Anhui Province (NO. 2008085QB59), the National Natural Science Foundation of China (NO. 22006002) and Key research project of Ecological environment in Anhui Province HKYKY05 -2021.

References

1. P. Zhang, L. Wang, K. Du, S. Wang, Z. Huang, L. Yuan, Z. Li, H Wang, L. Zheng, Z. Chai and W. Shi Effective removal of U(VI) and Eu(III) by carboxyl functionalized MXene nanosheets. J. Hazard. Mater., **396**, 122731-122739 (2020)
2. S. Huang, S. Jiang, H. Pang, T. Wen, A.M. Asiri, K.A. Alamry, A. Alsaedi, X. Wang and S. Wang, Dual functional nanocomposites of magnetic MnFe2O4 and fluorescent carbon dots for efficient U(VI) removal. Chem. Eng.., **368**, 941-950 (2019)
3. P. Gu, S. Zhang, C. Zhang, X. Wang, A. Khan, T. Wen, B. Hu, A. Alsaedi, T. Hayat and X. Wang, Two-dimensional MAX-derived titanate nanostructures for efficient removal of Pb(II). Dalton Trans., **48**, 2100-2107 (2019)
4. G. Cheng, A. Zhang, Z. Zhao, Z. Chai, B. Hu, B. Han, Y. Ai and X. Wang, Extremely stable amidoxime functionalized covalent organic frameworks for uranium extraction from seawater with high efficiency and selectivity. Sci. Bull., **66(19)**, 1994-2001 (2021)
5. C.L. Xiao, Z.H. Fard, D. Sarma, T.B. Song, C. Xu and M.G. Kanatzidis, Highly Efficient Separation

of Trivalent Minor Actinides by a Layered Metal Sulfide (KInSn2S6) from Acidic Radioactive Waste. J. Am. Chem. Soc., **139**,16494-16497 (2017)

6. W. Mu, S. Du, X. Li, Q. Yu, H. Wei, Y. Yang and S. Peng, Removal of radioactive palladium based on novel 2D titanium carbides. Chem. Eng. J., **358**, 283-290 (2019)
7. L. Wang, L. Yuan, K. Chen, Y. Zhang, Q. Deng, S. Du, Q. Huang, L. Zheng, J. Zhang, Z. Chai, M.W. Barsoum, X. Wang and W. Shi, Loading Actinides in Multilayered Structures for Nuclear Waste Treatment: The First Case Study of Uranium Capture with Vanadium Carbide MXene. ACS Appl. Mater. Interfaces., **8**, 16396-16403 (2016)
8. L. Wang, H. Song, L. Yuan, Z. Li, P. Zhang, J.K. Gibson, L. Zheng, H. Wang, Z. Chai and W. Shi, Effective Removal of Anionic Re(VII) by Surface-Modified Ti2CTx MXene Nanocomposites: Implications for Tc(VII) Sequestration. Environ. Sci. Technol., **53**, 3739-3747 (2019)
9. D. Yang, X. Wang, N. Wang, G. Zhao, G. Song, D. Chen, Y. Liang, T. Wen, H. Wang and T. Hayat, In-situ growth of hierarchical layered double hydroxide on polydopamine-encapsulated hollow Fe3O4 microspheres for efficient removal and recovery of U(VI). J. Clean. Prod., **172**, 2033-2044 (2017)
10. X. Wang, N. Wang, D. Yang, X. Wang, S. Yu, H. Wang, T. Wen, G. Song and Z. Yu, Highly efficient Pb(II) and Cu(II) removal using hollow Fe3O4@PDA nanoparticles with excellent application capability and reusability[J]. Inorg. Chem. Front., **5**, 2174-2182 (2018)
11. L. Yin, Y. Hu, R. Ma, T. Wen, X. Wang, B. Hu, Z. Yu, T. Hayat, A. Alsaedi and X. Wang, Smart construction of mesoporous carbon templated hierarchical Mg-Al and Ni-Al layered double hydroxides for remarkably enhanced U(VI) management. Chem. Eng. J., **359**, 1550-1562 (2019)
12. P. Gu, J. Xing, T. Wen, R. Zhang, J. Wang, G. Zhao, T. Hayat, Y. Ai, Z. Lin and X. Wang, Experimental and theoretical calculation investigation on efficient Pb(II) adsorption on etched Ti3AlC2 nanofibers and nanosheets. Environ. Sci.: Nano, **5**, 946-955 (2018)
13. Y. Cai, L. Chen, S. Yang, L. Xu, H. Qin, Z. Liu, L. Chen, X. Wang, and S. Wang Rational Synthesis of Novel Phosphorylated Chitosan-Carboxymethyl Cellulose Composite for Highly Effective Decontamination of U(VI). ACS Sustainable Chem. Eng., **7**, 5393-5403 (2019)
14. L. Yin, Y. Hu, R. Ma, T. Wen, X. Wang, B. Hu, Z. Yu, T. Hayat, A. Alsaedi and X. Wang, Smart construction of mesoporous carbon templated hierarchical Mg-Al and Ni-Al layered double hydroxides for remarkably enhanced U(VI) management. Chem. Eng., **359**, 1550-1562 (2019)
15. J. Wang, Y. Shen, S. Liu and Y. Zhang, Single 2D MXene precursor-derived TiO2 nanosheets with a uniform decoration of amorphous carbon for enhancing photocatalytic water splitting. Appl. Catal. B., **270**, 118885-118861 (2020)
16. P. Gu, S. Zhang, X. Li, X. Wang, T. Wen, R. Jehan, A. Alsaedi, T. Hayat and X. Wang, Recent advances in layered double hydroxide-based nanomaterials for the removal of radionuclides from aqueous solution. Environ Pollut., **240**, 493-505 (2018)
17. Y. Zou, Y. Liu, X. Wang, G. Sheng, S. Wang, Y. Ai, Y. Ji, Y. Liu, T. Hayat and X. Wang, Glycerol-Modified Binary Layered Double Hydroxide Nanocomposites for Uranium Immobilization via Extended X-ray Absorption Fine Structure Technique and Density Functional Theory Calculation. ACS Sustainable Chem. Eng., **5**, 3583-3595 (2017)

AUTHOR INDEX

EDP Sciences – Web of Conferences
17, Avenue du Hoggar
Parc d'Activité de Courtabouf
BP 112
F-91944 Les Ulis Cedex A
France

ISBN 978-1-7138-5245-2

9 781713 852452

CURRAN ASSOCIATES INC.
proceedings.com

9th International Conference on Engineering, Project, and Production Management (EPPM2018)

MATEC Web of Conferences Volume 312 (2020)

Cape Town, South Africa
24-26 September 2018

edp sciences